SYSTEMS TECHNOLOGY
APPLIED TO SOCIAL
AND COMMUNITY PROBLEMS

SYSTEMS TECHNOLOGY APPLIED TO SOCIAL AND COMMUNITY PROBLEMS

by

Robert L. Chartrand
Library of Congress

SPARTAN BOOKS

New York   Washington

FOREWORD

As the Nation enters the 1970 decade, we are faced with an environmental crisis of unprecedented severity and complexity. Our society is struggling to exist amid unrelenting change, much of which has resulted from our technological achievements. Experience is showing us that many of our "advances" are mixed blessings, and that successful solutions usually are a combination of political acumen and technical know-how.

Within the United States Senate, there has been a determination to move with the times, utilizing every possible resource so that future generations could enjoy a higher standard of living. One critical consideration was that of planning for manpower capable of the best understanding and knowledge of the techniques of solving problems. A Special Subcommittee on the Utilization of Scientific Manpower was established in the 89th Congress, and as its chairman I tried to explore two significant areas. First, the personnel skills and experience required to design, test, and operate new "systems" were to be examined. Secondly, the record of the aerospace industry in building better systems would be investigated, with an eye to adapting already proven technology to domestic problems.

Through a series of hearings, we gained a respectful understanding of the impressive potential of systems analysis, computer technology, planning-programming-budgeting techniques and standards, and operations research in dealing with our non-defense, non-space problems. The comments and questions of the testifying academicians, government representatives, and industrial specialists helped convince the subcommittee that there was hope for improving our living conditions. It was also apparent that a much greater use of systems technology can be made at all levels of government. The experience of State and local governments in using automatic data processing and systems analysis also was explored through two questionnaires sent out in 1966 and 1968. The governors of our States and the mayors of more than twenty large cities cooperated in describing where and how they were and were *not* utilizing systems technology. The hearings and questionnaire data were so valuable that I felt it was important to make our findings available to concerned groups and individuals across the country.

A report was prepared by Robert L. Chartrand, the Specialist in Information Sciences for the Legislative Reference Service in the Library of Congress. As a member of the senior staff of the Science Policy Research Division, he has been involved over a period of years in the hearings and questionnaire activity, and is responsible for a continuing study of our civil sector problems. Originally published as a Senate committee print, "Systems Technology Applied to Social and Community Problems"deserves wide distribution. The publisher has taken a vital step in marketing a commercial version of this important reference work.

V

The record is far from complete as we begin to marshall our resources, both political and economic, to create a better environment for all to share. Foremost in the public conscience is the need to maintain an understanding of these innovative devices and processes which can be helpful to mankind. The exertions demanded will be of monumental proportions and will necessarily require a strong public commitment and direct participation by the President, the Congress, and every sector of our society. I am confident that the ability of the American people to make such an effort will allow us to successfully meet the challenges of the future.

GAYLORD NELSON
United States Senator

CONTENTS

VII

LIST OF FIGURES

Page

I. SUMMARY AND CONCLUSIONS

A. Summary

The contemporary scene is filled with discussion and examples of the application of systems technology to various problem areas—environmental pollution control, transportation planning, housing redevelopment, health and welfare services, law enforcement and administration of justice, manpower and employment planning, and educational programs. In order to cope with these problems, new tools and techniques are being created and adapted; e.g., automatic data processing, simulation and modeling, critical path scheduling, and revamped planning-programing-budgeting methodology.

The U.S. Congress has perceived the seriousness of these interlocking social and community problems and the imperative need for effective policies and programs originating within the Federal Government. Congressional concern has led to an array of public laws to provide direct funding support, grants-in-aid, and technical assistance to State and local governments. In addition, legislation has supported private sector research and development which could lead to the creation of better devices and man-machine techniques for combatting societal problems.

The potential of systems technology for solving these nondefense, nonspace public problems was examined by the Senate Committee on Labor and Public Welfare. In 1965, the Special Subcommittee on the Utilization of Scientific Manpower had been created, under the chairmanship of Senator Gaylord Nelson, of Wisconsin. In order to acquire pertinent factual and interpretive information on the utility of systems technology in a new role, the special subcommittee held four series of hearings. In addition, questionnaires were sent to the 50 States, 22 large cities, and selected regional development groups to survey the extent to which systems analysis and ADP were being employed.

The principal objective of the special subcommittee, as expressed in formal statements of purpose, news releases, discussions in the Congressional Record, and commentary relevant to key legislation being considered—i.e., S. 430 and S. 467—was that an overall strategy must be conceived. This would allow a marshaling of all necessary national resources to meet the nondefense, nonspace problems affecting the people and institutions of the United States. Systems tools and techniques in particular should be utilized to the fullest extent.

Four critical requirements for future action were identified

First, pertinent information must be acquired regarding those social and community problems where systems methodology and automatic data processing had proven useful, or where it was believed that such techniques and devices would have a high potential as corrective mechanisms. This type of information is fragmentary at best, and has

(1)

never been collected and arranged so that the researcher and planner might use it. Cognizance also has been taken of the need for regular reporting of such program and project activity, with criteria and standards for the inclusion and formatting of the essential elements of information.

The second vital requirement is a careful review of the state of the art, to be performed in the light of civil sector requirements, both present and future. Experience has shown that many promising techniques, computer programs, and equipments cannot be produced within a reasonable price range, endure the conditions under which they might have to function, nor meet the exigencies of some of the problem situations which prevail today. The time and costs required to create or convert technological components responsive to the special needs of the Nation must be measured carefully. Consideration also must be given to simply upgrading existing manual systems through better organization and revised operational procedures. The findings of this survey of equipment and techniques will be of high importance to those responsible for devising corrective programs. An advisory role by private sector representatives, as an evaluation of this sort is performed, would seem both proper and of substantive value.

The third requirement is for a widespread, sustained program of orientation and education for State and city officials, regional functionaries, Federal program managers, and selected leaders from the industrial and academic world. Not only must their technical literacy be improved, but a deeper understanding of the changing environment—with its impact on government and the people—must be instilled. Also, the utilization by State and local government of private resources—as well as the more apparent forms of Federal assistance—must be examined and encouraged. The responsibility for carrying out such educational endeavors does not reside in one place, but will require a well-thought-out, coordinated program on a nationwide basis.

The fourth and final requirement involves legislative action both at the Federal and State levels which can lead to strengthened research and development programs and the initiation of experimental ("pilot") projects designed to apply innovative man-machine techniques and various types of devices (analog measurement, computational, communications) to a selected problem. In many cases, public laws have encouraged the development and use of technology but with no actual provision for funding nor the rendering of technical assistance which might be critical to the continuance or success of a given undertaking. It is recognized that jurisdictional conflicts may develop as plans emerge or that a promising project may be impaired due to asynchronous budgeting cycles.

Executive branch activity in promoting use of systems technology

The Federal Government has played a growing role in promoting the use of systems technology in connection with meeting the problems of our times—pollution, urban deterioration, transportation chaos, etc. The Department of Housing and Urban Development provides support to State-local or private sector projects in such diverse areas as the use of computers in housing design and construction, and employing systems analysis in water resource management planning.

The Department of Health, Education, and Welfare, in its recent study "Toward a Social Report," emphasizes the need for "social indicators" so that programs for change may be measured for effectiveness. Developmental efforts range from air pollution control measures to planning for the utilization of computer technology in education. Information exchange also has been stressed by HEW; e.g., the MEDLARS (Medical Literature Analysis and Retrieval System) network. The Office of Economic Opportunity also has organized the collection, storage, retrieval, and utilization of Federal assistance information through a computer-oriented information center, and provides key information to its own and other executive branch decisionmakers. Heavy reliance on external consulting groups in the early days of the organization allowed OEO to perceive many social and community problems in a way which allowed counteractions to be mounted quickly.

The Department of Transportation has cooperated with HUD in the preparation of the study on "Tomorrow's Transportation: New Systems for the Future." DOT then took over direction of the nationwide effort. The establishment of national transportation objectives, based on cost-performance planning and drawing heavily on systems technology in their implementation, has caused DOT to consult regularly with all governmental echelons as well as the carriers and labor. The Department of Commerce long has had a charter to collect, maintain, and process demographic, economic, business, scientific, and environmental information. The Economic Development Administration pursues its function of long-range economic development and programing for areas and regions of persistent underemployment. The National Bureau of Standards (through the Institute for Applied Technology) has been among the first to employ computers and systems analysis in manpower planning. The Office of State Technical Services has undertaken to encourage State and local private and public groups to apply science and technology to resources preservation, business welfare, information exchange, and industrial development. Other executive branch organizations include the Bureau of the Budget, the National Science Foundation, the Department of the Interior, and the National Aeronautics and Space Administration.

Congressional committee activity regarding the role of systems technology

Within the Congress committees and subcommittees have worked to delineate the issues in a given problem, prepare a course of investigation, schedule hearings through which to solicit advice from recognized experts in the field, and consider proposed legislation on the subject. The pressures of time and public demands for action are often a deterrent to thorough study.

During the past decade, the full potential of systems analysis, mathematical modeling, critical path scheduling, and computer technology has become better understood by congressional elements. Perhaps most importantly, electronic computers and systems methodology have been demystified and the continuing control by the human being of all systems' decisions reaffirmed.

Two broad areas concerning the role of systems technology have received attention: the planning-programing-budgeting process, and specific problem-oriented system developmental activities. In the former case, the Joint Economic Committee (Subcommittee on Economy in Government) held a series of hearings on "The Planning-Programming-Budgeting System: Progress and Potentials." The PPBS approach was also studied by the Senate Special Subcommittee on the Utilization of Scientific Manpower and the Senate Committee on Government Operations (Subcommittee on National Security and International Operations). The latter group has issued seven committee prints dealing with various aspects of PPBS, ranging from "PPBS and Foreign Affairs" to "Budget Bureau Guidelines of 1968."

Extensive hearings on the "Federal Role in Urban Affairs" were held by the Subcommittee on Executive Reorganization of the Senate Committee on Government Operations. Another pertinent action was the series of hearings on "Creative Federalism" covered by the Subcommittee on Intergovernmental Relations. The application of management science to social problems preoccupied the Subcommittee on Government Research during its hearings, which included one special seminar, on the Full Opportunity and Social Accounting Act (S. 843, 90th Cong., second sess.).

The Senate Select Committee on Small Business, concerned about the impact of technology on the Nation's economy, commissioned the preparation of two special studies by the Science Policy Research Division of the Legislative Reference Service (Library of Congress): "Policy Planning for Technology Transfer" and "Automatic Data Processing and the Small Businessman."

The Joint Economic Committee, aside from its scrutiny of the ways in which PPBS is impacting on Federal authorization-appropriations procedures, has been actively looking at employment and manpower problems in the cities, and through the Subcommittee on Urban Affairs has held hearings on "Urban America: Goals and Problems." The Subcommittee on Economic Progress published a compendium of papers on "Federal Programs for the Development of Human Resources" in 1968.

Environmental pollution has received concentrated attention by the Subcommittee on Science, Research, and Development of the House Committee on Science and Astronautics. In addition to urging establishment of "systems analysis and management capability" within the Federal Government in order to effect pollution abatement, the subcommittee also has devoted itself to technology assessment, the purpose being to "enable decisions for the public good." In July 1968, a unique "Joint House-Senate Colloquium To Discuss National Policy for the Environment" was sponsored jointly by the House Committee on Science and Astronautics and the Senate Committee on Interior and Insular Affairs.

Another House group concerned about the role of the Federal Government in planning and implementing programs such as sewage treatment, transportation network planning, and pollution control has been the Subcommittee on Research and Technical Programs of the House Committee on Government Operations. In particular, this subcommittee has voiced concern about the small sums being set aside for research and development in these critical areas.

In some instances, hearings were held related to specific legislation; e.g., the hearings conducted by the Senate Special Subcommittee on the Utilization of Scientific Manpower to consider the grants-in-aid approach contained in the Scientific Manpower Utilization Act (S.

430, 90th Cong., first sess.) and the establishment of a National Commission on Public Management (proposed in S. 467 and H.R. 20, 90th Cong., first sess.). The commission device allows the creation of a group which can focus exclusively on a specified problem area.

The Advisory Commission on Intergovernmental Relations has prepared such reports as "Urban and Rural America: Policies for Future Growth." A more recently established body, the National Commission on Urban Problems, has explored in depth the problems of the cities, and has utilized numerous university, industrial, and government agency consultants in preparing its final report, "Building the American City." Presidential action has created such groups as the President's Committee on Urban Housing, and Federal-private steps have placed in operation the Intergovernmental Task Force on Information Systems.

The effect of public laws in encouraging the use of systems technology

Legislation passed by recent Congresses in an effort to improve the standard of living throughout the Nation reflects the determination to see that all possible use is made of innovative tools and techniques. Sometimes the terminology will encourage the application of existing advances in technology (e.g., in housing construction); the establishment of information-handling systems using automatic data processing; or the research, development, and testing of "new and advanced technologies." Often the wording is oblique, or only intended to urge an action group to employ better planning methods or more sophisticated equipment. More recently, however, the encouragement is increasingly direct, and may specify that certain action ensue; for example, the duties of the Board of the newly founded Federal Judicial Center include studying and determining "ways in which automatic data processing and systems procedures may be applied to the administration of the courts of the United States."

An analysis of significant legislation featuring provisions for the encouragement of the use of advanced technology was prepared at the direction of the special subcommittee, and comprises one section of this report. Among the public laws resulting from action by recent Congresses, especially the 89th and 90th bodies, featured in this study are the following:

Demonstration Cities and Metropolitan Development Act (Public Law 89–754).

Housing and Urban Development Act (Public Law 90–448).

Water Quality Act (Public Law 89–234).

Clean Water Restoration Act (Public Law 89–753).

Solid Waste Disposal Act (Public Law 89–272).

Air Quality Act (Public Law 90–148).

Highway Safety Act (Public Law 89–564).

Urban Mass Transportation Act (Public Law 88–365).

Law Enforcement Assistance Act (Public Law 89–197).

Omnibus Crime Control and Safe Streets Act (Public Law 90–351).

Elementary and Secondary Education Act (Public Law 89–10).

Manpower Development and Training Act (Public Law 87–415).

Economic opportunity amendments (Public Law 90–222).

In addition, special attention was given to that legislation which created new agencies responsible for dealing with the varied social and community problems.

Private sector activity in exploring the potential of systems technology

Corporate underwriting of R. & D. projects which show promise of developing new devices and man-machine interactive processes has been on the upswing, and numerous universities have encouraged interdisciplinary studies and experiments with similar objectives in mind. One proven form of presenting new approaches and discussing alternative technical plans has been the conference, symposium, or seminar. The proliferation of these meetings is testimony to the desire for dialog between involved university, industrial, and government managers, planners, researchers, and implementers. The scope is reflected in the written products; in some instances a special volume such as "Science, Engineering, and the City" may be written as the result of a symposium on the interaction of technology with urban problems. A second form of record may be the publication of proceedings, as was done in the case of the "Conference on Science, Technology, and State Government," jointly sponsored by the National Science Foundation and the Southern Interstate Nuclear Board. A symposium may concentrate on a particular social ill (e.g., the IBM Scientific Symposium on Water and Air Resource Management). Yet another type of written expression may be a book, such as "The Year 2000," which gives evidence of the ongoing work of the Commission of the Year 2000 and also indicates the nature of continuing studies by the Hudson Institute.

As the plethora of public and private forums, writings, proposals, and action mechanisms is surveyed, a question which naturally arises is this: *Should there be a single point of coordination, either in a private entity or situated in a chosen governmental agency, so that a master plan of meetings, topics, and speakers might be created, thus utilizing more effectively the talents and time of the Nation's key personnel?*

In the past, most of the expertise has been applied to the problems of defense and space. President John F. Kennedy was well aware of this imbalance, and in his 1963 Economic Report reminded the Nation that:

* * * in the course of meeting specific challenges so brilliantly, we have paid a price by sharply limiting the scarce scientific and engineering resources available to the civilian sector of the economy.

For all of the exploratory and evaluative efforts exerted thus far by governmental and private sector groups to determine the benefits and limitations from applying systems technology to social and community problems, there still remains an acknowledged need for further, intensive study of the Nation's needs, its institutions and laws, the shifting patterns of living, and the role of systems technology in this ever-changing environment.

B. CONCLUSIONS

The extensive information comprising the basis for this report has been analyzed at the direction of the Senate Special Subcommittee on the Utilization of Scientific Manpower. The incentives for determining how, where, and when to substitute innovative tools and techniques for traditional forms of program planning, management, and operation are very real. The following conclusions suggest themselves as a result of the analysis and data compiled in the report.

1. The essential elements and interacting components of social and community problems must be monitored by some responsible public institution.

2. A comprehensive survey of activities featuring the use of systems tools and techniques must be conducted, and the results formatted and disseminated so that there is widespread cognizance of their nature and implications.

3. The state of the art of operations research, systems analysis, automatic data processing, and related techniques and equipments is swiftly advancing and must be reviewed regularly in the light of established programs and projected needs.

4. Information exchange mechanisms need to be developed. These should be capable of providing, both on a regular and ad hoc basis, salient narrative and statistical data on project findings, technical proposals placed before Federal and State agencies, literature citations, and equipment and software development and applications.

5. A master plan, under joint Federal-State sponsorship, for the orientation and education of key personnel regarding the potential of systems technology should be prepared. Participants in the training would include selected Federal Government personnel, State and local officials, and private sector representatives involved in urban planning and program performance.

6. A special evaluative capability at the Federal level to prepare, on a continuing basis, cost-benefit comparisons for proposed technological change is required. This would allow planners to be apprised in advance of the implications of their budgetary and program recommendations.

7. The Congress should consider authorizing and directing Federal departments and agencies to develop systems analysis and ADP capabilities specifically tailored to the requirements of the States and localities.

8. The Congress should determine the usefulness of a formal requirement that State and local governments utilize systems technology in implementing various programs—i.e., housing, highways, pollution control—where Federal funding is involved.

9. The Congress should explore the advantages and disadvantages of granting tax incentives for the electronics and communications industries and "think" groups who develop systems methodologies for the needs of State and local governments.

10. Federal agencies should make available to the States and municipalities their expertise and findings regarding systems technology and its applications to social and community problems, either by deliberate dissemination procedures or through a policy of active cooperation.

II. INTRODUCTION

The basic purpose of this report is to submit to the Senate Special Subcommittee on Scientific Manpower Utilization of the Committee on Labor and Public Welfare selected analytical and interpretive information which it can utilize in its study of the role of systems technology in social and community problems. The kaleidoscopic nature of these problems—transportation, urban renewal, pollution control, and others—has created crises for governmental leadership at all levels. The special subcommittee is concerned with the ways in which systems analysis, operations research, and the application of automatic data processing can be put to use in combating these burgeoning problems. Of equal importance is the reexamination of the role which human decisionmaking plays in the new man-machine environments. Finally, the special subcommittee intends to examine the application of systems technology to the problem areas within the context of existing Federal programs and State and local developmental efforts.

A. OBJECTIVE OF THE REPORT

In order to assist the special subcommittee in its assessment of a significant series of national problems, the Science Policy Research Division of the Legislative Reference Service in the Library of Congress has been requested to prepare this report. The objective of the report is to explore the present and potential application of systems technology to nondefense, nonspace public problems; describe selected programs and projects which have been undertaken at the Federal, regional, State, or local level; and review those legislative proposals concerned with the creation of a mechanism to apply innovative tools and techniques to the full range of social and community problems.

As a result of scrutinizing the facts, trends, and findings implicit in an examination of the problems facing the Nation and the array of corrective methods, standards, and devices available to be employed in their amelioration or solution, the Congress may determine that additional legislative action is required which can help affect future developments.

Illustrative of the growing concern about technology and society was the passage of Public Law 88–444 which provided for the establishment of a National Commission on Technology, Automation, and Economic Progress. In reflecting upon the relationship between technology and the needs of our civilization, this observation was made:

Technology has the potential, whose beginnings we already see, to realize a persistent human vision: to enlarge the capacities of man and to extend his control over the environment.[1]

[1] National Commission on Technology, Automation, and Economic Progress. Technology and the American economy, v. 1. Washington, U.S. Govt. Print. Off., February 1966, p. 73.

A realization is developing, as reflected in testimony before the
ecial subcommittee (see section VI), that the private and public sec-
⸱ ⸱s must join forces in the effort to meet and overcome the problems
ᵥ ich perplex society today. In many instances, only a joint delibera-
tᵢ ᵢ can establish those guidelines and priorities which should deter-
m ᵢe the way in which the Nation expends its not unlimited resources—
hᵤ nan, financial, and material.

B. The Significance of the Problem

Perhaps the significance of the problem facing contemporary America may be found in the compromises which must be struck between
the beneficial and detrimental aspects of technology. Mankind is
commencing to understand that the problems generated by technological activity are not simply "technological" in nature, but that
the "people problems" often are far more critical. Marshall McLuhan
has captured the essence of the problem which confronts our people:

Innumerable confusions and a profound feeling of despair invariably emerge in
periods of great technological and cultural transitions. Our "Age of Anxiety" is,
in great part, the result of trying to do today's job with yesterday's tools—with
yesterday's concepts.[2]

The spectrum of problems facing society and its leadership includes:
air and water pollution, solid waste disposal, transportation planning,
urban development and renewal, housing, health and welfare services,
manpower availability and education. In addition, such vital tan-
gential problems as information exchange and the need for improved
planning, programing, and budgeting procedures have risen to prom-
inence. Throughout its span of investigatory activity, the special
subcommittee has sought to develop approaches whereby systems
technology can be applied to these problems.

Decisionmakers must have comprehensive, timely, accurate information on the status of a given situation and possible alternatives for
improving it. President Johnson noted the vital nature of having such
information in these words:

Increasingly, as our Great Society Programs move forward, new types of information are demanded in such areas as education, health, housing and urban
development. This information must be provided. New information systems, in
which all levels of government cooperate, are needed.[3]

And upon another occasion, in introducing the governmentwide
Planning-Programming-Budgeting System (PPBS), President Johnson
stressed the importance of using management tools in a way that will
benefit the entire populace:

The objective of this program is simple: to use the most modern management
tools so that the full promise of a finer life can be brought to every American at the
least possible cost.

This program is aimed at finding new ways to do new jobs faster, better, less
expensively; to insure sounder judgment through more accurate information;
to pinpoint those things we ought to do less; to make our decision-making process
as up to date as our space-exploring equipment.[4]

One important aspect of improving decisionmaking systems through
the application of systems methodology and the power of the electronic

---

[2] McLuhan, Marshall. The medium is the message. New York, Bantam Books, 1967, p. 8.
[3] Johnson, Lyndon B. Telegram to the First National Conference on Comparative Statistics, Washington
D.C., February 24–25, 1966.
[4] Johnson, Lyndon B. Introduction of new government-wide planning and budgeting system. Statement
by the President to members of the cabinet and heads of agencies, August 25, 1965. *In* Weekly compilation
of Presidential documents, v. 1, n. 5, Monday, August 30, 1965. p. 141.

computer has centered about the availability of competent manpower to conceive and implement the new systems. Awareness of this manpower shortage caused the 89th Congress to establish the Special Subcommittee on Scientific Manpower Utilization of the Senate Committee on Labor and Public Welfare, with Senator Gaylord Nelson as chairman. As the special subcommittee initiated inquiries into the manpower needs of the Nation, its scope of activities broadened to encompass the area dealt with in this report.

While analysts in universities, foundations, industry, and government have enunciated diverse approaches to the central problem and possible solutions, there has been a growing consensus that effective action can result only from an "all-out" effort involving all of our resources. In the words of Senator Nelson:

We face an environmental crisis of almost unimaginable proportions right now * * * Nothing short of a massive effort by industry and government at every level will solve this problem. This can be done only after a total analysis of the problem and development of an overall program of action.[5]

The approach proposed by Senator Nelson (in S. 430, the 90th Congress) was to provide grants-in-aid to States and public or private institutions.

Another viewpoint was registered by Senator Hugh Scott, the chief sponsor in the Senate of S. 467, calling for the creation of a National Commission on Public Management. While attesting that significant achievements had been registered in pollution control, planned urban development and renewal, housing and education, and other key problem areas, Senator Scott reminded his colleagues that:

* * * the responsible authorities at all levels of government—local, State, and National—face a host of serious problems caused by a society which is characterized by its increasing complexity, rapid rate of growth, and high mobility.

The magnitude of the problems leave [sic] no doubt that traditional solutions alone will not provide relief.[6]

Representative F. Bradford Morse, the initial sponsor of the national commission concept in the House of Representatives (in H.R. 20), agreed that Congress has enacted many laws designed to solve our socioeconomic problems but pointed out that:

* * * the dimensions of our remaining problems are staggering: 10,000 of our Nation's communities will face serious problems of air pollution; the demand for water consumption may exceed the available supply before the end of this century; there are 9 million substandard housing units in the United States, most of them in urban areas; traffic jams cost the Nation over $5 billion each year; and scientific and technical information is doubling every 15 years.

It is clear that problems of this magnitude are not susceptible to the traditional solutions. We must reach beyond our history for new ways to manage the public business effectively and economically.[7]

The theme is iterated in many places and by many thinking people: New devices and man-machine procedures are needed if the Nation is to sustain its pattern of improved living conditions for an increasing number of inhabitants. In "The Year 2000," predictions

[5] U.S. Congress. Senate. Committee on Labor and Public Welfare. Special Subcommittee on the Utilization of Scientific Manpower. Scientific manpower utilization, 1965–66. Hearings before the Special Subcommittee (89th Cong., 1st sess., November 18, 1965). Washington, D.C., U.S. Government Printing Office, 1967, p. 6 [Remarks by Senator Gaylord Nelson]. Hereafter referred to as Scientific manpower utilization, 1965–66.

[6] U.S. Congress. Senate. Committee on Labor and Public Welfare. Special Subcommittee on the Utilization of Scientific Manpower. Scientific manpower utilization, 1967. Hearings before the Special Subcommittee (90th Cong., 1st sess., January 25, 1967). Washington, D.C., U.S. Government Printing Office, 1967, p. 70 [Remarks by Senator Hugh Scott]. Hereafter referred to as Scientific manpower utilization, 1967.

[7] Morse, F. Bradford. Managing the public business. Remarks in the House. Congressional Record (Washington), v. 112, Aug. 25, 1966. p. 20714.

concerning the course of events in the remaining third of this century are set forth with an emphasis on charting "alternative futures" as a condition for policy choices. Authors Kahn and Wiener focus upon:

* * * a new sophistication in methodology. We have begun to assemble statistical time-series both to plot trend lines and to extrapolate likely development * * * Along with time-series, we have begun to construct models or likely combinations of trends and developments in order to uncover the connections and casual relations between variables.[8]

Thus, the Nation is faced with demands for new services, revised policies reflecting enhanced governmental understanding of the needs of the citizens, and programs designed to educate all persons about the changes which are affecting their work, play, communicating, and planning for the future. While some of the subject problems have only intruded upon the public consciousness in the past few years, many of the predicaments have existed for a long time. Similarly, corrective techniques applicable to some of the most exasperating problems are neither new nor unknown. For example, systems analysis as it is termed today might equally well be called organized commonsense; that is, logic appropriately applied. In another instance, some veteran budgeteers note that certain principles and some of the procedures heralded in the "Planning-Programming-Budgeting System" were in being a few decades ago both in industry and selected governmental projects. A number of the basic calculating operations performed by the powerful computers of today were achieved through much more modest devices down through the ages.

The information gathered and studied by the special subcommittee may prove useful to those congressional and State legislative elements and program agencies at all levels of government as they strive to develop controls, forecasting procedures, and evaluation mechanisms capable of insuring the continued vitality and stature of the United States.

C. SCOPE AND METHODOLOGY

A brief chronology of events leading to the commissioning of this report by the Senate Special Subcommittee on the Utilization of Scientific Manpower is of value at this juncture. Early in 1965, the Employment and Manpower Subcommittee of the Senate Committee on Labor and Public Welfare addressed itself to the impact on national scientific and technical manpower of Federal research and development policies.[9] This subcommittee, under the chairmanship of Senator Joseph S. Clark, conducted hearings on the subject in June and July of 1965;[10] a "Report and Recommendations" was published a year later.[11] In the autumn of 1965 the Special Subcommittee on the Utilization of Scientific Manpower was formed, chaired by Senator Gaylord Nelson. At the same time, Senator Nelson—supported by Senators Clark and Jennings Randolph—introduced the Scientific

Kahn, Herman, and Anthony J. Wiener. "The Year 2000." New York, The MacMillan Co., 1967. p. xxvi-xxvii.

[9] Nelson, Gaylord. Hearings on the impact on scientific and technical manpower of Federal research and development policies. Remarks in the Senate. Congressional Record (Washington), v. 111, Apr. 13, 1965. p. 7835.

[10] U.S. Congress. Senate. Committee on Labor and Public Welfare, Subcommittee on Employment and Manpower. Impact of Federal research and development policies on scientific and technical manpower. Hearings before the subcommittee (89th Cong., first sess., June 2-4, 7-10, and July 22, 1965). p. 954.

[11] U.S. Congress. Senate. Committee on Labor and Public Welfare, Subcommittee on Employment, Manpower, and Poverty. The impact of Federal research and development policies upon scientific and technical manpower. Report and recommendations of the subcommittee (89th Cong., second sess., December 1966). Washington, U.S. Government Printing Office, 1966. p. 69.

Manpower Utilization Act of 1965 (S. 2662), a bill to mobilize and utilize the scientific and engineering manpower of the Nation to employ systems analysis and systems engineering to help to fully employ the Nation's manpower resources to solve national problems.[12]

The newly formed special subcommittee then embarked upon a series of hearings during the period from November 1965 until March 1967; except for the first series, which was held in California, all sessions were conducted in Washington, D.C.[13] As testimony was received from industry, universities, foundations, and all echelons of government, an expanded frame of reference evolved. Emphasis was placed on understanding the mission, objectives, procedures, and organizational considerations involved in a planning-programing-budgeting system; one facet investigated dealt with the relationship between manpower planning and PPBS. A concurrent topic featured in the hearings was the discussion of the strengths and limitations of systems analysis methodology, particularly in developing and testing alternate programs to meet goals. Consideration was given to the effect of PPBS on executive and legislative branch elements active in the authorization-appropriation cycle. A detailed discussion of the various hearings appears as section VI in this report.

One significant development which occurred midway in the hearings was the determination that S. 430, the 90th Congress version of S. 2662, which featured a grants-in-aid to States or other private and public institutions, might be considered jointly with S. 467. This latter bill, introduced by Senator Hugh Scott, called for the establishment of a National Commission on Public Management and stressed the importance of surveying extant programs and projects where systems analysis and automatic data processing were being applied to public problem areas. The two sets of hearings in 1967, therefore, were conducted within the context of the approaches set forth in these two bills. A discussion of these bills and related items of legislation are contained in section V.

In support of the information collected for the special subcommittee hearings, two questionnaires were sent to the Governors of all States, the mayors of the 22 largest cities, and the heads of selected regional development commissions. The first questionnaire, sent in 1966, was responded to by over 80 percent of the recipients; an analysis of the questions appears in the second volume of the hearings entitled "Scientific Manpower Utilization, 1967." [14] A similar questionnaire, repeating certain of the key interrogations but including additional inquiries, was disseminated in mid-1968. A comparison of the two questionnaires, in addition to an analysis of the 1968 questionnaire material, is discussed in section VII.

Certain salient points emerged from the testimony, questionnaire responses and ancillary studies:

    1. A "systems approach" to a problem by definition implies comprehensive consideration of all facets impinging on that problem;

---

[12] Nelson, Gaylord. A space age trajectory to the great society. Remarks in the Senate. Congressional Record (Washington), v. 111, Oct. 18, 1965. p. 27244.
[13] Scientific manpower utilization, 1967. *Op cit.*, pp. 6–13.
[14] Ibid., pp. 362–368.

2. When institutions attack regional problems such as transportation planning or pollution control, the modification of traditional jurisdictional prerogatives becomes imperative;

3. If optimum benefit is to be derived from systems tools and techniques already developed, the results of experimental and operational projects must be made available;

4. Since many of the necessary corrective programs require action by more than one level of government, a requirement exists for the careful examination of relevant Federal, State, and local statutes, and the impact which any new program will have upon them;

5. It is mandatory that both public and private resources be brought to bear upon the social overhead problems; and

6. Technological change has sharpened our realization that the role of the human being in the more sophisticated system becomes even more critical, for the element of control must reside with the manager of the system.

This report presents selected interpretive commentary and relevant background information reflecting the present role and future potential of systems technology in American society. A basis for understanding of systems terminology and applications will be established early in the report. A frame of reference for the reader concerned with governmental development of legislation and community action programs pertinent to a single problem or interlocking problems is supplied through a discussion of congressional consideration through legislation and a concise review of appropriate executive branch activity. In addition to viewing the situation from the vantage point of Federal endeavor, the report elucidates upon State and metropolitan use of ADP, operations research, and other systems development techniques. Candidate areas for possible future congressional, Federal, State, or local agency, or private organizational activity are identified.

This report contains the following major sections. Part I, summary and conclusions, presents an overview of the study prepared for the special subcommittee with an enumeration of the salient issues and items for possible action. The introduction (part II) sets forth the objectives of the report and reflects upon the significance of the problem; the path of action taken by the special subcommittee is traced, followed by the definitions of selected terms which appear throughout the document. Finally, the origins and development of systems technology are described, with emphasis on the current role of the new tools and techniques in helping to solve social and community problems. Part III features a series of questions to be considered by the Congress or its elements, including inquiry into the types of incentives and technical assistance which should be offered to State and local governments, and the important question of public and private organizations' accessibility to centrally stored data.

Part IV offers a perspective of governmental interest in applying systems technology to the problems of the Nation; activities within the executive and legislative branches, as well as intergovernmental endeavors, are noted. Significant legislation designed to encourage the use of systems technology in coping with our social and community problems is discussed in part V. Analysis of the hearings on S. 2662, S. 430, and S. 467 conducted by the Senate Special Subcommittee on the Utilization of Scientific Manpower appears in part VI. Questionnaire responses on State and local use of systems technology are analyzed in part VII, including a comparison of the data received in 1966 and that gathered in 1968. Part VIII contains descriptions of exemplary State and local programs and projects, sometimes supported by Federal funding, where the innovative tools and techniques are employed. In addition, the report contains a series of appendixes (e.g., a listing of pertinent Public Laws indicating major legislation on the topic of the study) and a list of selected references.

D. TERMINOLOGY FOR UNDERSTANDING

Technological developments during the past quarter century have created, in addition to numerous benefits and problems for society, a perplexing problem for those who demand precise definitions of concepts and methods. Many experienced observers have derived definitions which seemed to them to impart the essential elements of a given concept. Oftentimes, not surprisingly, other experts have offered alternative definitions. The working definitions which follow will provide a point of reference for the users of this report.

Systems technology refers to the application of scientific tools and techniques in the analysis and design of decisionmaking, information handling, or product-preparation functions. The merging of traditional problem-solving procedures with the flexibility and exactness of innovative man-machine techniques is featured.

The systems approach—

1. Is a body of highly developed capabilities for the solution of complex problems;

2. Looks at problems through their interrelationships in contrast to the more traditional view, which solely sees large problems consisting of separate parts;

3. Discards the trial and error method, and does not solely rely on the inductive method (i.e., gathering all facts and then analyzing them since the very data chosen can limit the view of the problem and thus restrict the number of possible solutions); and

4. Uses the inductive approach to gather and analyze information while giving paramount importance to the objective through the deductive process.[15]

One element of the systems approach is systems analysis. For many years, there has been a continuing debate about the difference between systems analysis and operations research. Charles J. Hitch of the University of California saw no need to distinguish between the two terms, but has characterized the former in this way:

Systems analysis at the national level * * * involves a continuous cycle of defining military objectives, designing alternative systems to achieve these objectives, evaluating these alternatives in terms of their effectiveness and cost, questioning the objectives and the other assumptions underlying the analysis, opening new alternatives, and establishing new military objectives.[16]

---

[15] A Strategy for planning. A report to the National Governors' Conference by the Committee on State Planning, Oct. 18, 1967, p. 11.
[16] Hitch, Charles J. Plans, programs, and budgets in the Department of Defense. In Operations research, January–February 1963. p. 8.

Systems engineering, has been seen as "the application of scientific and engineering knowledge to the planning, design, evaluation, and construction of man-machine systems and systems components." [17] Systems management consists of directing and controlling a system development program to completion, on schedule, and at the prescribed cost. [18]

One common misconception is the role of automatic data processing in systems analysis. There is no intrinsic connection between the application of computers and the analytic function. Dr. Alain C. Enthoven underscored the importance of understanding the proper role of ADP in his dialogue with Senator Nelson during the special subcommittee hearings in 1966:

One of the primary advantages of the computer to the systems analysis function is to permit us to examine a much larger number of alternatives in a shorter period of time than would be otherwise possible. This is especially important in the case of very complex and interrelated systems where hand calculations would limit the time available for the more important work of analysis. [19]

At this point, a definition of automatic (or electronic) data processing is included, it being the concept

Whereby a machine or computer can accept information or "input data," process the data according to a predetermined "program," and provide the results in a useable form. In an automatic data processing system, the electronic computer is the heart or focal point of the system. An ADP system consists of a number of components including input, processing, storage, and output devices. [20]

As the work of the special subcommittee progressed, the complexity of its task became more evident. A concomitant development was an identified need for meaningful definitions, criteria for deciding when and how to apply the new technology, indicators of progress or its lack, and standardized procedures for communicating selected information on projects' products and status from one governmental entity to another. Commentary on these needs will appear in the statements of witnesses before the special subcommittee (sec. VI) and in excerpts from State and local responses to the subcommittee questionnaires (sec. VII). A glossary of selected terms, featuring additional definitions, constitutes appendix A.

E. The Emergence of Systems Technology

Late in the 19th century, a few farsighted individuals became concerned with the improvement of human performance in a given work environment. Frederick W. Taylor, often called "the father of scientific management," introduced a new philosophy and approach to management. He sought to place management in the role of controller of the operation, as well as the organizer of the work force and the planner of all activities. The time studies undertaken by Taylor and companion efforts by Frank and Lillian Gilbreth in motion study analysis proved to be the first phase in an ever-widening endeavor which has benefited industry, government, and the Nation as a

---

[17] Scientific Manpower Utilization, 1967. Op. cit., p. 35 [remarks of Karl G. Harr, Jr.].
[18] Ibid.
[19] Scientific Manpower Utilization, 1965–66. Op. cit., p. 149 [remarks of Alain C. Enthoven].
[20] U.S. Congress. Senate. Committee on Government Operations. Report No. 938, (89th Cong., 1st sess., 1965). p. 6.

whole.[21] It should be noted that Taylor perceived and emphasized the importance of creating an organizational element dedicated to operational analysis.

During the early decades of the 20th century, management began to realize that not only production, but sales, engineering, and fiscal operations could be improved by applying some of the innovative techniques. With profit as a motive and corporate performance as a basis for evaluation, management decisionmakers invented, tried out, and sometimes abandoned such prediction and review mechanisms as econometric graphs, statistical charts and synoptic narratives, trend lines, general studies, and quantitative intrasystem analyses. Among the larger firms to apply the new approach were Du Pont, Westinghouse, and General Motors. One key development was that executives began viewing the corporate line and staff functions as interrelated elements of the whole.

Management at the Westinghouse Electric Corp. initiated an "operation analysis" program, with performance improvement linked to these 10 steps to a more effective mode of operation:

1. Make a preliminary survey.
2. Determine extent of analysis justified.
3. Develop process charts.
4. Investigate the approaches to operation analysis.
5. Make motion study when justified.
6. Compare the old and the new method.
7. Present the new method.
8. Check installation of the new method.
9. Correct time values.
10. Follow up the new method.[22]

In the years just prior to World War II, scientists involved in the development of military technology (e.g., radar) commenced to examine industry's use of systems technology. In Great Britain, the first interdisciplinary teams were formed, and began to apply "operational research" to numerous military problems. By the end of the war, a significant number of British, Canadian, and American personnel had received training and gained experience in various aspects of systems analysis and operations research.

Peacetime reconversion and the almost immediate inception of the "cold war" presented myriad opportunities for applying the new concepts and skills to private or joint public-private endeavors. A concurrent development in the midforties was the invention of prototype computers. These machines gave the pioneering analysts and operations research specialists unanticipated capabilities in the form of high-speed arithmetic operations, large memories, and an ability to work with alphanumeric information in ways never before possible.

Among those recognizing the importance of harnessing the new tools and techniques was Gen. H. H. Arnold. He and others foresaw the need for a new type of institution which could undertake complex, sometimes long-term, and often innovative studies of high consequence to the Nation and the world. In this setting the RAND Corp. came into being in 1946.[23] Its scientists and technologists studied problems related to defense, arms control, and other critical areas.

[21] Barnes, Ralph M. Motion and time study. London, John Wiley & Sons, 1955. Pp. 12–16. Also see Taylor, F. W. The principles of scientific management. New York, Harper & Bros., 1911. 77 p.
[22] Niebel, Benjamin W. Motion and time study. Homewood, Illinois, Richard D. Irwin, Inc., 1955. p. 3.
[23] Smith, Bruce L. R. The RAND Corporation. Cambridge, Harvard University Press, 1966, p. 332.

From this institution came the men who would one day dominate the national planning, programming, and budgeting activities: Charles J. Hitch, Alain C. Enthoven, and Henry S. Rowen, to name a few. Their impact on the national scene is exemplified by the establishment of the Federal Planning-Programming-Budgeting System (PPBS) and continuing emphasis on improving authorization and appropriations functions within the Congress.

The systems methodology and supporting equipment which had been nurtured in the private sector became an acknowledged facet of Federal management during the administration of President John F. Kennedy. The Department of Defense (DOD) had need of more efficient and expeditious means of handling the varied problems within that establishment. Secretary of Defense Robert S. McNamara had had experience in industry with quantitative analysis, and thus could furnish the leadership needed to execute a changeover to the new tools and techniques. Charles J. Hitch became the Comptroller, with the authority and responsibility for modernizing DOD planning, budgeting, and operational practices and procedures.

The new, systems-oriented mode of focusing on problems and resources was called the Planning-Programming-Budgeting System or simply "PPBS." The system features four distinctive characteristics:

1. It focuses on identifying the fundamental objectives of the Government and then relating all activities to these (regardless of organizational placement);
2. Future year implications are explicitly identified;
3. All pertinent costs are considered; and
4. Systematic analysis of alternatives is performed. This is the crux of PPBS. It involves (a) identification of Government objectives, (b) explicit, systematic identification of alternative ways of carrying out the objectives, (c) estimation of the total cost implications of each alternative, and (d) estimation of the expected results of each alternative.[24]

While there were cases of procedural and attitudinal impediments to change, the diverse DOD components gradually integrated the systems approach into many facets of their operations. Attention was given at the highest levels to the ways in which the decisionmaking function could be refined and upgraded, and the intricacies of the evolving man-machine system were examined continuously.

After PPBS had been in operation for a sufficient period of time to assess its strengths and shortcomings, the decision was made by President Lyndon B. Johnson to place the system in effect in all major executive agencies and establishments. A listing of agencies required, and those encouraged, to establish a formal planning-programing-budgeting system appears as figure 1. The directive by President Johnson to the Bureau of the Budget appeared as Bulletin No. 66–3 in October of 1965.[25]

[24] Hatry, Harry P. *and* John F. Cotton. Program planning for state, county, city. Washington, State-Local Finances Project of the George Washington University, January 1967. p. 15.
[25] Executive Office of the President. Bureau of the Budget. Planning-programing-budgeting. Bulletin No. 66–3 to the heads of executive departments and establishments. Washington, October 12, 1965. 13 p.

19

A. AGENCIES TO BE COVERED BY THE PREVIEW

Department of Agriculture
Department of Commerce
Department of Defense—separate submission for:
    Military functions (including Civil Defense)
    Corps of Engineers, Civil functions
Department of Health, Education, and Welfare
Department of Housing and Urban Development
Department of Interior
Department of Justice
Department of Labor
Post Office Department
Department of State (excluding Agency for International Development)
Treasury Department
Agency for International Development
Atomic Energy Commission
Central Intelligence Agency
Federal Aviation Agency
General Services Administration
National Aeronautics and Space Administration
National Science Foundation
Office of Economic Opportunity
Peace Corps
United States Information Agency
Veterans Administration

B. OTHER AGENCIES FOR WHICH A FORMAL PLANNING-PROGRAMING-BUDGETING
SYSTEM IS ENCOURAGED

Civil Aeronautics Board
Civil Service Commission
Export-Import Bank of Washington
Federal Communications Commission
Federal Home Loan Bank Board
Federal Power Commission
Federal Trade Commission
Interstate Commerce Commission
National Capital Transportation Agency
National Labor Relations Board
Railroad Retirement Board
Securities and Exchange Commission
Selective Service System
Small Business Administration
Smithsonian Institution
Tennessee Valley Authority
United States Arms Control and Disarmament Agency

FIG. 1.—AGENCIES REQUIRED AND ENCOURAGED TO ESTABLISH A FORMAL PLANNING-PROGRAMMING BUDGETING SYSTEM

(Originally appeared as Exhibit 1 in Bulletin No. 66–3 to the heads of executive department and establishments)

The specific procedures to be utilized in implementing PPBS were contained in a supplement to Bulletin No. 66–3, which was disseminated early in 1966.[26] Two forms of documentation were identified and explained in detail: the program and financial plan, and the program memorandum. The program and financial plan provides a listing of program outputs—such as "services" or "equipment"—and program inputs, which are denoted as items of cost or related financial data. The program memorandum contains detailed narrative and statistical data on each program identified in the program and financial plan, and is issued on an annual basis. Additional guidance to

[26] Executive Office of the President. Bureau of the Budget. Planning-programing-budgeting. Supplement to Bulletin No. 66-3. Washington, February 21, 1966. 31 p.

29–364—69——3

departmental heads is contained in Bulletin No. 68–2,[27] issued July 18, 1967 to replace Bulletin No. 66–3 and its supplement.

Since the PPBS operation affects many categories of line and staff personnel, a major effort has been initiated to orient and educate managers, program professionals, and key support personnel. In addition to courses organized by individual agencies, the U.S. Civil Service Commission sponsors a continuing series of seminars, structured by content and length for selected types of executive branch employees. The courses feature material on the forerunner activities of program budgeting—from the recommendations of President Taft's Commission on Economy and Efficiency to the establishment of the Tennessee Valley Authority to the activities of the War Production Board in 1942—as well as the problems faced today by PPBS personnel.

During the past three decades scientific discovery and the extensive utilization of technology in space penetration, nuclear energy development, and defense activities have forced leadership in Government and industry to seek new management controls and operational monitoring systems. In addition to drawing on the methodology and tools of the pre-World War II era, all sectors now rely heavily on the new capabilities made possible through the development of automatic data processing. Today, more than 70,000 computers are in use in the United States; [28] within the Federal Government alone the number exceeds 4,200.[29] With this array of tools, the Nation must now face up to its newest challenges: the specters of pollution, substandard services to the populace, and civil discord caused by inadequate governmental response to the needs of the people.

The importance of arriving at answers to these pressing problems is challenged in no quarter. But, the determination of which way to go, which ameliorative action to take, which resources to expend, has stirred unparalleled controversy. There are numerous questions which require answers, or at the very least a series of alternatives for consideration. The accuracy and applicability of the answers may determine the strength and stability of the Nation in years to come. This report will provide facts, selected commentary from both the public and private sectors, analyses and interpretations germane to the individual and intermeshed problems of the civil area. The emphasis will be increasingly upon the role of systems technology in the alleviation or solution of these nondefense, nonspace national problems.

[27] Executive Office of the President. Bureau of the Budget. Bulletin No. 68–2 to the heads of executive departments and establishments: Planning-programing-budgeting (PPB). Washington, July 18, 1967. 25 p.
[28] Computers and Automation. Newtonville, Mass., Berkeley Enterprises, Inc., October 1968 [See Monthly Computer Census, pp. 70–73.]
[29] General Services Administration. Federal Supply Service. Inventory of automatic data processing equipment in the U.S Government. Fiscal year 1968. Washington, U.S. Government Printing Office, 1969. p. 17, chart 1. [The actual number of computers for the end of fiscal year 1968 is 4,232].

III. QUESTIONS BEFORE THE CONGRESS

During the preparation of this report, a number of questions has been suggested which could warrant further consideration either in public forums, discussions with university and industrial consultants or staff seminars. The question categories have not been arranged in any priority nor has any connotation of public policy determination been set forth. In some cases, the questions transcend the aegis of the Senate Special Subcommittee on the Utilization of Scientific Manpower. The significance of the report as reflected in these questions, however, may merit future reflection and review by the Congress to the end that the needs of Government at all levels, and the citizenry throughout the Nation, be recognized and that successful corrective action be carried out.

A. ROLE OF THE FEDERAL GOVERNMENT

1. In what manner should the Federal Government, through one or more of its departments:

(*a*) Strive to bring together the Federal, State, and local governmental groups, with their range of serious problems, and the technological community with its array of new tools and techniques?

(*b*) Underwrite orientation conferences and problem-oriented workshops for State and local public and private groups?

(*c*) Review past efforts at educating and motivating industrial, academic and citizen groups in the use of systems technology in priority problem areas?

2. Should quality and performance criteria for public systems (i.e., transportation, sewage, etc.) be established as a prerequisite to systems analysis? What is the Federal role in identifying such criteria?

3. Is there a viable substitute to surveying the Federal, State, and local projects employing systems technology—as proposed in S. 467 and H.R. 20? What should be the Federal Government role in identifying alternatives? Other potential areas of application?

4. Should Federal computer facilities be made available on a time-charge basis by State and local governments? Under what terms? How would undesirable or unfair competition with private concerns be prevented?

5. How adequate is present Federal regulation of the communication of information between computers in interstate commerce? Is additional regulation needed? Should an existing or a new Federal agency perform this regulation?

B. MANPOWER CONSIDERATIONS

1. What criteria can be established to assure a more equitable distribution of trained manpower for automatic data processing and

systems analysis among the Federal, State, and local governments and supporting industrial and academic institutions?

2. How can State and local personnel experienced in public problem solution be brought to the Federal agencies which have a responsibility for furnishing funds for project development?

3. To what extent can visiting Federal experts provide technical assistance to State and local authorities?

C. Incentives and Technical Assistance

1. What incentives could be offered to State and local governments (by the Federal Government) which have commenced using systems tools and techniques in various operations?

2. Should special tax incentives, either State or Federal, be offered the private sector manufacturer, retailer, or transporter, to modernize his management or operations through the use of systems technology?

3. What is the extent of technical support and assistance now provided to State and local governments in automatic data processing and systems analysis by Federal agencies? Are there deficiencies or gaps in such assistance by the Office of Economic Opportunity, the Department of Housing and Urban Development, the Department of Transportation, and the Department of Health, Education, and Welfare?

4. Could such assistance be expanded without adversely affecting these agencies in performing their assigned functions?

5. Will a grants-in-aid approach, as initiated at the Federal level, provide a lasting motivation for State and local implementation of pilot systems which incorporate state-of-the-art techniques and machinery? Should such grants be awarded directly to the private sector, without State (or local) review or approval?

6. What new institutions, perhaps jointly sponsored, might be created within a State or region to facilitate the development of systems methodology germane to specific public problems?

D. Education and Orientation

1. What type of meaningful education can be presented to the leadership of political jurisdictions as the employment of data collection and measurement devices on an areawide basis is considered? Can both technical and nontechnical considerations be portrayed by an outsider so that the responsible leadership can accept the situation as presented, and respond?

2. In what ways should Federal systems analysis experts be used in preparing and conducting ADP-supported demonstrations and special training courses for interested State and local governmental personnel and those private groups involved in coping with such public problems as urban renewal, et cetera?

3. How might the academic community be encouraged to develop special curricula to teach the applications of systems technology to selected problems of State and local government?

(a) Should the Senate Committee on Labor and Public Welfare or some other congressional body assert its encouragement to this end?

(*b*) Should some Federal agency not only encourage but also finance the development of such curricula? Would this require authorization by Congress?

4. In what ways can the professional organizations—Operations Research Society of America, Association for Computing Machinery, American Society for Information Science, The Institute for Management Science, et al—develop educational offerings which are aimed specifically at applying systems technology to the problems of society?

5. In what specific areas can not for profit and profit organizations offer theoretical and practical assistance to regional, State, and local governments?

E. Accessibility to Key Data

1. In what ways can problem-oriented information about the characteristics and applications of automatic data processing, systems analysis techniques, and systems-supporting tools and procedures be made available to key State and local governmental decisionmakers?

2. To what extent should the statistical information collected by various Federal agencies, such as the Bureau of Labor Statistics, the U.S. Employment Service, and the Bureau of the Census be made accessible to private and public users? Under what terms and conditions? Should the responsibility for deciding who can get access to such information reside in the individual agencies?

3. Should special emphasis be placed on implementing regional information handling systems (e.g., those proposed for the San Gabriel Valley cities or the network projected for the States of Louisiana, Texas, and New Mexico, with old Mexico included)? Should the information already stored in the St. Louis Regional Industrial Development Corporation (RIRA) economic data bank be available to all categories of users? Should such an information center discriminate in its dissemination of data?

4. Have market analyses been prepared which measure the extent to which State and local governments would utilize ADP hardware and softwear, intercomputer communications devices, and analog measuring equipment? Could this information be available in a nonproprietary form to public and private sector groups?

F. Intergovernmental Information Exchange

1. Is there a requirement to establish a mechanism for transferring information about problem-oriented systems management techniques from Federal agencies to State and local functionaries? From a State with sophisticated systems methodology to a developing State?

2. To what extent would the establishment of an information service system, as proposed in Senate Joint Resolution 110 (90th Congress), with its emphasis on utilizing modern technology in the exchange of priority project data fulfill the need of State and local government for this useful information?

3. Should the Federal Government prepare guidelines to determine the nature and functioning of a national information exchange? Should Congress rely upon the executive branch to take the initiative, or should Congress itself do so?

IV. A PERSPECTIVE OF GOVERNMENTAL ACTIVITY IN APPLYING SYSTEMS TECHNOLOGY TO SOCIAL AND COMMUNITY PROBLEMS

The range and intensity of the social and community problems within the United States have caused a host of governmental components to become involved in policy formulation, financial transactions, interjurisdictional arrangements, information exchange, research and development, and program implementation and monitoring. In this section, a concise review of the activities of selected congressional groups and Federal executive branch organizations will be presented, together with commentary on certain intergovernmental efforts. Pertinent background information concerning the origins and evolution of the systems approach already has been offered in section II.

The significance of the degree of governmental involvement in this area should not overshadow the rising interest and endeavor of private sector elements. The commitment of funds and personnel in systems-supported projects is reflected in a number of the innovative efforts described in section VIII, and discussed by various witnesses before the special subcommittee (see section VI). The following sections indicate the types of activity sponsored or conducted by Federal Government groups, and where appropriate intergovernmental organizations, both in the past and at present.

A. Senate Special Subcommittee on the Utilization of Scientific Manpower

With the creation of the Special Subcommittee on the Utilization of Scientific Manpower within the Senate Committee on Labor and Public Welfare in 1965, the Congress had for the first time a suitable mechanism for examining the full potential of systems technology in lessening or solving social and community problems. Under the chairmanship of Senator Gaylord Nelson, who had extensive experience at the State level in using the systems approach in diverse planning and program functions, the special subcommittee created a focus and a forum for the philosophers, the students, and the practical men who would consider how the new tools and techniques were to be used effectively.

The course of action pursued by the special subcommittee featured three major efforts:

1. A series of hearings were held, with the solicitation of testimony from 38 carefully selected witnesses representing Federal and State governmental units (including the Congress), universities, foundations, industry, associations, and labor. A review and analysis of these hearings are contained in section VI.

2. Two questionnaires were prepared and disseminated, in 1966 and 1968, for the special subcommittee. These were dis-

tributed to all States, more than 20 large cities, and a few regional development groups. The emphasis in this inquiry was on the areas in which systems analysis and automatic data processing were being used, the level (and type) of manpower resources allocated, and the funding levels (with a notation as to Federal support, where appropriate). The responses to these questionnaires are analyzed in depth in section VII of this report.

3. The decision to have this report prepared by the Science Policy Research Division of the Legislative Reference Service at the Library of Congress, so that a reference work on the subject could be issued by the Congress for use by all interested persons in the public and private sectors. The "Scope and Methodology" of the report are set forth in II. B.

The pattern of exploration into the applicability of systems analysis, operations research, and automatic data processing to the complex and often interlocking civil public problems featured a desire on the part of the special subcommittee to evaluate existing legislative proposals (i.e., S. 430 and S. 467) and to intensify public awareness of the fact that such problems were as critical as those related to the defense of the United States. At one point, Senator Nelson reaffirmed his belief that:

> The problems faced by this Nation at the State, regional, and national level in the field of education, welfare, conservation, transportation, pollution, urban planning, and land use control * * * are certainly no less urgent or complex than the problems faced by the Defense Department.[1]

B. Other Congressional Activity

The Congress has reflected a sense of urgency in recent years in its handling of legislation and associated activities concerned with social and community problems. The responsibility for policy development, action on planning guidelines and processes, commitment of resources to program initiation and continuance, and at least minimal monitoring of on-going Federal and joint projects has been evident in chamber debate, numerous hearings on key topics or bills, published reports, and the public statements of individual Members.

A review of selected portions of these activities is contained in this section, and should be used in conjunction with related material found in sections II.E. (on the creation of the Federal PPB system), VI.C. (featuring witness testimony on PPBS), all of V. (on significant legislation), and those projects at the State and local level involving Federal support which are discussed in VIII. Information included in appendixes H, I, and L also is pertinent.

As the congressional elements have begun to realize the full potential of the computer and its ancillary devices, mathematical models, and the various techniques for projecting cost and manpower needs and allocations, they have listened with increasing care to those who have originated and used (often experimentally) the new technology. Throughout government, this fact has become understood and accepted: systems technology can perform many useful functions but the ultimate responsibility for that which the system produces resides with the human being who must make the judgments. Dr. Donald N. Michael stated the situation in these terms:

---

[1] Scientific manpower utilization, 1965–66. Op. cit., p. 145. [Remarks of Senator Gaylord Nelson].

In all, top leadership in many organizations will find that while the computer relieves them of minor burdens, it will enormously increase the demands on them to wrestle with the moral and ethical consequences of the policies they choose and implement.[2]

Senate Subcommittee on National Security and International Operations

Cognizance of the impact of PPBS on the Federal authorization-appropriations activity has led the Congress to hold several hearings and prepare various reports on the role of PPBS. The Senate Special Subcommittee on the Utilization of Scientific Manpower, as discussed in sections V and VI of this report, took the lead in soliciting comment on the objectives, strengths, and shortcomings of the new system. Also active in conducting exploratory hearings in the PPBS area was the Subcommittee on National Security and International Operations of the Senate Committee on Government Operations. In opening the hearings in August, 1967, chairman Senator Henry M. Jackson of Washington stressed that " * * * our Senate subcommittee has had a continuing interest in the role of budgetary process in helping plan and control national security policy."[3] He went on to say that the subcommittee could assist in a "frank stock-taking of the benefits and costs of the planning-programing-budgeting system."[4] Extensive testimony was presented by Charles L. Schultze, then Director of the Bureau of the Budget, and Dr. Alain C. Enthoven, which proved useful in providing definitive material for the subcommittee on questions of PPBS policy, control, evaluation criteria, and individual agency experience.

Seven committee prints were prepared during 1967 and 1968 by the Subcommittee on National Security and International Operations dealing with various aspects of PPBS: "Official Documents," "Initial Memorandum," "Selected Comment," "PPBS and Foreign Affairs," "Budget Bureau Guidelines of 1968," "Program Budgeting in Foreign Affairs: Some Reflections," and "Interim Observations."

Subcommittee on Economy in Government (Joint Economic Committee)

In September of 1967, the Subcommittee on Economy in Government of the Joint Economic Committee convened a series of hearings entitled "The Planning-Programming-Budgeting System: Progress and Potentials." The chairman, Senator William E. Proxmire of Wisconsin, described the concern of his group in improving management in Government. Noting that Federal components are responsible for handling approximately $175 billion annually, he went on to say:

Certainly at a time when approximately 30 percent of our national income flows through the public sector, it is of the utmost importance that our policy-makers be armed with the best possible tools for evaluating the effectiveness of our public programs and expenditures.[5]

It was pointed out that the intrinsic worth of PPBS is debatable, with supporters of the system claiming "that for the first time it provides decision-makers with a rational basis for choosing between alternative policies" while critics "view PPBS advocates as a new breed

---

[2] Michael, Donald N. Some long-range implications of computer technology for human behavior in organizations. In the American Behavioral Scientist, v. 9, April 1966, p. 35.

[3] U.S. Congress. Senate. Committee on Government Operations. Subcommittee on National Security and International Operations. Planning-programing-budgeting. Hearings before the Subcommittee (90th Cong., 1st sess., Aug. 23, 1967), p. 1 (Remarks of Senator Henry M. Jackson).

[4] *Ibid.*

[5] U.S. Congress. Joint Economic Committee. Subcommittee on Economy in Government. The planning-programing-bugeting system: progress and potentials. Hearings before the subcommittee (90th Cong., 1st sess., Sept. 14, 1967), p. 1. (Remarks of Senator William E. Proxmire).

of technocrat who think [sic] the computer can take the politics out of decision-making." [6] The hearings featured testimony by representatives of such civilian agencies as the Department of Health, Education, and Welfare, the Department of the Interior, and the Department of Housing and Urban Development. State and local applications of PPBS—for example, Wisconsin, New York City, and Vermont—were discussed, and future applications of the PPBS approach were noted by Government and university witnesses.

In addition to the congressional review of PPBS through public hearings, pressure has been brought to bear on the executive branch agencies to use systems technology to attain "strengthened management practices." During the 90th Congress, terminology was included in an amendment to the Foreign Assistance Act of 1961 (see H.R. 12048) to the effect that:

The Congress believes that United States foreign aid funds could be utilized more effectively by the application of advanced management decision making, information and analysis, automatic data processing, benefit-cost studies, and information retrieval.[7]

The potential of management science in enhancing the national planning function and the monitoring of a broad cross section of agency programs *is* very great. Similarly, the legislative branch in performing its traditional role may well be affected. Dr. Richard F. Fenno, Jr. of the University of Rochester has this to say about the effect of PPBS on the role of the Appropriations Committee:

1. Will PPBS in any way keep appropriations subcommittees from having the kind of sampling-type conversations with executives which legislators find advantageous?
2. Will PPBS help subcommittee members to pursue their traditional mode of decisionmaking?
3. Will PPBS open up new types of legislative-executive conversation that will be advantageous to subcommittee members? [8]

Senate Subcommittee on Executive Reorganization

The question of the proper role of PPBS was only one of several which attracted congressional inquiry, as it considered the difficulties of governing in such a way that pressing social and community problems could be overcome. Broad investigation of the role of the Federal Government in urban affairs was undertaken in the 1966–67 period by the Subcommittee on Executive Reorganization of the Senate Committee on Government Operations. A series of hearings documents (in 21 parts) were published, containing more than 4,000 pages of testimony and supporting commentary on the subject of the "Federal Role in Urban Affairs." In opening the hearings, chairman Senator Abraham Ribicoff of Connecticut stated that the mission of the hearings was to "come up with the understanding necessary to insure that our efforts are organized in the most efficient, effective, and coordinated manner." [9] During the first session, Senator Ribicoff asked a series of

[6] *Ibid.*, p. 1–2.
[7] U.S. Congress. House. H.R. 12048. A Bill to amend further the Foreign Assistance Act of 1961, as amended, and for other purposes. (90th Cong., 1st sess., Aug. 2, 1967), p. 38.
[8] Fenno, Richard F., Jr. The impact of PPBS on the congressional appropriations process. *In* Information support, program budgeting, and the Congress (Chartrand, Robert L., Kenneth Janda, and Michael Hugo. *ed.*) New York, Spartan Books, 1968, pp. 184–185.
[9] U.S. Congress. Senate. Committee on Government Operations. Subcommittee on Executive Reorganization. Federal role in urban affairs. Hearings before the Subcommittee (89th Cong., second sess., Aug. 1966, 1), pp. 3–4.

critical questions which reflected the need for Federal policy in the complex problem area, and indicated that new approaches were called for:

Are the techniques of our city aid programs obsolete and limited? Do they reflect the needs and conditions of national life a generation ago, and not the needs of modern America?

Is the effectiveness of the programs that do exist hurt by division of authority among many agencies, and many levels of government?

Most serious of all, do the goals of major Federal programs conflict—some working to revitalize the central city, some encouraging new urban clusters, some causing regional sprawl?

In short, do we have a clear, constructive national strategy geared toward the improvement of our cities? If not, what steps must we take to obtain it? [10]

Throughout the massive documentation is commentary on Federal, State, and local efforts to alleviate the depressing effects of the various civil sector problems; articles, speeches, and editorials by concerned individuals ranging from Congressmen to ministers to social workers; descriptions of community action projects using innovative approaches; industrial proposals to plan better cities; specific recommendations by experts in urban problems; discussion of minority (Negro, Indian, Spanish-American) living conditions; legislative program packages (e.g., the "Ribicoff package"); and information particularly germane to the use of computers and systems analysis in certain problem areas.

Among the topics discussed by the members of the subcommittee and its witnesses were the "geo-coding" of urban data and the role of city and State data banks (with Richard M. Scammon, vice president of the Governmental Affairs Institute); [11] the establishment of a "unified, computerized employment center" as proposed by the National Commission on Technology, Automation, and Economic Progress (and commented upon by Walter P. Reuther of the AFL–CIO); [12] and the role of technology in designing new cities, which led witness John H. Rubel (vice president, Litton Industries) to comment: [13]

In short, if one could set up a project for the creation of a new city from scratch, and offer the job to private industry, and set up project goals in terms of the performance of a dynamic system, new industries would spring up within the framework of existing firms to meet the new needs. Soon the multi-disciplinary teams would be assembled. The relevant analytical techniques would be applied. The new methods, the new technologies, the new insights would begin to emerge. The new insights and the new technologies would include wholly new species of engineering and technology. Social engineering, the applied counterpart of social science, would take form. The social sciences, increasingly linked to agencies for action, would develop more swiftly and with greater certainty than before. Now they would have new experimental situations at their disposal and the means to test and to compare hypotheses and concepts.

One can predict that they would begin to shift away from the descriptive phase with which all sciences begin. They would become increasingly manipulative, as the physical sciences have. Wind tunnels and airplanes are needed in order to design the next airplane, and after that the next, in a never-ending series in which the products of an advancing technology become the tools used by scientists and technologists to make further advances. Likewise, it is necessary to be able to design and construct cities under the aegis of city-specifying, city-

[10] *Ibid.*, p. 2–3. [Remarks of Senator Abraham Ribicoff.]

[11] Federal role in urban affairs. (89th Cong., second sess., Nov. 29, 1966, pt. 7), pp. 1485, 1488. [Remarks of Richard M. Scammon.]

[12] Federal role in urban affairs. (89th Cong., second sess., Dec. 5, 1966, pt. 8), p. 1847. [Remarks of Walter P. Reuther.]

[13] Federal role in urban affairs. (89th Cong., second sess., Dec. 7, 1966, pt. 10), pp. 2152–2153. [Remarks of John H. Rubel.]

building project auspices so that the next city-project will reflect the changed objectives and the more certain insights evolved during the first experience, all this in a never-ending series. For it is out of such a series that a "city technology" will be created analogous to "space technology." Again, it will be done not by borrowing the technologies or methods intact from the weapons or the space field, but by (a) imitating the applicable features of the organizational approaches that have given us modern, large-scale technology, and (b) creating a market in the form of projects in a framework which encourages the evolution of a new articulation of public purpose with private means. If there is a key to drawing forth the swift evolution of a "city technology" that might some day inspire a measure of that same awe inspired by our burgeoning achievements in space, this may be that key.

Another witness of note was Dr. Jerome B. Wiesner, provost of the Massachusetts Institute of Technology and former Science Adviser to Presidents Kennedy and Johnson. During his lengthy testimony, he stressed over and over the importance of understanding how to utilize technology in the urban area. In commenting on the absolute necessity of having the right information on hand, when it is needed, he also placed in proper focus the role of the computer: [14]

* * * you could make models of a new system, estimate costs, anticipate where your bottlenecks are going to occur, and so forth. If a slum is to razed, we could predict, if we had the right data, where its people may be relocated, at what prices; whether we would be pushing rents upward by creating untenable demands on a tight housing market; whether there will be important changes in patterns of commuting; whether our efforts at desegregation will be affected; whether the apartments and homes which are available for relocation are in school districts which can absorb the children; whether teacher reassignments will help to solve the school crowding problems, and so on.

It is the availability of the new computer technology which makes possible getting answers to such questions, but the computers themselves are not going to provide the answers. The computer is only a tool, but it is a tool of extraordinary utility and capacity. The data upon which the computer analyses are based depend upon elaborate and painstaking research, research which today is being done at far too modest a level to provide us with the information we need to manage the very large problems we have undertaken.

Later in his remarks, Dr. Wiesner gave his views on man's changing relationship with his environment, and the role of information in allowing him to render comparisons and establish goals: [15]

Technology like the city is man's brain child. In the course of evolution, the emergence of man seems linked to the acquisition of two critical skills: a new level of tool-making and tool-using and an enormously enhanced ability for communication among the members of a species. Once it became possible to handle information via a powerful system of spoken (and later written) symbols, i.e., language, the basis for the development of both technology and cities was given. When man learned to add the symbols of mathematics, he became ready to investigate meaningfully the regularities of nature.

Our most recent addition of computers—with their enormous capabllities of abstracting, storing, handling, tracking and retrieving information enhances the ability of the human brain to deal with the complexities of the molecules of life, the mysteries of the cosmos, man's urban existence. Through his increasing knowledge of the natural world, through his technological muscle, man has intervened in the evolutionary process. This is a very basic change in his relationship with the physical environment. And now he must better appreciate this fact in order to develop the wisdom to direct the process in more desirable ways. As man learns to apply his understanding of the physical world for practical purposes, he is, in reality, substituting a goal-directed evolutionary process in his struggle against environmental hardships for the slow, but effective, biological evolution which produced modern man through mutation and natural selection. By intelligent intervention, man has greatly accelerated the evolutionary process and has greatly expanded the range of possibilities open to him.

[14] Federal role in urban affairs. (90th Cong., first sess., Apr. 19, 1967, pt. 16), p. 3250. [Remarks of Dr. Jerome B. Wiesner.]
[15] Ibid., p. 3255–3256. [Remarks of Dr. Jerome B. Wiesner.]

However, he has not altered the fundamental fact that the process still remains trial and error. This means that is is subject to the old hazards of the evolutionary process, such as the danger of going down dead end paths and even of producing dinosaurs—both individuals and states—unfit to survive satisfactorily in the world that evolves.

Any learning process requires a feedback of information for a comparison of the accomplishment with the goal. While a variety of measures are used to provide feedback, the principal ones for society at large are the profit and the vote. In our country, as in many others, we believe that the political system should reflect equally the goals of all, and the feedback systems in operation attempt to reflect this view. In fact, the vast social changes of the past quarter century are efforts to approach the goal of equality more fully.

Trial and error are basic to all learning, including that involved in technical and social progress. In this instance the process is not controlled by chance alone, as in the case of biological evolution. Man's intelligence intervenes, as I indicated, and directs the process which remains, nonetheless, basically experimental in nature.

One exhibit included in the hearings documentation (part 16) which discussed the "systems" nature of urban complexes is John P. Eberhard's "Technology and the City"; at one point he emphasizes that: [16]

What is becoming clear is that we need to recognize the "systems" character of the city. I know this term "system" is overworked, but the concept is so powerful that it is displacing earlier notions of how to view the city. It is particularly useful if we want to look at technological opportunities, because it shifts the emphasis from city components—such as a house, an automobile, or a garbage dump—to the larger context in which these components are placed. For example, improving the means of consuming garbage at a "dump" will not affect the larger system of storage at the house, collection from this place of storage, transportation means from storage to dump, etc. The aggregate of the men, material, and facilities associated with a reasonably well defined function need to be viewed in toto. The system can then be examined from three points of view: its technological sophistication, its contribution to the effectiveness of our cities, and the cost of providing this capability. In order to do this, we first will need to put into perspective what a city system is, and then we will need measures of effectiveness. The first task is relatively easy, but the second is very difficult because it involves value analysis for what we might consider "the good life" or "a worthy civilization" or a "Great Society."

Specific reference to the use of computers in the construction industry—for example, the "School Construction Systems Development" project discussed by Dr. Donald Schon in section VI. D—is made in an exhibit prepared by James Alcott of the Midwest Research Institute.[17] Further treatment of this application area is found in part 19 of the hearings coverage, where the use of computers in "design, production scheduling, subcontracting, and bid estimating" is noted.[18] Other examples of the use of computer technology in housing construction and redevelopment are described in section VIII. C of this report.

Senate Subcommittee on Intergovernmental Relations

Within the Senate Committee on Government Operations, as has been shown already, several active subcommittees have been investigating diverse aspects of intergovernmental relations and the role of technology in society. In some instances, hearings have been held on proposed legislation, such as the Intergovernmental Cooporation Act which became law (Public Law 90–577), the Intergovernmental Personnel Act, and the Intergovernmental Manpower Act. The former law is

[16] *Ibid.*, p. 3327. [Remarks of John P. Eberhard in exhibit 216].
[17] *Ibid.*, p. 3341–3342. [Remarks of James Alcott in exhibit 217].
[18] Federal role in urban affairs. (90th Cong., first sess., May 16, 1967, part 19), p. 4128–4129. [Bureau of Labor Statistics Bulletin No. 1474, February 1966, exhibit 266].

discussed in sections V. D, and E. The hearings on these bills were conducted within the aegis of the Subcommittee on Intergovernmental Relations (Senator Edmund S. Muskie, of Maine, chairman), which also held a series of hearings in 1966 on "Creative Federalism." One featured element of this activity was the investigation into the desirability of authorizing (as set forth in S.J. Res. 187, 89th Cong.) "a study and investigation of an information service system for States and localities designed to enable such States and localities to more effectively participate in federally assisted programs." The hearings documentation for the above activities is listed in the "Selected References."

Senate Subcommittee on Government Research

During 1967, the Subcommittee on Government Research, chaired by Senator Fred R. Harris of Oklahoma, held hearings (one in the form of a seminar) on the Full Opportunity and Social Accounting Act. In discussing the application of management science to social problems, sponsor Senator Walter F. Mondale of Minnesota emphasized that the proposed system of social evaluation would—

    1. Sharpen our quantitative knowledge of social needs;

    2. Allow us to measure more precisely our progress toward our social objectives;

    3. Help us to evaluate efforts at all levels of government;

    4. Help us to determine priorities among competing social systems; and

    5. Encourage the development and assessment of alternative courses without waiting until some one solution had belatedly been proved a failure.[19]

The activities of the Subcommittee on Employment and Manpower of the Senate Committee on Labor and Public Welfare have been discussed elsewhere (section II. B) in this report. The 8 days of hearings held in 1965, together with an array of contributory comments, are documented under the title "Impact of Federal Research and Development Policies on Scientific and Technical Manpower" (see listing in "Selected References").

Subcommittee on Urban Affairs (Joint Economic Committee)

The Joint Economic Committee, in addition to its scrutiny of the Federal Planning-Programming-Budgeting System (see above), has undertaken studies of priority problems. At the committee level, a report was prepared entitled "Employment and Manpower Problems in the Cities: Implications of the Report of the National Advisory Commission on Civil Disorders." The Subcommittee on Urban Affairs under the chairmanship of Representative Richard Bolling of Missouri held hearings in the autumn of 1967 on "Urban America: Goals and Problems." The two volumes of material issued by the subcommittee are listed in the "Selected References." Chairman Bolling, in preparing for the hearings, invited over 20 specialists to contribute papers on various aspects of urban problems, trends, and solutions. These were published in order to assist in stimulating "new thinking about urban goals and problems, and to find out the areas of disagreement among the experts about the nature of the most pressing problems and the

---

[19] Mondale, Walter F. Some thoughts on "stumbling into the future." In remarks of Hon. Fred R. Harris, Congressional Record. [Daily ed.] (Washington), v. 114, Jan. 30, 1968, p. S628.

approaches that should be taken toward their solution." [20] This subcommittee published, in addition, "A Directory of Urban Research Study Centers," containing two types of organizations: university-sponsored study centers and nonprofit research institutes.

Subcommittee on Economic Progress (Joint Economic Committee)

The Subcommittee on Economic Progress of the Joint Economic Committee, with Representative Wright Patman as chairman, published a compendium of papers on "Federal Programs for the Development of Human Resources" in 1968. While the role of technology is not stressed, there is a lucid discussion of various key problem areas such as health care and improvement, and housing and man's environment. Manpower and education considerations are featured, and complement much of the exploratory work of the Senate Committee on Labor and Public Welfare.

Senate and House Committees on Public Works

The Committees on Public Works in both Chambers have long records of legislative activity reflecting an awareness of the importance of systematic planning, adequate information, and new technology in carrying out operational programs. As early as 1952, for example, the House Committee on Public Works was exploring the methods of cost-benefit analysis and allocation in connection with water project planning.[21] Later, a Select Committee on National Water Resources was created in the Senate and embarked on a schedule of widespread hearings (961 witnesses testified) and data gathering. The recommendations of the select committee covered five policy areas; the basic recommendations were: [22]

> 1. A national effort to prepare and keep up-to-date, comprehensive, basin development plans for all major U.S. rivers;
> 2. Aid to States for long-range planning of water development;
> 3. Coordinated Federal research on water utilization;
> 4. Biennial supply-demand analyses of U.S. water resources; and
> 5. Measures to improve efficiency of water development and use.

Senate Select Committee on Small Business

The Senate Select Committee on Small Business also has been concerned with the role of technology in our developing economy. Following hearings in 1963 on "The Role and Effect of Technology in the Nation's Economy," the select committee continued to be active in the area of concern. In 1967, a report entitled "Policy Planning for Technology Transfer" was published, containing an analysis of issues in obtaining the maximum benefits from Federal investments in scientific research and development. A later study, "Automatic Data Processing and the Small Businessman" (1968) explored the potential of computer technology as it might aid one sector of the urbanized economy.[23]

[20] U.S. Congress. Joint Economic Committee. Subcommittee on Urban Affairs. Urban America: Goals and Problems. Hearings before the subcommittee (90th Cong., first sess., Sept. 27, 1967), p. 2. [Remarks of Representative Richard Bolling].

[21] U.S. Congress, House, Committee on Science and Astronautics. Subcommittee on Science, Research, and Development. Technical information for Congress. Report to the subcommittee (91st Cong., first sess., April 25, 1969), pp. 504–505).

[22] Ibid., p. 505.

[23] U.S. Congress, Senate, Select Committee on Small Business. Automatic data processing and the small businessman. (90th Cong., second sess., 1968), 148 p.

House Subcommittee on Science, Research, and Development

Within the House of Representatives, several committee elements have addressed themselves to the ways in which ADP and systems analysis can be of assistance in treating social and community problems. The Subcommittee on Science, Research, and Development of the House Committee on Science and Astronautics has explored policy issues in science and technology. In a committee print entitled "Environmental Pollution—A Challenge to Science and Technology," this recommendation is made:

> To place pollution abatement on a comparable basis with other national technology programs, systems analysis and management capability should be established within the Federal Government. This approach should be used along with the "planning, programing, budgeting" technique to organize both near and long term Federal research and operational efforts in pollution abatement.[24]

The subcommittee, chaired by Representative Emilio Q. Daddario of Connecticut, also held a "Technology Assessment Seminar" in 1967 which brought together a group of senior individuals dedicated to providing expertise and thought to this charge as expressed by the chairman: [25]

> The past few years have brought a change in attitude toward science and technology, both in the public which is more technically literate, and in the Congress with its enhanced understanding. Faith in science, and awe of technology, have been supplanted by a recognition of a grave responsibility for decision—that is, what should we do with what we know?
> Technology assessment is a major key to discharging that responsibility. We are now turning to the natural sciences and asking them to move further in achieving a collective wisdom with politics, law, economics, and social interests for the management of technology. The evolution of a successful assessment process will require careful thinking from many sectors within our culture. The technical community must introduce a concern for public policy impacts and methods of operation—at an early point and in considerable detail.
> We believe that technology assessment will involve the scientific method and will be largely accomplished by scientists and engineers. But the purpose of assessment is to enable decisions for the public good. Up to now, the most important work on the assessment concept has been done by persons and groups outside of the natural science fields. These projects are characterized by the terms "science and culture" and "technology and society."

The Subcommittee on Science, Research, and Development was able to exert an unusually broad impact in the area of environmental pollution control through its careful management of inhouse staff resources, and support available from the Science Policy Research Division (Legislative Reference Service, Library of Congress), the National Academy of Science, and the National Academy of Engineering. Finally, on July 17, 1968, a "Joint House-Senate Colloquium To Discuss National Policy for the Environment" was held under the joint sponsorship of the House Committee on Science and Astronautics and the Senate Committee on Interior and Insular Affairs.[26] The purpose of the colloquium was to produce—

> a crystallization of the elements of a national policy for the environment;

---

[24] U.S. Congress, House, Committee on Science and Astronautics; Subcommittee on Science, Research, and Development. Environmental pollution—a challenge to science and technology. (89th Cong., second sess., House Report, committee print), p. 7.

[25] U.S. Congress, House, Committee on Science and Astronautics; Subcommittee on Science, Research, and Development. Technology assessment seminar. Proceedings before the subcommittee. (90th Cong., first sess., Sept. 21, 1967), p. 2. [Remarks of Representative Emilio Q. Daddario].

[26] U.S. Congress, Senate, Committee on Interior and Insular Affairs. And House, Committee on Science and Astronautics. Joint House-Senate colloquium to discuss a national policy for the environment. Hearing before the two committees. (90th Cong., second sess., July 17, 1968), 233 p.

an identification of common interests and ideas among the participants;

an imparting to the executive branch of congressional attitudes toward environmental quality, regulatory implementation, and organization;

an increased public awareness of the need for environmental quality controls; and

a summary of the alternative institutional, economic, legal, and political mechanisms within which a realistic environmental policy can be carried out.

House Subcommittee on Research and Technical Programs

The Subcommittee on Research and Technical Programs (Representative Henry S. Reuss, of Wisconsin, chairman) of the House Committee on Government Operations has been pursuing a line of inquiry regarding the efficacy of Federal R. & D. expenditures in meeting national goals. Criticism of the small sums being set aside for programs such as sewage treatment, transportation network planning, and pollution control is indicated in comments by subcommittee members, witnesses, and in the ancillary material published as "The Federal Research and Development Programs: The Decisionmaking Process," a 1966 GPO volume. Representative Reuss raised the complaint that while the United States is spending $3 billion annually to place a man on the moon, nothing is being allocated for research on how to develop "entirely new systems of urban transportation that will transport people speedily, safely, economically, and without ruining our cities or polluting our atmosphere." [27] He continued in a similar vein, comparing the "$560 million for further development of nuclear reactors, but only $1 million for the development of a better sewage treatment plant needed by most cities in the country." [28]

Recurring again and again in the growing volume of commentary on the role of technology in helping reduce or solve some of the social and community problems is the theme that there are no easy answers. In every case, decisionmakers and their planning counsels take calculated risks. In his now classic piece on "Long-Range Planning: Challenge to Management Science," Peter F. Drucker suggests that—

* * * while it is futile to try to eliminate risk, and questionable to try to minimize it, it is essential that the risks taken be the *right risks*. [29]

The Congress, as the record shows, has worked assiduously to create mechanisms whereby expert opinion could be drawn—from the universities, foundations, industry, and all levels of government— which would be helpful to the legislators as they sought to formulate new, responsive policies. Furthermore, the committee and subcommittee endeavor has resulted in many cases in a definition of goals, clarification of the problems, and a gradual cohesion of approach. Congressmen have found themselves interfacing with representatives of regional commissions (such as those listed in appendix J), councils (appendix K), and those State and local officials with whom they

[27] U.S. Congress, House, Committee on Government Operations; Subcommittee on Research and Technical Programs. House Government Operations Subcommittee to undertake study on how well $16 billion annual research and development budget contributes to meeting our national goals. Press release, Nov. 30, 1965. p. 1.
[28] Ibid., p. 2.
[29] Drucker, Peter F. Long-range planning: challenge to management science. In "Management Science," vol. 5, Apr. 1959. p. 240.

would normally never come in contact. And perhaps most significantly, the intensive schedule of meetings, hearings, colloquia, and seminars has reinforced their perception of the enormity of the task facing them, and which can be achieved only by even more arduous examination of these national problems and the passage of timely, judicious legislation.

C. U.S. DEPARTMENT OF HOUSING AND URBAN DEVELOPMENT

With the creation of the Department of Housing and Urban Development in 1965 (by Public Law 89–174), it became possible for the Federal Government to commence centralizing selected functions related to the welfare of the communities of the Nation. In the quarter-century since R. Harold Denton first had suggested the need for a Federal urban research and development program,[30] the fragmentation of such activities had been impressive. A Bureau of the Budget survey (1963) revealed that 12 Federal agencies were responsible for a $40-million amalgam of 400 urban research projects.[31] The predecessor to HUD, the Housing and Home Finance Agency (HHFA) had urged that financial support be provided for a range of research endeavors. Support for such expenditures was sought through the medium of several reports (e.g., "A National Program of Research in Housing and Urban Development" prepared by Resources for the Future, Inc.) and conferences such as the one on the Rationalization of Research on Housing and Urban Problems held in November of 1960. The Congress did appropriate modest sums for urban studies and housing research starting in 1962, as shown in figure 2.[32]

Fiscal year or years	Appropriation	Statutory authority
1948 to 1953	$4,876,526	Housing Act of 1948, Housing Act of 1949.
1954	$125,000 (appropriated to liquidate the housing research program).	Independent Offices Appropriation Act of 1954.
1955	(No request)	
1956	do	
1957	($175,000 request for census study of housing denied.)	
1958	($920,000 request for "housing studies" denied.)	
1959	(No request)	
1960	do	
1961	($600,000 request for "housing studies" denied.)	
1962	$375,000 ($900,000 requested)	Housing Act of 1948.
1963	$375,000 ($1,450,000 requested)	Housing Act of 1949.
1964	$387,400 ($2,500,000 requested)	
1965	$387,400 ($1,500,000 requested)	
1966	$750,000 ($1,500,000 requested)	
1967	$500,000 ($750,000 requested)	
1968	$10,000,000 ($20,000,000 requested)	Housing and Urban Development Act of 1966.

FIG. 2.—APPROPRIATIONS FOR URBAN STUDIES AND HOUSING RESEARCH RECEIVED BY THE HOUSING AND HOME FINANCE AGENCY AND THE DEPARTMENT OF HOUSING AND URBAN DEVELOPMENT, 1948-68 [1]

[1] Does not include funds for demonstration activities and statistical studies.

Source, Department of Housing and Urban Development and Independent Offices Appropriations Acts, 1948–68.

[30] U.S. Congress, Senate, Temporary National Economic Committee; Toward more productive housing (76th Cong., third sess., 1940). Pp. 123–194. [Study by R. Harold Denton, "The relation of productivity to low-cost housing".]
[31] Bureau of the Budget, Urban research under Federal auspices; Prepared for the Senate Committee on Government Operations, Subcommittee on Intergovernmental Relations (88th Cong., second sess., 1964).
[32] Carroll, James D. "Science and the City: The Question of Authority. In Science, vol. 163, No. 3870, Feb. 28, 1969. p. 905, table 1.

Among the duties assigned the Secretary of the new Department of Housing and Urban Development were those related to conducting "research and studies to test and demonstrate new and improved techniques and methods of applying advances in technology in housing construction, rehabilitation, and maintenance, and to urban development activities," and encouraging and promoting "the acceptance and application" of these techniques and methods (see section V. D. for more detailed comment).

A forum for private sector opinion was offered through a joint HUD–Office of Science and Technology conference held at Woods Hole, June 5–25, 1966. This "Summer Study on Science and Urban Development" was performed by five working panels on: rehabilitation, new housing, environmental engineering, transportation, and health services.[33]

Within the following year, President Lyndon B. Johnson recommended: [34]

1. Legislation to authorize a new Assistant Secretary in the Department of Housing and Urban Development for research, technology, and engineering.

2. [that] the Secretary of Housing and Urban Development * * * encourage the establishment of an Institute of Urban Development, as a separate and distinct organization.

3. Twenty million dollars in fiscal 1968 funds [be] appropriated to the Department of Housing and Urban Development for general research.

Shortly thereafter, an Office of Urban Technology and Research was created within HUD, with staff responsibilities for coordinating on-going data collection, demonstration, and research activities, and preparing a comprehensive urban R. & D. budget. In addition, the new office was charged with administering a number of existing contract and grant programs. These included the preparation of a plan to demonstrate the use of systems analysis in urban planning by the International City Managers' Association, and the development of a bibliographic data system (URBANDOC) by the City University of New York, for use by urban technologists.

The view taken of systems analysis by HUD was expressed succinctly in these terms: "we define systems analysis in the context of comprehensive planning as follows:

"1. Formulation of community values and objectives.

"2. Formulation of alternative long-range plans (or strategies) intended to achieve community goals.

"3. Evaluation of the effectiveness of each alternative plan and the selection of the most effective plan for programing and implementation.

"4. Formulation of short-range programs, such as capital budgets and community renewal programs (CRP's), intended to guide public and private action." [35]

---

[33] Summer study on science and urban development. See transcript of summary report and individual panel reports of conference held at Woods Hole, June 5–25, 1966.

[34] Johnson, Lyndon B., "America's Unfinished Business: Urban and Rural Poverty". In Weekly Compilation of Presidential Documents, vol. 3, No. 11, Mar. 20, 1967. p. 469.

[35] Letter from William L. Slayton (Commissioner) to Hon. Gaylord Nelson, U.S. Senator, dated Feb. 21, 1966, p. 1.

The program responsibilities within HUD have been placed for the most part under Assistant Secretaries for Metropolitan Development, Renewal, and Housing Assistance, and Model Cities and Governmental Relations. The following discussion will indicate some of the programs, offices, and agencies which have been active in sponsoring the use of systems technology in a variety of social and community problem applications.

The Urban Transportation Administration, functioning under the direction of the Assistant Secretary for Metropolitan Development, was responsible for the issuance of a thorough study called Tomorrow's Transportation: New Systems for the Urban Future. The study, prepared in consultation with the Secretary of Transportation, offered "recommendations for a comprehensive program for national leadership in research, development, and demonstration of all aspects of urban transportation and of the basis upon which it was formulated." [36] Contributing to the study were 17 contractors from scientific research centers, universities, industry, and foundations. The time-phased program recommended involved these four areas: improved analysis, planning, and operating methods; immediate system improvements; components for future systems; and the development of entirely new systems for the future.[37] Figure 3 depicts a matrix of program elements and recommendations for using the transportation system to enhance and improve the total city system.[38]

Another effort underwritten by this group in HUD was the contracting with Alan M. Voorhees & Associates, Inc., to design and write a system of computer programs to assist in long-range mass transit planning. The 12 programs were grouped into three analytical categories:

1. Network analysis.
2. Modal split model development and application.
3. Passenger loading.

[36] U.S. Department of Housing and Urban Development, Tomorrow's transportation: new systems for the urban future. Washington, D.C., 1968, p. xi.
[37] Ibid., p. 2.
[38] Ibid., p. 78, table 3.2.

	Contributions to Solutions of Urban Problems							
Program elements	1. Equality of access to urban opportunity	2. Quality of service	3. Congestion	4. Efficient use of Equipment and facilities	5. Efficient use of land	6. Urban pollution	7. Urban development options	8. Institutional framework and implementation
1.0 Improved Analysis, Planning, and Operations								
1.1 Investment, financing, and pricing	○	○	○	○			⊘	●
1.2 Comprehensive planning	○	○	○	○	●	○	⊘	○
1.3 Operating efficiency	○	⊘	○	●				○
1.4 Social impact	●	○			○	⊘	○	○
1.5 Evaluation techniques	●	○	○	○	○	○	○	⊘
2.0 Immediate Systems Improvements								
2.1 Urban bus systems	⊘	●	○	○		○	○	○
2.2 Exclusive guideway systems		●	○	⊘		○		
2.3 Urban automobile innovations	●		⊘		○	○	○	○
2.4 Improvements for pedestrians	⊘	○	●		○		○	
2.5 Improvements of general application	○	⊘	○	●				○
3.0 Components for Future Systems								
3.1 Automatic systems controls	○	⊘	○	●	○		○	
3.2 Propulsion and power transmission		○		⊘		●		
3.3 Suspension and guideway components		●		○		⊘		
3.4 Elevated structure design			○		●	⊘	○	○
3.5 Tunneling			○		⊘	●	○	○
3.6 Goods movement			●	⊘	○	○	○	
4.0 New Systems for the Future								
4.1 Dial-a-Bus	●	⊘		○			○	○
4.2 Personal rapid transit systems	●	○	○			○	⊘	
4.3 Dual mode personal vehicle systems	●	⊘	○			○	○	
4.4 Automated dual mode bus	⊘	●	○	○			○	○
4.5 Pallet or ferry systems		⊘	○			○	○	
4.6 Fast intraurban transit links	○	○			⊘	○	○	
4.7 Systems for major activity centers	○	○	●	○	⊘	○		○

Key: ●PRIMARY—Project areas which are primarily related to or make a major contribution to solutions of urban problems.

⊘ SECONDARY—Project areas which make significant contributions to solutions of urban problems, but which are secondarily related to the problem.

○INDIRECT—Project areas which are indirectly related but which make substantive contributions to one or more problem areas.

FIG. 3.—SUMMARY PROGRAM RECOMMENDATIONS AND PROBLEM AREAS

Figure 4 shows the inputs, outputs, and interfaces of all of these programs.[39]

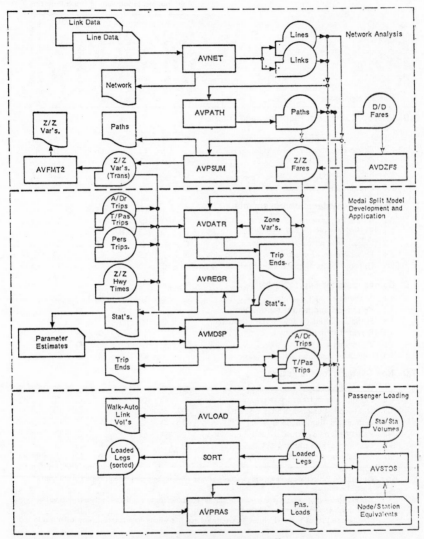

FIG. 4.—TRANSIT PLANNING SYSTEM

The Office of Intergovernmental Relations and Planning Assistance prepared a useful publication listing 100 outstanding programs and projects funded by title VIII of the 1964 Housing Act and title IX of the 1966 Demonstration Cities and Metropolitan Development Act.[40] The document serves two purposes: it singles out the innovative and superior proposals submitted by well-established agencies and the more modest, but well-conceived, efforts of newcomers to the field;

[39] Dial, Robert B. and Richard E. Bunyan, Public transit planning system; In Socio-econ. plan. sci., vol. 1, No. 3, July 1968, p. 348, fig. 2.
[40] U.S. Department of Housing and Urban Development. Community development training and urban information and technical assistance: 100 outstanding programs. Washington, D.C., June 1968, 48 pp.

and it serves to acquaint designated agencies, local officials, and other interested groups with the wide variety of activities which are ongoing under these two grant programs. For each grant program, the projects are subdivided by program category (e.g., automatic data processing, manpower analyses and surveys, information systems, and services) and identified by State. A listing of State administering agencies includes departments of community affairs, offices of local affairs, universities, State planning agencies, and municipal leagues. Related information on State and local activities is contained in section VII of this report.

The Midwest Research Institute often has been the recipient of contracts from HUD, and its contribution entitled "A National Clearinghouse Network for Urban Information"—discussed in detail in section VIII. D.—is representative of the work performed, in this case for the Office of Intergovernmental Relations and Planning Assistance. Another MRI project for HUD has been a survey of local needs in connection with Federal aid program information. A simplified picture of the information flow in this realm of activity is shown in figure 5.[41]

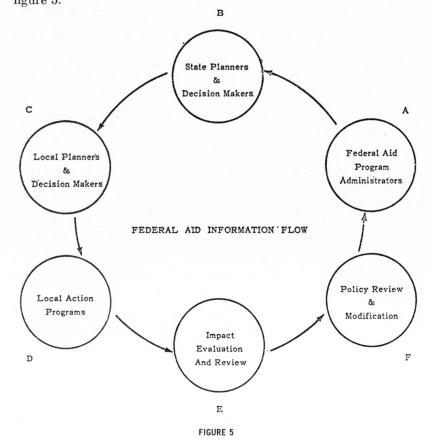

FEDERAL AID INFORMATION FLOW

FIGURE 5

[41] Federal aid program information: a survey of local government needs. Final report, prepared for the urban planning research and demonstration program, Department of Housing and Urban Development, September 1967. Kansas City, Midwest Research Institute, p. 2.

The Assistant Secretary for Renewal and Housing Assistance has been responsible for a variety of study projects, including a grant from the (then) Urban Renewal Administration for the State of Michigan on "Systems Analysis Techniques and Concepts for Water Resource Planning & Management," and a grant for a "computer-based cost-analysis and design system for low-income housing" from the low-income housing demonstration program for work on the North Carolina Fund Low Income Housing Demonstration Project. Both of these efforts are discussed in Section VIII. Originally prepared for the Housing and Home Finance Agency but reissued by HUD was a study on an information system for urban planning by a professorial team from the George Washington University, which was supported by the Maryland-National Capital Park and Planning Commission. Emphasis was placed in this effort on local needs. Numerical information in the urban planning information system was placed in three decks of punched cards (for use on an IBM 1401 computer):

> 1. A land use deck, divided into a properties subdeck and a supplemental parcels subdeck.
> 2. A family characteristics deck, a sample based on the land use deck.
> 3. An employment characteristics deck, also a sample based on the land use deck.

An example of the organization of the data is shown in figure 6, where the employment characteristics questionnaire form (coding sheet) is reproduced.[42] While the ADP handling of such key data was featured, consideration also was given to the utility of an automated visual data processing system which would feature aperture card containment of aerial photographs, property maps, and topographic maps.

The projected scope of the model cities program will encompass 150 cities, with continuous, multiple-faceted communications between Federal, regional, State, and local elements. Action has been taken to establish, in a number of phases, a city demonstration agency information system capable of collecting, formatting, transmitting, storing, processing, and retrieving selected narrative and statistical data on societal problem areas.[43] Every form of assistance is being offered the State and local components so that they may better understand the full potential available under laws like the Demonstration Cities and Metropolitan Development Act. A "Program Guide: Urban Information and Technical Assistance Services" recently was published by the Office of Demonstrations and Intergovernmental Relations within HUD containing explicit information on title IX of the above act.[44] These two activities, and many others, were carried out under the direction of the Assistant Secretary for Model Cities and Governmental Relations.

[42] Campbell, Robert D. and Hugh L. LeBlanc. An information system for urban planning. Prepared for the Maryland-National Capital Park and Planning Commission. Washington, D.C., U.S. Government Printing Office, 1967, p. 62, fig. 12.

[43] Chartrand, Robert L. The systems approach in social legislation. In Systems approach to social problems. Charles L. Vaughn, ed. Chestnut Hill, Mass., Boston College Press, 1968, p. 22.

[44] Solving urban problems through urban information and technical assistance. A program guide for urban information and technical assistance, title IX, Demonstration Cities and Metropolitan Development Act of 1966. Washington, D.C., U.S. Department of Housing and Urban Development [1968]. 46 p.

Coding Sheet No. _____

EMPLOYMENT CHARACTERISTICS QUESTIONNAIRE

Name of Employer _____

Street Address _____

Columns	Code	Description
1-4	☐☐☐☐	Employer (Standard Industrial Classification)
5-9	☐☐☐☐☐	Number of Employees
10-11	☐☐	Per cent of employees male
12-14	☐☐☐	Occupational title of largest employed group: _____
15-17	☐☐☐	Occupational title of second largest employed group: _____
18-20	☐☐☐	Occupational title of kind of employee most difficult to find: _____
21-23	☐☐☐	Median employment period, in months
24-27	☐☐☐☐	Median monthly salary (if applicable)
28-31	☐☐☐☐	Median hourly wage (if applicable)
32	☐	Employment peak: (1) spring; (2) summer; (3) fall; (4) winter; (5) none
33	☐	Employment low: (1) spring; (2) summer; (3) fall; (4) winter; (5) none
34-35	☐☐	Per cent "input materials" from within metropolitan area
36-37	☐☐	Per cent of "market" within metropolitan area
38-45	☐☐☐	Name code (firm) _____
46-47	☐☐	Number of locations within metropolitan area
48-50	☐☐☐	Months in business (life of firm)
51-53	☐☐☐	Months in metropolitan region
54-56	☐☐☐	Months at same location
9-79	☐☐☐☐☐☐☐☐☐	Location
80	☐	Deck Number

Date data gathered _____ Interviewer _____

FIGURE 6

During its brief existence, HUD has endeavored to evolve a consistent approach in coordinating the heretofore fragmented Federal
grants-in-aid. Former Secretary of HUD Robert C. Weaver, in testimony [45] before the Subcommittee on Intergovernmental Relations
of the Senate Committee on Government Operations on November 16,
1966, described the three programs which stressed the development
of improved manageability and coordination at the local level: the
demonstration cities program—where "the guiding principle is coordination of both Federal and non-Federal resources"—the workable
program for community improvement, featuring neighborhood anal-

[45] Weaver, Robert C. Statement before the Senate Committee on Government Operations, Subcommittee on Intergovernmental Relations, on management problems created by the increased number of Federal grant-in-aid programs, Nov. 16, 1966, p. 6.

yses, comprehensive community planning, administrative organization, and financing (using a PPBS-type approach); and the community
renewal program, where grants have been make to 162 communities
which have met the HUD policy requirements for (1) intensive
surveys of the need for renewal; (2) an analysis of the economic base;
(3) development of goals; (4) examination of resources available; and
(5) preparation of a program of renewal action.[46]

Perhaps the most difficult requirement laid upon HUD leadership
is to facilitate planning and implement programs against a highly
fluid environmental situation. It has been recognized that all possible
private sector support must be marshaled, and this has been undertaken through arrangements with nonprofit organizations such as
ACTION-Housing, Inc., or industrial firms like Armstrong Cork Co.
and Conrad Engineering, both involved in housing rehabilitation using
new techniques and products.[47] The special role of commissions, committees, and task forces will be discussed in section IV. I.

The Department of Housing and Urban Development, in its key
position as a corrective force to combat the Nation's social and community problems, has set a course which will cause its technical
expertise and funds to be applied to the most critical needs of States
and localities. Whether it is a grant to develop a Bay Area Transportation system—as described in section VIII and appendix D of
this report—or the establishment of a new (November 1968) Urban
Information Systems Inter-Agency Committee (USAC) with membership from several other Federal agencies,[48] HUD will continue to
press for the sustained, imaginative use of systems technology in the
planning, testing, and operations of those applications which can lead
to an improvement in social and community conditions.

D. U.S. DEPARTMENT OF HEALTH, EDUCATION, AND WELFARE

The farflung responsibilities of the Department of Health, Education, and Welfare have taxed the various major elements of that
executive branch entity—the Public Health Service, Office of Education, and others—to remain responsive to the needs of a society which
is exemplified by growth and an expectation of ever greater service
by national, tax-supported establishments. Created in 1953, the
Department has received strong financial support from the Congress
in the form of R. & D. appropriations. Indeed, as shown in figure 7,[49]
HEW received approximately one-half of the R. & D. obligations
to those 28 agencies other than the DOD-NASA-AEC complex.

---

[46] *Ibid.*, p. 9.
[47] Hincks, Joel B. The role of private enterprise in the solution of major urban problems. Washington,
D.C., Library of Congress, Legislative Reference Service, Nov. 30, 1967, p. 11–12.
[48] First meeting of urban information systems interagency committee (USAC). In Small-area data
activities, vol. 3, No. 4, December 1968, p. 2.
[49] National Science Foundation. Federal funds for research, development, and other scientific activities:
fiscal years 1967, 1968, and 1969. Vol. XVII. Washington, U.S. Government Printing Office, 1968. p. 35, table
16.

[Dollar amounts in millions]

Agency	Actual, 1967	Estimates	
		1968	1969
Total, all agencies	$16,529	$16,230	$17,300
Total, "other 28" agencies	$2,356	$2,580	$3,064
Percent distribution:			
Department of Health, Education, and Welfare	49	50	48
Department of Transportation	12	11	15
National Science Foundation	11	11	10
Department of Agriculture	11	10	9
Department of the Interior	7	8	7
Department of Commerce	3	3	3
Veterans Administration	2	2	2
Office of Economic Opportunity	2	2	2
Department of Housing and Urban Development	1	1	2
Other agencies	3	3	3

Note: Percent detail may not add to 100 because of rounding.

FIG. 7.—FEDERAL R&D OBLIGATIONS OF THE 28 AGENCIES OTHER THAN DOD, NASA, AND AEC (DOLLAR AMOUNTS IN MILLIONS)

Not only has HEW played an appreciable role in promoting research and development, but it has been directed to "search for ways to improve the Nation's ability to chart its social progress." In March 1966, the President asked the Secretary of HEW to:

* * * develop the necessary social statistics and indicators to supplement those prepared by the Bureau of Labor Statistics and the Council of Economic Advisers. With these yardsticks, we can better measure the distance we have come and plan for the way ahead. [50]

As a result of this directive, the document "Toward a Social Report," was prepared by the Office of the Assistant Secretary for Planning and Evaluation, with contributions from private sector representatives. Dealing with such aspects of the quality of American life as health and illness, social mobility, the physical environment, income and poverty, and public order and safety, the report was designed to represent "a preliminary step toward the evolution of a regular system of social reporting." [51] In commenting on the role of science and technology, and the need to consider carefully those policies affecting them, the report cautions:

If there is almost sure to be more heat generated by issues of science policy in the future, ways must be found to generate more light. Priorities in science could be laid out more systematically, and farther in advance. Issues involving such priorities could be exposed to wider public debate. The very unpredictability of scientific breakthroughs could be made the basis for more rational development of scientific manpower, institutions, and communications with an emphasis on keeping these resources flexible.[52]

It should be noted that the responsibility and authority assigned the HEW Assistant Secretary for Planning and Evaluation stress the coordination of Department activities in "economic and social analysis, program analysis, and long-range program planning"; and that the development and administration of the HEW Planning-Programing-Budgeting System will be accomplished by:

1. Developing long-range objectives;
2. Evaluating alternative means of achieving the objectives;
3. Conducting cost benefit studies; and

[50] U.S. Department of Health, Education, and Welfare. Toward a social report. Washington, U.S. Government Printing Office, 1969. p. iii.
[51] *Ibid.*
[52] *Ibid.*, p. 74.

4. Providing staff leadership in the conduct of economic and systems analyses on a departmental basis.[53]

In order to present ideas and data which may be helpful to members of the HEW staff, the Office for Planning and Evaluation selects and circulates various "occasional papers," such as the study on "Measuring the Effectiveness of Criminal Rehabilitation Programs."[54] This report was prepared by W. Michael Mahoney of the staff of the Resources Management Corporation for the Office of Juvenile Delinquency and Youth Development.

The Public Health Service has been instrumental in encouraging various State and local jurisdictions to cooperate in coping with such problems as air pollution. For example, the National Air Pollution Control Administration conducted a study of the San Francisco Bay area. This activity was related directly to the direction provided in the Air Quality Act of 1967 to HEW to designate "air quality control regions," which is discussed in section V. D of this report. Since cooperation with State and local authorities is a stipulated condition in such HEW efforts, this study involved either directly or indirectly the California Department of Public Health, Bay Area Air Pollution Control District, Association of Bay Area Governments, and the San Francisco Bay Area Council.

The evaluation of urban factors—land use, population, transportation, existing governmental organizations—was considered in terms of the influences exerted by the geographic characteristics of the San Francisco Bay region. The nine counties involved within the physical context have become socially and economically interrelated. Figure 8 presents a flow diagram for State action to control air pollution on a regional basis.[55] For purposes of comparison with another recent study, of the Los Angeles Basin, the highlights of a proposal for "Adapredictive Air Pollution Control for the Los Angeles Basin" are discussed in section VIII.C. Individual members of the HEW staff have been providing commentary on the role of technology in air pollution control; an instance of this is John H. Ludwig's contribution to the Law and Contemporary Problems' issue on "Air Pollution control." [56]

Sponsorship of conferences designed to increase understanding and expedite the exchange of salient findings between public and private personnel also has been stressed by the leadership of the Public Health Service. The Conference on Biomedical Communication: Problems and Resources (April 1966)[57] to discuss urgent health needs and critical manpower shortages, and the Conference on Quantitative Analysis for Planning Control (April 1968)[58] illustrate the type of jointly sponsored activity which the Public Health Service encourages

[53] General Services Administration. National Archives and Records Service. Office of the Federal Register. "U.S. Government Organization Manual: 1968–69." Washington, U.S. Government Printing Office, 1968. pp. 359–360.

[54] Mahoney, W. Michael. Measuring the effectiveness of criminal rehabilitation programs. RMC Research Document RD–015. Originally published as Occasional paper n. 5. Washington, U.S. Department of Health, Education, and Welfare, February 1968. 10 pages plus figures.

[55] Report for consultation on the San Francisco Bay area air quality control region. Washington, U.S. Department of Health, Education, and Welfare, Public Health Service, December 1968. P. 5, fig. 1.

[56] Ludwig, John H. Air pollution control technology: research and development on new and improved systems. In Law and contemporary problems, v. XXXIII, n. 2, Spring 1968. (Special issue on air pollution control). Pp. 217–226.

[57] This conference was jointly sponsored by the New York Academy of Sciences and the U.S. Public Health Service Audiovisual Facility, in New York City, April 4–6, 1966.

[58] This conference was presented by the Department of Medical Care Organization, School of Public Health, University of Michigan, under the sponsorship of the U.S. Public Health Service, Division of Medical Care Administration, at Ann Arbor, Mich., April 8–10, 1968.

in order to alert various key groups to the role of technology in the biomedical realm.

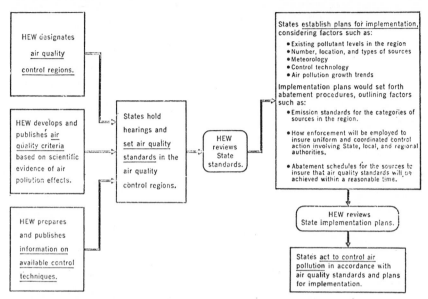

FIG. 8.—FLOW DIAGRAM FOR STATE ACTION TO CONTROL AIR POLLUTION ON A REGIONAL BASIS

The National Institutes of Health also have been deeply involved in the application of systems tools and techniques to problems of the component institutes. Typical contracts are those like the three for studies on the feasibility of a computer-based regional or national inventory system for whole blood, blood products, and blood donors. If feasible, such an inventory system might—

Make possible a more equitable distribution of the available blood supply and thus prevent local shortages due to unavoidable fluctuations in the blood supply and in clinical demands upon it.

Reduce losses of whole blood through outdating in storage by making possible redistribution of local surpluses.

Provide a means of summoning potential blood donors during slack donation periods.

Provide a means of locating rare-blood donors within a region. These could then be called upon to donate only when a need for a rare blood type arose, thus eliminating most rare blood losses due to outdating.

Provide means of screening and excluding donors who are probable carriers of infectious hepatitis.

Make possible closer coordination of civilian and military blood programs.[59]

Another area drawing the attention of the Public Health Service is that of prognostic epidemiology and here again the question of the utility of developing theoretical approaches (e.g., featuring model

[59] U.S. Department of Health, Education, and Welfare, National Institutes of Health, National Heart Institute, News release, July 15, 1968. First page.

building) or practical ones, such as the creation of a sample system, must be considered. Additional commentary on the tradeoffs possible through using a systems approach to certain problems in the health services area is contained in section VIII.C. of this report.

The National Library of Medicine, with the computer-supported MEDLARS (Medical Literature Analysis and Retrieval System) capability, also reflects the HEW determination to apply computer technology and systems analysis wherever possible in its daily operations. MEDLARS, developed with strong external consultant support, provides users with the "Index Medicus" and other indexes and bibliographies, as well as retrieving on demand citations to scientific publications. A sample card printout made as the result of a computer search is reproduced in figure 9.[60] Six MEDLARS search stations are situated at major medical libraries at the Universities of California (Los Angeles), Colorado, Alabama, Michigan, and Harvard and Ohio State Universities. Thus a computer search of the central holdings is possible through this sophisticated system.

SAMPLE CARD PRINTOUT

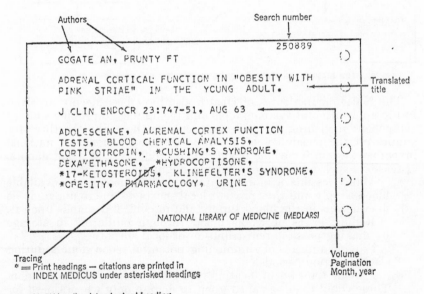

FIG. 9.—SAMPLE CARD PRINTOUT

The Office of Education within HEW has been quite aware of the need to employ technology wherever and whenever possible in the educational systems of the Nation. President Johnson in an address before the delegates of the Williamsburg Conference on October 8, 1967, put it in these terms: [61]

[60] U.S. Department of Health, Education, and Welfare. Public Health Service. National Library of Medicine. Guide to MEDLARS services. Washington, Public Health Service publication No. 1694 (1967). P. 10.
[61] Johnson, Lyndon B. An address before the delegates of the Williamsburg, Va., conference, October 8, 1967.

If the world's financial systems were forced to function with no better facts than these which educational systems live by, a financial panic would swiftly seize all capitals of the world * * *.

I can see no reason in the world why modern technology cannot, for example, permit the best professor in the world to teach students all over the world.

The national need, as set forth in the 1966 report, "Automation and Technology in Education" prepared by the Subcommittee on Economic Progress of the Joint Economic Committee of the Congress, expressed the prospective technological demands in this way:

It seems clear that rapid and effective application of these devices and new techniques will require important adjustments within the educational system. The role of teachers and other educational personnel may be altered. Application of the new technology will require much more specific planning for the teacher-pupil relationship, with some departures from past dependence on improvisations and intuitive insights.[62]

The report then continued, identifying the technological aids: educational television (open and closed); video tape; computerized instruction; use of computers in student testing, guidance, and evaluation, and the storage, retrieval, and distribution of information; programed courses of instruction, teaching machines, particularly the talking typewriter; the use of microfilm and microfilm viewing equipment; and language laboratories. Another important item was included: stressing that the systems approach be used in the development and utilization of educational technology.[63]

The great potential of educational technology caused the MITRE Corp. to prepare a detailed, system development study proposal entitled "Definition of a Planning and Design Study for a National Center for Educational Technology." [64] This document provided a distinctive, useful commentary on the legislation (see section V.D.) which has been enacted by a Congress anxious to improve the performance of our school systems. The Cooperative Research Act, as amended by the Elementary and Secondary Education Act, has for nearly 15 years supported educational research by the academic world, but it was not until the passage of the Elementary and Secondary Education Act that sufficient funds—$100 million over 5 years for construction of educational research facilities—were available to establish a system of regional educational laboratories. A listing of the top 20 educational establishments to receive Cooperative Research Act funds appears as figure 10.[65]

Public Law 89–10 (the Elementary and Secondary Education Act) also created a National Program of Education Laboratories. Twenty regional educational laboratories, each incorporated as a nonprofit organization, have been set up so that a widely representative group of public and private individuals and agencies can participate in the planning and implementation cycle. The MITRE study states that in spite of the expenditure of hundreds of millions of dollars of public funds, technical development in educational systems is extremely primitive,[66] and that the status of management understanding of the role of the new technology is no less unsatisfactory.

---

[62] U.S. Congress. Joint Economic Committee. Subcommittee on Economic Progress. Automation and technology in education. A report of the subcommittee. (89th Cong., 2d sess, August 1966.) Washington, U.S. Government Printing Office, 1966. P. 5.

[63] Ibid.

[64] Definition of a planning and design study for a national center for education technology. Washington, the MITRE Corp., May 1968. 60 p.

[65] Ibid., p. 10, table 1.

[66] Ibid., p. 14.

The proposal for a "National Center for Educational Technology" emphasizes two possible orientations: first, a center which could concentrate on national issues, including such areas as telecommunications interconnection policy, the development of national standards for computerized instructional programs, computer utility rate regulation, and the like.[67] The other alternative, chosen for further development in the MITRE study, concentrated on a national center oriented more toward the prosecution of pedagogical research employing advanced technology, toward the training of teachers and administrators in the use of educational technology, and toward the development and training of a nationwide series of regional, State, or metropolitan centers for educational technology.[68]

Institution	Amounts received under Cooperative Research Act, fiscal year 1965	Ranked by amount of funds received under Cooperative Research Act, fiscal year 1965	Ranked by total amount of Federal support received, fiscal year 1965
University of Pittsburgh	$1,351,494	1	22
Harvard University	1,333,280	2	10
University of Wisconsin	975,516	3	11
University of Oregon	667,076	4	36
University of Michigan	622,029	5	2
Stanford University	598,825	6	8
University of Illinois	533,974	7	6
University of Chicago	507,189	8	14
Ohio State University	418,911	9	19
Syracuse University	374,300	10	59
University of California (Berkeley)	313,670	11	7
University of Minnesota	313,603	12	9
Columbia University	250,489	13	4
Webster College (Missouri)	218,900	14	(1)
Northwestern University	203,218	15	38
Michigan State University	179,062	16	42
New York University	177,195	17	12
Pennsylvania State University	174,856	18	30
Purdue University	166,654	19	24
University of Texas	150,979	20	17
Total	15,878,177		

1 Not in 1st hundred.

FIG. 10.—COOPERATIVE RESEARCH ACT FUND ALLOCATION

The center for educational technology would contain several related laboratories and centers, such as the teacher training laboratory, a student education laboratory, an experimental counseling center, and experimental classrooms and libraries. The Center also would feature an operational proving ground for the technology applications developed in the Center's laboratories, and which would be comprised of a local school district, its classrooms, libraries, counseling centers, and administrative offices. Service to all of these elements would come from a Time-shared, Interactive, Computer-Controlled, Education Television (TICCET) system. The technical concept for the center is shown in figure 11,[69] and the organizational concept as figure 12.[70]

67 Ibid., p. 17.
68 Ibid.
69 Ibid., p. 24, fig. 3.
70 Ibid., p. 28, fig. 4.

51

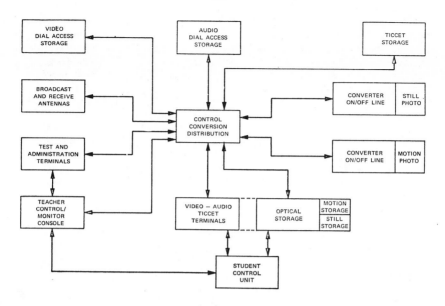

FIG. 11.—TECHNICAL CONCEPT—EDUCATIONAL TECHNOLOGY CENTER

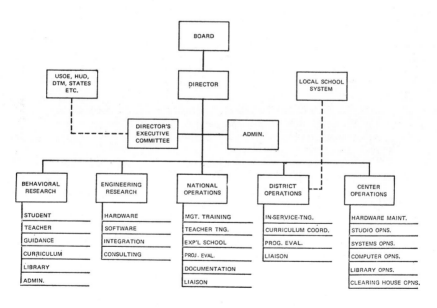

FIG. 12.—NATIONAL EDUCATIONAL TECHNOLOGY CENTER—ORGANIZATIONAL CONCEPT

The final section of the MITRE proposal is devoted to a description of a TICCET system in the Washington, D.C. area, including details of computer, television, storage retrieval, terminal and data software subsystems. Various applications for this type of man-machine system are noted, and the use of the system by administrators, teachers, researchers, and students explored.

The Office of Education, much like the National Library of Medicine, has worked to establish a national information network (called ERIC, for Educational Resources Information Center) which collects, abstracts, indexes, stores, retrieves, and disseminates the most timely and significant educational research reports and program descriptions. Certain regular publications are issued (e.g., Research in Education,[71] a monthly abstract journal), and the Office of Education also prepares one-time catalogs and conducts special literature searches.

As HEW works to serve the citizenry more effectively, and often accomplishes its objectives through the use of computer technology and systems techniques, its efforts are noted by private sector elements which then strive to apply the systems approach to certain health services problems. The IBM Corp., for example, prepared a manual, "Introduction to Federal Government Hospitals," which describes the organization and operations of Government hospitals, and then identifies the applications of ADP techniques in the various divisions and services.[72]

Sometimes, professional organizations perceive the needs of medical service organizations and are moved to undertake special symposia so that all may benefit from the exchange of ideas. One recent meeting of this nature, sponsored by the Operations Research Society of America, featured a discussion of management sciences and health planning, attracting a strong panel from the Federal Government (Dr. Robert N. Grosse of HEW), the city of New York (Dr. Howard J. Brown of the Health Services Administration), and the Kaiser Foundation Health Plan and Hospitals (Mr. Arthur Weissman).[73] The panel topics included:

> Procedures currently used to plan health programs;
> Measures of benefits of health programs;
> Adequacy of mathematical and statistical concept to model health programs;
> The availability of pertinent data to permit sophisticated health planning;
> The probable implications of further health planning to the Government and private sectors; and
> The professional background desirable for future health planners.

The Department of Health, Education, and Welfare has struggled to develop a program which would establish a cohesive and coherent set of departmental objectives and relationships. The need is to define human ecology and develop a strategy and tactics which allow research and development while still performing the established, day-to-day services. The responsibilities of HEW in providing social

---

[71] U.S. Department of Health, Education, and Welfare. Office of Education. Bureau of Research. Division of Information Technology and Dissemination. Research in education. A monthly abstract journal.

[72] Introduction to Federal government hospitals. IBM data processing application. Washington, IBM Federal Region, Applications Development Department, [no date], 23 p.

[73] This conference was sponsored by the Operations Research Society of America, in San Francisco, May 1–3, 1968.

and community services are enormous, and require that management
avail itself of every possible approach, technique, and device which
can help furnish enhanced assistance to the Nation's populace.

E. U.S. OFFICE OF ECONOMIC OPPORTUNITY

The Office of Economic Opportunity has been in the forefront of
those Federal Establishments which have seen the potential of systems
analysis and ADP and moved to utilize this technology. Immediately
upon being created in 1964, OEO could rely upon sophisticated data
handling support from its information center, featuring a powerful
computer configuration. A concerted effort to collect, format, and
make available key data—on the basis of State, county, and congressional district—on Federal assistance activity and related programs has been made by the Information Center. The first Catalog of
Federal Programs was published in December of 1965; subsequent
versions bearing the title Catalog of Federal Assistance Programs
have been expanded to include all domestic assistance programs.[74]

The Assistant Director for Research, Plans, Programs, and Evaluation has been deeply involved in preparing a PPB system for OEO,
and is charged with such tasks as developing a statement of national
antipoverty requirements.[75] The R.P.P. & E. Office has relied upon
university, corporation, and other Government agency specialists in
systems analysis to provide alternative approaches and/or answers to
many of the poverty program problems. James Ridgeway, writing in
the New Republic, opined that the poverty programs are an area
where there is a natural marriage between politicians and behavioral
scientists who design social systems.[76] An analysis of the R.P.P. & E.
contracts let in fiscal year 1966 to external consultants shows that of
$4.7 million spent on nearly 40 projects, $2.1 million went to universities, $1.8 to other governmental groups; that is, Bureau of the Census
and the remainder ($0.9 million) to private (industrial) corporations.

Many of the contracts with OEO featured computer simulation of
certain problem situations. For example, the Philco Corporation
proposed to use a computer to simulate a community action program,
with 500 statistical indicators reflecting the economic, political, geographic, and social parameters of a hypothetical but typical community. Another project where computer technology was instrumental
in preparing results was the analysis of survey data in New Haven to
determine needs of employment—related education, skill training,
placement services—to be performed by Community Progress, Inc.

A recent effort by OEO which has involved many executive branch
agencies and departments, and drawn attention from all quarters of
the Nation, has been the establishment of a Federal Information
Exchange System. This system collects information from all Federal
departments and agencies and develops a central data bank for use by
all Federal, State, and local agencies and private sector groups.
Computer manipulation of these data has been carried out by the
OEO Information Center, which has published Reporting Instructions

[74] Executive Office of the President. Office of Economic Opportunity. Catalog of Federal Assistance Programs. Second edition, June 1, 1967. p. 699.
[75] Office of Economic Opportunity. Establishment of the OEO planning-programing-budgeting system
(PPBS). OEO Instruction n. 72-2. Washington, July 6, 1966. p. II–1, II–2, VII–1.
[76] Ridgeway, James. Simulating poverty: input and output. In the New Republic, June 11, 1966. p. 9.

for the Federal Information Exchange System.[77] This document works 'to assure maximum uniformity of reporting from Federal agencies, including formats and procedures. The OEO development of this system also has involved providing technical assistance to selected States (five were in the pilot group) so that they could make optimum use of the data available. The role of the analyst group as it supports a Governor and his staff will be to—

1. Advise on reports and information already available through the Federal Information Exchange System; and

2. Assist the State organization to inventory, define, classify, and better utilize the information already available.[78]

Also, attempts to standardize the information being reported will be made.

As a new governmental group, OEO has been able to innovate and utilize new tools and techniques without many of the built-in deterrents, both human and procedural, which often impede similar action by more established agencies. The antipoverty mission, accorded the highest priority during the 1960's, often has required unprecedented measures in order to achieve the goals set for the responsible agencies. The encouragement and "hard" support provided to the States and local jurisdictions by OEO also is indicative of a new determination to utilize all available weapons in the battle against the formidable array of social and community problems.

F. U.S. DEPARTMENT OF COMMERCE

The Department of Commerce, long in point of service but with continually diversified responsibilities, has among its missions the development and dissemination of basic demographic, economic, business, scientific, and environmental information; assuring the fullest use of the Nation's scientific and technical resources; and the economic development of communities and regions with lagging economies. In order that the role of science and technology would be given the correct emphasis, a position of Assistant Secretary for Science and Technology was created, with cognizance over the research and development activities by all organizations of the Department.

The ways in which the Department of Commerce components are participating in the effort to improve the national standard of living are many. Systems technology in one or many forms is employed increasingly by the management and field operations of the Department. The Bureau of the Census, a pioneer in the development and utilization of automatic data processing and data collection, formatting, storage, analysis, retrieval operations, conducts major census on population and housing (once every 10 years), governments and transportation (once every 5 years). Data collected on a more current basis, ranging from weekly to annually, are used to provide estimates and projections of the population and housing, and State and local government finances and employment. Surveys include the National Health Survey, the survey of employment and unemployment, and the survey of expenditures for research and development. The Bureau of the Census maintains a large data processing facility and staff,

---

[77] Executive Office of the President. Office of Economic Opportunity. Reporting instructions for the Federal information exchange system, Washington, U.S. Government Printing Office, 1968. 36 pp.

[78] Office of Economic Opportunity. The Federal information exchange. Mimeographed broadside, June 29, 1967. p. 3.

which handle statistics and survey information, in addition to providing assistance to other Federal agencies.

The Economic Development Administration has as its primary function "the long-range economic development and programing for areas and regions of substantial and persistent unemployment and underemployment." [79] The Public Works and Economic Development Act of 1965, as amended, authorizes certain Federal aids for areas designated as redevelopment areas and centers.

The National Bureau of Standards has among its subordinate elements the Institute for Applied Technology, which controls the Clearinghouse for Scientific and Technical Information, operates a Center for Computer Sciences and Technology, and provides for "the use and application of technology within the Federal Government and within the civilian sector of American industry." [80] The Clearinghouse serves:

* * * as the national center for the dissemination of Government-generated information in the physical sciences, engineering, and related technology. In brief, the clearinghouse has been established as the single point of contact in the executive branch for supplying the industrial and technical community with unclassified information about Government-sponsored research and development * * * It thus makes readily available, at low cost, research information which may aid in the development of a new product, solve a processing problem, or increase productivity through technical improvement.[81]

Government R. & D. literature is available from the Clearinghouse at low cost to Federal departments and agencies, State and local governments, industries, universities, and other private sector organizations. A "fast announcement service" provides selected information from the 50,000 documents acquired each year by the Clearinghouse, thus saving the users money and time.

The Institute for Applied Technology has taken the lead in applying computer technology to manpower planning. A study was begun in July 1965 under the joint sponsorship of the Institute and the President's Commission on Automation, Technology and Economic Progress "to determine if the current techniques available in systems analysis and operations research could be of positive benefit in the planning and operation of programs relating to manpower utilization and adjustment." [82] The 6-month study made one complete pass at a gross systems analysis oriented toward current systems problems; this involved:

1. Identification of systems components;

2. Identification of objectives, constraints, key control variables and suitable criteria for systems performance;

3. Development of a crude model simulating the system performance;

4. Identification of alternative systems strategies for control and improvement;

5. Use of the model to test and evaluate these strategies; and

6. Evaluation of the results, assessment of model performance, and conclusions concerning the total effort.[83]

[79] U.S. Government Organization Manual: 1968–69, op. cit., p. 308.
[80] Ibid., p. 319.
[81] U.S. Congress, Senate, Committee on Government Operations; report on S. 1136. Committee print (89th Cong., first sess., 1965) pp. 60–61.
[82] U.S. Department of Commerce, National Bureau of Standards, Institute for Applied Technology. The uses of systems analysis in manpower adjustment. A summary of a 6-month study for the President's Commission on Automation, Technology, and Economic Progress. Washington, Jan. 25, 1966. P. ii.
[83] Ibid.

The Office of State Technical Services, created by Public Law 89–182, has served as a catalyst for encouraging public and private groups (often working together) to apply science and technology to the functions of American enterprise. This has been undertaken in the face of restrictive budget limitations. The annual report for fiscal 1967 includes brief descriptions of those OSTS activities where ADP and systems methodology are employed.[84] A diagram originally appearing in the first annual report (for fiscal year 1966) which shows the projected participation at successive levels of the OSTS operations is included in this report as figure 13.[85]

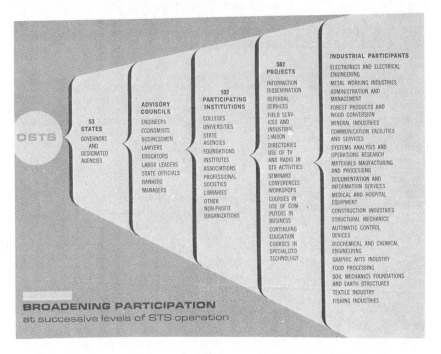

FIGURE 13

Illustrative of the type of proposals which the OSTS program of support has attracted is the "special merit proposal" submitted by the coordinating board of the Texas College and University System for "A Southwestern Regional Information Dissemination Network;" this document also is mentioned in section VIII. D. The role of OSTS has been noted with interest and growing enthusiasm by State and local governments and private sector institutions and organizations. In his remarks before the Conference on Science, Technology, and State government held in Louisville, Ky., in September of 1968, Wyatt M. Rogers, Jr. (associate director, Southern Interstate Nuclear Board) cited the importance of the matching fund provision for carry-

[84] U.S. Department of Commerce, Office of State Technical Services. Annual report, fiscal year 1967. Washington, U.S. Government Printing Office, 1968. Preface.
[85] U.S. Department of Commerce. Office of State Technical Services. Office of State Technical Services first annual report, fiscal year 1966. Washington, Department of Commerce, 1967. p. 2, fig. 1.

ing out such activities (called "technical assistance projects") as seminars, workshops, information centers, and field visits.[86]

Thus, the Department of Commerce carries a significant portion of the load insofar as providing technical assistance to sub-Federal groups concerned with social and community problems is a matter for Federal intervention and support. An overview of the State and local functions where systems methodology and ADP play a discernible role—25 activity areas are identified—and where Federal support may be a factor is presented in section VII both in narrative and tabular form.

G. U.S. DEPARTMENT OF TRANSPORTATION

As the requirement for a coordinated approach to the Nation's transportation problems became a cause for serious concern, the Congress moved to establish a separate entity to cope with this multifaceted area. Public Law 89–670 (the Department of Transportation Act of 1966) created a new cabinet level post, with responsibilities for directing the development of "national transportation policies and programs conducive to the provision of fast, safe, efficient, and convenient transportation at the lowest cost consistent therewith." [87] A heavy reliance on cost-performance planning and systems technology is implicit in other DOT charter elements:

1. [Assuring] the coordinated, effective administration of the transportation programs of the Federal Government;

2. [Facilitating] the development and improvement of coordinated transportation service, to be provided by private enterprise to the maximum extent feasible;

3. [Encouraging] cooperation of Federal, State, and local governments, carriers, labor, and other interested parties toward the achievement of national transportation objectives; and

4. [Stimulating] technological advances in transportation.[88]

In the light of the "sorting out" process which has continued during the past few years as regards assignment of various problems to different executive branch elements, DOT often has been in the position of "consulting" with another department—as in the case of the preparation of the report, "Tomorrow's Transportation: New Systems for the Urban Future" (with HUD)—and then later being the recipient of the full responsibility for the implementation of the national program. Also, there can be a requirement for following through on a systems-oriented project begun under another department's aegis; an example of this is the study prepared by The Pennsylvania State University Institute for Science and Engineering on the "Use of Computers in the Storage and Retrieval of Severance Effect Information" (discussed more fully in sec. VIII. D. of this report). The transfer of the Bureau of Public Roads from the Department of Commerce to DOT brought with it questions of residual support for such efforts.

Contractor support has been sought in order to study and preliminarily design some of the systems, both interstate and intra-urban, which have been given a high priority. The Office of the Assistant Secretary for International Affairs and Special Programs

[86] Background study report for the Conference on Science, Technology and State Government, held in Louisville, Ky., Sept. 19–20, 1968. Report prepared by the Southern Interstate Nuclear Board. [Opening remarks of Wyatt M. Rogers, Jr. on "Science and Technology in State and Local Affairs".] P. 4.
[87] U.S. Government organization manual: 1968–69, op. cit., p. 400.
[88] Ibid.

has commissioned Charles Rivers Associates to undertake a study of "Choice of Transport Technology under Varying Factor Endowments in Less Developed Countries" with the objective of developing:

* * * comparative analyses of costs and services characteristic of different transportation modes in freight and passenger movements for various terrains, trip distance, and traffic densities under alternative factor endowments (i.e., availability of labor, capital, foreign exchange, and other basic inputs), as experienced in the less developed countries.[89]

Another study being pursued is a "Feasibility Study on Improved Transportation Communication System" by Urban America, Inc. The importance of systems methodology (and often ADP) in the examination of these problem areas, the transfer of common elements and interactions from one environment to another, and the precision manipulation of the data elements cannot be overstated. The Assistant Secretary for Research and Technology not only is responsible for "scientific and technologic research and development relating to the speed, safety, and economy of transportation," but for the "improvement in the gathering, classification, accessibility, and use of transportation information." [90]

The maintenance of computer programs which may be used in a given task area has been a point of activity for some of the DOT groups. The Bureau of Public Roads, for instance, has an "Electronic Computer Program Library," and publicizes its holdings for external users. Included are such varied computer programs as:[91]

> Assignment of traffic to street and freeway systems;
> Forecasting interzonal traffic movements;
> Digital terrain model system horizontal alinement programs;
> Analysis of equipment costs and rentals; and
> Hydraulics of bridge waterways.

The critical state of the Nation's airways poses a problem for governmental leadership, many facets of which remain unsolved, and are causing DOT to reexamine established management procedures. The Federal Aviation Administration within DOT has shown a willingness to experiment with and apply management tools and techniques to many of its functions. In a publication produced in 1967 called "The Systems Approach to Management Decisionmaking," the FAA tells of its decision to adopt a carefully planned and organized approach to management, and describes the "management system" as a process which usually provides or requires:

> A formal statement of goals or end objectives;
> A specific, quantitative definition of what constitutes successful achievement of the stated goals so that results can, in fact, be measured against what the organization set out to accomplish;
> An integrated system for gathering, analyzing and reporting relevant data needed for decision-making at various managerial levels;
> A built-in requirement for careful analysis of alternatives at key decision points;
> An evaluation and feedback mechanism which measures progress against goals during operations as well as after action has

---

[89] Information provided by James A. McDevitt, Director, Office of Technical Assistance, Department of Transportation, 1969.
[90] U.S. Government organization manual: 1968–69, op. cit., p. 401.
[91] Electronic computer program library. Memorandum 11. Bureau of Public Roads, U.S. Department of Commerce, July 1965. 35 pp.

been completed, and which flags deviations from the norm so that operations or goals or the process itself can be adjusted accordingly; and

Safeguards which insure that components within the organization are consciously related to each other—that action by one component is only taken in full cognizance of its impact on other components. [92]

A concerted effort by FAA management to orient and educate its personnel has resulted in a better understanding on the part of all echelons of the steps which must be taken to develop a management system. An example of an "Integrated Facilities Maintenance Program Management System," developed as part of an FAA briefing package, reflects the approach being taken by this DOT component (see Figure 14). [93]

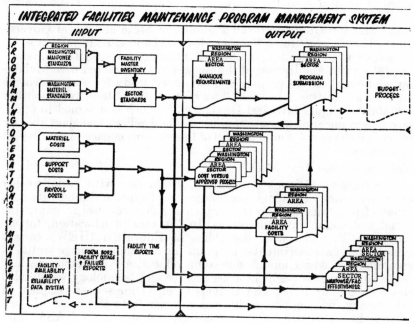

FIGURE 14

The activities described above give an indication of the role assigned to developing approaches, tools, and techniques which can be useful in overcoming the problems now besieging the transportation planners of the Nation, and prepare a sound foundation for better permanent mechanisms with which to shape the future mobility of a growing populace.

H. Other Executive Branch Activity

The executive branch of the Federal Government is faced today with a welter of social and community problems, and as has been shown there has been a reshuffling of responsibilities and where needed the

---

[92] The systems approach to management decisionmaking. Washington, Federal Aviation Administration, 1967. P. 1.
[93] Ibid., fig. 1 [no pagination].

creation of new agencies to cope with the public dilemmas. Although major rôles are being carried out by HUD, HEW, OEO, the Department of Commerce, and DOT, several other agencies and departments have given evidence of their action to use systems technology in working on civil sector problems. The money available for research and development, support of contractor studies and State or local experimental efforts, or the development of information exchange systems often is quite limited, but the even occasional project lends credence to the belief that Government leadership has come to understand the importance of employing modern technology in the management, planning, and conduct of key programs.

Bureau of the Budget

The U.S. Bureau of the Budget has been responsible for letting a series of contracts to the State-local finances project of the George Washington University which led to the preparation of a report entitled "Encouraging Improved Planning in State and Local Government: The Federal Role."[94] This is discussed and cited in section V. D. In addition to selective funding of management-oriented developmental system, the Bureau of the Budget has contracted with McKinsey and Company to prepare a preliminary study for a management information system. Also, the Office of Statistical Standards publishes a useful "Federal Statistical Directory" [95] which is designed to facilitate communication among various Federal offices and does this by providing the names and locations of key persons engaged in statistical operations.

National Science Foundation

The National Science Foundation, interacting with various other executive branch elements, encourages and supports through grants a range of programs and projects which are instrumental in improving social and community conditions. In the area of education, for example, NSF works with the Office of Education (HEW) toward establishing an experimental program for developing the potential of computers in education. A milestone pilot study to determine how State governments might best use science and technology was funded (for $85,000) by NSF. The University of Tennessee, as recipient of this money, will receive additional support from the State of Tennessee, the Tennessee Valley Authority, and through a NASA "Sustaining University Award" to the university. The specific objectives of this effort are to: [96]

 1. Describe how government now uses science and technology to promote Tennessee's economy;

 2. Identify and appraise the factors influencing the structure and innovative capacities of the State's socioeconomic institutions;

 3. Appraise from the State's viewpoint the relevant Federal programs and processes that affect the State and its communities;

 4. Explore ways to link public and private organizations with technological information and activities; and

[94] Mushkin, Selma J., Harry P. Hatry, and Marjorie C. Willcox. Encouraging improved planning in State and local government: the Federal role. Washington, D.C., State-local finances project, the George Washington University, September 1968. 54 p. plus appendix.
[95] Executive Office of the President, Bureau of the Budget. Federal statistical directory, 21st edition, 1967. Washington, U.S. Government Printing Office, 1967. 233 p.
[96] NSF supports Tennessee study of State science policy. In State technical services newsletter, vol. IV, No. 2, February 1969. Washington, U.S. Department of Commerce, Office of State technical services. P. 3.

5. Recommend actions for the formulation and implementation of a State science and technology policy.

Department of the Interior

Among the departments concerned with improving the handling of information in a circumscribed problem area is the Department of Interior. In 1966, a Natural Resources Scientific Information Center was established by the Secretary of Interior, with the first announced component to be a Water Resources Scientific Information Center. The requirement to bring together the splintered intradepartmental activities could be met by this center, which serves as a focal point for national water resource technical information activities, and will be responsible for publishing an abstract bulletin and initiating a selective dissemination of information (SDI) system.[97] Another, earlier activity originating within the Department of Interior was the publication of a "Water Quality Management Data Systems Guide," prepared as the result of a joint effort by Federal and State representatives serving on the Joint Committee on Water Quality Management Data of the Conference of State Sanitary Engineers.[98]

National Aeronautics and Space Administration

The Office of Technology Utilization of the National Aeronautics and Space Administration, established in 1963, has stressed the identification, reporting, processing, and dissemination of selected information about technological achievements throughout the world. More than 300,000 technical documents on a broad spectrum of subjects have been indexed on computer tapes, and the number is increased by approximately 75,000 annually.[99] In addition, NASA and NASA contractor efforts are monitored and their reports screened for applications which may prove helpful to nonaerospace activities. Through a network of regional dissemination centers, information is made available to outside users through an SDI-type operation; also, retrospective searches can be requested. NASA also is supporting several experimental centers, some of which are computer-equipped.

A unique type of service supported by NASA is the Computer Software Management and Information Center (COSMIC) at the University of Georgia Computer Center (at Athens, Ga.). Computer programs (source decks) may be purchased in card or tape form, with fees for a computer program on this schedule: $75 for a nonreturnable tape, and the same price for a 2,000 (or less) card deck. The fee covers a copy of the program statements on either cards, a nonreturnable tape, or by listing, and program documentation. A separate schedule of fees applies for documentation only.[100] A users' index is prepared in a Keywork-in-Context (KWIC) format and referenced to short descriptions of each program. The contents of the COSMIC library also are published regularly in International Computer Programs Quarterly,[101] together with hundreds of other available programs owned by governmental and private organizations.

---

[97] Department of the Interior activates water information center as first component of total info system on natural resources. In Scientific information notes, vol. 9, No. 6, December 1967–January 1968. p. 1.

[98] Water quality management data systems guide. Published by the Pennsylvania Department of Health, [no date]. 69 p.

[99] National Aeronautics and Space Administration. NASA's technology utilization program. Washington, U.S. Government Printing Office, 1966. p. 5–9.

[100] Directory of computer programs available from COSMIC, vol. 1, Athens, Ga.; COSMIC computer center, July 1, 1967. pp. 1–2.

[101] ICP quarterly. Indianapolis, Ind., International Computer Programs, vol. 3, No. 1, January 1969. pp. 63–89.

The Office of Technology Utilization, in its recognition of the value of mathematical management models and systems analysis, has taken the initiative in having Abt Associates prepare a survey of "Applications of Systems Analysis Models," [102] which has proven to be of considerable aid to administrative and operations personnel alike.

The efforts of other agencies and departments have been mentioned elsewhere in this report, such as the studies prepared for the Arms Control and Disarmament Agency in applying aerospace resources in the civilian sector (see sec. VIII. E.). Then there are special publications like "Recreation Information Management," [103] prepared by the U.S. Department of Agriculture Forest Service, which describes a computer-oriented system for the management of "information about people, places, and things over periods of time."

This section has indicated the breadth of executive branch activity in moving to meet the challenges of social and community problems with the latest, most responsive types of management techniques, machine support, and problem-centered systems methodology. In the analysis of the hearings conducted by the Senate Special Subcommittee on the Utilization of Scientific Manpower, contained in section VI, there is a redundant theme of the awakening of those individuals and agencies responsible for the welfare of the Nation.

I. Special Commissions, Committees, and Task Forces

The American democratic process seems to rely in part on the contributions to be made by special commissions and task forces, and in the area of social and community problems this has become increasingly the case. A list of commissions, councils, task forces, and committees which have been established as the result of congressional or presidential action, or by a responsibile officer in the executive branch comprises appendix J. Of these, a few are particularly noteworthy because of their efforts to report on current conditions in rural and urban areas, or to point up desirable corrective action which should be taken by Federal officials, State and local governmental elements, or on a joint basis.

Advisory Commission on Intergovernmental Relations

The Advisory Commission on intergovernmental Relations came into existence in 1959 with a broad range of duties. As set forth in Public Law 86–380, the Commission will— [104]

> (1) bring together representatives of the Federal, State, and local governments for the consideration of common problems;
>
> (2) provide a forum for discussing the administration and coordination of Federal grant and other programs requiring intergovernmental cooperation;
>
> (3) give critical attention to the conditions and controls involved in the administration of Federal grant programs;
>
> (4) make available technical assistance to the executive and legislative branches of the Federal Government in the review

---

[102] ABT Associates, Inc. Applications of systems analysis models: a survey. Washington, D.C., Office of Technology Utilization, National Aeronautics and Space Administration, 1968. 69 p.

[103] U.S. Department of Agriculture, Forest Service. Recreation information management. An in-service training guide. Washington, U.S. Government Printing Office, 1968. 127 p.

[104] The Advisory Commission on Intergovernmental Relations. Statutory and administrative controls associated with Federal grants for public assistance. Washington, U.S. Government Printing Office, 1964. P. iii.

of proposed legislation to determine its overall effect on the Federal system;

(5) encourage discussion and study at an early stage of emerging public problems that are likely to require intergovernmental cooperation;

(6) recommend, within the framework of the Constitution, the most desirable allocation of governmental functions, responsibilities, and revenues among the several levels of government; and

(7) recommend methods of coordinating and simplifying tax laws and administrative practices to achieve a more orderly and less competitive fiscal relationship between the levels of government and to reduce the burden of compliance for taxpayers.

In a series of reports, the Commission examined various facets of Federal-State, Federal-local, and State-local, as well as interstate and interlocal relations. Among the reports issued were:

"Statutory and Administrative Controls Associated with Federal Grants for Public Assistance"

"A Handbook for Interlocal Agreements and Contracts"

"State Legislative and Constitutional Action on Urban Problems in 1967"

"Urban and Rural America: Policies for Future Growth"

The latter report explores in considerable detail the pattern of urbanization in America; regional, State, and local economic growth; new communities and their objectives; and policy considerations in planning for controlled growth. The need for greatly enlarged services by States to local communities is discussed, and reflects the growing emphasis on improved planning, orientation, and public services (see fig. 15).[105]

---

[105] The Advisory Commission on Intergovernmental Relations. Urban and rural America: policies for future growth. Washington, U.S. Government Printing Office, 1968. P. 98, table 46.

				Planning and technical services								
State	Land use	Zoning	Assessment and taxation	Transport and communications	Power and fuel	Water and waste	Housing and health	Legal	Engineering	Offers special courses	Sponsors conferences	Make special studies
Alabama	X	X				X	X					
Arizona	X	X				X	X	X		X	X	X
Arkansas											X	X
California	X	X	X	X	X	X	X	X			X	X
Colorado	X	X	X	X	X	X	X	X		X	X	X
Connecticut	X	X	X			X	X	X		X	X	
Florida	X	X	X	X	X	X	X	X			X	X
Georgia											X	
Hawaii												
Idaho	X	X		X		X	X	X			X	X
Illinois	X	X	X	X		X	X	X			X	X
Iowa			X					X	X	X	X	X
Kansas	X	X	X	X		X	X	X			X	X
Kentucky	X	X	X	X		X	X	X			X	X
Maryland	X	X	X	X		X	X		X	X	X	X
Michigan	X	X	X	X		X	X				X	X
Minnesota			X			X	X	X				X
Mississippi												
Montana	X	X	X	X	X	X	X	X	X		X	X
Nebraska												
Nevada	X	X	X	X	X	X	X	X			X	X
New Jersey	X	X	X	X	X	X	X	X		X	X	X
New York	X	X	X	X	X	X	X	X			X	X
North Carolina	X	X	X	X	X	X	X	X			X	X
North Dakota	X		X	X		X	X		X		X	X
Ohio	X	X		X	X	X	X	X		X	X	X
Oklahoma	X	X		X		X	X				X	X
Oregon	X	X	X	X		X	X	X		X	X	X
Pennsylvania	X			X		X	X	X			X	X
Rhode Island	X					X	X	X		X	X	X
South Dakota						X	X				X	X
Tennessee	X	X		X		X	X	X		X	X	X
Texas	X	X		X	X	X	X	X	X		X	X
Utah										X	X	X
Vermont	X	X	X	X	X	X	X	X			X	X
Virginia	X	X		X		X	X			X	X	X
Washington	X	X		X		X	X	X			X	X
West Virginia	X	X		X	X	X	X		X		X	X
Wisconsin	X	X	X	X		X	X	X		X	X	X
Puerto Rico	X	X										

Source: The Council of State Governments, "Economic Development in the States" (Chicago, Ill., 1967), p. 25.

FIG. 15.—STATE SERVICES TO LOCAL COMMUNITIES (1966)

In discussing the need for new communities, the report notes that urban experts have told of the requirement for "50 to 100 standard-setting models;" [106] these could be useful in assessing where and how public services should be developed within communities old and new in order to support a scattered, unorganized population.

National Commission on Urban Problems

The National Commission on Urban Problems, appointed by the President on January 12, 1967, has performed to carry out the purposes defined in section 301 of the Housing and Urban Development Act of 1965 (see a description of this act in app. L). In addition to its final report, "Building the American City," [107] the Commission has published five volumes of hearings and a series of 14 research reports, ranging from "The Challenge of America's Metropolitan Population Outlook—1960–85" [108] to "Local Land and Building Regulation." [109]

President's Committee on Urban Housing

The President's Committee on Urban Housing, created in June 1967, has followed a course of meetings, consultations, and special studies—plus a group of technical studies—in order to determine how housing costs can be reduced, production increased, and decent housing built.

We have learned that no single new development in technology or in social and economic organization will solve at a stroke this pressing problem. We have learned, also, that although the responsibility of the Federal Government is great, Federal action alone cannot build the needed housing. Instead, there must be creative new action by many institutions and agencies, by government at the State and local level as well as in Washington, and especially by private enterprise.[110]

The report treats the role of technology by looking at Federal R. & D. expenditures and calls out the HUD funds expended (see section IV. C. above), with subsequent discussion of the activities of other Federal and some State elements in this regard: [111]

Some R&D on housing problems is supported by agencies for the Federal Government other than HUD, perhaps in part because of past Congressional reluctance to provide its housing agencies with R&D funds. The Department of Defense is now encouraging the development of innovative construction systems through its purchases of military family housing. In addition, the Army Corps of Engineers plans to devote some $2,000,000 a year to a construction research laboratory at the University of Illinois. Data on the housing industry and on the condition of the housing stock have historically been collected by the Department of Commerce, which is also generally responsible for preparation of standards for the Federal Government. The Department of Health, Education and Welfare has financed most of the significant studies of the effects of housing conditions on the health and social performance of its occupants. The Bureau of Labor Statistics in the Department of Labor has done a modest amount of research over the years on housing costs, and on productivity trends in housing. Other R&D on housing problems has been financed by the Department of Agriculture, the Office of Economic Opportunity, the Advisory Commission on Intergovernmental Relations, and other agencies. Despite this lengthy list of Federal agencies which have played a part in housing R&D in the past, the sum of their efforts has been insignificant in comparison to the needs A few states, such as California and New York, have

---

[106] Ibid., p. 103.
[107] The National Commission on Urban Problems. Building the American city. Washington, D.C. U.S. Government Printing Office, 1968, 504 p.
[108] Hodge, Patricia Leavey and Philip M. Hauser. The Challenge of America's Metropolitan Population Outlook. Washington, D.C., U.S. Government Printing Office [no date]. 99 p.
[109] Manvel, Allen D. Local Land and Building Regulation, Washington, D.C., U.S. Government Printing Office, 1966, 48 p.
[110] The President's Committee on Urban Housing. A decent home. Washington, U.S. Government Printing Office, 1969, p. i–ii.
[111] Ibid., p. 197.

extended some support to housing-oriented research. But all government efforts put together have not yet been sufficient to establish a sound base for a systematic and growing body of knowledge on housing problems.

Intergovernmental Task Force on Information Systems

The problem of improving the flow of information within and among Federal, State, and local governments led to the establishment of an Intergovernmental Task Force on Information Systems. This action was arranged by the U.S. Bureau of the Budget, Council of State Governments, National Association of Counties, National League of Cities, U.S. Conference of Mayors, International City Managers Association, and the Advisory Commission on Intergovernmental Relations. The purpose of the study was twofold: to identify impediments to attaining an effective flow of information between governments, and to recommend actions that could be taken at various levels of government.

Twenty specific recommendations were set forth, varying from endorsement of legislation (the Joint Funding Simplification Act, the Intergovernmental Manpower Act) to the creation of an Intergovernmental Information Systems Exchange (under the auspices of the Advisory Commission on Intergovernmental Relations) and a Federal Information Center on Assistance Programs. The emphasis on planning and the utilization of systems and computer technology permeates the discussion and recommendations, and the need for realistic mechanisms to allow the exchange of representative views on a regular, coordinated basis is underscored. "The Dynamics of Information Flow" [112] is a thought-inspiring document, and it strives to develop a compound of recommendations—legislation, new action mechanisms, revised procedures for interaction—that gives the governmental leadership at all levels a number of alternatives from which to choose.

J. Summary

During the past decade, the Federal Government has taken the initiative in establishing policies, designing and implementing direct assistance and other types of corrective programs, and creating improved channels of communication with State and local governments—all of which could lead to more effective handling of social and community problems. Science and technology, as adapted from the aerospace environment or applied in an innovative way (without prior testing elsewhere) to some specific civil sector problem (e.g., pollution control), have begun to fulfill an increasingly key role as the Nation moves to solidify its gains and advance even further toward a better existence for its citizens.

Leadership in large part has been provided by the Congress, as it has sorted out the options for legislative action and then compressed often complex and paradoxical situations into meaningful law. The activities of the Senate, House, and Joint committees (and subcommittees) have represented, often in almost a classical fashion, the difficulty of dealing with a problem, the enormity of which may be threatening the fabric of our civilization. In this section, the activities of several committees and subcommittees have been noted, and in many cases linked appropriately with legislation—discussed separately in section V—which ensued from their deliberations.

---

[112] The Intergovernmental Task Force on Information Systems. The dynamics of information flow. Washington, Executive Office of the President, April 1968, 31 p.

Within the executive branch of the Federal Government a band of departments and agencies has been assigned the responsibility of devising programs capable of removing or reducing various threats to the public welfare. The missions of such establishments as HUD, HEW, OEO, the Department of Commerce, and DOT have been reviewed, and a selection of projects involving the utilization of ADP, operations research, and other forms of systems methodology has been presented. Descriptions of yet other projects conducted either under completely private, joint governmental-private, or State-supported funding are contained in section VIII.

The entire question of how to provide technical assistance and monetary support to States, local units, and private sector groups has been a major focus of the Senate Special Subcommittee on the Utilization of Scientific Manpower. The grants-in-aid approach, suggested by Senator Gaylord Nelson in S. 430, continues to be used by many agencies so that imaginative, technology-oriented efforts may be tested and evaluated. The other major alternative under consideration during the hearings of the special subcommittee—section VI reviews and analyzes the four series during which nearly 40 witnesses offered commentary on the strategy and tactics of Federal aid—centered about the creation of a National Commission on Public Management (as proposed in S. 467 and H.R. 20). The work of several key commissions and task forces in the area of studying and recommending solutions to the nondefense, nonspace problems of the country is noted in section IV. I.

Finally, the part played by professional, university, and industrial groups in the process of studying and trying out corrective programs has been considered. As the Nation grows larger, and the greater competence of disciplinary groups is degraded by the dilemma of how to communicate within and between specializations, the importance of symposia and seminars grows. From these may emerge useful proceedings or topical treatments of some special problem. As technology and its multiple-pronged impact stirs the thinking of scientists, technologists, and administrators, the frustrations brought about by communications difficulties must be lessened and responsive exchange mechanisms established. Toward this end, the public servant and the professional from industry or academe must join forces and agree upon an achievable plan of action.

V. SIGNIFICANT LEGISLATION

The legislation of prime concern to the special subcommittee has been S. 430, the Scientific Manpower Utilization Act, and S. 467, calling for the creation of a National Commission on Public Management. These bills and their predecessor versions (in the 89th Congress) as well as H.R. 20, the counterpart bill to S. 467, served as the focal points for witness testimony during the four hearings (see section VI), and were an instrument for attracting public attention to the need to utilize our systems capability in new areas.

In addition to the legislation identified above, other bills and resolutions were introduced during the 90th Congress which set forth the requirement for employing automatic data processing, systems analysis, and related tools and techniques in the handling of certain priority programs. These proposals, together with selected public laws treating various social and community problems, are reviewed in this section.

A. COMMENTARY ON S. 430 (AND S. 2662)

The Scientific Manpower Utilization Act first was introduced in the Senate in October 1965 by Senator Nelsons and Clark. It (S. 2662) was

. . . designed to foster the application of systems analysis at the regional, State, and local levels as well as to provide for an adequate base of trained manpower.[1]

Reintroduced early in the 90th Congress by the two original sponsors and Senator Randolph, S. 430 had this purpose:

. . . to facilitate and encourage the utilization of the scientific, engineering, and technical resources of the Nation in meeting urgent problems facing the Nation or localities within the Nation, by promoting the application of systems analysis and systems engineering approaches to such problems.

The problems referred to included but were not limited to the areas of education, unemployment, welfare, crime, juvenile delinquency, air pollution, housing, transportation, and waste disposal. The responsibility for carrying out the purposes of the act was placed with the Secretary of Labor, who was directed to—

1. Make "appropriate grants to States," and
2. Enter into "appropriate arrangements" (grants, contracts, or other agreements) with universities or "other public or private institutions or organizations."

Such grants or contracts would be "for the purpose of causing systems analysis and systems engineering approaches to be applied to national or local problems."

The burden for determining that such grants were used properly and the results disseminated to appropriate Federal and State groups was vested with the State (see sec. 4 of the text of S. 430 below). In particular, the grant was to be employed in such a way that the

---

[1] Nelson, Gaylord. Systems analysis. Remarks in the Senate. Congressional Record. [Daily ed.] (Washington), vol. 112, Oct. 21, 1966, p. 27301.

knowledge and experience gained would have *"substantial relevance"* [italics added] to problems within the purview of the act and which exist in other States.

The bill also provided for two or more States to combine to apply for one or more grants jointly to carry out the purposes of the act. The need for regional funding is discussed in section VI of this report.

The full text of S. 430 follows:

[S. 430, 90th Cong., first sess.]

A BILL To mobilize and utilize the scientific and engineering manpower of the Nation to employ systems analysis and systems engineering to help to fully employ the Nation's manpower resources to solve national problems.

Be it enacted by the Senate and House of Representatives of the United States of America in Congress assembled, That this Act may be cited as the "Scientific Manpower Utilization Act of 1967".

SEC. 2. It is the purpose of this Act to facilitate and encourage the utilization of the scientific, engineering, and technical resources of the Nation in meeting urgent problems facing the Nation or localities within the Nation, by promoting the application of systems analysis and systems engineering approaches to such problems. The problems referred to in the preceding sentence include, but are not limited to, problems in the area of education, unemployment, welfare, crime, juvenile delinquency, air pollution, housing, transportation, and waste disposal.

SEC. 3. The Secretary of Labor (hereinafter referred to as the "Secretary") shall carry out the purposes of this Act by—

(1) making appropriate grants to States, and

(2) by entering into appropriate arrangements (whether through grants or contracts, or through other agreements) with universities or other public or private institutions or organizations,

for the purpose of causing the systems analysis and systems engineering approaches to be applied to national or local problems of types which the Secretary, by regulations, designates as being within the purview of this Act.

SEC. 4. (a) Any grant made under section 3 to a State shall be used only for the purpose for which the grant was made, and may be used by the State for such purpose directly, or through the State's entering into appropriate arrangements for the carrying out of such purpose (whether through grants or contracts, or through other agreements) with universities or other public or private institutions or organizations.

(b) No grant under this Act shall be made to a State unless the Secretary finds that—

(1) the knowledge and experience expected to be gained from the employment of such grant would have substantial relevance to problems within the purview of this Act which exist in other States;

(2) the State has presented a plan setting forth in detail the purposes for and manner in which such grant is to be used, together with the objectives expected to be achieved from the use of such grant;

(3) the State has designated an officer or agency of the State who has responsibility and authority for the administration of the program in which such grant is to be employed; and

(4) the State agrees fully to make available to the Federal Government and to other States (and political subdivisions thereof) data and information regarding the employment of such grant and the findings and results stemming therefrom.

(c) There shall not be granted to any State under this Act amounts the aggregate of which exceed 20 percentum of the aggregate of the amounts which have been appropriated to carry out this Act at the time amounts are granted to such State hereunder.

(d) Two or more States may combine to apply for one or more grants jointly to carry out the purposes of the Act with respect to one or more of the problems which they have in common and which are within the purview of this Act, and in any such case, the provisions of subsection (b) shall be deemed to require the submission of a joint plan for the utilization of the grant and the designation of one or more officers or agencies having responsibility and authority to carry out the joint plan. Each State participating in such a joint plan shall be deemed, for purposes of subsection (c), to have received an amount equal to the amount produced by dividing the amount of the grant received to carry out such plan by the number of States participating in such plan.

SEC. 5. The Secretary, in awarding grants to States and in entering into arrangements with universities or other public or private institutions or organizations, shall follow procedures established by him for the purpose of assuring that the grants or other expenditures made to carry out the purposes of this Act will be equitably distributed among the various major geographic regions of the Nation.

SEC. 6. For the purpose of making the grants and entering into the other arrangements provided under section 3 of this Act, there is hereby authorized to be appropriated, without fiscal year limitation, not more than $125,000,000.

B. COMMENTARY ON S. 467

Recognition that a formal mechanism must be established to bring about a means of managing the public business more effectively and economically caused Senator Scott and several colleagues to introduce a bill (S. 3762) in August of 1966 which would establish a National Commission on Public Management. The 90th Congress version appeared as S. 467, with this stated purpose:

To provide for a study with respect to the utilization of systems analysis and management techniques in dealing with problems relating to unemployment, public welfare, education, and similar problems.

The instrument for accomplishing this task would be a 13-member commission with a tenure of $2\frac{1}{2}$ years, with a mandate to answer two fundamental questions: "Can the systems approach contribute to the solution of these problems? If so, how can it best do the job?" [2] The Commission, with a membership composed of representatives of government, business, labor, and education, would:

First, define a social and economic problem to which the application of the systems approach appears to hold promise.

Second, examine the many modern management techniques currently being used in the aerospace industry to ascertain which are best suited for application to these problems and what modifications may be required.

Third, determine the proper relationship between public and private investment in these areas, including the degree of public involvement and the best procedures for governmental support and funding.

Fourth, decide on the optimum organizational relationships among several levels of governmental authorities.

Fifth, explore the roles of small business and organized labor in the application of·these new management techniques.

Sixth, assess the potential contributions of the universities toward resolving public management problems.[3]

These six actions reflect the nine specified duties of the National Commission on Public Management (see the full text of S. 467 below).

While only a preliminary report, at the end of the first year of the Commission's activity, and a final report are required, interim reports—e.g., consultants' reports, transcripts of testimony, seminar reports—could be published. Interface with executive branch elements for the purpose of obtaining "information, suggestions, estimates, and

[2] Scott, Hugh. National Commission on Public Management. Remarks in the Senate. Congressional Record. [Daily edition] (Washington), vol. 113, Jan. 18, 1967, p. S. 432.
[3] Ibid.

statistics" is provided for, and the Commission is authorized to hold seminars or informal conferences in order to provide a forum for the discussion of applying the new technology to the Nation's community problems.

Senator Dominick, a cosponsor of S. 467, signified his belief that the new Commission would "revolutionize our political approach to vital national problems" and went ahead to inform his peers that: "We must set forth our desire to honestly meet and solve the problems and encourage our citizens that finally we are going to employ the tools which are necessary." [4] The emphasis, then, in the words of Senator Scott, was to create a means to cope with the management problems of today:

All levels of government—Federal, State, and local—are finding it increasingly difficult to solve their complex management problems on a piecemeal basis, to a large extent because they lack the management techniques and skills that have been applied so successfully in private industry.[5]

The emphasis on private sector involvement in the problems of public management is reflected in several sections of S. 467, the full text of which appears below:

[S. 467, 90th Cong., first sess.]

A BILL To provide for a study with respect to the utilization of systems analysis and management techniques in dealing with problems relating to unemployment, public welfare, education, and similar problems.

Be it enacted by the Senate and House of Representatives of the United States of America in Congress assembled,

CREATION OF COMMISSION

SECTION 1. In order to study and recommend the manner in which modern systems analysis and management techniques may be utilized to resolve problems relating to unemployment, public welfare, education, and similar national and community problems in the nondefense sector, there is hereby established a National Commission on Public Management (hereafter in this Act referred to as the "Commission"), under the general supervision and direction of the Secretary of Labor.

MEMBERSHIP OF THE COMMISSION

SEC. 2. The Commission shall be composed of a Chairman, a Vice Chairman, and eleven other members to be appointed by the President, by and with the advice and consent of the Senate. Members of the Commission shall be individuals concerned with the subject matter to be studied by the Commission, including individuals with experience derived from Government, business, the labor movement, or from teaching and research.

DUTIES OF THE CHAIRMAN AND VICE CHAIRMAN

SEC. 3. (a) The Chairman shall be responsible for calling regular quarterly meetings of the Commission and other special meetings as he deems necessary. The Chairman shall determine the time, place, and agenda for each regular or special meeting.

(b) The Vice Chairman shall act in the Chairman's absence.

QUORUM

SEC. 4. Seven members of the Commission shall constitute a quorum.

COMPENSATION OF MEMBERS OF THE COMMISSION

SEC. 5. (a) Members of the Commission, other than officers or employees of the Federal Government, shall receive compensation at the rate of $75 per diem while engaged in the actual performance of duties vested in the Commission, plus

[4] Ibid., p. S 433 [remarks of Senator Dominick].
[5] Scott, Hugh. Congressional interest in systems approach spurs legislation for national commission. In Aerospace Management, fall-winter 1966, p. 12.

reimbursement for travel, subsistence, and other necessary expenses incurred by them in the performance of such duties.

(b) Any members of the Commission who are officers or employees of the Federal Government shall serve on the Commission without compensation, but such members shall be reimbursed for travel, subsistence, and other necessary expenses incurred by them in the performance of duties vest in the Commission.

STAFF

SEC. 6. (a) The Commission may appoint an Executive Director and such other personnel as it deems advisable. The Executive Director shall be the chief staff member of the Commission and shall be responsible to the Commission for the direction of its staff. The annual compensation for the Executive Director shall be $28,500.

(b) The Commission may procure temporary and intermittent services in accordance with section 3109 of title 5, United States Code, but at rates not to exceed $75 per day.

EXPENSES OF THE COMMISSION

SEC. 7. There are hereby authorized to be appropriated such sums as may be necessary, but not exceeding $500,000 in the aggregate, to carry out the provisions of this Act during the initial year of Commission operation.

DUTIES OF THE COMMISSION

SEC. 8. The Commission shall concern itself with the management of the public business and shall give attention to the development, dissemination, and implementation of modern management technology and analysis of the systems interrelationships involved in public business problems. The Commission, in the performance of its duties, shall:

(1) Develop information on the methodology of the systems approach and its applications.

(2) Analyze the possible application to public programs of such recognized management planning and control techniques as operations analysis and research, econometrics, mathematical programing and modeling, simulation, project management, and the utilization of automatic data processing devices and procedures for program control and information systems.

(3) Determine and categorize the national and community problems to which the application of such techniques offers the greatest promise of solution.

(4) Assess the proper relationship between governmental and private investment to obtain the most effective application of the techniques involved.

(5) Make recommendations to the executive and legislative branches of the Federal Government regarding data requirements, management techniques, and systems interrelationships in the formulation of legislation.

(6) Conduct studies into unemployment, public welfare, education, and other specific problem areas and make recommendations.

(7) Schedule seminars, symposia, and prepare publications to expand public knowledge of and stimulate the use of modern management technology.

(8) Encourage the Nation's best talent in Government, labor, university, and private enterprise to study public management problems and to participate in the improvement and extension of modern management technologies and their application to public problems.

(9) Analyze alternative methods and make recommendations of Federal, State, and local governmental support and encouragement of the application of modern management technology to public problems through the use of various contracting procedure, grants, loans, cost allowances, and tax incentives.

REPORTS

SEC. 9. (a) Within one year after the fist meeting of the Commission it shall submit to the President and the Congress a preliminary report on its activities with particular emphasis on the plan for the study and investigation provided for in section 8 and any activities undertaken to carry out such plan, including an estimated budget for the remainder of the life of the Commission.

(b) Within thirty months after such first meeting the Commission shall submit to the President and the Congress a final report on its study and investigation which shall include its recommendations and such proposals for legislation and administrative action as may be necessary to carry out its recommendations.

(c) In addition to the preliminary report and final report required by this section, the Commission may publish such interim reports as it may determine, including but not limited to consultants' reports, transcripts of testimony, seminar reports, and other Commission findings.

POWERS OF THE COMMISSION

SEC. 10. (a) The Commission or, on the authorization of the Commission, any subcommittee or member thereof, may, for the purpose of carrying out the provisions of this Act, hold such hearings and sit and act at such time and places, administer such oaths, and require, by subpena or otherwise, the attendance and testimony of such witnesses and the production of such books, records, correspondence, memorandums, papers, and documents as the Commission or such subcommittee or member may deem advisable. Subpenas may be issued under the signature of the Chairman of the Commission, of such subcommittee, or any duly designated member, and may be served by any person designated by such Chairman or member. The provisions of section 102 to 104, inclusive, of the Revised Statutes (U.S.C., title 2, secs. 192–194), shall apply in the case of any failure of any witness to comply with any subpena or to testify when summoned under authority of this section.

(b) The Commission is authorized to secure directly from any executive department, bureau, agency, board, commission, office, independent establishment, or instrumentality information, suggestions, estimates, and statistics for the purpose of this Act; and each such department, bureau, agency, board, commission, office, establishment, or instrumentality is authorized and directed to furnish such information, suggestions, estimates, and statistics directly to the Commission, upon request made by the Chairman.

(c) The Commission is authorized to hold seminars or informal conferences as it deems appropriate to provide a forum for discussion of the application of modern systems analysis and management techniques to the solution of national community problems.

TERMINATION

SEC. 11. On the sixtieth day after the date of its submission of its final report to the President, the Commission shall terminate and all offices and employment therein shall expire.

C. COMMENTARY ON H.R. 20

Within the House of Representatives a sizeable number of Members among the Republicans introduced bills in support of the Morse bill (H.R. 20) to create a National Commission on Public Management. A complete listing of these sponsors, together with an enumeration of Senators sponsoring S. 430 and S. 467, is shown in figure 16. The concept of a national commission, as described in H.R. 20, was endorsed by the United States Chamber of Commerce.

CONGRESSIONAL SPONSORS OF S. 430, S. 467, AND H.R. 20

SENATE

S. 430 Gaylord Nelson (D–Wis.)

Co-Sponsors

Joseph S. Clark, Jr. (D–Pa.)   Jennings Randolph (D–W. Va.)

Bill referred to Committee on Labor and Public Welfare.

To provide for a study with respect to the utilization of systems analysis and management techniques in dealing with problems relating to unemployment, public welfare, education and similar problems.

S. 467 Hugh Scott (R–Pa.)

Co-Sponsors

Gordon Allott (R–Colo.)
Wallace F. Bennett (R–Utah)
Clifford Case (R–N.J.)
Peter H. Dominick (R–Colo.)
Paul Fannin (R–Ariz.)
Hiram Fong (R–Hawaii)
Robert P. Griffin (R–Mich.)

Jacob K. Javits (R–N.Y.)
Len B. Jordan (R–Idaho)
Thomas H. Kuchel (R–Calif.)
Thruston B. Morton (R–Ky.)
George Murphy (R–Calif.)
Winston L. Prouty (R–Vt.)
John G. Tower (R–Tex.)

Bill referred to Committee on Labor and Public Welfare.

To mobilize and utilize the scientific and engineering manpower of the Nation to employ systems analysis and systems engineering to help to fully employ the Nation's manpower resources to solve national problems.

HOUSE OF REPRESENTATIVES

To establish a National Commission on Public Management, and for other purposes.

H.R.   20 F. Bradford Morse (R–Mass.)
       101 John B. Anderson (R–Ill.)
       102 Mark Andrews (R–N.D.)
       103 William H. Bates (R–Mass.)
       104 Alphonzo Bell (R–Calif)
       105 William S. Broomfield (R–Mich.)
       106 Clarence J. Brown, Jr. (R–Ohio)
       107 Elford A. Cederberg (R–Mich.)
       108 Don H. Clausen (R–Calif.)
       109 James S. Cleveland (R–N.H.)
       110 Glenn Cunningham (R–Neb.)
       111 Thomas B. Curtis (R–Mo.)
       112 Edward J. Derwinski (R–Ill.)
       113 Robert Dole (R–Kan.)
       114 John J. Duncan (R–Tenn.)
       115 Florence P. Dwyer (R–N.J.)
       116 John N. Erlenborn (R–Ill.)
       117 Paul Findley (R–Ill.)
       118 Peter H. B. Frelinghuysen (R–N.J.)
       119 Seymour Halpern (R–N.Y.)
       120 Craig Hosmer (R–Calif.)
       121 Theodore R. Kupferman (R–N.Y.)
       122 Robert McClory (R–Ill.)
       123 Joseph M. McDade (R–Pa.)
       124 Charles M. Mathias, Jr. (R–Md.)
       125 Chester L. Mize (R–Kan.)
       126 Charles A. Mosher (R–Ohio)
       127 Albert H. Quie (R–Minn.)
       128 Charlotte T. Reid (R–Ill.)
       129 Ogden R. Reid (R–N.Y.)
       130 Ed Reinecke (R–Calif.)
       131 Howard W. Robinson (R–N.Y.)
       132 Herman Schneebeli (R–Pa.)
       133 Richard S. Schweiker (R–Pa.)
       134 Garner E. Shriver (R–Kan.)
       135 Henry P. Smith III (R–N.Y.)
       136 Burt L. Talcott (R–Calif.)
       517 John T. Duncan (R–N.Y.)
       708 Frank J. Horton (R–N.Y.)
       793 Odin Langen (R–Minn.)
    1215 William F. Ruppe (R–Mich.)
    1459 John W. Wydler (R–N.Y.)
    3911 Harold D. Donohue (D–Mass.)
    3997 Donald Rumsfeld (R–Ill.)
    4280 Sam Steiger (R–Ariz.)
    4514 George M. Rhodes (D–Pa.)
    4742 Hastings Keith (R–Mass.)

4761 Clark MacGregor (R–Minn.)
5655 William A. Steiger (R–Wis.)
7220 Harold R. Collier (R–Ill.)
7404 Edward J. Gurney (R–Fla.)
10348 Marvin L. Esch (R–Mich.)
10383 William L. St. Onge (D–Conn.)
All Bills referred to Committee on Government Operations.

FIG. 16.—CONGRESSIONAL SPONSORS OF S. 430, S. 467, AND H.R. 20

The bill prepared by Representative Morse—and its 89th Congress version, H.R. 17310—was similar to its Senate counterpart (S. 467). Among the differences was the statement of purpose, which was somewhat more general:

In order to study and recommend the manner in which modern systems analysis and management techniques may be utilized to resolve national and community problems in the nondefense sector there is hereby established a National Commission on Public Management.

Four congressional Members are included in the composition of the Commission, two each being appointed by the Speaker of the House of Representatives and the President of the Senate. The total membership of the Commission, as in S. 467, is 13.

Of the nine specified duties, all are the same as S.467 except that No. 6 reads: "Conduct studies into specific problem areas and make recommendations." The reports required are the same in both the House and Senate versions.

In commenting on the role of science in society, and the importance of insuring the rights of the individual, Representative Morse told his colleagues:

We should not ignore the caution urged by those who see modern technology only as the precursor of a society of robots, where individual identity is subsumed in the mass production of everything from man's comforts to his personality. We can avoid that world, the world of '1984,' only if we make the new technology our servant, and not allow it to be our master. The best assurance that the progress of science will not mean the insignificance of man is to recognize the revolution in technology, to anticipate its growth, to assure adequate personal safeguards from its excesses, and to employ it for the betterment of man. We must not merely be awed by science; we must be inspired by it to summon equal creativity in the political and economic fields.[6]

Prior to the introduction of H.R. 20, Representative Morse stressed the criticality of involving more deeply the various elements of the private sector in public management. In an article prepared for the Harvard Business Review, he identified several reasons why business should assume a responsibility in the new area:

The attractions for business in this field are several. First, there is the profit motive, and the federal government of course should make certain that its private partners are sufficiently compensated. Then there is the chance for large firms to diversify, and possibly find and open up new 'civilian' markets. Finally, there is the opportunity to take maximum advantage of the heavy investment already made in manpower and equipment necessary for systems techniques.[7]

Thus, a more effective Government-industry partnership was called for as the needs of the Nation began to force a reassessment of the roles and allocation of resources of the assertive forces within our society.

[6] Morse, F. Bradford. National Commission on Public Management. Remarks in the House. Congressional Record. (Daily ed.) (Washington) vol. 113, Jan. 10, 1967, p. H40.
[7] Morse, F. Bradford. Private responsibility for public management. In Harvard Business Review, vol. 45, No. 2, March/April 1967, p. 178.

D. Commentary on Selected Problem-Oriented Public Laws

While the propositions contained in S. 430 and S. 467 were the subject of intensive examination and discussion by the Special Subcommittee on the Utilization of Scientific Manpower, there was also an increased awareness of provisions for attention to technology and its application to the social and community problems. The terminology found in the public laws often was quite general or ambiguous, but it did reveal congressional determination to encourage the use of new technological methodology, automatic data processing, and analytical planning processes.

The legislation discussed below was passed, for the most part during the 88th, 89th, and 90th Congresses. A listing of pertinent public laws is found in appendix L of this report. The nondefense, nonspace public problems where public laws reflect the encouragement for using innovative technology include: urban planning, housing, environmental pollution, transportation, law enforcement, health services, recreation, education, manpower, and (in a related sense) the creation of new agencies.

Demonstration Cities and Metropolitan Development Act (Public Law 89-754)

The Demonstration Cities and Metropolitan Development Act of 1966 features "comprehensive city demonstration programs" jointly funded by grants from the Department of Housing and Urban Development and State or local moneys. Public Law 89-754, title II (planned metropolitan development), specifies that grants may be made for development projects in metropolitan areas, including "sewer, water, and sewage treatment facilities; highway, mass transit, airport, and other transportation facilities; and recreation and other open-space facilities." Title III (urban information and technical assistance services) authorizes Federal financial assistance to State and local agencies to:

* * * enable them to (1) make available information and data on urban needs and assistance programs through centers established for such purpose, and (2) provide technical assistance to small communities.

It should be noted that section 1010 is intended to encourage application of existing advances in technology created by private research and development efforts so as to assist industry in lowering the costs and raising the quality of housing construction. The Secretary of HUD is directed to:

(1) conduct research and studies to test and demonstrate new and improved techniques and methods of applying advances in technology to housing construction, rehabilitation, and maintenance, and to urban development activities; and

(2) encourage and promote the acceptance and application of new and improved techniques and methods of constructing, rehabilitating, and maintaining housing, and the application of advances in technology to urban development activities, by all segments of the housing industry, communities, industries engaged in urban development activities, and the general public.

In discussing the objectives of research and studies, this law states that such efforts "shall be designed to test and demonstrate the applicability to housing construction, rehabilitation, and maintenance, and urban development activities, of advances in technology relating to (1) design concepts, (2) construction and rehabilitation methods, (3) manufacturing processes, (4) materials and products, and (5) building components."

Finally, Public Law 89–754 deals with urban environmental studies:

The Congress finds that, with the ever-increasing concentration of the Nation's population in urban centers, there has occurred a marked change in the environmental conditions under which most people live and work; that such change is characterized by the progressive substitution of a highly complex, man-contrived environment for an environment conditioned primarily by nature; that the beneficent or malignant influence of environment on all living creatures is well recognized; and that much more knowledge is urgently needed concerning the effect on human beings of highly urbanized surroundings. It is the purpose of this section to authorize a comprehensive program of research, studies, surveys, and analyses to improve understanding of the environmental conditions necessary for the well-being of an urban society, and for the intelligent planning and development of viable urban centers.

The powers and duties of the Secretary of HUD in carrying out the various assignments in this new area of activity are specific; he is authorized and directed to—

(1) conduct studies, surveys, research, and analyses with respect to the ecological factors involved in urban living;

(2) document and define urban environmental factors which need to be controlled or eliminated for the well-being of urban life;

(3) establish a system of collecting and receiving information and data on urban ecological research and evaluations which are in process or are being planned by public or private agencies, or individuals;

(4) evaluate and disseminate information pertaining to urban ecology to public and private agencies or organizations, or individuals, in the form of reports or otherwise;

(5) initiate and utilize urban ecological information in urban development projects initiated or assisted by the Department of Housing and Urban Development; and

(6) establish through interagency consultation the coordinated utilization of urban ecological information in projects undertaken or assisted by the Federal Government which affect the growth or development of urban areas.

Housing and Urban Development Act (Public Law 90–448) and Earlier Laws?

The lawmakers and executive branch decisionmakers concerned with *housing and urban renewal* long had recognized the importance of careful planning. As early as the passage of the Housing Act of 1949 (Public Law 81–171), the Federal Government required certain analytical documents to be prepared either during or after the planning period. Specifically, an "Urban Renewal Plan" and a "Workable Program for Community Improvement." This legislation also charged the Housing and Home Finance Agency with encouraging and assisting "the use of new and standardized methods, materials, and equipment." Later legislation such as the Housing and Urban Development Act of 1968 (Public Law 90–448), featuring amendments to section 701 of the Housing Act of 1954 (Public Law 83–560), also stressed the analytical elements of planning: an inventory of current status, a projection of need, and a general statement on standards of evaluation. This law also directs the Secretary of HUD "to institute a program under which public and private organizations shall submit plans for lower income housing using *new and advanced technologies*" [italic added].[8] Another section:

Permits grants for developing and testing urban renewal demonstration grant program and authorizes funds necessary for urban information and technical assistance program, and studies of advances in technology in housing and urban development.

Other indications of the awareness that the application of new technology must be encouraged are found in Public Law 88–560 (Housing Act of 1964) which broadens experimental programs if the housing "involves the utilization or testing of new design, materials, or similar items."

Water Quality Act (Public Law 89–234)

Legislation concerning the control and improvement of *environmental pollution* has been enacted only after a consideration of many highly diverse proposals, extended hearings, and numerous compromises involving the forms which these laws finally assumed. The Water Quality Act of 1965 (Public Law 89–234) created a Federal Water Pollution Control Administration within HEW, and provided for the establishment of water quality standards. Provision also was made for grants for "projects which conform to comprehensive metropolitan or regional planning." The urgency of developing "new or improved methods" of controlling pollution activities is implicit in the terminology of this and related laws. The Secretary of HEW is charged with promoting "the coordination of, research, investigations, experiments, demonstrations, and studies relating to the causes, control, and prevention of water pollution." Emphasis is placed on the collection and exchange of information on projects (research, investigatory, demonstration) related to the problem, and to the issuance of grants-in-aid to public or private agencies and institutions or individuals for research and training projects. In addition, research fellowships in HEW are supported.

Clean Water Restoration Act (Public Law 89–753)

Evidence of the Federal Government's desire to consider such a problem as water pollution within the larger context of public well-being is reflected in the statement of this major goal in title II of the Clean Water Restoration Act of 1966 (Public Law 89–753):

"(g) (1) The Secretary shall, in cooperation with the Secretary of the Army, the Secretary of Agriculture, the Water Resources Council, and with other appropriate Federal, State, interstate, or local public bodies and private organizations, institutions, and individuals, conduct and promote, and encourage contributions to, a *comprehensive study* of the effects of polution, including sedimentation, in the estuaries and estuarine zones of the United States on fish and wildlife, on sport and commercial fishing, on recreation, on water supply and water power, and on other beneficial purposes. Such study shall also consider the *effect of demographic trends*, the exploitation of mineral resources and fossil fuels, *land and industrial development*, navigation, flood and erosion control, and other uses of estuaries and estuarine zones upon the pollution of the waters therein.

[8] Public Law 90–448 further authorizes HUD to support five major plans for new housing technology in developing housing for lower income groups on surplus Federal lands to be made available to HUD by other Federal agencies upon request. The use of such lands for research and development purposes allows bypassing some of the impediments (zoning regulations, housing codes) encountered in utilizing private property.

80

"(2) In conducting the above study, the Secretary shall assemble, coordinate, and organize all existing pertinent information on the Nation's estuararies and estuarine zones; carry out a program of investigations and surveys to supplement existing information in representative estuaries and estuarine zones; and identify the problems and areas where further research and study are required.

"(3) The Secretary shall submit to the Congress a final report of the study authorized by this subsection not later than three years after the date of enactment of this subsection. Copies of the report shall be made available to all interested parties, public and private. The report shall include, but not be limited to—

"(A) *an analysis* of the importance of estuaries to the economic and social well-being of the people of the United States and of the effects of pollution upon the use and enjoyment of such estuaries:

"(B) a discussion of the major economic, social, and ecological trends occurring in the estuarine zones of the Nation;

"(C) recommendations for a comprehensive national program for the preservation, study, use, and development of estuaries of the Nation, and the respective responsibilities which should be assumed by Federal, State, and local governments and by public and private interests.

[Emphasis added.]

Solid Waste Disposal Act (Public Law 89-272)

Title II of Public Law 89–272, the "Solid Waste Disposal Act," contains the results of congressional study followed by a brief statement of the purpose of the act. Here again, the problem is presented in terms of the needs of society and the corrective programs which must be pursued.

FINDINGS AND PURPOSES

SEC. 202. (a) The Congress finds—

(1) that the continuing technological progress and improvement in methods of manufacture, packaging, and marketing of consumer products has resulted in an ever-mounting increase, and in a change in the characteristics, of the mass of material discarded by the purchaser of such products;

(2) that the economic and population growth of our Nation, and the improvements in the standard of living enjoyed by our population, have requested increased industrial production to meet our needs, and have made necessary the demolition of old buildings, the construction of new buildings, and the provision of highways and other avenues of transportation, which together, with related industrial, commercial, and agricultural operations, have resulted in a rising tide of scrap, discarded, and waste materials;

(3) that the continuing concentration of our population in expanding metropolitan and other urban areas has presented these communities with serious financial, management, intergovernmental, and technical problems in the disposal of solid wastes resulting from the industrial, commercial, domestic, and other activities carried on in such areas;

(4) that inefficient and improper methods of disposal of solid wastes result in scenic blights, create serious hazards to the public health, including pollution of air and water resources, accident hazards, and increase in rodent and insect vectors of disease, have an adverse effect on land values, create public nuisances, otherwise interfere with community life and development;

(5) that the failure or inability to salvage and reuse such materials economically results in the unnecessary waste and depletion of our natural resources; and

(6) that while the collection and disposal of solid wastes should continue to be primarily the function of State, regional, and local agencies, the problems of waste disposal as set forth above have become a matter national in scope and in concern and necessitate Federal action through financial and technical assistance and leadership in the development, demonstration, and application of new and improved methods and processes to reduce the amount of waste and unsalvageable materials and to provide for proper and economical solid-waste disposal practices.

(b) The purposes of this Act therefore are—

(1) to initiate and accelerate a national research and development program for new and improved methods of proper and economic solid-waste

disposal, including studies directed toward the conservation of natural resources by reducing the amount of waste and unsalvageable materials and by recovery and utilization of potential resources in solid wastes; and

(2) to provide technical and financial assistance to State and local governments and interstate agencies in the planning, development, and conduct of solid-waste disposal programs.

Air Quality Act (Public Law 90–148)

In the Air Quality Act of 1967 (Public Law 90–148), the importance of treating such problems as air pollution on the basis of regional impact and corrective action is repeated, thus echoing many of the statements made during the hearings of the Special Subcommittee on the Utilization of Scientific Manpower. The mechanism in this case is to be "regional air quality commissions" which shall pursue a systematic approach to abatement through the establishment of air quality standards, "taking into consideration * * * the concentration of industry, other commercial establishments, and population and the technological and economic feasibility of achieving such quality."

While the number of public laws passed by the 89th and 90th Congresses in an attempt to improve living conditions has been impressive, and indicative of the Members' deep concern over the problems of environmental pollution, the private sector also has begun to commit more money to waste treatment facilities. Senator Edmund S. Muskie, the sponsor of many pieces of legislation in this area, gave the keynote address at the IBM Scientific Computing Symposium on Water and Air Resource Management in 1967, and during the course of his remarks pointed out the value of systems technology to private corporations:[9]

Systems analysis techniques can be extremely valuable to individual companies in determining the best combination of control techniques to achieve a given level of control. They can lead to improvements in processing which reduce waste and cut costs as well as curtail pollution.

Beyond the individual factory, stack emission control, or treatment system problem, computer technology is essential to the development of any rational and economic environmental quality program. As we delve deeper into the mysteries of watersheds and airsheds, we are aware of the vast range of data we need to understand the complicated interactions within these ecosystems. Without computer technology, we would be faced with what Dr. Kenneth Boulding has called the "pollution of our information system."

Senator Muskie then spoke of the responsibility of the assembled private sector representatives, stating that "you who are charged with the responsibility of applying new technology, developing systematic methods, and predicting long-range trends will determine the policy decisions in environmental protection."[10]

Highway Safety Act (Public Law 89–564) and related legislation

The mobility of the American public is dependent in large part upon highway networks, reliable urban mass transportation (busses, trains), interurban vehicles (airplanes included), and traffic control systems. *Transportation* planning needs of the Nation have led to the creation of the Federal Department of Transportation in 1966 (Public Law 89–670) and a series of legislation encompassing interstate highway apportionment, highway safety (Public Law 89–564), urban mass transit, and regional transportation authority establishment. Controls

---

[9] Muskie, Edmund S. Computers, environmental planning, and the quality of life. Keynote address *in* proceedings of IBM scientific computing symposium on water and air resource management. White Plains, N.Y., IBM Data Processing Division, 1968. P. 4.

[10] Ibid., p. 5.

on funds granted by the Bureau of Public Roads to States, and through them to cities and counties, are identified in Public Law 85–767 ("Highway Planning and Research Program") as residing with the State highway departments. Planning in urban areas (with a population of 50,000 or more) must be coordinated with the State's "comprehensive planning process." The elements of the planning process are noted in the volume Encouraging Improved Planning in State and Local Government (The Federal Role), prepared by the State-local finance project of The George Washington University: [11]

(a) Setting objectives.

(b) A continuous inventory of roads.

(c) An implicit consideration of need (now to be based upon projections until 1990).

(d) Implicitly required is a consideration of alternatives, and a ranking of road construction priorities which are formulated in a project plan for the year.

The role of interstate highway development, both in terms of binding distant population centers more closely and providing major conduits for heavy traffic flow within a megalopolitan area—such as the Boston-New York-Baltimore-Washington, D.C., conglomerate—has not been ignored by the Congress. From the passage of the Federal Aid Highway Act of 1956 to the successor legislation 10 years later (Public Law 89–574), large sums have been spent in acquiring rights-of-way, performing preliminary engineering tasks, and actual construction. Where land use was changing from agricultural to urban functions, the Secretary of Agriculture (through Public Law 89–560) was authorized to cooperate with States and other public agencies "by providing soil surveys and other technical assistance in planning for use changes in rapidly expanding areas including farm and nonfarm areas."

The requirement to insure adequate national highway safety was considered by Congress in 1966, with the result that a Highway Safety Act was passed. This law encourages the States to create safety programs in accordance with uniform standards established by the Department of Transportation, and that these standards are to be such as to—

(1) improve driver performance;

(2) provide for an effective record system of accidents;

(3) provide for accident investigations;

(4) insure current vehicle registration, operation, and inspection;

(5) affect highway design and maintenance;

(6) establish traffic control;

(7) provide for vehicle codes and laws;

(8) establish surveillance of traffic for detection and correction of high or potentially high accident locations; and

(9) provide for emergency services.

Certain categories of data, such as vehicle registration and accident records may already be stored in the computer-supported files of such cities as St. Louis, or similar systems under development in Los Angeles and Alameda County, Calif. The maintenance of current, cross-referenced files in State department of motor vehicles is allowing intersystem queries. As discussed in section VIII of this report, the

---

[11] Mushkin, Selma J., Harry P. Hatry, and Marjorie C. Willcox. Encouraging improved planning in State and local government: the Federal role. Washington, D.C., State-local finance project, The George Washington University, September 1968. P. A–35.

New York State Identification and Intelligence System (as one example) is able to recall valuable written and graphic data ranging from stolen motor vehicles to auto registration forgeries.

Urban Mass Transportation Act, (Public Law 88–365) and related legislation

The problems related to the development of urban mass transit have been numerous and complex, and each of the last three Congresses has wrestled with some aspect of the dilemma. The Urban Mass Transportation·Act (Public Law 88–365) stressed assistance to State and local bodies and "mass transportation companies" in order to provide "mass transportation facilities necessary for the orderly growth and development of urban communities." This law allowed States to enter into interstate compacts in order to carry out the necessary programs. During the 89th Congress, consent was given for a compact between Missouri and Kansas to create the Kansas City Area Transportation District and the Kansas City Area Transportation Authority (Public Law 89–599). Similar legislation was passed to create a Washington Metropolitan Area Transit Authority (Public Law 89–774) and encourage "the prosecution of a transit development program for the National Capital region" (Public Law 90–220).

As the difficulties of implementing a modern mass transportation system became better understood, more explicit action as taken by the Congress "for the planning, engineering, and designing of urban mass transportation projects and for studies concerning the nature, utilization, and economic feasibility of facilities, equipment, and systems" (Public Law 89–562). A special category of graduate fellowships was created so that managerial, technical, and professional personnel already in the transportation field could learn about innovations in machines and techniques. Research and development support in highspeed ground transportation "to improve the national transportation system" was authorized in Public Law 89–220, and amplified in Public Law 90–423 which made the Secretary of the Department of Transportation responsible for contracting "the construction of two suburban rail stations in furtherance of a demonstration program."

Throughout the remarks of the sponsors of the above legislation, in public discussion, and in dialog between technologists and public authorities has emerged a theme of increased reliance on systems analysis in the identification of feasibility projects, the determination of useful criteria and developmental check points, the measurement of basic and applied research purportedly contributory to the development of an improved operating system, and the establishment of a set of "hard" evaluation parameters by which the utility of a piece of equipment or a new systems procedure can be measured.

Law Enforcement Assistance Act (Public Law 89–197)

Significant strides have been made in the application of systems technology to *law enforcement and the administration of justice,* Sophisticated communications and computer processing capabilities have been established, as discussed in section VIII of this report. As the crime rates have increased in many areas, law enforcement officers, judges, and the citizenry have initiated action to employ every possible advanced tool and technique. The Law Enforcement Assistance Act

(Public Law 89–197) authorizes the Attorney General to make grants to "any public or private nonprofit agency for the purpose of improving the quality of State and local law enforcement." Provision also is made for studies of "matters relating to law enforcement organization, techniques, and practices."

Omnibus Crime Control and Safe Streets Act (Public Law 90–351)

The Omnibus Crime Control and Safe Streets Act of 1968 (Public Law 90–351) which provides funds to States and local governments requires that a comprehensive State plan be prepared—and this process may include local participation—which will identify objectives, identify the resources available and needed, consider the costs and benefits of alternatives within a multiyear framework. The law also features, in the interest of improving law enforcement throughout the Nation, an authorization for the Attorney General to:

. . . conduct research and evaluation studies, to collect, evaluate, publish, and disseminate statistics and other information on the condition and progress of law enforcement and criminal justice in the several States.

Federal Judicial Center (Public Law 90–219)

During the preparation of legislation (H.R. 5385) and others designed to create a Federal Judicial Center, language was included which called for the use of ADP technology in connection with the improvement of judicial administration. There had been effort in this application area as the result of the work of the President's Commission on Law Enforcement and the Administration of Justice, and congressional interest soon led to the preparation of a general study on the topic.[12] In the section of Public Law 90–219 which defines the duties of the Board of the Federal Judicial Center, this directive is found:

(5) study and determine ways in which automatic data processing and systems procedures may be applied to the administration of the courts of the United States.

Legislation concerning health services (Public Law 89–749 and others)

The terminology related to the use of systems technology in the realm of *health services* tends to be general, with occasional references to "modernization" of facilities and provisions for "studies and demonstrations." Planning requirements for medical facilities include overtones of systems-oriented estimation of needs, as included in the Hospital and Medical Facilities Amendment of 1964 (Public Law 88–443), with the maintenance of a current inventory specifically noted. Grants to States and nonprofit institutions for training, studies, planning, and services are included in the Comprehensive Health Planning and Public Health Services Amendments (Public Law 89–749). Perhaps the most descriptive wording to encourage modern methods of improving health facilities and services is contained in Public Law 90–174 (partnership for health amendments), which:

Authorizes the Secretary of Health, Education, and Welfare to make grants and contracts for projects for the conduct of research, experiments, or demonstrations aimed at developing new methods or improvement of existing methods or organization, delivery, or financing health facilities and services.

[12] Chartrand, Robert L. Systems technology and judicial administration. In "Deficiencies in Judicial Administration," hearings before the Senate Subcommittee on Improvements in Judicial Machinery (90th Cong., 1st sess., 1967), pp. 455–470.

This law, amending Public Law 89–749, asserts positively that planning at both the State and local levels should consider—[13]

1. Goals and objectives;
2. Measurements of health status;
3. Inventories of existing programs and facilities;
4. Assessment of their adequacy; and
5. Consideration of alternatives and priorities.

There is little detailed discussion of these elements nor any guidelines as to their fulfillment, however.

Outdoor recreation planning assistance (Public Law 88–578)

The Bureau of Outdoor Recreation of the Department of the Interior is empowered by Public Law 88–578 (outdoor recreation planning assistance) to furnish funds which may be used in the development of a "State comprehensive plan." Such a recreation plan is deemed acceptable if it examines needs, includes an inventory of outdoor facilities and areas, and has a "plan of action" for the acquisition and development of lands and facilities.

Elementary and Secondary Education Act (Public Law 89–10)

Several major pieces of legislation were passed during the 89th and 90th Congresses in an effort to assist in the solution of community *educational problems*. The Elementary and Secondary Education Act of 1965 (Public Law 89–10) stressed the development of supplementary educational services and the creation of exemplary programs as models for regular school programs. Federal funds were made available for specialized instruction and equipment for students in advanced courses, "developing and conducting exemplary educational programs," educational radio and television; in addition, grants could be made for educational planning on a statewide basis, collection-analysis-dissemination of educational data, programs fostering or aiding educational research programs and projects. The planning guidelines were twofold, featuring (a) the preparation of a budget-plan document necessary to receive funds allotted to the States, and (b) a general planning process. The State-local finances project, directed by Dr. Selma Mushkin, in reviewing this and associated legislation, commented on the nature of the planning document: [14]

The document called a "state plan" is prepared in order to show how money allotted to the states under this title will be used. It must show educational needs and how this money will be used to meet the needs. In addition, provision is necessary to adopt procedures for evaluating the effectiveness of projects which the state funds. The purposes of exemplary programs, one of the thrusts of this title, are to try out educational innovations which in itself can be considered part of an overall planning process. The administrative funds can be used to shape the whole of a state's educational program by planners who wanted to guide the use of innovative monies. The money could be used to fund a variety of planning activities. These are not required activities, but the U.S.OE is attempting to promote them as parts of a planning process.

Higher education amendments (Public Law 90–575)

The Higher Education Amendments of 1968 (Public Law 90–575) placed heavy emphasis on strengthening education for public service.

[13] Mushkin, op. cit., p. A–20.
[14] Ibid., p. A–44.

Among the new programs were those to create "Networks for Knowledge" whereby colleges and universities would "share their technical and other educational and administrative facilities and resources through cooperative arrangements." The testing of such multi-institutional arrangements would be made either by the colleges or universities involved or by established public or nonprofit private organizations. The law also provided for Federal assistance in the purchase of special equipment for education in the natural or physical sciences. The Vocational Educational Amendments of 1968 (Public Law 90–576) concentrated on improving occupational educational programs, including "making available modern vocational education equipment and specially qualified personnel." This law provided, as well, for:

. . . support or services for the comprehensive and compatible recording, collecting, processing, analyzing, interpreting, storing, retrieving, and reporting of State and local vocational education data, including *the use of automated data systems.* [Italic added.]

The impact of these and related (e.g., the National Defense Education Act) laws has been great, for university-situated laboratories—the University of Pittsburgh, Harvard University, the University of Wisconsin, Stanford University, to name a few—have been enabled to explore the application to education of technology. The establishment of a national program of educational laboratories (NPEL) by Public Law 89–10, and the further provision within that act for a "cooperative research program" resulted in the creation of a national capability of unprecedented potential. The Elementary and Secondary Education Act of 1965 also includes a charter for an advisory committee on supplementary educational centers and services responsible for reviewing grant applications, policy matters related to program administration, and the delineation of management criteria. A more detailed discussion of the role of the U.S. Office of Education is contained in section IV. C. of this report.

U.S. *Employment Service Wagner-Peyser Act of 1933*

Beginning with the administration of President Franklin Delano Roosevelt, there has been recurring action by the Congress to deal with the problem of *unemployment and manpower resources.* For the most part, the application of the systems approach to this problem has been through the inclusion in legislation of terminology requiring "systematic planning." In the U.S. Employment Service Wagner-Peyser Act of 1933, emphasis is placed on coordination of and communication among federally sponsored agencies which furnish manpower services at the national, regional, State, and local levels. The instrument for this action was the cooperative area manpower planning system. A concentrated employment program—situated in 90 major urban areas with high incidences of manpower problems—was charged with the responsibility for channeling all manpower services controlled by the Department of Labor. Analytical planning of the demand-supply ratio, with the critical implications of accurate forecasting, is commented upon in encouraging improved planning in State and local government: [15]

The problem in manpower is primarily to try to match the demand and supply of labor, and much of the planning involves forecasting these quantities. To provide funds for these purposes the states have been required annually to submit

[15] Ibid., p. A–48, A–49.

formal budget type justifications to the BES for increases in funding over the base in the existing year.

Under a new regulation, an annual document must now be developed from the local offices to provide a one-year work plan with an indication of how existing resources will be used, the results expected, and estimates of what additional resources would be required to provide effective service. The state, in turn, develops a state plan based on these local plans. The federal administrators hope to develop information on five-year needs.

Analyses are primarily of labor market supply and demand conditions and not of the form which services should take. The form of services is largely predetermined by the legislatively authorized mechanisms for an employment exchange function. Indeed, this legislative rigidity has meant that other types of manpower services have to be done by separate organizations.

Manpower Development and Training Act (Public Law 87–415)

The Manpower Development and Training Act (Public Law 87–415) stressed in like fashion the coordination of all planning activities of manpower programs. With grants being jointly administered by the Manpower Administration (Bureau of Employment Security) of the Department of Labor and the Office of Education (Bureau of Adult and Vocational Education) of HEW, the need for precise planning information is very important; these elements of information are required: [16]

1. A summary of existing economic conditions, labor market supply and demand, and anticipated economic developments.
2. Identification of manpower problems (needs).
3. Inventory of program resources available.
4. One-year operating plans for the most effective use of available program resources.
5. An evaluation of the differences between needs and resources available to meet them.

Economic opportunity amendments (Public Law 90–222)

Another directive causing community action agencies to adopt a systematic approach to the achievement of purposes and the utilization of funds is contained in Public Law 90–222, the Economic Opportunity Amendments of 1968 (to Public Law 88–452). This law also "requires careful and systematic evaluation of the Job Corps program," one of the rare instances when such terminology is featured. This evaluation is to be achieved directly or through independent contracts (some of which are noted in section IV. C. of this report) or through consultation with other agencies. The goal: "to compare the relative effectiveness of Job Corps and other programs." The concern of the Senate Special Subcommittee on the Utilization of Scientific Manpower is reviewed in sections II and IV. A., with additional commentary by Senator Gaylord Nelson in the section (VI) which analyzes the several hearings conducted by the special subcommittee.

Legislation creating new agencies: HEW, OEO, OSTS, HUD, and DOT

In the period since the Second World War, pressures upon the populace and the overwhelming display of a need for a new mechanism to provide service in some public problem area have spurred the Congress to create new agencies, departments, and offices within the Federal executive branch. The Department of Health, Education, and Welfare (created by the Reorganization Plan of 1953), the Office of Economic Opportunity (established by the Economic Opportunity

[16] Ibid., p. A-51.

Act of 1964, Public Law 88–452), the Office of State Technical Services (established by departmental order 7A of November 19, 1965, pursuant to Public Law 89–182), the Department of Housing and Urban Development (Public Law 89–174), and the Department of Transportation (Public Law 89–670)—all of these organizations have prescribed responsibilities for working with the States (and local authorities, in most instances) to combat our social and community problems.

Intergovernmental Cooperation Act (Public Law 90–577)

Experience has shown that often special legislation, allowing multistate action, is required, as in the passage of the Public Works and Economic Act of 1965 (Public Law 89–136) which provides for the establishment of regional planning commissions. The Intergovernmental Cooperation Act of 1968 (Public Law 90–577) goes even further in solidifying arrangements between the Federal Government and the States. Governors of the States are to receive full information on all grants to their States, a "coordinated intergovernmental urban assistance policy" requiring local government review of certain applications for Federal aid in urban problems is established and congressional review of future grant programs, "to insure that such programs are systematically reexamined and reconsidered in the light of changing conditions," is required.

The sum of the legislation discussed, then, is a more powerful (though often fragmented) capability for coping with the problems of America in the midsixties and beyond. The role of systems technology, increasingly identified and specifically assigned to some facet of a problem, is growing. Public law amendments to earlier legislation reflect the urgency attached to the creation of new policies and organizations which can and will address the encroaching problems. Perhaps most importantly, the decisionmakers in the legislative and executive branches have evinced a willingness to work together in effecting a complement of forces equal to the task of preserving our civilization.

E. Related 90th Congress Legislative Proposals

There is much evidence, as this report indicates, that man is exerting every energy to maintain control of his environment. But there is an equal measure of proof that the accomplishment of this goal will depend in good part upon his ability to invent and apply new management techniques. Discussion continues to be heavy about the extent to which "creative federalism" should be sanctioned. There is concern on every hand about the proliferation of Federal grant programs, with their often disruptive influence on State and city operations. And there is a growing opinion that intergovernmental cooperation is the key to the survival and continued rise in the standard of living of the people of the United States.

During the 90th Congress, several proposals were introduced in both chambers which addressed the questions of intergovernmental cooperation and planning, the exchange of certain information both laterally and vertically between governmental elements, and the need for an effective mechanism by which to measure the efficacy with which Federal program funds were being expended. While a number of thoughtful concepts did not receive approval by the Congress, their introduction and the subsequent discussion of the merits of each served to alert the leadership of the Nation to the significance of the subject areas.

As noted earlier in this section, the Intergovernmental Cooperation Act of 1968 became law (Public Law 90–577) and established a series of strong guidelines for Federal-State cooperation. In a similar vein, with decided emphasis on the need to develop "comprehensive, policy-based, action-oriented planning processes and systems," was S. 799, sponsored by Senator Hugh Scott and others. While the wording of this bill did not specifically mention the systems approach, the repeated underscoring of the importance of effective planning, programing, and budgeting and the "collection on a systematic basis" and analysis of demographic and economic data indicates the author's awareness of the role of systems technology. The Comprehensive Planning and Coordination Act also strives to:

. . . give an incentive to State and sub-State areas to participate in the national program budgeting system (PPBS) and thereby provides a State input into the Federal budget-making process.

Senator Edmund Muskie took the lead in sponsoring several bills and resolutions in the area of governmental activity: S. 671 called for the establishment of a National Intergovernmental Affairs Council in the Executive Office of the President; S. 699 suggests a Federal assistance program to improve State and local personnel and training programs, by focussing on four prime problem areas (merit systems, personnel management, in-service training, and interchange of Federal, State, and local employees). The latter bill, called the Intergovernmental Personnel Act of 1967 together with S. 671, and Senate Resolution 68 which called for the creation of a Select Committee on Technology and the Human Environment, reflected congressional concern not only with existing problems, but the projected enlargement of problem situations. The proposed select committee would be responsible for conducting:

. . . a comprehensive study and investigation of (1) the character and extent of technological changes that probably will occur and which should be promoted within the next fifty years and their effect on population, communities, and industry, including but not limited to the need for public and private planning and investment in housing, water resources (including oceanography), education, automation affecting interstate commerce, communications, transportation, power supplies, welfare, and other community services and facilities; and (2) policies that would encourage the maximum private investment in means of improving human environment.

Congressional concern for improving the flow of key information between Federal, State, and local governmental units was manifested in Senate Joint Resolution 110 which was introduced by Senator Edward Kennedy. He noted that modern technology allows for the expeditious exchange of selected data, and that there is a need for "a study and investigation of information service systems for States and localities;" in this connection, he states that:

. . . scientific advances in computer and information retrieval technology represent a major new capability which may have important applications to the development of a modern intergovernmental information system.

As the number of Federal assistance programs has mounted, congressional desire to know more about how the money is spent, and with what results, has grown apace. The need for a reliable, penetrating evaluation process has not been questioned, and while many Members have indicated a willingness to let the cognizant executive branch agencies monitor their own programs, other Congressmen have come forward with propositions for new assessment mechanisms. The concept of a Government Program Evaluation Commission (S. 2032) was introduced by Senator William Proxmire. The Commission would:

> . . . make a full and complete study and evaluation of existing Federal programs and activities (old and new) and of projected expansions of such programs and activities for the purpose of determining, in the light of the fundamental needs of the Nation and its vital objectives—
>> (1) the effectiveness of each such program or activity in terms of its present and projected costs,
>> (2) whether such program or activity should be continued and, if so, the level at which it should be continued, and
>> (3) in the allocation of Federal funds, the relative priority which should be assigned to such program or activity.

Another suggestion which took the form of legislation was that to establish an Office of Program Analysis and Evaluation and a Joint Committee of Congress on Program Analysis and Evaluation. Introduced by Senator Scott (S. 3322) and with counterpart bills in the House of Representatives, this approach featured the use of "objective, scientific, and empirical analysis" in determining the effectiveness of Federal programs and activities.

F. Summary

The Members of the U.S. Congress, acting as individual legislators or in their capacity as members of committees, have been active during the 1960 decade in initiating legislation designed to improve social and community living conditions. One aspect of the bills and resolutions presented in the two Chambers for consideration was that of applying systematic planning and systems technology to these problems. The terminology differed, but the underlying aim was the same: to encourage and in many cases finance the design, testing, implementation, and evaluation of innovative tools and techniques which could at least, improve serious problem situations such as environmental pollution and urban renewal.

The Senate Special Subcommittee on the Utilization of Scientific Manpower, in hearings and through questionnaires, has sought to elicit comments from experienced persons in the public and private sectors. The focus has been a dual one: specific critique of S. 430 and S. 467, which are discussed at length earlier in this section and in section VI, and the larger context of the potential of the systems approach in the service of society.

While an analysis and comparison of the two subject bills (together with H.R. 20) has been of prime interest, the significance of legislation providing for the application of systems methodology to civil sector problems seemed to justify an examination. Also, there has been a number of bills and resolutions submitted both in the Senate and the House of Representatives which embody the new approach to reducing crisis conditions or providing for their solution, but which did not receive favorable (or final) action during the 90th Congress.

The narrative discussion of the above legislation is complemented by appendix L, a collection of the synopsized contents of those public laws which relate to specific social and community problems, and the selected references which contain listings of relevant congressional documentation.

VI. ANALYSIS OF HEARINGS ON S. 2662, S. 430, AND S. 467

A. Purpose and Organization of the Hearings

The Special Subcommittee on the Utilization of Scientific Manpower, under the chairmanship of Senator Gaylord Nelson, conducted four series of hearings from 1965 to mid-1967. Senator Nelson stated that the hearings had two purposes:

* * * first, to explore systems analysis and systems approaches to find the best way to make them available; and second, to develop a hearing record that will make clear to local officials and national officials, with administrative rather than technical responsibilities, what the systems approaches are all about.

As the hearings progressed, special areas of emphasis evolved—an example of this was the examination of the impact of PPBS both at the Federal and State levels—due to new perspectives provided by witnesses and the development of new approaches to the subject by members of the special subcommittee. Testimony was solicited from and provided by university and foundation specialists in social and community problems; Federal and State officials with responsibilities in the planning, programing, and budgeting realm; managerial and systems analysis personnel from industry; and representatives from civic and trade organizations.

In most instances, the special subcommittee generated selected questions which were sent to participants in the hearings so that an opportunity existed for acquiring specific information. Included in these questions were requests for comments on the proposed legislation (S. 430 et al.) and the potential for adapting aerospace systems technology to the nondefense, nonspace problem areas. In addition to those persons actually making an appearance before the subcommittee, contributions in the form of papers were accepted from interested individuals and organizations. The testimony presented during the course of the four sets of hearings, together with ancillary documentation, has been published in two volumes.[1] A listing of selected witnesses is found in appendix B of this report, and the "Selected References" section contains many of the papers submitted to the special subcommittee for inclusion in the hearings documentation. A more detailed description of each of the four hearings' series is set forth in the sections which follow.

B. First Hearing, Los Angeles, 1965

The first hearing was held in Los Angeles, Calif., on November 19, 1965, with Senator Nelson presiding.[2] The Scientific Manpower

---

[1] U.S. Congress, Senate Committee on Labor and Public Welfare, Special Subcommittee on the Utilization of Scientific Manpower, Scientific Manpower Utilization, 1965–66, hearings before the special subcommittee (89th Cong., first sess., Nov. 18, 1965, and second sess., May 17–18, 1966), 213 pages.
   U.S. Congress, Senate Committee on Labor and Public Welfare, Special Subcommittee on the Utilization of Scientific Manpower, Scientific Manpower Utilization, 1967, Heatings before the special subcommittee (90th Cong., first sess., Jan. 24–27 and Mar. 29–30, 1967), 377 pages.
[2] Scientific Manpower Utilization, 1965–66, op. cit., p. 1–143.

Utilization Act (S. 2662) served as a focal point for many of the comments from public and private sector witnesses. There was a consensus of support for the Nelson bill, with considerable discussion about the basis for granting Federal funds to State or local governmental groups, or directly to private organizations such as universities. Detailed reports on the systems analysis studies performed for the State of California by four aerospace companies were presented.

Gov. Edmund G. Brown, of California, appeared as the initial witness, and expressed his full support for S. 2662. He stressed the phenomenal rate of growth throughout the Nation and in California in particular, pointing out that by the year 2000:

* * * there will be 150 million more people in America than there are today. Where there is one house today, there will be two. Where there is one hospital, one school, they will need two. They will need five times as much electricity as we generate today. They will drive 240 million cars—more than three times as many as we have on the roads today.

Governor Brown then reviewed the heavy emphasis on aerospace work in his State, and told of convening a panel of industrial executives to assist him in identifying courses of action which might be desirable and feasible if a major conversion to consumer production were necessitated. During these discussions, two facts asserted themselves:

First, fully one-half of all of the engineers and scientists trained in space research and development live and work in this State.

Second, these men shared a talent that might well be applied to the development of systems to solve nearly any problem of any kind that might be presented to them. And when we began thinking in terms of the public sector—the many unmet needs of our State—we began to realize that we had a precious and unused resource.

The next step described by Governor Brown was the letting of four $100,000 contracts to aerospace firms, each of which would study one priority problem area:

    1. North American Aviation was asked to determine whether systems engineering could help plan and build a transportation complex sufficient for the long range (i.e., 50 years) needs of the State.

    2. Aerojet-General was asked to investigate methods of waste disposal, including air and water pollution and the solid waste problem.

    3. Space-General was asked to study crime, which costs California $600 million each year.

    4. Lockheed was asked to develop a better statewide system for information handling.

The reports of these four firms are presented in volume 1 of the special subcommittee hearings, with salient aspects discussed below. Governor Brown offered the opinion that perhaps the " 'cold mathematical' approach is the only way that we can afford the compassionate goals of our society during the crucial decade just ahead." He then ended his testimony by pointing out that:

One factor—cost—makes these studies more national than local in scope. No State can afford to carry the burden of a fully engineered system in very many problem areas. And even those who could do so should not be asked to, since these studies will be useful in nearly all of the States.

Both Governor Brown and his deputy director of the California Department of Finance, Harold R. Walt, stressed the savings which had been projected through applying systems technology to certain problems. In waste management alone, an annual savings (in terms of pollution alleviation) of $3 billion would ensue.

In discussing the need for additional research to develop new tools and techniques, Senator Nelson reminded his hearers that "air pollution is doing $11 billion damage" per year, so preventive research is more than justified.

In the dialog which followed Governor Brown's statement, several questions were raised by Senator Nelson and Roy Millenson, a staff member of the Senate Committee on Labor and Welfare. In response to the query "Are there funds for systems analysis available from any source of the Federal Government now?", Governor Brown cited the Aid to Education Act which allows research within certain criteria. He then noted that the Clean Air Act provides for research in the field of waste management; the Law Enforcement Assistance Act also was mentioned in terms of allowing limited usage of systems analysis. Another question—by Senator Nelson—concerned the availability of funds for grants to industry to engage in research in the development of new devices for waste disposal. Mr. Walt replied that the State Technical Services Act offers a hope for obtaining funds to develop salt-waste disposal plans and water desalinization facilities.

Finally, in regard to the funding aspects of S. 2662, the State of California witnesses urged allocation of grants on the basis of "open and free competition," but not on a matching fund arrangement. Governor Brown also felt that awards should be on a statewide basis but that New York City might require special handling. After a discussion of the needs of large cities for Federal support, Senator Nelson iterated the theme of S. 2662 in these words:

> It was my intention in drafting the bill that the problems tackled were problems that had a universality about them, and grants can be made under the bill to States or groups of States, but not for the sole purpose of solving that State or group of States problems, but that the answer they came up with would be applicable to the solution of the similar problems in all of the rest of the States.

During the testimony of State Senator Thomas M. Rees, the importance of attacking "regional problems with regional thinking and planning without subverting the concepts of home rule and local government" was stressed. The creation of a basinwide flood control district—jointly supported by city, county, State, and Federal elements—was noted as an example of regional problem-solving. After noting such major problems as air pollution, weather control, and hardship areas (such as Watts), Mr. Rees proceeded to comment upon the role of the Federal Government, which should be:

> * * * that of a coordinator, and since the Federal Government will be supplying some funds, with these funds they can say we will only cooperate if this is done on a regional basis. We will not single-shot it by hitting separate jurisdictions.
> I don't think the Federal Government should come in and help the community unless they are willing to help themselves.

Presentation of the reports by the four aerospace firms commissioned by the State of California to study selected problem areas next occupied the special subcommittee. Representing the Lockheed Missiles & Space Co. were Dr. J. P. Nash, vice president; Kenneth T. Larkin, direct of special programs; and Louis J. Lauler, project leader, special programs. The objective of the Lockheed study was to "determine what the informational process was in the State of California." More than 800 interviews were held throughout the State and five sample counties plus five sample cities were selected for intensive scrutiny.

Following this information acquisition phase, the study team then developed a concept for an "optimum" State information system, and an implementation plan. The latter phase involved such questions as: "What would the program cost? How should it be phased? What steps should take place first?" Finally, the report was fully documented.

Information usage was a matter of prime concern to the Lockheed analysts. Several types of basic data were visualized:

    1. "People" data—information about people themselves, e.g., location, identification, descriptive data, etc.

    2. Data concerning parks, water use, and urban renewal.

    3. Data associated with vocational skills, industrial requirements for plant location and labor force.

    4. Data in the area of public transportation—highway locations and timing.

    5. Data related to the physical facilities of the State such as prisons and hospitals.

The actual functions of collecting, storing, retrieving, and exchanging information (within the various levels of jurisdiction within the State) and the ways in which these data are used are shown in graphic form in figure 17. The witnesses noted that the Federal Government would be involved in the information handling process in the future.

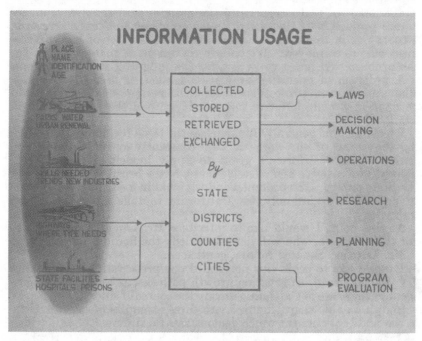

FIGURE 17

One facet of the State's information handling problem which caused concern to the analysts was the lack of coordination between those planning statewide ADP systems and the organizations projecting computation centers. Information *exchange*, a subject often to appear during the special subcommittee hearings, was not being provided for in the State of California. A special graphic, tracing data flows on a county-to-State basis, reflects the study team emphasis on thoroughly understanding intergovernmental information exchange. (See fig. 18.)

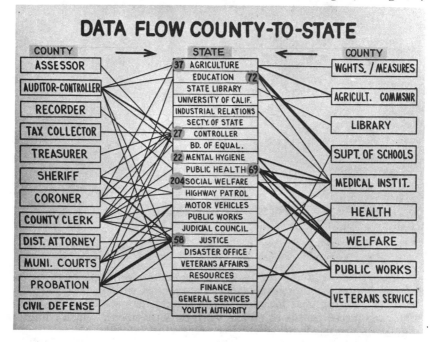

FIGURE 18

The approach taken by the Lockheed investigators in delineating the information needs of the State of California was to identify the various reports emanating from each governmental entity, how they were indexed, stored, retrieved, and disseminated. Four major needs (as shown in fig. 19) were stressed:

1. A common identifier, assigned on a geographical basis.

2. An information locator, indicating where certain data was maintained.

3. A single data entry, allowing the transmission of selected data in a specific form to all organizations having the same need.

4. Paper flow reduction through the development of a modern data communication system.

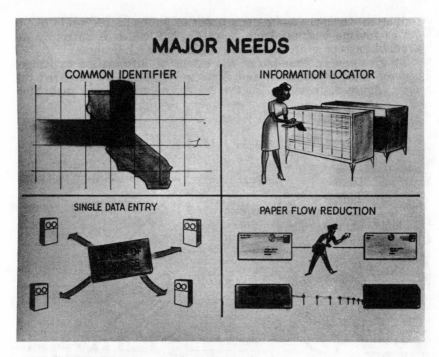

FIGURE 19

The witnesses for Lockheed ended their presentation by underscoring the savings possible through the implementation of an improved, ADP-supported system for handling the State's rapidly expanding volumes of information. It was stated that $183 million annually could be saved with such a system, while the general benefits—cost reduction, program evaluation enhancement, service improvements, and the creation of new capabilities—would allow the State to continue effectively in the decades ahead. Examples of the listed general benefits appear in figure 20.

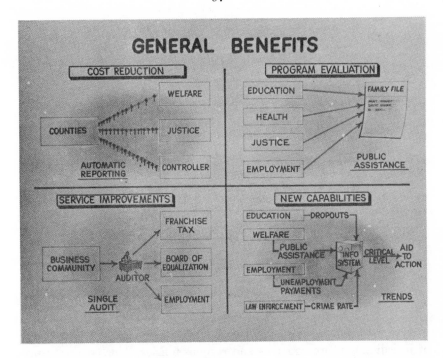

FIGURE 20

At this juncture in the hearing, Dr. John V. Zuckerman of the University of Southern California injected the telling statement that:

While we do have the scientific knowledge and the technical skills to deal with these problems we feel that these have to be integrated with the conceptual and practical understanding of management processes.

Dr. Zuckerman specifically recommended that S. 2662 be modified to include an emphasis on what he termed "the management of change."

The second study to be described to the special subcommittee was that performed by Aerojet-General Corp. in the waste management area. Dr. B. Dwight Culver, manager of the Life Support Systems Division, made a lengthy presentation which stressed the need for a unified approach. Political and technical problems are so intertwined that they must be analyzed carefully by professional groups appreciative of the changing state of our environment and the increased pressures of a future civilization. The study team approached the assignment by projecting the effects of three different systems: the "expanded existing system," the "state of the art" system using all present technology in the most efficient way, and the "development system" which would "incorporate new and novel means of transporting, treating, reclaiming, and disposing of wastes." The dissemination of gaseous, liquid, and solid wastes in shown in figure 21.

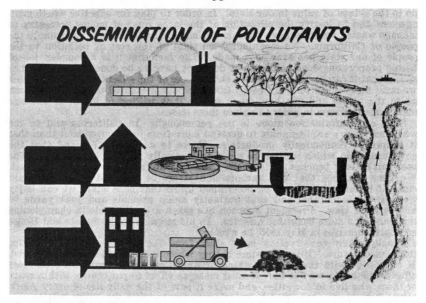

FIGURE 21

One of the chief impediments to controlling pollutants is the question of jurisdictional prerogatives and responsibilities. This condition prevails in virtually all metropolitan areas, and in many of the State or multistate situations. A graphic rendering of the fragmented responsibility for waste control is contained in figure 22.

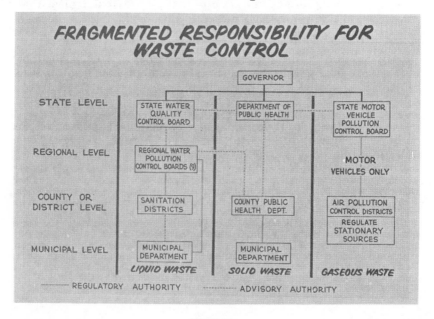

FIGURE 22

Dr. Culver's report also discussed at length the need to establish meaningful environmental standards. These would include esthetic guidelines and the specific limits placed on air or water pollution conditions. Figure 23 reflects present water environmental standards and selected present values. The Aerojet-General study recommended that a feasibility study be undertaken; this would last from 3 to 5 years and cost between $8 and $10 million. The recommendations included an identification of the need for a number of models and computer programs, and the creation of a national environmental simulation facility for the study of the complex problems associated with environmental processing capacity and the relationship between pollution and effect.

PRESENT WATER ENVIRONMENTAL STANDARDS AND SELECTED PRESENT VALUES

CONDITION	WHO INTERNATIONAL (1958) PERMISSIBLE LIMIT	U.S.P.H.S. (1962) RECOMMENDED LIMIT	PROPOSED CALIF. WATER PLAN	SACRAMENTO RIVER MAX. PERMISSIBLE CONCENTRATION	METROPOLITAN WATER DISTRICT
DISSOLVED SOLIDS (ppm)	500	500	400	525	704
SULFATE (ppm)	200	250	100	130	297
CHLORIDE (ppm)	200	250	100	130	92
SODIUM (%)	—	—	50	60	63
HARDNESS AS $CaCO_3$ (ppm)	—	—	160	200	195 (SOFTENED FROM 330)

FIGURE 23

The North American Aviation statement was made by Jack Jones' assistant to the President, who commenced his discussion of the transportation study by pointing out that this study alone involved a Federal sponsor-monitor (the Department of Commerce). The study team had a threefold assignment: to survey the total field of transportation, obtaining a feel for the size of the problems; to see if systems analysis and management could be applied fruitfully and, if so, in what fashion. The study team looked at all transportation modes—tube trains, air-cushion machines, sky buses, electric automobiles—within a single frame of reference. In addition, the interrelated ramifications of population, land use, economy, and technology as they affect transportation demand and opportunities were studied.

In looking forward to the year 2015, land utilization planning must be considered a critical item, as population, personal income, and productivity combine to saturate the transportation capacity of the State. During the study, regional, interurban, and urban transportation needs were considered, and a series of models were conceived for moving people and produce, as well as viewing land use and various economic activities. The planning device proposed by the North American Aviation team would feature six submodels to be completed and exercised during a 52-month period, on this schedule:

Basic submodels

1. Land use: Developed and checked out at the end of the 10th month. Can be used independently to investigate land-use trends and their attendant effects upon industrial and population distributions.
2. Population: Developed and checked out at the end of the 11th month. Can be used independently to investigate population growth and demand trends.
3. Econometric: Developed and checked out at the end of the 15th month. Can be used independently to project interindustry flow, productivity and employment trends, and gross State product.

These basic submodels would be integrated with each other and with the data base at the end of the 21st month, and could be exercised to measure and project in a number of areas.

Transportation submodels

4. Transportation simulation: Developed and checked out at the end of the 25th month. Using independently generated inputs, can be used to assess adequacy of links by mode and to examine new or proposed transportation schemes.
5. Evaluation: Developed and checked out at the end of the 26th month. Can be used to evaluate the outputs of the independently operated transportation simulation submodel as well as the information generated by the simultaneous operation of the population, land use, and econometric submodels.
6. Transportation demand: Developed and checked out at the end of the 29th month. Serves as the bridge between the basic and transportation submodels.

Figure 24 portrays the relationships between the elements of the California transportation model.

THE CALIFORNIA TRANSPORTATION MODEL

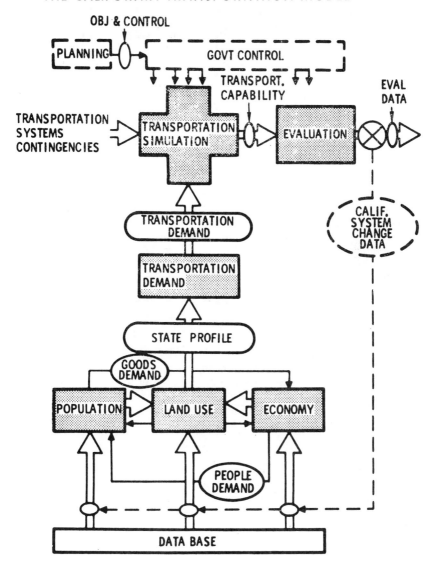

FIG. 24.—THE CALIFORNIA TRANSPORTATION MODEL

The fourth study commissioned by the State of California was that concentrating on crime and delinquency problems. Representing Space-General Corp. was Frank Lehan, its president, and John Kuhn, project manager. The 6-month effort was a systems analysis and cost-effectiveness study of the California system of criminal justice, including local law enforcement, the courts, probation, juvenile and adult institutions, and parole. Noting that California spends $600 million per year for the prevention and control of crime and delinquency, a program was outlined which would lead to an improved system of criminal justice. The major elements of the proposed program:

A continuing systems engineering analysis of the management and the effectiveness of the California system of criminal justice.

The development of an information system linking together various agencies of criminal justice and being capable of evaluating program and system effectiveness through collection, storage, and processing of appropriate data.

A systematic study of persons involved in criminal activity and identification of crime susceptible groups.

Carefully selected prevention programs directed toward the susceptible offender groups.

Technical assistance in the apprehension and processing of offenders.

Development of more effective methods in the management and treatment of offenders with attendant studies of subsequent behavior and costs.

The development and training of manpower to carry out the program.

The development of public support and understanding through information and community education programs.

A comprehensive master plan which projects over a 5-year period the scheduling and costs of the program.

The technical presentation by Mr. Kuhn emphasized that the problems had been approached analytically, with every effort made to demonstrate that the quantitative and numerical techniques used in engineering could be valuable in coping with crime and delinquency. Recent trends in felony crimes and arrests, juvenile delinquency, and age distributions for offenders were supplemented by appropriate population projections. Graphic representation of California's "crime susceptible group" (ages 14–29) is shown in Figure 25. The size of this age group is growing at twice the rate of the overall State population.

CALIFORNIA POPULATION HISTOGRAM

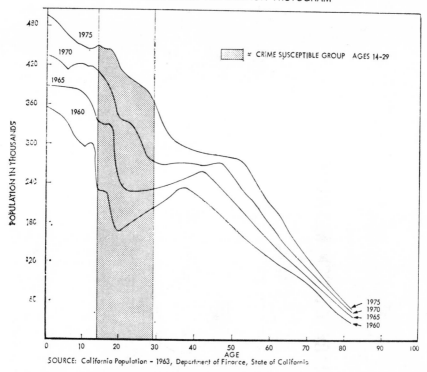

SOURCE: California Population – 1963, Department of Finance, State of California

FIG. 25.—CALIFORNIA POPULATION HISTOGRAM

In studying the system of criminal justice, a decision network was constructed so that key point for human action can be identified. Figure 26 shows such a "Juvenile Summary Decision Network:"

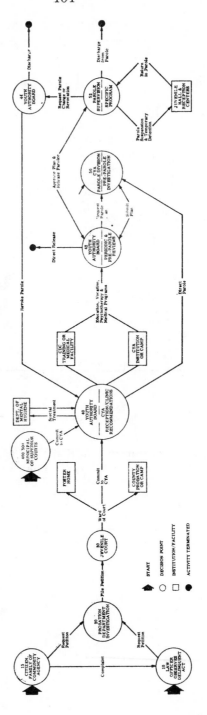

JUVENILE SUMMARY DECISION NETWORK

FIGURE 26

'The program presented by the study team, embodying 34 separate program proposals and research techniques, would cost $122 million over a 5-year period. Featured would be the development of prevention programs and the measurement of their effect; also apprehension, case management, police work, the problem of crime reporting, and the element of potential offender identification.

Although the special subcommittee was occupied during the first hearing largely with the reports by the four aerospace companies, other worthwhile commentary was forthcoming from other California groups. The Friends Committee on Legislation in California, represented by Robert A. Mang, evinced a concern about rewording the Nelson bill to reinforce the fact that social sciences resources also would be utilized in combating the social and community problems.

The testimony and informal commentary of Thomas C. Rowan, vice president and manager of Advanced Systems Division of the System Development Corp., served to give the special subcommittee background information on the underlying concepts of the systems approach and the gradual development of new devices and techniques during the 20th century. The importance of creating a truly interdisciplinary team "interacting in a common environment on a day-to-day basis with a common problem-oriented focus" was cited. Mr. Rowan pointed out that perhaps the most important development within the aerospace community in regard to systems development was the creation of a "dramatic capability for organizing and managing the manifold resources involved in large-scale, complex, development projects conducted under severe performance and time constraints."

The dialog then turned to the need for national long-term commitments, with Mr. Rowan stressing the fact that "political tenure seldom coincides with the duration of the proposed efforts." (Concern over this problem and a discussion of steps taken to lessen the impact of such uncoordinated circumstances are found in sec. IV.) Education of the Nation's leadership and citizenry also was seen as a matter for serious consideration, since many of the social and community problems are felt rather than understood. Another facet of the educational process drew this comment:

That is the problem involving the glamour that often surrounds the application of technology to new areas. I want to stress that the determination of system program objectives cannot be left to the technologists alone. This must be a joint effort between the elected officials who are responsible to the public and the specialists who are attempting to develop the program. The purposes, uses, and goals of any program must derive from the values and mores of the public. Thus, some of the trade-offs implicit in the judgment to follow or not to follow a particular course of action, are not necessarily subject to rational analysis. Therefore, it is imperative to remember that system analysis is a technology that applies the scientific method to the allocation of limited resources among a variety of competing demands, given a set of objectives. In and of itself, it cannot nor is it intended to determine those objectives.

The final portion of the Nelson-Rowan discussion featured opinions about the transferability and generalizability of the product resulting from the application of systems technology to a particular problem area. Mr. Rowan was reserved in his optimism about solution transfer from one geographic area to another, ending his contribution with a realistic statement about the potential of systems technology in the civil sector:

In summary, I want to emphasize that system analysis is not a panacea. It is not a black box into which one drops problems at one end and automatically receives solutions at the other. While I do not want to detract from the value of all the knowledge and experience that can be transferred from the defense community, I believe we should recognize that problems of the civil sector are no less difficult, and perhaps are more so, and require the utmost creativity. It is clear that one of the reasons that the four aerospace contracts look so promising is that these companies have applied some of their most creative people to the effort.

The first hearing of the special subcommittee resulted in the acquisition of firsthand information about actual studies into the application of systems technology to nondefense, nonspace public problems, and the reaction by public and private sector individuals and groups to the proposed Scientific Manpower Utilization Act.

C. SECOND HEARING, WASHINGTON, D.C., 1966

The second hearing of the special subcommittee took place on May 17 and 18 in Washington, D.C.[3] The first day featured testimony by Dr. Alain C. Enthoven, Assistant Secretary for Systems Analysis at the Department of Defense, and Henry S. Rowen, Assistant Director of the Bureau of the Budget. In addition to Senator Nelson who presided, Senators Jacob K. Javits, Claiborne Pell, and George Murphy were present.

Since both Dr. Enthoven and Mr. Rowen had been instrumental in developing the Federal PPBS concept and implementing it within the Defense establishment, much of the discussion between them and the Congressmen centered on the underlying principles of the systems approach and its applicability to the new spectrum of civil sector problems.

Dr. Enthoven set the stage for his later commentary on the DOD experience by selecting several key points which had appeared in an anthology called "A Modern Design for Defense Decision":

(1) Systems analysis is a reasoned approach to problems of decision, accurately described as "quantitative commonsense."

(2) Systems analysis is an application of scientific method, using that term in its broadest sense.

(3) There are limitations in the application of systems analysis, although these have often been overstated.

(4) In 1961, the Defense planning and budgeting system had to be changed to permit the application of systems analysis.

(5) Systems analysis is a regular working contributor to the annual Defense decisionmaking cycle.

(6) Two necessary conditions for the successful applications of systems analysis as a working part of an operating organization are that it be used by decisionmakers, and that it be fed with ideas by a broadly based interdisciplinary research program.

(7) Systems analysis can be applied to the problems of State and local government, including programs for social welfare.

After reminding his listeners that "systems analysis is *not* [italic added] synonymous with the application of computers," Dr. Enthoven carefully noted the support role of ADP (also see sec. II.D. of this report). Pointing out that systems analysis is "at once electric and unique," he stated that it is not "physics, engineering, mathematics, economics, political science or military operations," and yet it involves elements of all of these. Next, the method of science was defined and discussed:

[3] Ibid., pp. 145-206.

First, the method of science is an open, explicit, verifiable, self-correcting process. It combines logic and empirical evidence. The method and tradition of science require that scientific results be openly arrived at in such a way that any other scientist can retrace the same steps and get the same results. Applying this to weapon systems and to strategy would require that all calculations, assumptions, empirical data, and judgments be described in the analysis in such a way that they can be subjected to checking, testing, criticism, debate, discussion, and possible refutation. Of course, neither science nor systems analysis is infallible. But, infallibility is not being claimed; it would be worse than unscientific to do so. However, scientific method does have a self-correcting character that helps to guard science from persistence in error in the long run.

Second, scientific method is objective. Although personalities doubtless play an important part in the life of the physics profession, the science itself does not depend upon personalities of vested interest. The truth of a scientific proposition is established by logical and empirical methods common to the profession as a whole. The young and inexperienced scientist can challenge the results of an older and more experienced one, or an obscure scientist can challenge the findings of a Nobel Prize winner; and the profession will evaluate the results on the basis of methods quite independent of the authority of the contenders, and will establish what is the correct conclusion. In other words, the result is established on the objective quality of the physics and not on the reputations of the persons involved * * *

Third, in scientific method in the broadest sense, each hypothesis is tested and verified by methods appropriate to the hypothesis in question. Some are tested and verified logically, some experimentally, some historically, etc. Some sciences, of course, can reproduce experiments cheaply and they tend to emphasize experiment. This is notably the case with the physical sciences. In others, particularly some branches of medicine and the social sciences, one cannot experiment readily, if at all, and the detailed analysis of available historical data is most appropriate. In this respect, they resemble military science very closely. In choosing weapon systems some experimentation is possible, but a great deal of analysis is also required. In fact, in the development of weapon system analysis, one is more handicapped than in most of the sciences, for fully realistic tests come only at infrequent intervals in war, while the development of new weapon systems also takes place in peacetime. But this argues for better analysis and more heavy reliance on analysis where fully relevant experience is not generally available.

Fourth, quantitative aspects are treated quantitatively. This is not to say that all matters can be reduced to numbers, or even that most can be, or that the most important aspects can be. It is merely to say that the appropriate method for dealing with some aspects of problems of choice of weapon systems and strategies requires numbers. Non-quantitative judgment is simply not enough. What is at issue here really is not numbers or computers versus words or judgment. The real issue is one of clarity of understanding and expression * * *

Dr. Enthoven candidly informed the special subcommittee that:

It was a possible misconception when we brought systems analysis into the Pentagon that this somehow was going to be a substitute for and an end to debate.

He continued—

it has not turned out that way at all. Rather, I think it might be better described as a sophisticated, logical set of ground rules for conducting a debate, ground rules which help the debate to converge on some useful conclusions and to eliminate the unsupportable arguments as you go along.

The discussion then turned to the often-raised questions about the limitations of systems analysis, such as the emphasis on quantitative analysis or the tendency to "favor the old" when using this technique. And of course the complaint that "systems analyses oversimplify complex problems." Dr. Enthoven admitted that these are real considerations, but that none is a crippling indictment, and that balanced judgment on the part of the managers of any system is the key to an effective operation.

The discussion then turned to the "rolling 5-year plan" introduced in 1961 in the Department of Defense. Mr. Rowen interjected the comment that some nonmilitary Federal components long had used multiple year planning, but with this difference:

* * * these earlier plans have lacked something very important. They lacked tying the specific accomplishments which were spelled out in those plans to the finances required to accomplish them in a very specific, precise, and detailed way. They weren't really operational, they were sort of dreams for the most part, things that would be nice to have.

Senator Nelson made a telling point when he commented that:

* * * the system itself then has a built-in factor that compels you to utilize all the information you have gained this year in revising your predictions for the 5th year away or the 10th year away.

After reviewing the way in which the Systems Analysis Office in DOD functions, Dr. Enthoven then noted three conditions necessary to the successful development and functioning of a systems analysis group within a policymaking organization; that:

1. The responsible decisionmakers make use of systems analysis and take it seriously.
2. The systems analysis operation be fed with ideas growing out of a broadly based interdisciplinary research program.
3. The studies be conducted with continuity and depth.

In closing, Dr. Enthoven stressed his conviction that systems analysis can assist in the "design, development, and consideration of alternative approaches" to a broad range of public problems; he then drew this line of comparison for his listeners:

It is often suggested that these problem areas will be resistant to systematic analysis because they do not lend themselves to quantification. In commenting on that, I would like to point out that we, in the Defense Department, also have our own imponderables to deal with. We try to measure those things that are measurable, and, insofar as possible, to define those things which are not, leaving to the responsible decisionmakers the job of making the difficult judgments about the imponderables. It has been our experience that in those areas most difficult to quantify, years of research and the application of a good deal of ingenuity will often yield ways of measuring and making comparisons that were not available at the outset.

Ultimately, policy decisions will be based on judgments about relative values, the likelihood of uncertain future events, which risks we should and should not run, et cetera. But, in defense, and in these other areas as well, good analysis can do a good deal to sharpen the issues, clarify the alternatives available to the decisionmakers, and narrow substantially the range of uncertainty, thus freeing the responsible officials to concentrate their attention on the crucial judgments.

Senator Nelson discussed the value of collecting and using information in the social field—for example, the problem areas of coping with the mentally retarded and transportation planning—and Senator Murphy noted that planning figures often are outdated very quickly, as in the instance of developing the new Irvine campus where a 20-percent deviation occurred. Senator Javits then asked that a sample outline of how the systems approach would be applied to a specific area be prepared; the fields of medical services, education, and unemployment were identified. Such an outline never was provided to the special subcommittee.

The testimony of Mr. Rowen concentrated on three major topics:
1. The development of the Federal PPB System;
2. The kinds of analysis underlying PPBS and which are crucial to it; and
3. The essence of the problems and the opportunities for the future in broadening the applications of systematic approaches to Government operations.

During the discussion on the first topic, Senator Nelson raised the question of locating trained systems analysts, and Mr. Rowen agreed that good ones are in scarce supply. This description of an analyst was mentioned: "The sort of person who has a good academic foundation and who is interested in these broader social problems and who is an organizer." Senator Nelson recalled that during his tenure as Governor of the State of Wisconsin he had encountered only one "first-rate" analyst.

So I am wondering, where do you get this fellow and are there programs for training people in the concept of systems analysis or is it just something that people learn in industry or in Government just because at some stage they are compelled to start looking at the broader problem and happen to be broad-gage people who understand the broad concept and have a capacity for leadership and understanding what discipline is necessary to meet this total problem?

The reply from Mr. Rowen emphasized two kind of training. First, in certain academic disciplines and quantitative methods; second, on-the-job training with a group that is doing actual systems analysis. The establishement of special curricula, some with Department of Defense support, at universities was considered an important facet of building the Nation's analytical work force.

In developing his second topic, Mr. Rowen iterated that PPB is not a computer operation, though ADP may be useful upon occasion. He refuted the "black box syndrome," saying that the thinking that must underlie agency plans cannot be produced by machinelike analysts.

The real nature of the analysis which major social problems require is a good deal more commonsensical. First, it requires a precise statement of objectives; not, for example, "the development of a safer civil aviation system," but rather, for example, "the reduction of the civil aviation accident rate by 10 percent annually over the next 5 years." That has some bite to it. That can mean something. That leads one directly into programs to see if this can be done. This is just one example.

Other key factors to be remembered:

Systems analysis involves the better measurement of costs.

Good analysis requires the development and examination of alternative methods of reaching program objectives.

The final step is the review of the original objectives in the light of the costs and of the probable effectiveness of the programs which can be designed to attain them.

Thus, the utilization of systems analysis requires dedication to the spirit and the procedures of dissecting and assessing a problem situation. If this is done, this Rowen maxim holds:

This leaves the decisionmaker essentially where he has always been and where he always will be: Forced to use his own judgment, intuitions, and faith in determining what to do and what not to do. But that judgment and faith can be far better informed than they now typically are, better aware of costs, more sensitive to alternatives, equipped with estimates of likely benefits, and conscious of the relevant criteria.

The final topic discussed by Mr. Rowen involved the role of the systems approach in State and local programs. Management systems can do much to help the sub-Federal governmental elements meet the problems of the times. The usefulness of such a system at the State and local level should be reflected:

1. In a better, more effective utilization of State and local resources;

2. In an improved utilization of Federal resources;

 3. In an enhancement of our understanding of the problems to be solved;

4. In an increased ability to estimate the probable costs and effectiveness of the various corrective measures; and

5. In more realistically relating local efforts to Federal efforts.

Pilot efforts such as the 5–5–5 project (see sec. IV. D.) and those undertaken by individual States are a key to our future as the lessons learned are passed on to other troubled States and localities.

The necessity of using more effectively the analytical talent now in being was stressed by Mr. Rowen, who noted that while the aerospace firms have a sizable manpower pool, the nonprofit organizations and universities also possess competent analytical talent. The importance of having defense oriented centers (RAND, RAC, IDA, CNA, etc.) undertake to study social issues or as an alternative the creation of new study centers also became a consideration for the special subcommittee.

Senator Nelson and Mr. Rowen then discussed the possible need for establishing a congressional analytical support capability, with the problems inherent in doing this. Should an independent systems analysis agency for the Congress be formed? Should each committee have its own qualified analysts who are able to evaluate with some independence the work of the executive agency? Should the Appropriations Committee and the appropriate committees of the Congress receive all the benefit of *all* [italics added] the background and the reasons for the decisions made in their recommendations by the executive branch? The substance of Mr. Rowen's response: that the agencies will make all possible information available to the Congress, but that the shortage of good analysts might make it impossible for congressional committees to build adequate expertise in the key areas of concern.

The role of regional planning commissions and their importance in coping with areawide problems was the final point of discussion, with Senator Nelson citing the recent establishment of the Southeast Regional Planning Commission in Wisconsin, involving about a million and a half people. Mr. Rowen noted that the Bureau of the Budget was working with the Council of State and Local Governments, the League of Cities, and comparable county organizations. The importance of creating lines of communications with relevant government jurisdictions throughout the country has been recognized by the Department of Housing and Urban Development and the Economic Development Administration, thus implementing the Nelson emphasis on a regional approach to many of the social and community problems.

The second day of the second hearing also featured representatives of the Federal Government with responsibilities for carrying out the PPBS. Appearing before Senators Nelson and Murphy were William Gorham, Assistant Secretary for Program Coordination for the Department of Health, Education, and Welfare; Joseph Kershaw, Assistant Director of the Office of Economic Opportunity; and William Ross, Deputy Under Secretary for Policy Analysis and Program Evaluation of the Department of Housing and Urban Development.

The comments of Mr. Gorham about HEW's use of systems analysis were made within the context of a rapidly expanding department, with a significant amount of legislation passed by a concerned Congress. He noted that "the focus in social programs has been so much on getting attractive new programs through, that insufficient attention has been paid in the past to the balance of programs." The HEW staff, now in excess of 90,000 persons, is responsible for programs totaling almost $10 billion, or 20 percent of the U.S. nondefense budget. As the programs to combat poverty, disease, and related problems increase in number and scope, the need for a coordination mechanism within the Federal Government becomes mandatory. Several agencies may find their social programs interlocked or impacting one another. Lines of demarcation become blurred, and a struggle for precious resources inevitably ensues.

A new line of discussion was initiated by Senator Nelson, who differentiated between the two aspects of using systems analysis in a given departmental environment. First, he alluded to its use for the purpose of making a total evaluation of programs of a particular department, including the objectives and costs of the programs. But of equal importance is the question of the internal management of the programs themselves. In the latter area of endeavor, he explained that teams of experts were used to evaluate programs in the State of Wisconsin, with good results.

Mr. Gorham then proceeded to describe the six major program categories (e.g., human investment programs) of HEW, but cautioned that the sponsorship of various programs and projects by various Federal and private elements sometimes had caused an imbalance in the Department's schedule of corrective action. He then described the five program analysis groups which had been established to work in these areas: programs which increase future earning power, college student aid programs, income and benefit programs, selected disease control programs, and child health care programs. The program analysis groups will:

    1. Specify the objectives of each program and ways of measuring them; and

    2. Prepare a description of alternative programs which would lead toward the objectives, an analysis of the costs and benefits of each alternative, and a discussion of the basic assumptions upon which the conclusions depend and the uncertainties which affect the estimates.

In support of the HEW in-house analysts are university consultants and experts from OEO, Department of the Treasury, and the Council of Economic Advisers. Mr. Gorham stressed that the persons from other agencies acted in the capacity of "individuals," not as official representatives of said agencies.

An intensive look at the need for a clearly presented overview of programs involving various (or multiple) agencies was the focus of the next exchange between Senator Nelson and the witness. In reply to the Nelson query "If Congress wanted an analysis of all the manpower functions that are performed at OEC, Labor and elsewhere and what their objective is, how much they cost, who the people are, is that information available now?" Mr. Gorham expressed doubt that such information could be obtained. Senator Nelson:

You mean as to all the agencies, all the functions that do cut across agency lines, when your analysis is over with and the analysis of all the other agencies is over with, that there will have been some careful evaluation of what is being done in all aspects of any program that cuts across agency lines and what the objectives and what the cost is? HEW will be able to say we are performing such and such function affecting some class of people, here is the objective and in each agency there will be an overall analysis of these functions that do cross agency lines so that a total picture will be available?

Mr. Kershaw of OEO, also formerly at RAND, began his testimony by expressing his rationale for using a PPBS-type approach:

The value of these techniques is that they require us to look beyond some of our traditional conceptual limitations and in doing so to sharpen the judgement and intuition of decisionmakers at various levels of Government. PPBS principles and techniques require us first to look beyond the limits of traditional classifications of resources and effort. This is their structural contribution. Secondly, they require us to look beyond the limits of traditional forms of information needed for decisions between alternatives. Finally, they require us to look beyond the limits of our traditional systems for the analysis of alternatives. This is their analytical contribution.

He went on to say that a new framework for classifying and considering alternative proposals has been developed, with emphasis in the planning process on (a) a classification of effort or resources that looks to the end products; (b) on goal-oriented functional services which cut across organizational lines; and (c) on efforts which are directed toward eliminating the causes of the poverty problem rather than its symptoms. The close relationship between some of the HEW and OEO categories was noted, and Mr. Kershaw then reflected on the need for certain types of critical information. For example, he pointed out that the Council of Economic Advisers has estimated that 30 percent of the poor population moves in and out of poverty from year to year. His questions: Are the same people crossing the line in both directions? How frequently? Are different people moving in and out? Any analysis of this complex problem is contingent upon having appropriate data, but this does not exist for the most part.

Another instance when systems analysis may prove essential in determining the future of key social programs:

Both Job Corps and Neighborhood Youth Corps are designed to serve disadvantaged youth and to give them the necessary work experience, education, and vocational training to permit them to get and hold jobs in the private sector. We ought to compare these two systems by examining their respective costs and outputs.

We know that Job Corps costs about four times as much as Neighborhood Youth Corps; if it produces five times the results it is a better program; if it produces three times the results, it is the inferior program. The results in this case appear to be fairly easy to compare, although this ease of comparison disappears if, as is likely, we decide that Job Corps should serve more difficult youngsters than Neighborhood Youth Corps. It may be that for these difficulty youth, Job Corps is the only program which can reach them effectively, and thus its benefits are infinitely greater than any alternative. But in any case with some additional experience in the two programs and some knowledge about what happens to youngsters who go through each of the programs, which we do not yet have, we will be well on our way toward making effective comparisons.

Mr. Kershaw, in closing, indicated that one innovation that he considered extremely valuable, and which had been adapted from the DOD environment, was "the program change system":

This is a system by which changes from the officially adopted program are made. It requires the person proposing the change to submit costs of his proposal, and it requires a staff in the agency to examine alternative ways of financing the new proposal. When the package reaches the director for decision, the system requires him to recognize that a decision to spend more money on a good idea also requires at the same time a decision to spend less money on something else. With proper staff work, this system forces the decisionmaker, all during the year and not only at budget time, to make choices on a rational basis with as much information as possible at hand, and it forces the recognition of the fact that every decision is a choice among alternatives.

The final comment by the witness, in response to Senator Murphy's question about effective utilization of ADP equipment within the Federal Establishment, was that OEO has an information center employing 82 persons, and that its computers are used to capacity so that additional time is rented on external machines.

Reflecting the responsibility of HUD to cope with several major community problems, William Ross chose to look at the underlying relationships which can effect analysis, thus allowing the analyst to "spot a change for the benefit of the decisionmaker—not after he has gotten himself into trouble by applying an obsolete response to a new version of a problem." He selected trash collection as a point of reference for further comments, "because this is a very pervasive local government problem and also one to which analysis has relevance." The question may be whether the city manager should try to get an incinerator constructed or whether he should continue to use the system of hauling trash out to a sanitary landfill:

In looking through the kinds of system analysis which Mr. Gorham, Mr. Kershaw, Mr. Rowen, and Mr. Enthoven have mentioned, in looking at the total systems cost just for the definition of that problem, or that subsystem of the problem, we have a tradeoff between the operating cost of hauling, the extra time involved in the collection and hauling of trash to distant points—and as these points become more distant year after year, hauling costs become higher—or the question of developing and building an expensive incinerator which permits the city to then reduce the volume of trash substantially and make a much less frequent or much more efficient form of hauling away that amount of waste which cannot be handled by incineration. This is the kind of tradeoff between capital cost and operating cost for which our knowledge is completely adequate. It is the kind of problem on which the local city manager may tend to make continually the wrong decision, if he does not look at total systems cost because he does not have to go for a bond issue in order to continue to pay the extra cost for hauling trash back and forth unnecessarily. He does not even have to defend that decision. He is carrying on traditional practices, and no one questions it.

An engineering analysis will tell the city manager many things, according to Mr. Ross: the physical relationships, the analysis of capital costs, the differing efficiencies of incenerators by size, the relative wage costs of incinerator operation versus trash hauling and collection, and so forth. The exercise will give him the basis for a "totally new decision," one which may serve 50 States, 18,000 cities, and hundreds of counties, Mr. Ross informed the special subcommittee that HUD had received solicitations from the National Association of City Managers, the Council of State Governments, the Conference of Mayors, and specific cities for financing systems analyses that can be used in their cases.

The city of Detroit was identified as a future recipient of a research contract to help them work on trash collection and police manpower utilization problems. (For other developmental efforts in Detroit, see sec. VIII.D.) Also, the city of Oakland, Calif., has outlined studies which are needed in police and fire department use of manpower. Senator Nelson then asked whether problems of cities of comparable size could be approached by using the efforts in similar locales. Mr. Ross opined that the gathering of statistics and the development of alternative responses—i.e., the process of analysis—were transferable, but that the results probably were not. Senator Nelson: "Would you see any value in a proposal [in S. 2662, for example] which would make grants available to various class cities, classed by size?" The reply was that such a grant would be relevant, but perhaps additional legislation is not necessary. Mr. Ross pointed out that title VIII of the Housing Act of 1965 "gives us authority to make grants for training purposes to States and State instrumentalities in response to plans that the States submit to us." Both training and research are included and that while the authority was not presently funded, several States have submitted proposals in anticipation of such funding. Another authorization was cited: Title I of the Higher Education Act of 1965 provides for grants by the Office of Education to State centers for training and educational work. Also, the Secretary of Commerce under Public Law 89–182 can make technical service grants for the diffusion of the results of scientific knowledge to business and industry. Finally, NASA has been making grants to various universities which allow them to develop demonstrations of aerospace technology spinoffs. (Additional discussion of Federal programs and funding assistance may be found in sec. IV.)

Testimony received during the second hearing was especially valuable to the special subcommittee, because the senior Federal officials appearing as witnesses were responsible both for the design of the planning-programing-budgeting system and its implementation in individual agencies. A focus was established regarding some of the problems encountered during the transition from traditional budget and planning operations to a PPBS-type activity. A consensus of support for S. 2662 emerged during the formal presentations and question-and-answer periods. Concern was expressed by more than one witness, however, regarding the form which any Federal funding mechanism might take, and also whether the large cities should be the recipients of grants-in-aid on the basis of special wording in the proposed legislation. Finally, the commentary by the witnesses revealed the difficulty of getting responsible groups and individuals to graduate from a perception of any single problem or group of problems in terms of broad generalities to that of specificity allowing the initiation of tangible corrective action.

D. Third Hearing, Washington, D.C., January 1967

Four days, January 24–27, 1967, comprised the third hearing.[4] A preponderance of the witnesses was from the private sector, but there were executive and legislative branch representatives, as well as State government officers. Senator Nelson presided over the hearing, with Senator Jacob Javits and Senator Peter Dominick (who had co-sponsored S. 467 with Senator Hugh Scott) in attendance. This hearing marked the beginning of a scrutiny of S. 430, the 90th Congress version of S. 2662, and S. 467 which called for the creation of a National Commission on Public Management. Senator Nelson noted that a "bipartisan bill" was a possibility, and that the merging of the grants-in-aid approach of S. 430 with the Scott-Dominick-Morse Commission concept (as set forth in S. 467 and H.R. 20) merited serious consideration. The first day of the hearing featured the testimony of Representative F. Bradford Morse; Karl G. Harr, Jr.,

---

[4] Scientific manpower utilization, 1967. *Op. cit.*, p. 1–192.

president of the Aerospace Industries Association; and E. R. Roberts, vice president for development of the Aeroject-General Corp.

Representative Morse opened his remarks by endorsing the Nelson approach to the utilization of scientific manpower in coping with nondefense, nonspace public problems. He then commented upon the working nature of the National Commission on Public Management—which is treated at length in section V of this report—and stressed that one of the specific Commission tasks would be to assess the relative merits of various financing techniques. Greatest emphasis was placed on learning more about the problems and techniques under consideration:

Initially this Commission must define the problems involved. Then it must determine the applicability of the many systems analysis and management techniques as they relate to a myriad of problems ranging from those relatively simple and local in nature to extremely complex national issues. Finally, it must recommend optimum means for developing Government-private enterprise cooperation to encourage and support the techniques found to be applicable. Unless these crucial points are investigated thoroughly, we will not be able to pi ovide guidelines for the many future applications of the systems approach, and instead we will be limited to trial and error procedures. We require a "conceptual definition" phase—to borrow from Department of Defense terminology—before we can move full speed into the implementation phase. This, in fact, is merely using good systems analysis procedure.

Several on-going efforts then were mentioned, such as the New York State Identification and Intelligence System (NYSIIS), and follow-on studies to the four California aerospace contracts. One of these, the California regional land use information system performed by TRW Inc., is described in appendix G.

Congressman Morse addressed also the question of jurisdictional control or responsibility:

The geographic boundaries and historic charters that once created obvious administrative divisions are no longer of central importance. For example, the problem of pollution in the Merrimac River is one involving at least two States and several local communities. They must find new means for working together to solve this problem, irrespective of jurisdictional lines.

Senator Nelson and the witness proceeded to discuss the differences between S. 430 and S. 467; these are analyzed in section V. Congressman Morse underscored the importance of "taking a comprehensive look at the 'state of the art' " so that a coordinated approach can be pursued. The impact of the Nelson-Scott-Morse legislation was reflected in this comment:

We are on the threshold of an entirely new approach to the management of the public business. Last year an esteemed political scientist told me that the concepts which were included in the bill which I filed and the bill which the chairman of this distinguished subcommittee filed were as much of a dramatic departure in the field of political science as the discovery of the atom bomb was in the physical sciences.

Action within the Federal Government to deal with the growing national problems was alluded to in terms of the passage of the Demonstration Cities and Clean Rivers Restoration Acts, and in the steps taken by various agencies and departments (e.g., Department of State) to acquire systems personnel. He also noted that the Department of Commerce had let a systems analysis contract rather than a systems management contract in the northeast corridor high-speed rail transportation developmental area. Representative Morse concluded his comments by applauding the S. 430 strategy of grant awards to groups of States. Senator Dominick interjected with his ideas on the value of regional awards that would allow several States to confront jointly a major problem. He then restated the approach contained in the Scott-Dominick bill:

It does seem to me that the major difference at the moment between the approach that you suggest in your bill, S. 430, and the one that we suggest in S. 467 is that you have more or less solidified on the system of grants to the States. We are asking that a commission be appointed to determine whether this is the best way to approach the problem.

Senator Nelson, during the course of the Morse testimony, raised the point that each State should be encouraged to develop a "comprehensive" plan which would allow it to analyze its current situation, projecting resource allocation in terms of population growth and mobility and related identifiable problems. In this way, educational and transportation facilities can be planned, land use can be undertaken on a rational basis, and the interests of succeeding generations may be protected.

The second witness on the first day was Karl G. Harr, Jr., who presented a set of basic definitions for systems analysis, systems engineering, and systems management (see sec. II D), and then pointed out the importance of having an "institutional framework" able to exploit the advantages of the systems approach. During his discussion of governmental involvement in public problems, Senator Nelson took issue with the Harr statement that:

The only significant respect in which applying the modern systems approach to socioeconomic and human environmental problems will differ from its application to defense and space problems lies in the fact that, of necessity, the public authority will be primarily at a level other than the Federal Government.

The importance of obtaining Federal resources for, and in many cases management control of, problems affecting entire regions was stressed by Senator Nelson, who used the example of pollution in the Mississippi River. The problem would be more political than technical, and obviously would involve dozens of political units:

But to apply the systems concept to analyze the problem and come up with a proposed solution would be a relatively simple one. What we would have to do is to analyze or compute all the factors that are polluting the Mississippi River, and then make an analysis of what it would cost to require compliance with certain water quality standards.

After Mr. Harr agreed that experimentation would be a necessary facet of any approach, whether of the S. 430 or S. 467 variety, he was asked by Senator Javits to contribute a special statement which would be responsive to these questions:

You are urging your industry, the private sector, to dovetail closely with Government. Will you undertake for your association to study these two bills and give us in detail precisely what you think you can do in the framework with the merged plan which the chairman has outlined?

If you find there are any deficiencies or any areas which would disable you from cooperation or close coordination and contribution because of the terms of the law, will you tell us?

I want to know from you what you are going to do to help this work, and whether the legal framework which our chairman has laid out accommodates your concepts of how the aerospace industry can contribute to this effort.

I bar nothing, not even money, facilities, or management. You tell us what you want to do in this respect, and whether what you want to do is accommodated by the statutory scheme our chairman has laid out.

The material requested by Senator Javits was not delivered to the special subcommittee.

At this point in the hearing, Mr. Harr introduced Ward Dennis, of the Northrop Corp., and H. L. Wheeler, of North American Aviation, who conducted a dual presentation on the systems process. The sequential steps involved in using systems techniques as a basis for or executive decision are shown in figure 27.

Concept definition—systems analysis provides:
    Alternative approaches to satisfy requirements.
    Relative effectiveness of each.
    Probable extent of Government obligations.
Program definition—systems engineering provides:
    Preferred system designs with alternatives.
    Cost effectiveness of each.
    Firm plans and schedules or design and development.
Program implementation—systems management provides:
    Plans and costs for orderly system development, production and operation.
    Timing and nature of obligations.
    Assurance of cost and program control.

FIG. 27.—SUMMARY—SYSTEM TECHNIQUES AS ONE BASIS FOR EXECUTIVE DECISION

In order to orient the decisionmakers regarding the time phasing of any systems development effort, a chart showing the time increments which should be anticipated was presented (as fig. 28).

SYSTEM PROCESS PHASES

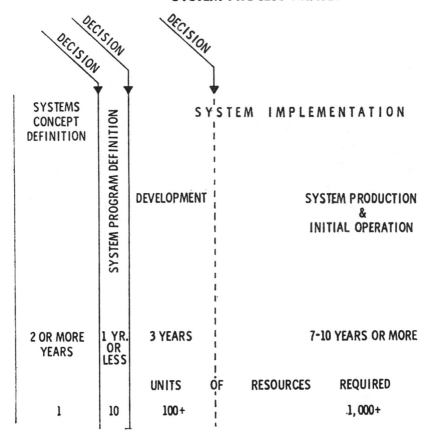

FIG. 28.—SYSTEM PROCESS PHASES

118

An example of a tactical logistics support system, which has many parallels in the civil sector, then was presented. This hypothetical example carefully depicted the areas for decision on the part of DOD management, DOD analysis and evaluation groups, Air Force headquarters and systems commands, and the contractor aerospace companies. The development of strategies, concepts, and data, and the subsequent modular systems activity are shown in figure 29.

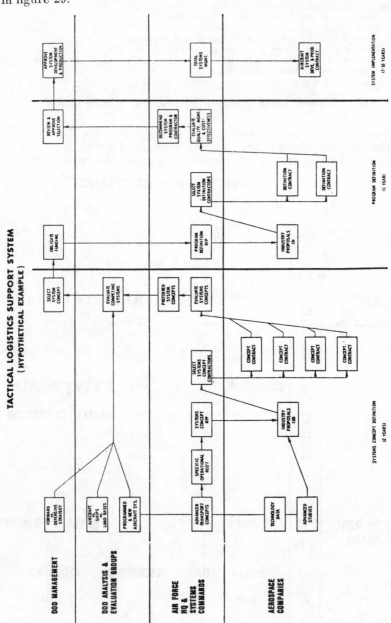

FIG. 29.—TACTICAL LOGISTICS SUPPORT SYSTEM (HYPOTHETICAL EXAMPLE)

Following this presentation, Senator Nelson requested of Mr. Harr that a narrative presentation such as the one just given the special subcommittee be developed for the area of pollution, so that it could be used to inform Members of the Congress in a direct, succinct fashion. This request was not fulfilled.

The final witness on the first day was E. R. Roberts of the Aerojet-General Corp. After showing how systems methodology was useful in providing a better insight into such problems as crime, social welfare, and waste administration, he described two major obstacles which were encountered, both of which "may stem from the * * * lack of truly planned socioeconomic systems:"

1. The uncertain nature of goals and objectives.
2. The lack of necessary data to find cause and effect relationships.

Mr. Roberts offered the opinion that a typical military weapon system can be assessed quantitatively in terms of affecting the total system performance. In contract, the socioeconomic systems have "evolved without the setting of firmly defined goals."

The problems with which they have to deal exist against a backdrop of social, political, legal, and economic influences.

Since the socioeconomic systems today have no quantitatively established objectives, the magnitude of the problems they face cannot be measured in terms of system performance. Thus, a logical solution to the problem is obstructed until we remove this obstacle.

Further, in order to come up with solutions, at least reasonable understanding of cause and effect relationship between the more significant contributing factors must be developed. This is particularly important since solutions to socioeconomic problems must evolve from the present environment. This environment, regardless of how bad it may be, cannot be swept away like an obsolete weapon system when it is replaced by a more modern one.

Two areas—crime and delinquency, and waste management—were chosen by the witness as activities which could be addressed in the short-range time frame. His delineation of the things which could be accomplished realistically and the time required to obtain both essential data and other specific benefits warrants iteration:

In the area of crime and delinquency, we believe there are activities which we can profitably undertake immediately. For example, substantial benefits could be derived by improving the information flow within the system and assure the availability of maximum meaningful data at the decision points in the system. Further, we could improve the system to effect a faster processing of the criminals. In undertaking this effort, which would take from 3 to 5 years, we are dealing primarily with methods of handling the problem, rather than with the problem itself. More germane would be our contribution to the solution of the problem, Mr. Chairman, by designing, with the help of social scientists, behavioral scientists, penalogists, psychologists, psychiatrists, and so forth, well designed and monitored statistically significant sociological experiments to yield data which would permit us to verify or reject hypotheses previously made. The very nature of this iterative process and the long leadtime required for feedback takes at least 5 to 15 years before meaningful data can be expected.

On the basis of the data resulting from such experiments, we might be enabled to intelligently use our resources to cause the basic problem to recede. Finally, as far as the criminal problem is concerned, we can contribute to pinpoint areas of basic research and help to evaluate data obtained in the course of the research. This activity would be directed to understand what makes a human become a criminal and would have to go on, like cancer research, until the problem is totally resolved.

In the waste management area, our immediate contribution should be directed to establish useful environmental quality standards. We believe that within 3 years, through a program of basic research, involving environmental sciences and system analysis techniques, we could fix useful quality standards for air, water, and land. Further, we believe that within 5 years we could design concepts capable of meeting these standards. Paralleling this effort, within 10 years, these concepts could be reduced to hardware, and in 15 years, we could start the implementation of the system and really solve the problem of pollution. We are fully aware of the fact that no socioeconomic system can be implemented in a vacuum, and we must take utmost care to avoid or at least minimize adverse effect on interfacing systems of our society.

The final point in the Roberts testimony was a statement of belief that the responsibilities of existing institutions should be redefined or realigned. He stressed that State and local governments should sponsor meaningful research and development pertaining to local problems, and should be responsible for implementing proposed solutions. Universities should continue to derive support for research in such a way that their traditional freedom to think and act would be unimpaired.

The second day's activity included testimony by Senator Hugh Scott, chief sponsor of S. 467; Dr. Simon Ramo, vice chairman of the board of TRW, Inc.; and Dr. Robert Lekachman, chairman, Department of Economics, State University of New York.

Senator Scott, in his statement, gave witness to the growing significance of systems technology in our society. He also reminded his listeners that the resources of the Nation must be expended with care. In particular, he pointed out that "No longer can we afford to expend 90 percent of all Federal research and development funds on defense, space, and nuclear energy developmental projects." Senator Nelson, in thanking Senator Scott for his participation in the hearing, iterated the nonpartisan nature of the legislation thus far introduced and said that finding a solution is "a matter of equal concern to all people, regardless of party." The special subcommittee chairman then stressed the quandary which often arises when a program must be implemented, and cited the difference between the DOD situation and the social problems identified in the hearings:

The difference in applying it to a social problem is when you get to the decisionmaking stage, it isn't just a question of DOD and the systems engineering group to say, "We will now take this money that has been appropriated and spend it." When you get to the social side, then it has to go to the city councils, the county boards, the regional planning commissions, and the Congress of the United States.

The political decision to implement a program which has been evaluated and alternative solutions proposed is a political decision. They, I think, found it difficult to visualize how that would work. I don't think it is quite so difficult as they do.

Senator Nelson noted that several regional planning commissions—such as the one in Dade County, Fla., the Wolf River Planning Commission in Wisconsin, and others—are employing successfully the innovative tools and techniques.

Dr. Simon Ramo, noted leader in creating and applying systems methodology and a variety of man-machine computing and communications devices, early expressed the belief that systems engineering and systems analysis are broadly applicable to social engineering problems. Within a frame of reference of his corporation's experience in several project areas including health, transportation, information systems, and land use (for the latter, see app. G), Dr. Ramo described briefly TRW's work with the Los Angeles Police Department and the study and design of a $100 million medical center complex in Alberta, Canada.

The System Development Corporation, it should be noted, also has been involved during the past 3 years with the Los Angeles Police Department. Figure 30 depicts the projected "LAPD Operational System." [5]

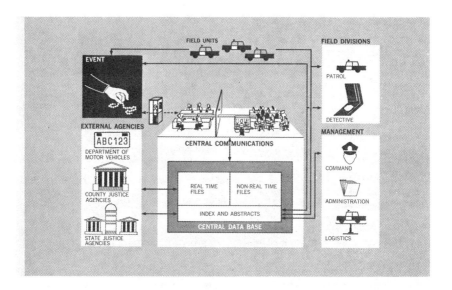

FIG. 30.—LAPD OPERATIONAL SYSTEM

In discussing the study of the medical center, Dr. Ramo emphasized the importance of spending "a great deal of time living and working with those who have been in hospital areas," and then stated that:

A good system job depends on understanding the nature of the problem well enough to identify the kinds of specialists as well as generalists one needs. A good systems engineering team for a problem is one that is properly interdisciplinary in nature, not just is interdisciplinary in sociology, but in the total socioeconomic sphere involved.

The witness and Senator Nelson then proceeded to discuss the problem of actually carrying out an innovative design effort, with the latter referring to the baffling of systems engineers where "political implementation" is involved. Pursuing the example of the purging of the Mississippi River of pollution raised earlier in the hearings, Senator Nelson noted that "somebody else politically is going to make the decision as to whether or not and how you would implement the work that has been done by the team of systems analysts."

Once again the question of recruiting or training systems analysts became a focus of conversation, with Senator Nelson posing these four questions about the elusive systems analyst: What is he? What qualifications does he need? Can he be trained in an educational institution? Can he only learn by experience? Dr. Ramo replied that there is "quite a lag between the recognition of the need for and potency of intellectual disciplines and the arranging of the teaching of them."

[5] Los Angeles Police Department. Administration of Justice. In SDC magazine, v. 8, n. 10, October 1965. p. 17.

The universities are beginning to teach certain tools (e.g., computer programing) and explore some of the problem areas. The difficulty is in finding the person interested both in the socioeconomic and the technical aspects of the field. In closing, Dr. Ramo presented brief descriptions of several projects which TRW, Inc. was engaged in where systems technology played a major role:

HEALTH

Edmonton health sciences centre

This $100 million medical center is being developed by the Province of Alberta, Canada. The 6-year project will involve the design and construction of ten buildings, in the development of a major medical complex that will provide patient care, education and research.

In a project of this magnitude and complexity, systems analysis is an essential tool to aid in planning, evaluating alternatives, and preventing schedule slippages. TRW is also designing a fully integrated communications-logistics-data handling system, so that ultimately anything that moves—information, supplies, etc.—in the complex will do so as part of this system.

Our systems analysis effort should reduce construction costs, provide continuing review and examination of alternatives, optimize such factors as site location, and identify—in advance—potential delays or other problems.

Our design and information systems effort is expected to lead to substantial reductions in operating costs over the life of the complex, and to enable operating efficiencies never before realized in a medical center of this type.

Regional medical program

This Federal program evolved from the deBakey Presidential Committee on Heart Disease, Cancer, and Stroke. The program contemplates methods for making available on a broad basis the medical excellence that now exists within the Nation's medical schools and related teaching hospitals.

The program is federally funded, but administered on a regional basis. We are under contract to the Northern New England Region, with the University of Vermont as its centroid. Our role is to provide analytical and technical support and to work with local and regional agencies as the "engineering member-of-the-team," to plan and implement the regional objectives.

This activity demonstrates one method of employing engineering innovation towards the solution of civil or social problems. Our engineers are working together with regional and local representatives of the medical profession, hospital administrators, public health authorities, and educators, to create a harmonious ensemble that can bring to play all the resources necessary to a frontal attack on the Nation's three killer diseases: heart, cancer, and stroke.

Department of Defense hospital feasibility study

Several months ago TRW completed a hospital systems study for the Department of Defense—in which we were teamed with Beckman Instruments and the architectural/engineering firm of Daniel, Mann, Johnson & Mendenhall. Purpose of the study was to examine the feasibility of applying advanced engineering techniques to the operation of existing military medical facilities, and to the design of future military hospitals. Feasibility was established; the studies clearly demonstrated that significant operating cost reductions could be realized through the employment of the systems approach in the initial design of a medical facility.

Automated medical examination systems

Another happy marriage of engineering and medical profession is in the making with the advent of the speedy, inexpensive—but comprehensive—medical examination. Here again, the application of advanced engineering technology can lead to a systems design and improved methods of integrating the various subsystems, so that the product will be a superior means of providing medical examinations and a more effective method of utilizing the latest advances in medical diagnosis.

TRANSPORTATION

High speed ground transportation (Northeast Corridor)

The problem of moving people within the so-called "Northeast Corridor" is one of the Nation's most critical transportation problems. The Office of High Speed Ground Transportation (HSGT), which will be an element of the new Department of Transportation, has contracted with TRW to perform analytical services and technical support, leading to the system definition of an optimum High Speed Ground Transportation System for use in the densely populated corridor between Boston and Washington, D.C.

Under this program, we will analyze the various subsystems and provide technical syntheses that will permit the HSGT office to select the system that will best meet the intercity transportation requirements. Our engineering analysis will be supported by specialized research and cost studies, as necessary, to assure the successful completion of the project.

Airline transportation problems

Although TRW's participation in the solution of the numerous airline problems is still relatively minor, this is an area in which advanced technology is essential to help overcome the congested airport situation, airline scheduling, flight separation, and terminal-to-city ground transportation.

Work is progressing on technical solutions to these technologically caused problems. These solutions will involve such advanced techniques as the employment of communications satellites for relaying voice communication between aircraft and ground, improved air traffic control systems, and the development of computer scheduling techniques.

URBAN SYSTEMS

California land use information system

TRW is currently under contract to the State of California to develop a regional improved land use information method. The project has the specific objective of providing the means for more effective sharing of information between departments within a government jurisdiction, and between jurisdictions.

The project requires the determination of needs for land use data, the development of criteria to satisfy system requirements, and the design of a land use information system within the framework of the prospective federated information concept in the State of California.

Then, using Santa Clara County (Calif.) as the base, TRW will demonstrate the operation of the proposed system.

Again, through the application of sophisticated engineering techniques, and with the aid of the big computer, we will be able to handle problems, and mountains of data in a manner that will vastly improve efficiency and reduce costs. Another example of the benefit of applying modern technology to government-oriented problems.

Law enforcement

Authorities in the law enforcement field say the only way they will be able to cope with the growing urban crime problem is through the employment of the latest technological advances, particularly in communications, electronics and information storage and retrieval.

Take, for example, the deployment of police patrol cars and other emergency vehicles. We now know that a mobile system's efficiency can be improved three-fold, through the use of modern communications, display devices and information systems. As I mentioned earlier, the Los Angeles Police Department hopes to solve some of its problems along these lines. Other police departments around the country are also exploring ways of obtaining more law enforcement at less cost, by utilizing advanced engineering techniques.

WATER

Ground water simulation

TRW is currently under contract to the San Bernardino Valley Municipal Water District, in California, to perform computer simulations and analyses of ground and surface water flows in the San Bernardino Valley. The simulations will portray water distribution and movement, to assist municipal, industrial and agricultural users in identifying underground watercourses and locating potential sites for wells, etc.

Another objective of this computer-based effort is to explore, analyze, evaluate, and present a range of possible plans for the coordinated use of ground and surface water resources to meet the increasing water needs of the area.

Desalination

For the Office of Saline Water (U.S. Department of the Interior), TRW is studying membrane processes such as electrodialysis and reverse osmosis, to desalinate water economically. Information derived from our studies of the biological mechanisms of ion transport is expected to lead to the development of more efficient membranes capable of operation for prolonged periods and low temperatures, low pressures, and with minimal energy requirements.

A systematic and comparative investigation of certain salt transport characteristics is also underway to determine the relationship between the divergent salt tolerance properties of these bacteria and membrane structure and function.

The testimony of Dr. Robert Lekachman of the State University of New York reinforced some of the points made in the Rowen and Enthoven statements:

Intelligent advocates of systems analysis are frank to concede limitations upon the capacity of their technique to assign numerical weights to some of the important variables in social or, for that matter, military problems.

What they very properly prescribe is the measurement of the measurable, the identification of the unmeasurable, and the evaluation of what has been measured and what has resisted measurement by the appropriate decisionmaker whether he be the Secretary of Health, Education, and Welfare, the mayor of New York, the chief of a local police force, or, possibly, a member of a legislative body.

Dr. Lekachman continued by saying:

Numbers imply precision and certainty, the resolution of doubt, and the end of indecision. Unless we take care we shall imperceptibly move toward a condition in which numbers and techniques determine the kind and the quality of the decisions which we are capable of making.

He then emphasized that in his opinion "systems analysis can unquestionably aid us in achieving ends upon which we are *reasonably united* as a society." [Italic added] His final caution was that while systems analysis often is useful, and that he supports the legislation being considered by the special subcommittee, that we must avoid complacency which may come from adopting some "technique of social choice." The "emphasis on tidiness" must be made carefully, particularly by the social planner.

Testimony on the third day of hearings was presented by Dr. Charles N. Kimball, president of the Midwest Research Institute; Dr. Donald A. Schon, president of the Organization for Social & Technological Innovation; and Paul Grogan, Director of the Office of State Technical Services, Department of Commerce.

Early in his commentary before the special subcommittee, Dr. Kimball sought to place the not-for-profit research institutes in a "time perspective," and to note some of the major ones:

There are 10 or 15 institutes that are usually named when one lists the independent not for profits. In the aggregate they employ about 7,000 people and their total research volume now reaches about $230 million annually. Thus, they account for roughly 1 percent of all the R. & D. work funded in the United States.

The largest in terms of staff and research volume are Battelle Memorial Institute at Columbus, Ohio, and Stanford Research Institute at Menlo Park, Calif. Then come a group of middle-sized institutes—our own at Kansas City; IIT Research Institute at Chicago; Cornell Aeronautical Laboratories; Southwest Research Institute at San Antonio; Franklin Institute Laboratories at Philadelphia; Mellon Institute at Pittsburgh; Southern Research Institute at Birmingham; and Research Triangle Institute at Durham, N.C. There are some recent additions to the list such as Gulf South Research Institute in Louisiana; Spindletop in Kentucky; North Star at Minneapolis; and a new institute which MRI helped to organize just last year in Rhode Island. You will note from this list that we have pretty well covered the U.S. map.

After citing the shortage of trained personnel in the systems analysis field, and the need for a "period of initiation and incubation" during which sub-Federal governmental groups can adapt to systems approaches, Dr. Kimball then turned to the problems of implementing the new ideas and plans in such complex areas of concern as environmental control and city management.

The witness then addressed a topic which had arisen several times during the hearings and to which Senator Nelson had paid especial attention: the need for a regional approach to many of the critical problems of our times. Dr. Kimball did so within the context of the regional value and impact of the research institutes. He said:

I believe these are some of the means to the ends we seek:

We can create regional data centers to collect, analyze, and make available the information needed for more effective government throughout our region.

We can do much more of the kinds of work we presently do for State and local government, either by scaling up the size of the problem area we work with, or by adding to depth and continuity of our research and support services.

We can develop continuing research and consulting arrangements with various governmental groups, on a drawing account basis.

We can provide the vitally needed link between private industry and the public sector in critical social problem areas like housing, city renewal, job training, or medical care. We find that the companies are willing if their risks can be minimized, and we know that State and local governments are ready to experiment. The missing link is the matchmaker who creates a workable marketplace in objective, impartial, and apolitical terms.

In collaboration with certain universities, we can create highly pragmatic training programs for State and local government people, to help speed the transfer of systems management know-how and other techniques. This is a logical extension of the technology transfer programs which we have successfully developed for NASA and the Office of State Technical Services.

Senator Nelson and Dr. Kimball discussed next the 701 grants by the Department of Housing and Urban Development, which have induced the creation of a number of regional development commissions. The witness went ahead to mention some of the projects being undertaken by the Midwest Research Institute, including the development of computer planning models and demand studies for State highway departments, and the building of a socioeconomic model to show the impact of certain railroad mergers. Other MRI efforts are included in the "Selected References" section of this report.

After dialog on pesticide residue observation and control and comment on a recent project to analyze all forms of Federal aid available to a large city, Dr. Kimball mentioned that MRI was the first regional dissemination center established by NASA (in 1961) to "interpret and disburse to the private sector advanced technology developed by NASA for another purpose." The point then was made that PPBS or critical path scheduling may be too costly for a State or city to establish "from scratch," but that implementation from an intermediate point can allow many localities to benefit. Transfer of technology is both possible and desirable; what is needed is "more coupling from the source of knowledge to the people who need to have it." One such mechanism is the recently formed Mississippi Research and Development Center which obtains sustaining funds from the State legislature but also draws upon Federal and other State funds for project support.

At the end of Dr. Kimball's testimony, Senator Nelson said that "We are probably going to merge the two concepts one way or another [referring to S. 430 and S. 467] * * * and we expect to offer a bipartisan bill." He then requested that Dr. Kimball's staff prepare a critique of the two bills, with any "suggestions upon how it ought to be done, who ought to allocate the money, what kind of advice should they have in allocating the money, what objectives should we seek to accomplish, and how should we accomplish these objectives." The request was not fulfilled.

The initial focus of Dr. Donald Schon's testimony was upon the schools construction system development (SCSD) project in California, under the direction of Ezra Ehrenkrantz. Dr. Schon chose to define the term "systems approach" in terms of this project, the rationale for which is described as follows:

The why of SCSD is apparent in the gap between the increasingly complex, constantly changing demands being made on our schools, and the ability of traditional building practices and products to meet them. New teaching methods and equipment call for new ways of arranging new types of instructional space.

Changes in curriculums, teaching techniques, organization, and grouping of students and staff, require corresponding changes in buildings. And change is beginning to be recognized by educators as a continuing part of the educational scene. Upgraded educational standards point to an upgraded environment—good lighting, effective sound control, air conditioning, even carpeting. At the same time, the student population grows and shifts; budgets remain tight. In short, we are asking for more variety, greater flexibility, higher quality, and lower costs—a combination the schoolhouse can seldom provide.

Collectively, schools form a building market second only to housing; but because they are built one at a time, schoolhouses do not offer the manufacturer enough volume to spur product development to meet new educational requirements. As a result, school architects must select from products which are developed independently, often for other building types, and therefore do not fit perfectly either the school's physical needs, its budget or one another. Too much of the architect's time is spent fitting together bits and pieces of material, instead of grappling with vital problems of program and design.

In 1961, when SCSD was established, it was abundantly clear that such procedures—inefficient educationally as well as economically—were not the best answer to the demands of a decade in which taxpayers would buy $27.3 billion worth of primary and secondary public schools, and in which change would be the only constant. And it was becoming evident that current attempts at reform—stock plans, prefabs, portables, and so on—offered only limited solutions and were winning only limited acceptance. Certainly, no latter-day Henry Ford was in the offing, ready to start rolling identical schoolhouses off an assembly line.

It was described how a market was identified, and then the user requirements for a school—who would use it, teach in it, interact in it—were established:

1. Freedom in overall planning, from the single, large loft building to the multi-unit, campus-style school.

2. The simple and economical arrangement of a variety of spaces in a variety of ways for a variety of purposes. For many of the districts the self-contained classroom for 30 students was no longer the basic teaching space.

3. Altering and rearranging these spaces as the need arises. In fact, one of the assumptions underlying the design criterial was that an average of 10 percent of the interior partitions would be changed yearly.

The next step was to divide the basic design of the school into a set of component systems: structural, ceiling lighting, air conditioning, and the moving and operable portions. Special union problems caused the exclusion of exterior brickwork and plumbing. Significant savings from using this "systems approach" soon could be ascertained: long span construction with interior flexibility now can be obtained for $1.81 per square foot (compared to a shorter span cost of $3.24 per square foot). Ceiling lighting under the SCSD plan will cost $1.31 per square foot in contrast to the traditional $1.67 per square foot. Dr. Schon said that five such schools were under construction, and that the importance of his example resided in the illustration of some of the key features of a systems approach:

1. The object of inquiry was the whole system—the school and all its functions—not just a part or component of that system.

2. In the course of inquiry, the system was divided into interconnected subsystems.

3. For these subsystems performance criteria were developed.

4. A process was set in motion which led to the making of a variety of alternative inventions meeting these performance criteria.

5. Through the consortium of school systems, markets were created which served to attract technological innovation.

6. The whole building process, including its social and political problems, was taken as a subject of study and an attempt was made to design that process.

Dr. Schon then discussed the impediments to improving certain public sector conditions, such as the long-standing networks of companies and other institutions which provide "sets of interlocking products and services." The production of low cost housing, attempted by several powerful firms (Monsanto, Johns Manville), has not succeeded because the total cost of overcoming existing regulations and practices has proven to be too high. Today, innovation in housing has been channeled into the production of prefabricated trailers, which now comprise about 20 percent of the market.

The discussion then turned to a consideration of several activity areas where the systems approach is playing a role:

1. "City business"—more than 70 cities are in various stages of development and construction as the result of efforts by companies like General Electric, Sunset Petroleum, and others.

2. Crime control and prevention—this includes a series of projects ranging from the Arthur D. Little study of the Philadelphia courts to the work of Dr. Schon's firm with the District of Columbia and the National Crime Commission.

3. Community improvement—support has been forthcoming from the Department of Housing and Urban Development in this area, as the physical rehabilitation of dwellings and new construction are emphasized. The Bedford-Stuyvesant project in New York City typifies the efforts now underway.

4. Man-and-job matching—occupational requirements and manpower resources need matching, and this area was worked on by IBM and the Commission on Automation, Technology, and Economic Progress.

5. Health services—as discussed in section VIII of this report, several metropolitan areas are undertaking systems studies in this problem area; also, the health problems on Indian reservations are being studied using systems techniques. Dr. Simon Ramo referred to work by his firm (TRW, Inc.) during his appearance before the specia subcommittee.

6. Educational systems—systems endeavors in this area have taken several forms: design, management, and implementation of a junior college system, or the total skill and training needs of a community.

7. Transportation—in this instance, Dr. Schon mentioned the Institute for Applied Technology (Department of Commerce) study, under his direction, of the Northeast Corridor. The major research activities connected with this project are shown in figure 31: [6]

[6] ABT Associates, Inc. Applications of systems analysis models: A survey. Washington, D.C., Office of Technology Utilization, National Aeronautics and Space Administration, 1968. p. 25, fig. 2.

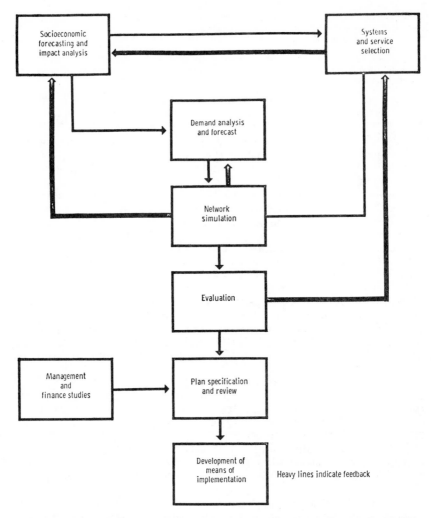

FIG. 31.—MAJOR RESEARCH ACTIVITIES OF THE NORTHEAST CORRIDOR TRANSPORTATION PROJECT

Following a discussion by the witness and Senator Nelson regarding the institutional, regulatory, and administrative barriers to improving societal conditions, the former gave the opinion that national legislation could be effective only in certain cases. He then addressed two other points. First, that a few universities (e.g., Nassau County in New York State) are beginning to provide curricula designed to develop systems skills. Second, that there must be "clients who are prepared to understand" the systems approach. In essence, he concluded, we are trying to "propagate a frame of mind throughout the country," and it would be useful to set up a small munber of demonstration projects across the spectrum of public systems needs. In this way, teaching and training would occur simultaneously.

The final witnesses on the third day were Paul Grogan, Director of the Office of State Technical Services, and Martin Robbins, Assistant Director for Special Programs. After some opening observations about the impact of technological change on the Nation, Mr. Grogan informed the special subcommittee that under the State Technical Services Act (STSA) each State must designate an agency to prepare a 5-year plan that outlines the technological and economic conditions of the State. The problem areas identified describe a wide range of conditions extant among many States; the conditions may be categorized thusly:

1. Existing industries fail to apply modern methods.
2. Shortages of experienced managers and qualified technical personnel.
3. Economy in transition is seen as unfavorably influencing economic growth and development.
4. Location, resource, and climate factors seen as limiting development.

The States of Nebraska and Massachusetts were given as examples of conscientiously developed 5-year plans; the plan for the former State is listed in the Selected References section. Mr. Grogan then said that under the "technical services" covered by the act are information centers, referral centers, field visits, demonstrations, and educational programs in a variety of formats. Nearly 600 such activities were supported in 24 State programs during fiscal year 1966. Several representative examples were noted:

1. Michigan: A program at Ann Arbor, May 24, 1966, on how to apply computer technology in business was attended by more than 200 businessmen. In another pair of programs on the numerical control of machine tools, 69 participated at Ann Arbor and 44 at Traverse City.
2. Wisconsin: The State technical services program in Wisconsin is sponsoring a series of conferences around the State in which the applications and the economic benefits to be derived from new technologies, such as fluid power, adhesives, welding, powder metallurgy, metal forming and cutting (removal), and computer applications in industry are presented in such a way that the top manager has a better basis for judging the relevance of the new technology to his company products or processes.
The most recent STS newsletter from Wisconsin lists more than 30 engineering short courses in Madison and Milwaukee the first 3 months of this year. Topics include: Computer-aided solid circuit design, manufacturing cost analysis, critical path method, adhesives for wood and paper, adhesives for metal and plastics, matrix methods of structural analysis, and nondestructive testing techniques.
3. In Iowa, where there is more of a plastics industry than people realize, they ran a course on plastics for 2 days, attended by nearly 100 persons. An industrial engineering workshop at Ames was attended by more than 80 persons.
4. Illinois: The Illinois Institute of Technology Research Institute has designed a series of 1-day seminars on numerical control techniques aimed at the needs of small- and medium-sized "tool and die shop" type operations. Firms wishing to go deeper into the subject may then engage in a practical working experience in the application of numerical control to various manufacturing processes.
5. New York: Syracuse University is operating a technical information service sensitive to the scientific and technical needs of central New York business, commerce, and industry. Interest profiles of local industry have been compiled, with inventories of technical information resources established and made available to industrial users.
6. Texas: Conference'and workshop activities served a total of 100 persons in the three specialized areas of professional engineering, production management, and nondestructive testing techniques using neutron activation analysis.

Among the many projects supported by the OSTS were those courses on the use of computers by industry and business, and a lesser number of seminars and symposia on operations research. The following listing reflects the range of orientation and educational activities made possible through the use of STSA funds:

COURSES IN USE OF COMPUTERS IN BUSINESS

Electronic computers in business.—A 1-day conference will feature topics such as electronic data processing (EDP), integrated systems technology, the processing cycle, computer units, programing, the feasibility study, payout applications, long-range contributions of EDP, and personnel aspects of conversion to EDP.—University of Arkansas, Little Rock, Ark.

Application of computer simulation to problems of physical distribution models.—A program of study on the use of TRANSIM in the solution of physical distribution problems will be offered to managers of transportation and distribution organizations. TRANSIM is a method of simulating transportation and/or distribution problems on a computer.—University of California at Los Angeles, Los Angeles, Calif.

Computer sciences for small business.—The rational, methodical, and effective application of computers in small businesses will be developed at a 1-day seminar.—Brevard Engineering College, Melbourne, Fla.

Management science.—Seminars will be conducted on the application of inventory management, computer simulation, and linear programing to management decisionmaking.—University of Georgia, Athens, Ga.

Computers, computations, and computional mathematics.—A course for research scientists and engineers will be offered on the organization of modern computer systems, including the structure of current and future programing systems. The meaning and results of an algorithmic procedure, as well as the basic underlying mathematics of numerical algorithms, will be discussed.—University of Illinois, Urbana, Ill.

Industrial applications of the computer.—A seminar on computer applications to the manufacturing and packing industries will be held for middle and upper management personnel.—Clark College, Dubuque, Iowa.

Digital computers in industry.—This course will provide industrial management and engineering personnel with an understanding of computer technology and will include the language limitations and data requirements. The computer's role in management will be especially stressed.—Iowa State University, Ames, Iowa.

Electronics and manpower in banking.—This seminar will be designed to explain to banking management the benefits obtained through the use of computers and electronic equipment in the banking profession.—Morningside College, Sioux City, Iowa.

Computer technology for business usage.—A workshop for the industrial community of northern Louisiana will acquaint management with the values of analog and digital computers and with the impact of computers on industry. An opportunity to practice computer techniques will be provided to participants.—Louisiana Polytechnic Institute, Ruston, La.

Data processing for small business.—A 1-day conference will be directed at motivating management of small- and medium-sized companies to examine the potentials of electronic data processing. The conference will be followed by a series of closed-circuit television presentations, training films, and taped portions of the conference. It is estimated that approximately 20 half-hour programs will be developed.—Northern Michigan University, Marquette, Mich.

Computer technology.—The topics in this 1-day conference of interest to small- and medium-sized businesses will include potential profitability of computers, evaluation of the benefits of electronic data processing (EDP), acquisition of an EDP capacity, computer operation and organization, communication with a computer and computer time-sharing operations.—University of Michigan, Ann Arbor, Mich.

Computer graphics and time-shared computers.—This 2-day workshop will be a followup program of the preceding computer technology conference. Its contents will be determined largely by the specific interests and qualifications of the conference attendees.—University of Michigan. Ann Arbor, Mich.

Information systems and computers.—An 80-hour course will be devoted to managerial considerations in systems design, the role of operations research and econometrics within an information system, and recent developments in computer hardware.—University of Minnesota, Minneapolis, Minn.

Use, value, and dangers of computers in small business.—This short course will cover the applications and values of the computer for sales analysis, accounting, production, inventory, and payroll. Discussions will evaluate the strengths, weaknesses, and dangers of computer systems. No previous knowledge of automated data systems will be necessary.—Washington University, St. Louis, Mo.

Remote terminal computer systems.—This institute for engineers and applied scientists will be concerned with the acquisition, financing, operation, programing, and other aspects of remote terminals coupled to large computer systems.—Washington University, St. Louis, Mo.

Introduction to digital computing.—Businesses in the southern tier area of New York will be offered a short course on digital computing that will cover basic instruction in programing techniques; and include a discussion of current and potentially available computer equipment and computer applications. A summary of the capabilities and economics of digital computers in support of research, engineering, and industrial administration will also be presented.—Cornell University, Ithaca, N.Y.

Business and industrial computer use.—A conference for small business management will survey existing computer programs to demonstrate the feasibility of using computers by firms presently unable to afford the ownership of computer equipment.—Tennessee Technological University, Cookeville, Tenn.

Computers and industry.—The content of this 3-day course will include computer technology and its limitations.—College of the Virgin Islands, St. Thomas, V.I.

Applications of data processing.—A 10-week institute for small businessmen in the Spokane area will be conducted to provide training and skill in management decisionmaking. A computer simulation game will be run, along with workshops on the following subjects: (1) functions of management and business leadership; (2) electronic data processing as it affects modern business operations; (3) scientific decisionmaking techniques; (4) cash flow analysis and use of funds statements; (5) sources of capital; (6) financial planning and control for small business; (7) problems in buying; (8) inventory modeling; and (9) forecasting and planning for profits.—Eastern Washington State College, Cheney, Wash.

Operations research

Operations research.—A 1-day seminar will be offered which is directed primarily at utilizing operations research techniques in establishing parameters for the solution of management and technical problems in the electronics and aerospace industries.—Brevard Engineering College, Melbourne, Fla.

Operations research.—A series of 8-day symposia for management in large firms will be held at monthly intervals involving fields such as quantitative models, linear programing, transportation models, inventory models, production scheduling, program evaluation review technique (PERT), and critical path method (CPM) techniques, organization theory, communications in the firm, dynamic programing, and sequential decision analysis.—University of Missouri, Columbia, Mo.

Operations research.—A 3-day conference will be presented to disseminate basic knowledge of operations research techniques as applied to business and industry.—Oklahoma State University, Stillwater, Okla.

Next, Mr. Grogan discussed some of the limitations of the modern university in serving society. For example, he pointed out that the "conventional system does not offer commensurate incentives for activities that encourage the application of new knowledge." Also, a barrier to full university use of its manpower is the "traditional discipline orientation and departmental structure of the university." His final comments were concerned with the lessons learned in critiquing the DOD study of major defense systems, called Project Hindsight.

The Hindsight study shows conclusively that the application of new technical ideas requires that three major criteria be satisfied:

(1) There must be a well-defined problem;
(2) There must be an organized group of scientists and engineers capable of working on the problem; and
(3) Sufficient resources must be committed.

Other significant findings of Project Hindsight that seem pertinent to these hearings include:

(1) Defense systems are built upon 20- to 30-year-old science, not new science. This would relate to the second point;
(2) That the continuing professional education seems critical and necessary if practicing engineers and scientists are to become familiar with and therefore apply the new science;

(3) Defense systems involve the combination of large numbers of innovations, the aggregate effect of which often is greater than expected from the total of the individual contributions. They call this the synergistic effect;

(4) The systems approach to the design of defense materiel results in concepts and breakthroughs that would not have been possible if attention merely had been focused narrowly upon continued product development;

(5) The barrier to the transfer of technology in either direction between the industrial and military sectors is very great, although the reason behind this is not clearly understood.

Mr. Grogan explained that OSTS is taking the lessons of Project Hindsight and the experiences of foreign and domestic organizations with similar missions, as well as "the current literature on the subject of our responsibility very much to heart." Senator Nelson, in asking for comments from Mr. Robbins, was told that in certain States, all possible resources are not brought to bear on a problem because of political decisions. In many instances, outside organizations are brought in to carry out the systems analysis functions.

Witnesses appearing before the Nelson subcommittee on the fourth day were Henry Rowen, president of the Rand Corp. (making his second appearance, but in a new capacity); Prof. John C. Geyer, Professor of environment engineering, at the Johns Hopkins University; Prof. Kathleen Archibald, assistant director of the Public Policy Research Organization, the Univeristy of California at Irvine; and Frank Wallick, Washington legislative representative for the United Automobile, Aerospace & Agricultural Implement Workers of America, AFL–CIO.

In opening his testimony, Mr. Rowen sought to establish certain boundaries for the systems approach; for example, he pointed out that it is not "a method of understanding 'everything' about, or even 'everything important' about a given subject." He then suggested that "perhaps the most useful, brief way to think about or to describe systems analysis is to regard it as a way of making discoveries * * * not of things so much, but about objectives or values, or relationships, or facts." The procedure for doing this kind of thinking, he went on, is to—

1. Be clear about objectives and criteria;
2. Clarify what the issues and purposes are, or might be;
3. Think about objectives, which may provide useful insights;
4. Identify the key factors and describe the relationships among these factors;
5. Look at the full range of factors as an "entire system";
6. Develop alternatives and assess them; and
7. Take the uncertainties into account in any assessment.

Mr. Rowen proceeded to explore briefly the role of the private sector elements in solving our domestic problems. He pointed out that "the better managed firms" can make "much more significant contributions *than has been realized so far*" [emphasis added]. Government must figure out how to use the talent and energy of the private sector groups. Universities, possessing a wide range of skills, must posture themselves to undertake interdisciplinary research.

Local governments must be modernized, and while many cannot afford the more sophisticated tools and techniques, others are able to do so. The need for new institutions was signified, and they should possess the features of diversity, continuity of effort, and the creation of a "well-designed and imaginative program for continuing study both at the 'fundamental' and 'applied' levels." A sampling of some of the programs and projects being undertaken by States and cities is found in section VIII B. and D. of this report.

A discussion of the steps being taken by certain Federal agencies— HUD, HEW, OEO—to encourage various types of research followed:

For example, in the Department of Housing and Urban Development, $20 million has been allocated for the purposes of urban research. This is a real departure for that Department. It has never had a serious research program.

In the Department of Health, Education, and Welfare, a substantial amount, I cannot recall the total, has been put into the budget for the purpose of evaluation, not just research in education or health in general, certainly not biomedical research in this instance, but research in the programs, and the same is true in a number of other areas.

Now, I think this is going to be a big step forward. I believe it has left open the extent to which this work would be done in-house, to what extent it would be done outside.

I believe, for example, in the case of the Office of Economic Opportunity that substantial sums have been provided for analysis, evaluation of community action programs in communities, at least the principal cities, so this clearly would be done at the local level.

The possibility of assigning "block grants" to the States, as provided for in S. 430, was supported by Mr. Rowen, who recalled that many Federal grants are channeled into State departments with no appreciable research tradition. "Pushing money on people * * * does not solve problems," he remarked.

In closing, the witness stressed the problem of so many localities in coping with such basic predicaments as snow removal, garbage collection, and street cleaning. He opined that there are no institutions to "design work on problems, to work on solutions that might have wide applicability." At this juncture Senator Nelson asked if a national commission, consulting with private systems groups, could not address effectively the problems in focus, identifying those problems that are "of significance and common to a number of jurisdictions." The chairman then commented, as he had so often in other hearings, on the need for more skilled systems personnel.

The initial emphasis in Professor Geyer's remarks was on the fact that while the application of operations research techniques to civil engineering problems has grown rapidly, similar activity in the social problems area had been much slower. Deterrents to cities' use of the systems approach included lack of money and trained personnel, and the "measurement, quantification, and understanding of the forces at work where human behavior is involved" are much more difficult than many of those in the "world of physical things."

Endorsing the approach contained in the Scientific Manpower Utilization Act of Senator Nelson, the witness said that providing money to the States for the kinds of studies being discussed was good for three reasons:

1. First, it provides decentralization which always seems desirable;

2. Second, it brings the State into areas of research and study which have been neglected at the State level; and

3. Third, it assures that the Governor and his agencies know what is going on and can participate in and take full advantage of the work.

Professor Geyer then presented his views regarding the States' use of the money available through the grants-in-aid mechanism contained in S. 430:

The States should be encouraged to use the support in any of a wide variety of ways designed best to promote the use of systems analyses in attacks on any important public problem. Education and training might be promoted by establishment of fellowships, by development of inservice training programs or by sending personnel to school.

The research needed to supply facts for the analyses could be carried out by an appropriate government agency or could be done through grant to or contract with public or private educational or research institution. The actual systems analyses might be by a central group or they could be farmed out as best suited any situation.

The objective of all these activities would be to take maximum advantage of all available talents and institutions in a concerted effort to use systems technology to understand and come to grips with problems of modern society.

I would put no strings on what a State might do with the money except that (1) the activity supported contribute to the application of systems approaches, and (2) the results be made available for the use of others. I think this would come closer to optimum utilization of manpower than any other way of going about the job.

Next, he commented upon a key point which had not been remarked upon by other witnesses, State versus local management responsibilities:

Most States do not have any management obligations relative to municipal services such as water supply, waste disposal, air pollution control, drainage, metropolitan transit, and the like. Improvement of the management of these public services is a prime area for application of systems analyses. It is to be hoped that money going to the States could be passed on to cities or to any other appropriately constituted group for making systems studies of their problems.

Professor Geyer then recalled that prior testimony had underscored the need to recognize the boundaries, the interfaces, the inputs, and the outputs of a system. Furthermore, it is imperative that we also understand "(1) the system from which the inputs arrive; (2) the interrelationships within the system itself; and (3) the effects on the system which receives the output."

In discussing new institutions and mechanisms, the witness suggested that a State Governor might have an office of science and technology supported by a science advisory group as is done at the Federal level. A discussion between Professor Geyer and Senator Nelson on the lack of information about university training and research in applying systems technology to public problems followed. The Johns Hopkins University, the witness explained, has a department of operations research, and is being funded by the Public Health Service to conduct systems analyses studies of solid waste disposal systems. Senator Nelson asked whether there was a "national register" of the schools engaged in training systems—none was signified—and went on to say that "we need an analysis to find out what is going on in institutions." [This course of action is an essential component of S. 467.] The assignments won by Johns Hopkins University graduates, including the management of the Water Pollution Control Administration laboratory activities, were noted. Senator Nelson's final point was to inquire whether universities could establish short courses (e.g., 3 months) which would furnish an initial orientation and training of professional personnel, such as budget analysts, in systems analysis.

Testimony next was offered by Prof. Kathleen Archibald, assistant director of the public policy research organization on the Irvine campus of the University of California. Miss Archibald explained the role of the public policy research organization as "falling somewhere between the traditional university research institute and the nonprofit advisory corporation:"

It has two primary objectives: (1) to undertake long-term programs of policy research and analysis in several substantive fields, such as education, international development, and social welfare; and (2) to establish educational and training programs in the field of policy research and analysis. The training program will be connected both to the academic curriculum within the university and to ongoing policy research programs within the public policy research organization.

The hope is to develop a "better articulation" between systems analytic approaches and the social sciences, and with this in mind, Ex-President Clark Kerr had established an ad hoc committee under the chairmanship of Roger Ravelle. This group was charged with exploring the desirability, feasibility, and possible design of a policy research organization. The letter of appointment to committee members stated:

A structure for policy research should be complementary to, rather than competitive with, the teaching and basic research functions of the university. It may need to be quite different from our usual research institute in its relations, both to the world outside the university and to the people within the university. At the same time, it should be designed in such a way that it can help and not hinder the research of individual faculty members and the education of graduate students.

The witness then proceeded to outline the basic requirements important to the conduct of good policy research:

1. Independence: This was defined not simply as "the freedom to tell a client something different than he wants to hear, nor is it necessarily the freedom from having any client," but it is the freedom to select jobs, or even to initiate research when no client has appeared.

2. Stability: The policy research organization should be in a position to think in terms of 5- or 10-year programs in certain areas.

3. Flexibility: Being able to modify the size and composition of the professional staff as the situation dictates is very important. The University of California public policy research organization hopes to maintain a small, versatile staff skilled in the methodology of policy research, while drawing upon the talents of outside expertise (including the faculty from all campuses of the University of California).

4. Significance: This should mean being able to do work that contributes "something of importance to the policy issues of our day, or of tomorrow":

There is no way of guaranteeing significance but there are ways of encouraging it: by having first-rate research people; by maintaining good external relations, particularly with responsible decisionmakers; and by providing a setting that facilitates significant work. Two facets of such a setting are (a) to have the flexibility to move into new lines of inquiry when problems are anticipated but before they are salient as policy issues, and, (b) to maintain several major interdisciplinary programs concerned with long-term policy issues or problem areas.

5. Interdisciplinary work: Translating what is desirable—and agreed to—in principle to practice is recognized as being difficult, but an objective to be met.

6. Access: A policy research program must have both formal and informal access to policymakers. In the first instance, cooperative research can involve the policy officials, and in the latter case, rapport of a personal nature is valuable.

Miss Archibald then asked two questions reflecting the difficulty of pursuing policy research in a university setting: first, Why should policy research and analysis be connected with a university? Second, what are the benefits to policy research and to the university from such a connection? She identified four things which could comprise answers to these queries:

I think probably the most important thing is, as seems to be very clearly the case, there is this need for trained personnel and the logical place to train people is within the university. And it is difficult to train people unless you are actually doing this kind of work within the university.

The second thing is that major academic institutions—and particularly State universities supported by public funds—have a responsibility to contribute to more effective public policy.

A third thing that at least entered into our decisions in thinking about the policy research organization was that it seemed that a university setting might be able to minimize some of the problems that the nonprofit corporations experienced. If, from the point of view of policy research, the traditional university institute is tied too closely to the academic structure, then it could be said that, on the other hand, the nonprofit corporations have sometimes suffered from being tied too closely to governmental agencies. The nonprofits have had some problems in attaining and maintaining both independence and significance for this reason.

A fourth factor suggesting a university connection is that perhaps the best way of assuring the increased use and further development of systems analytic techniques and related approaches is to securely link this field to the institutions of higher education. Here, I think, it is not just a matter of training people but of increasing the acceptance and legitimacy of the field.

The policy research organization connected to a university is able to resist certain pressures brought to bear by its clients, and also can give its attention to long-term and fundamental problems.

The physical presence of government personnel, on an exchange basis, in the research organization was offered as a possible way of developing a governmental in-house capability and appreciation of how the researchers work. A concomitant consideration should be the placement of certain researchers in the Government agency. Training persons in systems analysis and in appreciating the role— limitations as well as benefits—of systems analysis led to this comment by the witness:

There seems to be, as of now, general agreement on two of the ingredients needed for training: (1) a strong and quantitatively oriented training, perhaps not solely in one discipline but in relevant aspects of several disciplines; (2) involvement in the actual conduct of systems analysis and related research.

A third possible ingredient would be some experience in working in a governmental agency. Even if these are the right ingredients, we still do not have a recipe that tells us now to mix them or in what amounts. When should a student start on such a program? At the beginning of his graduate career, in the midst of it, or perhaps at the postdoctoral level? How much time would be required? Can the time be shortened by use of such devices as case studies, possibly simulation, and other such techniques?

Miss Archibald's final comments concerned the approaches contained in S. 430 and S. 467; in regard to the first bill she urged using "the first grants to prime the pump in order to get continuing programs and continuing interest at the State level." She echoed Professor Geyer's admonition to create an intragovernmental systems design and monitoring capability, not worry too much about institutional arrangements, and initiate short-term training programs for public agency personnel.

The final witness for the day was Frank Wallick, Washington legislative representative for the United Automobile, Aerospace & Agricultural Implement Workers of America (AFL–CIO). After stating his group's endorsement of the S. 430 approach, Mr. Wallick indicated the interest of labor in maintaining a high level of employment, and that any reduction in the Nation's space program should be offset by new programs in social problem areas. He then urged the inclusion in any systems-oriented efforts of representatives from the ranks of organized labor, saying that:

Since S. 430 involves the systems approach in dealing with social problems, it is particularly important to get people with direct and immediate insight into the various problems.

Obviously, trained scientists must be relied on for the major technical work. But there may be many instances where poor people, working people, and people with ordinary, even humdrum experience can make a significant contribution to a systems team.

Mr. Wallick then described briefly an OEO-financed project which featured the use of electronic planning and budget control techniques in connection with Detroit's poverty program. Most of the work was performed by Touche, Ross, Bailey & Smart, a consulting firm. (See sec. VIII. B. for an account of this company's work for the State of Washington.) A critical element of this project was the development of "a reliable method for quantifying client characteristics in a way that permits measurement of their progress from heavy reliance on public services to relative independence." The statement prepared for the special subcommittee entitled "Managing the poverty program in Detroit" described the efforts made by the professional staff to involve and educate the representatives of the poor in the operation of the system. Gradually, the quantification scale was developed with the aid of the "people off the street," social workers, and academicians. Figure 32 shows a worksheet such as that used in the creation of a quantified "dependency" scale:

Types	Characteristics										
	Health	Education	Work experience and skills	Family stability	Urban adjustment skills	Total income	Income from public assistance	Alienation	Personal appearance, speech, behavior	Housing	Employment status: Fully employed, underemployed, etc.
Negro, male, 1 to 4											
Negro, female, 1 to 4											
White, male, 1 to 4											
White, female, 1 to 4											
Negro, male, 4 to 9											
Negro, female, 4 to 9											
White, male, 4 to 9											
White, female: 4 to 9											
9 to 13											
13 to 18											
18 to 25											
25 to 40											
40 to 65											
Over 65											

FIG. 32.—WORKSHEET FOR PREPARATION OF QUANTIFIED "DEPENDENCY" SCALE

The care taken to establish meaningful quantification and guidelines for use of the innovative models and programs is reflected in this commentary:

Each "consultant" was given a chart similar to the one attached and asked to work across, line by line, rating the comparative importance of each characteristic for each type. When all results were in and synthesized, T.R.B. & S. worked out a numerical scale which theoretically measured each client's dependency. The scale ran from 1 to 10, with 10 representing self-sufficiency.

Information about each client is collected at various intake points by regular interviewers. It is analyzed and coded by specially trained staff to insure consistency. Periodic remeasurements, and comparison with earlier ratings, should enable evaluators to determine how quickly each client is progressing, what his current needs and past successes are, what overall impact the poverty program is having, and how much it costs to dispense each unit of service. By shuffling data around, they can also determine which component programs are contributing most, and which least, to the city's fight against poverty.

Using a computer, the system can recommend various strategies for dealing with the problems of poverty. It sets up mathematical models and, on the basis of present program performance, suggests the best mix for spending whatever money is available. It can also accept preconditions, such as the political need to invest heavily in highly visible services, and suggest how remaining funds should be handled. It can also predict how much time and money will be needed to finish the job.

Within the testimony of the witnesses appearing in the third hearing several major points began to emerge:

1. The grants-in-aid approach embodied in S. 430 should be scrutinized in the light of problems which required a multijurisdictional approach, and in terms of the needs of certain metropolitan entities.

2. A commission-type structure could perform a useful service in the collection and ordering of salient information on corrective action underway in a range of problem areas.

3. A serious look should be taken at the nature, operations, and utility of "new institutions," the need for which had arisen in discussion from time to time.

4. A determined effort to acquaint public and private management as to the potential and shortcomings of the systems approach, when applied to discrete or intermeshed problems, is an important "next step" in actually developing an improved system.

5. The creation of communications and experience channels, both through information exchange mechanisms and physical transfer of consultants and government personnel from their own environment to another, was mentioned by several witnesses.

6. Expectations regarding the transferability of technology from the aerospace milieu to the civil sector should not be too optimistic.

7. Many of the basic problems of the city and State should be confronted from the point of view of improving the existing system through relatively mundane changes, rather than always thinking in terms of "quantum" jumps through the introduction of sophisticated man-machine wizardry.

As the special subcommittee assimilated the factual and interpretive offerings of the various witnesses, its ability to appreciate the enormity of the task facing the Congress and the Nation increased. Also, the "magic" of the systems approach was replaced in many instances by a realization of the hard thinking and time-consuming effort which is an integral part of any significant systems improvement.

E. Fourth Hearing, Washington, D.C., May 1967

The fourth and final hearing, on March 29 and 30, 1967,[7] featured university, industry, and State government witnesses. Senator Nelson presided on both days. Witnesses on the first day were Arthur W. Barber, Deputy Assistant Secretary (Arms and Trade Control), International Security Affairs, Department of Defense; E. T. Weiler, dean, Krannert Graduate School of Industrial Administration, Purdue University; Daniel Alpert, dean, the Graduate College, University of Illinois; and Robert Nelson, director, public sector projects, the Raytheon Co.

In his opening remarks, Mr. Barber ascribed the "little or no progress" in applying the systems approach to civil sector problems in these words: "No one is buying!" He described the "cultural gap" between the managers of our corporate society and the managers of State and local governments:

The dynamic parts of American industry constantly reassess their goals and program objectives. They plan 5 years in advance. They analyze the total costs of programs, both in human and financial resources. they make rigorous analyses of private and public spending over the next 5 years, and longer. They outline and analyze many alternative solutions.

While dealing with this kind of professional management capability, the State government executive, however competent, is relatively helpless. He is rarely able to plan 3 to 5 years in advance. He almost never has "thinking money" at his disposal. He cannot make budget tradeoffs, to increase the training effort, for example, in order to reduce welfare costs, or to increase capital spending to decrease operating costs.

He can abolish a job, or create new offices only with very great difficulty. He has little budgetary discretion. For example, school superintendents cannot make major changes in their curricula, and most can allocate only 5 percent of the budget available to them.

The government program manager cannot begin to achieve the effectiveness that is possible through government-industry cooperation until he is permitted a functional analysis of his job, establishing its goals and programs. When this is done there is no question that dramatic programs will begin to appear across this Nation.

Next, the witness noted that while governmental units have expanded their budgets, and often their staffs, by 400 to 500 percent in the past 20 years, there has been no commensurate reorganization nor "reexamination of the purposes or functions of government."

At this point, Senator Nelson observed that while Mr. Barber had been emphasizing the shortcomings of the State and local governments, "Probably the Federal Government is worse off." A quotation of Alfred North Whitehead then was used by Mr. Barber to lead into his next area of comment:

The greatest invention of the 19th century was the invention of the method of invention. We must concentrate on the method itself; that is the real novelty which has broken up the foundations of the old civilization.

In America, we lack "effective local political institutions to manage change," Mr. Barber commented. "The result is that we react to change rather than reflect upon it and shape it to our purpose. *The future is not inevitable as long as we are willing to develop a set of goals and work toward them.*" [Emphasis added.]

---

[7] Scientific manpower utilization, 1967. Op. cit., p. 193–274.

This question then was asked of the special subcommittee: "What can we do to encourage a society organized to meet human goals with all their complexity and variety, rather than the goals of bureaucracy? And these possible actions were identified:

1. We must recognize that no national program or series of programs can solve the problem.

2. New programs * * * should be determined and managed at the lowest possible levels in our society.

3. Convince the people in our great cities and in the suburbs that they have a role in shaping their own future.

Mr. Barber then suggested that the Federal Government should consider "dramatic demonstration programs" to set standards of excellence in education, urban renewal, and other problem areas. In addressing himself to Senator Nelson's question about the need for new institutions, the witness replied: "Frankly, I think the primary problem is to introduce both flexibility and planning into existing public institutions."

* * * efficient management, while necessary, is not enough. There are values above and beyond cost effectiveness. What we badly need is a wider and broader public discussion and participation in the choosing of the alternatives open to American society.

Mr. Barber then expressed the opinion that there is a need for institutions "free from both Government and industry influence" which can "bring to the attention of the public the issues and alternatives that lie before society." He then cited the potential public service which a multi-band equatorial satellite could provide, and that such a system could be created if the public wanted it enough.

The discussion then turned to the nature of S. 430 and S. 467, with Mr. Barber making three points:

1. The bill on creating a public commission would be strengthened by providing for two or three demonstration programs;

2. These bills would provide the means to educate the public; and

3. The bills would be useful in aligning private interests with the public interest.

He then drew a distinction between:

* * * public education, public reporting and programs of this kind, which I consider to be basically sound which educate the public, and a program in which there actually would be a two- or three-block area that would be dramatically changed, or one in which a school system was dramatically improved—in other words, something outside the intellect and the printed page, in the outside world, that was improved dramatically—a demonstration of excellence and hope.

Senator Nelson interjected his feeling that only a modest demonstration program could gain congressional acceptance at this time. He then stressed, as he had done in earlier hearings, the importance of analyzing the needs of a city of a million (for example) and transfering the methodology used to comparable metropolitan areas. The enunciation of his goal merits iteration:

But what I am seeking to find out is the best approach to put some seed money into the area in order to develop some understanding of the concept of systems engineering and also to do some analysis of some problems around the country.

The final point of discussion between Mr. Barber and the chairman concerned the creation of a new framework within which public problems are considered and corrected. Senator Nelson chose to talk

about the role of regional planning commissions, and reminisced about
his experience in Wisconsin:

I can recall very well back in 1959, 1960, 1961, when I was Governor of Wisconsin and I was advocating and did create several regional planning commissions
whose objective was to do precisely this, that in southeastern Wisconsin there was
almost no interest in regional planning commissions, and there was a great sus-
picion among the counties that they would lose some of their sovereignty and
rights.

In a short time, after these commissions were created, they have become very
popular, and they are doing a systems engineering job of evaluating problems in the
area in which they are involved. They are doing a very good job.

I appeared before every county board in the area, and some of them two or three
times, to try to persuade them to join in a planning commission. The same with the
planning commission in central Wisconsin, but now there is political acceptance,
and now they are making substantial headway.

Before welcoming the next witness, Dean Weiler of Purdue University, Senator Nelson said that he had given thought to providing an
appropriation to the National Science Foundation which then could
consult with private profit and nonprofit organizations, and perhaps
jointly evaluate how to best "train people in the utilization of the tools
and educate people as to the efficacy of the concept."

Dean Weiler, after stating his support for the two bills being considered by the special subcommittee, turned his attention to the universities, and how they might utilize systems techniques in helping
to solve complicated public problems. He described how 11 universities in the Midwest had formed the committee on institutional
cooperation (CIC) which created a council of economic growth,
technology, and public policy. This council has called for a program
of studies and systems analyses aimed at increasing understanding
of the problems of midcontinent America, and which would suggest
alternatives for decisions that would "involve human values and the
political process."

Planning is underway for this experiment in which a consortium of universities
would cause pertinent academic disciplines to interact on major socioeconomic
problems through the common techniques of systems science. The task involves
mastery of asking precise questions of the appropriate disciplines to obtain beneficial answers and feasible alternatives. It means coupling science—in the broad
sense—with the social studies to develop viable policy end products.

Such a consortium would be a new departure. It not only would attempt to
recruit working teams of problem solvers from the various disciplines, but it
would gather the team members from two, three, or more great universities. It
not only would try to order many knowns, unknowns, and variables in a small
area, it would attempt to do that for a geographic region of several States. It
would take the tenets of a technique demonstrated to be successful in the aerospace industry, revise them as necessary, and apply the resultant techniques to
a complex net of social, economic, and governmental interrelationships.

The witness then turned to the problem of organizing the university
personnel to meet the new challenges. Unfortunately, universities
traditionally have operated within disciplines, and—as Henry Rowen
of RAND expressed it—"problems just don't come cut this way."
Dean Weiler quoted another witness before the special subcommittee,
Dr. Ramo, who had said that the universities should be producing
"politico-econo-socio technologists." Special hybrid training, perhaps
with Federal support, is needed, and at this point Dean Weiler cited
the recommendation made by the National Commission on Technology, Automation and Economic Progress, as stated by its chairman,
President Howard R. Bowen of the University of Iowa in this proposal:

* * * that the Federal Government "experiment with the formation of university institutes or interdisciplinary programs, adequately financed and fully integrated with the educational function of the university, which would serve as laboratories for urban problem analysis and resources for local communities that would want to use their advice and services."

The next focus by the witness was on the subject of regional approaches to problem-solving; here he noted the creation of the Upper Great Lakes Regional Commission—embracing the northern tiers of counties in Minnesota, Wisconsin, and Michigan—and the Great Lakes Basin Commission, which was recommended to the President by the Federal Water Resources Council under the Water Resources Planning Act of 1965 (see app. L for related public laws).

Dean Weiler was succeeded by Dean Alpert of the University of Illinois who pointed out that a problem situation can be corrected not just by the establishment of a good analysis team, but that a "good client" also must be present. He expanded upon his concept of a client, noting that individual municipalities may have to turn to the megalopolitan whole in order to improve transportation or air pollution conditions. Thus, a "joint client" becomes a reality, and grants may have to be sought by this type of new entity. He then offered these comments on the proposed legislation, with especial emphasis on S. 467:

In talking about the Commission which is proposed in S. 467, a commission for the study of systems analysis, it seems to me that this would be valid if the Commission itself were viewed as a client. In many of our national committees and commissions, we call together a group of very busy people who can meet, let us say, once a month for a year or 2 years, and what they are actually trying to do is to carry out a systems analysis of a pretty tricky problem.

It seems to me that above all, a commission that studies systems analysis should use the techniques of systems analysis that are available, and in the role of the client. It might well be that one could merge the thrust of these two bills by supplying sufficient funds to several commissions or to subcommissions, in different problem areas, which themselves would seek out valid client relationships and find sample areas where systems analyses could be done, always in the context that there existed a potential client or an actual client who could make use of the systems analysis.

Having heard the testimony of the two academic representatives, Senator Nelson embarked upon a line of questioning regarding the way in which additional systems personnel can be obtained. In referring to the role of the CIC—comprised of the Big 10 universities and the University of Chicago—he asked what expertise was available in its membership to "establish a program of giving 1-week training to heads of departments, 2 or 4 or 6 weeks training to deputy assistants to heads of departments on systems engineering?" When asked point-blank whether there are "specific programs aimed at reaching people to become systems analysts?" Dean Weiler replied that he knew of none. Dean Alpert later added that "outside help" might be needed in order to conduct such a training program. Senator Nelson ended the discussion by reflecting that while there are people in the country who are prepared to analyze the biggest problems, "We cannot find anybody prepared to analyze the little ones."

The final witness on the first day was Robert Nelson of the Raytheon Co. In discussing mechanisms for prompting private groups to train personnel in analyzing civil sector problems, Mr. Nelson pointed out that one very effective means had been used in the procurement for the California studies (see discussion in sec. VI. B). He reminded his listeners that 50 companies had prepared proposals, but only four had been successful:

What of the other 46? The fact is that each company assigned its best talent and developed its approach on the subjects of crime, information, transportation, or waste systems. Each prepared and submitted its proposal. Where a company lacked the requisite skills, it teamed with others and employed consultants. As a result, the knowledge of the subject and the analysis of the applicability of the systems technology to these problems now exists in a reservoir of 50 companies, not just four.

There is an additional advantage. Simultaneously, the State administrators had the opportunity to review and evaluate the best ideas of the 50 companies and in the process to increase their understanding of the problems and their understanding of the administration of these new techniques for their solution.

After endorsing the proposed National Commission on Public Management which could "provide a focus for the evaluation of prospective programs and * * * assemble data on actual experience," the witness related some of the in-house research projects and contract responsibilities currently pursued by his company. Water resource planning (e.g., the Merrimac River) and land-use analysis, including the exploration of mineral and oil resources, were receiving corporate funding. Mr. Nelson told of work being done for several municipalities in educational systems "where we are relating curriculums development, the media required to meet the curriculums, and communications techniques." He also mentioned a contract in which Raytheon was teamed with a university to design an improved traffic control system for Detroit, with the Highway Research Board monitoring the developmental effort.

Four witnesses appeared before the special subcommittee—with Senator Nelson presiding and Senator Javits in attendance—on the second day of the fourth hearing: Michael Michaelis, manager, Washington office, Arthur D. Little, Inc.; Vincent J. Moore, assistant director, office of planning coordination, State of New York; Roger Schrantz, director, policy planning and program development, Bureau of Management, State of Wisconsin; and Richard M. Cyert, dean, Graduate School of Administration, Carnegie Institute of Technology.

With such credentials as the former Executive Director of the White House Panel on Civilian Technology, Mr. Michaelis could advise the special subcommittee about some of the ramifications of using systems methods to guide governmental actions. His initial subject was the critical need to increase the rate of utilization of new knowledge; he pointed out that the need had not changed in the past 5 years and that some new mechanism was required to control experimentation with new applications of management techniques. He went on to say that while the *procedures* of systems analysis vary widely, "reflecting the inherent uncertainties of the problem under review, as well as the difference between organizations," the *structure* is quite consistent from case to case and will involve these successive steps:

1. Understanding the objectives of the desired system in the context of its working environment.

2. Stating the interrelations between the objectives and the variables of the system that are chosen for analysis, thus constructing a model.

3. Quantifying functional relationships between elements of the model and its outputs, often described as the benefits.

4. Quantifying functional relationships between elements of the model and inputs or resources needed, often called the costs.

5. Combining (3) and (4) into an input-output or cost-benefit relationship that flows through the whole model.

6. Determining from the input-output relationships that choice of all possibilities of systems characteristics and manner of operation that produces the most desired result, and operating rates that correspond to that optimum.

Mr. Michaelis observed that while the Department of Defense was "a centralized and integrated decisionmaking mechanism," the civil sector had a multitude of customers, competitive producers, and service organizations. A common language—systems methodology—can be created to link managers in government, industry, labor, and the private sector centers of knowledge generation. Next, the witness discussed research and development (R. & D.) vis-a-vis "innovation":

R. & D. is only a comparatively small part in the whole chain of innovation from the conception of an idea to the sale of a new product or service in the marketplace. R. & D. generally costs no more than 10 percent of the total cost of innovation; it occupies a comparably short timespan in the total chain of innovative events and, surprisingly enough, it often carries comparatively less financial risk than some of the other functions in the chain, such as marketing and distribution.

This is worth repeating: Innovation is not the same as R. & D. R. & D. creates new knowledge. Innovation is the process of utilizing such knowledge. Innovation thus entails the challenging of wisdoms and beliefs that are often long held and cherished by the body politic. Innovation, therefore, is a "political" process in which the systems method can provide assistance but can, by itself, provide no solutions. Personal values and attitudes enter significantly into the picture. Fears and ambitions joust with each other; labor and management seek compromises between apparently opposing viewpoints; the public and the private sectors attempt to do likewise—and each, whether an individual or an organization, proceeds from his own formulated set of objectives and assumptions.

In this situation, the systems method can provide a common language so that the battles that will be fought and the compromises that will be reached can derive from at least a common understanding. There are plenty of differences that will remain because we are, after all, human beings, and react differently even to the same stimuli.

At this point, Mr. Michaelis commented at length upon the school construction systems development project in California—citing it as an example of the usefulness of the systems method in a "decentralized, fragmented, and traditionally adversary-prone situation"—which had been described originally in Dr. Schon's testimony (see sec. VI. D).

The concept of a National Council for American Progress, which had been proposed by Mr. Michaelis when he served as the chairman of a committee of the U.S. Chamber of Commerce concerned with industry-Government cooperation in innovation, was placed before the special subcommittee. The general scope and objectives of this council. (1) will seek and manifest consensus on the means of achieving national goals; (2) will illuminate the opportunities for living up to our potential; and (3) will identify actions we must take if we are to realize our expectations.

In order to achieve these ends, the council must be representative of both the private and the public sector of our society. It will need the fullest support and active participation from Government (executive, legislative, and judiciary), from industry and finance, from labor, and from the academic and professional community. It will be the purpose and responsibility of the council to confront these groups on critical issues pertaining to America's progress. Such confrontation and collaboration between these groups will be undertaken with the help of an interdisciplinary staff, skilled in the most advanced management methods, including those of systems analysis, systems engineering, and systems management. Such confrontation and collaboration—

on a professional level—should lead to a better mutual understanding of the opportunities and obstacles before us, and thus to greater unity of purpose.

Mr. Michaelis assured Senator Nelson and Senator Javits that the council would not "seek to supplant, impede, or circumvent any of our democratic processes and institutions," and that its multisectoral representation would be advantageous. After discussing his support for both S. 430 and S. 467, and that the proposed council might offer a means for combining the elements of each, the witness then was asked by Senator Nelson how such a body would be created and funded. The reply: initially by foundations and eventually by matching grants from Government, industry, and labor. In response to Mr. Michaelis' statement that the council should be "a self-generating mechanism," Senator Nelson said that he had never seen "a good idea spontaneously combust." The chairman then returned to the rationale behind his bill (S. 430) and his feelings about the proposed council:

> The reason that I drafted one of these bills in the first place, was to try to figure out a method, a feasible technique, for generating the interest in utilization of the concept of broad interdisciplinary planning, and that is the purpose of the bill.
> The reason for drafting the bill was that it was my judgment that it was not going to originate on any local level because the patient does not know what is wrong with him. As a matter of fact, he does not even know in most cases that he is not feeling good.
> He does not know what the problem is. I think it is very important, your idea of involving the decisionmaker. Without the decisionmaker involved, whether it is the mayor, or the budget director or the police chief, or the motor vehicle director, or the superintendent of schools, or whoever it may be, without having the interest and cooperation of the actual person who makes the decisions, it is not possible to implement a proposal no matter how good it is.
> I like the idea of the American Council for Progress, and it reads very well. I just do not know how you will get it started.
> That is the reason for introducing these bills, to find some technique for introducing some funding and exciting some interest in the concept both from the standpoint of educating people as to the need and demonstrating by some projects the results that can be accomplished.

The discussion of the nature of the mechanism to achieve Senator Nelson's objectives continued, with the chairman commenting that he was "looking for an established Federal institution that already exists to administer funds." Also, he repeated an option voiced earlier in the hearings: that of placing the responsibility with the National Science Foundation, and empowering that group to hire consultants when necessary. Another possibility: to call in the American Management Institute, which could work with the aerospace and nonprofit experts in grappling with the problems of society. In closing, the witness cited several agencies which might play a role in the furtherance of the principle of using systems technology to solve the burgeoning social and community problems; included were the General Accounting Office, the Department of Transportation, the Office of Education, and the Department of Housing and Urban Development.

Representing the State of New York before the special subcommittee was Vincent J. Moore, assistant director of the office of planning coordination. In his opening remarks, Mr. Moore reviewed the creation in 1961 by Governor Nelson Rockefeller of the office for regional development, which in 1964 produced the noteworthy report entitled "Change/Challenge/Response." Among the 15 recommended actions of the report were two which formed the basis of the current New York State systems effort:

RECOMMENDATION NO. 4

Formulation of comprehensive statements of statewide development factors and needs pertaining to urbanization, transportation, resource development, public facilities and other such fields. Such functional statements would be prepared by the concurrent efforts of interested State departments, private consultants, university faculties and other private or public agencies. The statements would be coordinated by the division of the budget and office for regional development and would be subject to periodic review.

RECOMMENDATION NO. 6

Annual preparation of and submission by the budget director to the Governor, as part of the proposed executive budget, of a specific physical program and a financial program for meeting such immediate and long-term State requirements are deemed feasible.

The formalized planning-programing-budgeting system for the State was initiated in April of 1965 and featured such special provisions as:

1. Adding a 3-month period for long-range programing in advance of the initiation of the annual budget preparation by the agencies.

2. Having the program planning activity of the departments overlap the beginning of the next immediate fiscal year, thereby permitting the agencies to take into account the legislative adjustments to their programs.

3. During the summer, division of the budget and Office of Planning Coordination staffs meet to review and evaluate the program plans and adjust and coordinate them on the basis of 10 major functional areas.

4. Establishment of a uniform report format, which requires, in sequence:

I. The identification of exogenous factors influencing major program areas. These are factors such as changes in the population structure, potential scientific breakthroughs or technological improvements, changes in behavioral patterns, and changes in economic structure.

II. The projection of indicators of future levels of need for program services — at this point without respect to the responsibilities of the various Government sectors or private sectors for meeting these needs.

III. The analysis and evaluation of the relative roles and responsibilities of the State government and the other levels of government and private enterprise for meeting the needs.

IV. The analysis and evaluation of the relative roles and responsibilities of various departments of the State government if the major program activities are split between two or more agencies.

The above four topics establish the policy base upon which specific program plans developed.

V. The establishment of specific program goals and plans, including the appropriate multiyear projections of program achievement and workload.

The final four topics, also on a multiyear format, present the projected requirements for the resources need to carry out the program plans.

VI. The projection of the personnel required to operate each program.

VII. The projection of any capital facilities required.

VIII. The projection of the overall fiscal support required.

IX. The projection of the basic reasearch support required to improve the program operation.

The guidance for the State PPBS is found in the "Guidelines for Integrated Planning-Programing-Budgeting" manual. Mr. Moore next turned his attention to three problems of importance in the development of the new system:

1. Recruitment of planners and public administration experts;
2. Improvement of the basic information base for planning; and
3. Intensification of the research into domestic problem areas.

In connection with the acquisition of "well-educated, inquisitive, creative planners and public administration specialists who are people-oriented rather than machine-oriented," Mr. Moore highlighted a feasibility study on midcareer education for comprehensive planning recently completed by him and soon to be published by a group of foundation-supported consultants in connection with a detailed plan for implementing the President's proposal for a National Urban Design Institute.

My recommendations call for an early identification, after some post entry experience, of potential leaders in comprehensive planning and public administration, and offering such people the opportunity to participate in an intensive 2-year work-study program which would be designed to both broaden their field of knowledge beyond their basic specialty, and to update and train them in the most modern techniques of planning and public administration. The "students" of such an institute would spend part of their time working on actual problems for actual governments or other agencies contracting with the Institute for such services. They would be paid a salary commensurate with their potential earning power as a full-time employee of the Government or corporation from which they were selected.

It appears that such a program could be financed almost completely by both contributions from the "student's" sponsoring Government (contingent upon his agreement to return to that Government for a specified time following his completion of the program) and from the proceeds of the Institute's contractual agreements with governments and other agencies desiring its services. Some small additional foundation or Federal grant assistance would probably be required.

The witness also noted that technical manpower support for applying computer technology to planning and decisionmaking is needed, including professionals from such disciplines as economics, sociology, geography, education, and penology.

The creation of the new PPB System revealed to the Office of Planning Coordination the inadequacies of the information available for its purposes. Mr. Moore cited the requirement for current data on population composition and distribution, economic structure, land use, and housing conditions; he also recalled that Governor Rockefeller recently had proposed the establishment of a central planning data bank to be utilized for systems analysis approaches to planning by the State and local government units.

The third problem area—the need for more intensive research in domestic problem areas—drew the comment from Mr. Moore that while millions of dollars are spent on State and local development plans, better research information must be used in their creation. Since State and local government funds are insufficient to bring about research at the needed level, the Federal Government could do two things:

1. Insure that the tremendous research activity of the Federal departments and agencies is coordinated and communicated to State and local government planning and budgeting staffs.
2. Provide greater recognition and more fiscal support for research activities related to the preparation of the State plans required by various Federal grant programs.

After underscoring the need for a "comprehensive planning agency at the national level," the witness outlined the shortcomings of existing agencies in any such endeavor, and gave the opinion that either a permanent national commission or an agency within the Executive Office of the President should be established.

The testimony and discussion then focused on the problem of training intelligent, educated persons in systems techniques. Mr. Moore set forth a new approach in this manner:

At this point I would like to insert that I think there is a role here for the universities and private industry in developing new techniques for programed instruction. These are self-instructing techniques whereby manuals are prepared which individuals can study and use to test themselves using automated teaching procedures. Systems techniques are particularly adaptable to programed instruction.

I think we ought to try to bring together some of our major data processing firms, such as International Business Machines and other such groups in the country, with the government officials and the university faculties concerned with producing this type of individual and see if some basic curriculum material and program instruction methods couldn't be worked out.

Mr. Moore, in response to Senator Nelson's query about New York's plans to commence such an effort, applied affirmatively, and went on to amplify the scope and nature of the special educational effort:

Basically what we want to do is have a series of what you might call brainstorming sessions related to various functional or a specific functional problem area, to discuss what the relationships should be between the variables contributing to the development of such a system, such as a park and recreation system, and determine exactly what goals and objectives we would like to accomplish in terms of planning an integrated State recreation network.

We would have people trained in computer applications to analysis audit these sessions and later respond with a discussion of where the computer techniques could be applied to this functional area.

We think that this training has to take place related to the real world problems which our staff planners are facing.

Senator Nelson complimented the witness on the idea, calling it creative and valuable.

Senator Javits entered the discussion, mentioning that he had been agitating for "a national commission on national goals," a prototype of which had been in existence during President Eisenhower's administration under the chairmanship of Walter Wriston. He questioned whether the commission spelled out by Mr. Moore involved "compulsion." The witness said that he did not think this was necessarily the case, and went ahead to say that in the postwar era, "the concept of planning has changed radically from one of static, master planning to one of what you might call process planning which seeks to improve the decisionmaking process." Senator Javits then asked if Mr. Moore had analyzed S. 430 and S. 467. The detailed statement of the witness follows:

STATEMENT ON S. 430 AND S. 467

To review, S. 467 would create a Federal commission to conduct a feasibility and applications development study of systems analysis and other management techniques to the problems of major functional areas of government. The commission would exist under this statute for some 32 months and its first year of operations would be supported by $500,000 appropriated by this bill.

Bill S. 430 appropriates $125 million to be administered by the Secretary of Labor for grants to the States and direct contracting for developing and implementing systems analysis applications to national and local matters which are defined by the Secretary as "problems." No program duration is specified in the bill.

We often hear it said that "ever since there have been men there have been governments which make decisions. Why should systems analysis be called in to assist with such decisions? What reason is there for such a radical change in the habits of governments?" One thoughtful reply given by researchers at Johns Hopkins is: "The necessary and sufficient cause is the exponent growth of the progress of humanity." They also emphasize the acceleration of the casual sequence of research, innovation, and economic growth and demonstrate the absolute necessity of preliminary scientific reflection before major decisions are taken whether at the National, State, or local level. The progress of humanity is marked by the necessity to accommodate to the technological changes that are occurring at an accelerating rate. This puts a severe strain on the decisionmaking process. The management sciences including systems analysis and the computer can bear part of the burden. Government must assimilate these new approaches because of its key role in society, and should provide the leadership and ingenuity in using modern technology positively to design the social institutions which the complex problems of today demand. Technological progress in the past has not produced catastrophe, but has instead resulted in a higher standard of living.

This country has shown a talent for social innovation and it is important that the powerful tool of computer-based automation for production, paperwork, research, decisionmaking, and planning an invention be utilized to adapt to the changing times. However, research is costly. Valid perceptions of efficient, effective systems analysis do not just happen. They are the product of planning, testing, experience, observation, and imagination. Very little has been done in this area at the State or municipal level. Pioneering efforts include the New York State planning-programing-budgeting system and the SOGAMMIS (South Gate municipal management information system) project being undertaken by a University of Southern California team. The approach in both of these cases is a combination of functional planning, systems theory, and decision theory as it relates to resource allocation (e.g., program budgeting) and a dynamic conception of data storage (continuous input and random access, not a static data "bank"). Systems analysis incorporates the concept of information feedback which reflects both the performance of the governmental organization and the effect of the programs on the community being served. The significance of recent theory on the decisional processes for allocating resources (program budgeting, etc.) is particularly relevant because most if not all such theory deals with how to improve the definition of organizational goals and the optimal allocation of resources to them.

The expression "systems analysis" has not a single denotative universal definition, but rather it possesses several different connotations. For example, the authors, McMillan and Gonzalez, in "Systems Analysis" (Irwin 1965) include such analytic techniques as linear programing and queuing theory in the subject while many educational institutions continue to present these techniques in operations research courses. The U.S. Government Organization Manual lists an Assistant Commissioner for Planning and Research in the Treasury Department with responsibility for unspecified "systems development." The manual lists a systems research and development responsibility in the Federal Aviation Agency, and specifies the systems concerned are air traffic control and navigation. We infer, therefore, that both bills are using "systems analysis" in its most all-embracing sense. Because other organizations do not use the same definition of the term, it might be wise to spell out Senate meaning in the bill.

The Commission of S. 467 has a large bearing upon the aims, scope, and finances of the program called for in S. 430. In fact, the S. 467-authorized study is a necessary prelude to effective program planning to implement S. 430. But a final report is not called for from the Commission on Public Management (S. 467) for some 30 months following their first meeting. In order for the output from the Commission study to be rational input to the program of systems analysis of governmental problems the Commission, Labor Department, states, and other interested groups will require more than the informal coordination we infer from the bills.

The program which would be authorized by S. 430 is only in part concerned with technical manpower. The implications of this program for intergovernment and intragovernment planning, management, and organization are undoubtedly of greater short-range import.

Ideally, a study of systems analysis feasibility, such as that which would be authorized by S. 467 *might* become, should precede the program planning required to implement that proposed in S. 430. However, if both study and program are to proceed simultaneously there are clear needs to be satisfied by program funds. A moderate investment in regional intergovernment OR technical meetings could offer an exchange of experience and methodology which is not provided by any organization at present.

Development of programed learning materials for specific analytic methods and techniques would afford government at all levels with a valuable and permanent training resource. Beyond these suggestions, the development of program requires better feasibility measures.

Clearly the need exists for the kind of encouragement on the part of the Federal Government which is evidenced in U.S. Senate bills S. 430 and S. 467. The problems of employment, public welfare, education, mental health, water and air pollution, urbanization, crime, juvenile delinquency, housing, etc., can and must be solved. Just as the modern private organization can continue as a viable institution only if a substantial portion of its funds are systematically channeled into research and development activities related to its goal, public needs, and the mechanics essential to program achieving, so too must government respond to this need.

U.S. Senate bills S. 430 and S. 467 are in essence two approaches to the same problem. It would seem to us that S. 430 would better accomplish the important goal of getting personnel at the State and local level involved in utilizing new techniques to solve problems. The result could very well be a system which would require basic changes in the way the State or local government has traditionally conducted its business. Systems planning might prove to be an important change agent in tooling government for the increasingly complicated governmental programs growing out of our urban society. Hence, State participation in the research efforts as provided in S. 430 would have considerable practical advantages. It is important that abstract theories be subjected to rigorous analysis and test. The New York State PPB system is an excellent example of the evolution of a concept which, once refined, can be the nucleus for a similar program in every State in the Union. There is a substantial amount of literature about systems planning, program planning, etc. What is needed here is not a general study, but an attack on specific problems using the new analytic tools. For this reason, it would seem that the Department of Housing and Urban Development would be the logical agency to dispense these kinds of planning funds.

In essence we see this as a "demonstration" program quite analogous to the HUD model city program which seeks "to test whether we have the capacity to understand the causes of human and physical blight, and the skills and the commitment to restore quality to older neighborhoods, and hope and dignity to their people."

Another thorough description of a State government's efforts to modernize its management and information handling was provided by Roger Schrantz, director for policy planning and program development (bureau of management) for the State of Wisconsin. In welcoming the witness, Senator Nelson recalled that that State had the first "comprehensive State planning in the United States" and the first planning-programing-budgeting program. Mr. Schrantz began his testimony by recounting these achievements, which began with the then Gov. Gaylord Nelson's establishment of a consolidated department of administration in 1959. The statutory purpose of the department pointed to the development of a management and decision process that would eventually embrace the use of system analysis techniques. The directives included:

Present clearly defined alternatives and objectives of State programs and policies so that the State's agencies, the Governor, and legislature may plan cooperatively and finance the services which the State will provide for its citizens.

*   *   *   *   *   *   *

Help the State's agencies furnish the agreed upon services as efficiently and effectively as possible.

*   *   *   *   *   *   *

Assure the Governor and the legislature that the services are being provided to the public at the agreed upon quantity, quality, and cost.

*   *   *   *   *   *   *

Anticipate and resolve administrative and financial problems faced by the agencies, Governor, and legislature of the State.

The next step was to modernize the State's budgeting techniques and concurrently develop pilot program budgets for several of the State agencies. The latter effort allowed a sharpening of the concepts for a practical application of program budgeting theory, and also tested the marketability of program budgeting with the legislators. In 1961, Wisconsin embarked on a comprehensive State planning process, to be carried out in two phases:

1. Phase 1, an overall framework of comprehensive planning data and techniques in which specific functional plans could be prepared; this was completed in 1963.

2. Phase 2, the preparation of specific long-range plans, including a freeway plan for 1990, a library facilities plan, an airport system plan, and economic plans for eight State regional areas.

In 1963, with full bipartisan support, Wisconsin commenced "the mammoth task of developing a comprehensive program budgeting process for all State government activities and agencies." In 1965, the results of this effort were evidenced in the State's budget, now framed in a program budget format. Later steps included the creation of a program planning component (in the department of administration), a management sciences unit which would "stimulate the use of modern scientific techniques in managing State government activities," and the preparation of a series of policy papers outlining the Governor's budget policies.

Mr. Schrantz enumerated some important characteristics which marked the Wisconsin policy decision efforts: There had been a continuing bipartisan effort with both the executive and legislative branches involved, and the focus had been on policy decisions rather than operations. He then commented upon the ways in which the State's need for systems analysis capabilities could be satisfied:

1. Orientation of all policy decisionmakers and key supporting personnel in the concepts and procedures of systems analysis.

2. Development of "at least a modest in-house systems analysis capability."

3. Reliance in large part for support from private sector—i.e., not for profit—organizations and the State university.

Additional details on the Wisconsin experience and plans are featured in appendix H.

The witness described several systems studies being conducted in cooperation with management consultant firms; such projects include a range of problem areas:

For example, the State highway department has contracted with one of the Nation's leading management consultant firms for a multiyear, $175,000 study of the department's planning and management system. Now well underway, the study is being done in conjunction with department staff efforts, and is aimed at developing an integrated operations and management information system for the department's management and decision process. The sheer magnitude and complexity of the study and systems implementation demanded skills and attention beyond those of the department staff.

Mr. Schrantz then stated his belief that the venerated "Wisconsin idea"—that of joining campus and capital resources in the solution of serious social problems—would take on "a new dimension" in the years ahead.

The concluding portion of Mr. Schrantz's time with the special subcommittee was occupied with a discussion between Senator Nelson and him about the need for special training courses in systems analysis. The witness mentioned the 9-month academic offering of the University of Wisconsin for Federal agencies on PPBS, and agreed that 1-week or 10-day courses for executive level decisionmakers should be instituted. Mr. Schrantz then reviewed several suggestions which resulted from his examination of S. 430 and S. 467. Included in these were an endorsement of Federal encouragement and funding for the education of policymakers, the requirement for Federal support in developing State government in-house systems analysis staffs, further Federal sponsorship of private enterprise capability enhancement, and establishment of an exchange mechanism whereby systems analysis studies' results can be made available to many States with the same problems.

The final witness to appear before the special subcommittee was Dean Richard M. Cyert of the Graduate School of Administration, Carnegie Institute of Technology. Part of his prepared statement contained these reflections on the proposed commission:

There must be a concerted and systematic effort made to apply the best analytic methods to the solution of pressing social problems. In particular I like the idea of forming a commission to develop operational approaches to handling the important problems. There is always the danger that another commission generates another report which occupies space and results in no action. On the other hand, there is no approach which guarantees constructive action. I would like to see a commission composed of working scientists in mathematics, statistics, economics, sociology, political science, psychology, civil engineering, mechanical engineering, and chemical engineering given the task of proposing in detail, action and research plans designed to utilize analytic methods in deriving solutions to a set of social problems. The commission would serve to encourage research in social problems by giving "seed grants" and by giving individuals a particular place in government to which to come with their ideas and the fruits of their work. Many imaginative ideas for dealing with society's problems have been lost simply because there was no recognized place to send the ideas. The commission should be viewed as an architectural commission laying out the paths to be followed. At the completion of such a set of plans the Congress and the Executive would be in a better position to decide which departments should be given specific tasks. We would have a coordinating plan which could fit the diverse activities together into a unified whole. We would also be in a position to view our problems as a system and would be able to take into account the impact of a proposed solution in one area on the solution in another area.

Senator Nelson and Dean Cyert then discussed at length the aim of a "public commission" or "public work group"; its composition also raised concern that the correct mixture of generalists and specialists be found, and that their energies be tapped for a sufficient period of time so that meaningful recommendations could be forthcoming. This series of questions was posed by Senator Nelson:

How do you select them and, for example, should it be full time? You can't—I am assuming—expect to set up a full-time commission of highly paid people who would take a year, or 2 years, or what have you. Would it be feasible in your judgment to create such a commission as a working commission composed of all the disciplines you mention here, assign them the task of doing exactly what you mention, pick some problem areas and probably decide, if they are going to use outside consultants, which they will, what profit or nonprofit organizations should be employed to do what kind of projects, what universities, or what have you could they contract with, what projects ought to be done? Could this be done by a group who were hired as consultants themselves on a part-time basis?

In other words, could you have them meet for a week or so as a preliminary and create some subcommittees to tackle some problems, meet again then over a period of 6 months or so and analyze the area and come up with a proposal?

Once again the possibility of assigning the National Science Foundation the responsibility for working with private consultant groups to meet the needs of the age was raised. Also, the question of full-versus part-time workers on any such project was brought up for discussion, but not resolved.

Dean Cyert then talked about the five-course curriculum established by the Carnegie Institute for Federal employees, with the goal of improving the PPBS capability within the Federal Government. The courses, taken over a 9-month period, include cost-benefit analysis, planning and budgeting, and computers in Government. He noted that a calculus background is required for this course of study. The witness also mentioned a 9-week course at his institution for people in private management.

In adjourning the final hearing, Senator Nelson said:

* * * We certainly will change the two bills that were introduced and our objective will be to really have a bipartisan bill that accomplishes what we are seeking to do here.

The fourth hearing was significant because of the valuable testimony on State experiences in establishing PPBS capabilities—as presented by Mr. Moore and Mr. Schrantz—and the emphasis on the time required to build a modern management structure. Second, specific proposals for new types of institutions were presented to the special subcommittee, such as Mr. Michaelis' Council for American Progress. Third, a more complete discussion of the role of "special education" in the creation of a greater national capability in systems analysis ensued.

F. SUMMARY

The multiple hearings conducted by the Senate Special Subcommittee on the Utilization of Scientific Manpower proved to be instrumental in focusing the attention of public sector entities and private sector institutions—foundations, universities, profit and not-for-profit firms—alike on the requirement to explore the potential for applying systems technology to social and community problems. With two distinct legislative proposals in being, the witnesses could present critiques within a broader context of the Nation's needs for new approaches to long-standing but newly critical problems.

The "systems approach" was defined, discussed, and delimited at great length, and as the hearings progressed, the sophistication of all of the participants increased. The terminology which grew up about the systems analysis-operations research-computer technology cult became more familiar, and the cycle of activity involving the steps in the process (as shown in fig. 33)[8] became more comprehensible.

[8] Thome, P. G. and R. G. Willard. The systems approach: a unified concept of planning. In Scientific manpower utilization, 1967. Op. cit. p. 301.

154

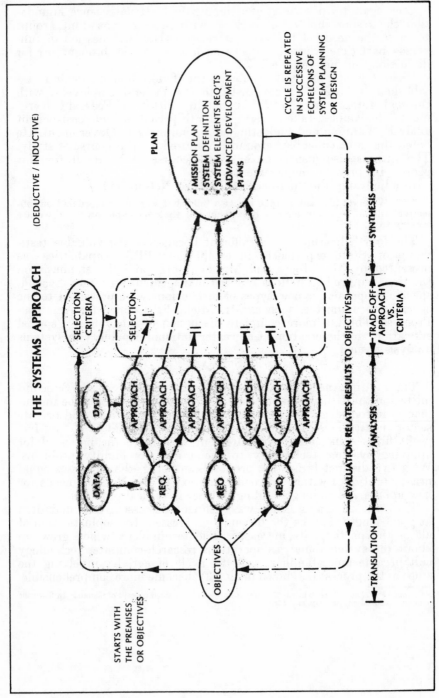

FIGURE 33

In addition to establishing an overview—indeed, a common point of reference—for the members and support staff of the special subcommittee, the penetrating discussions of the beginnings and step-by-step evolution of the systems approach were important since they provided a foundation for the consideration of applying the new techniques and devices to a new set of problems. As each step within the systems approach was examined and examples of its implementation brought forth, the relationships between the elements became a matter for thoughtful study. The "absolute" and "relative" parameters and the interactions between the steps within the systems approach are shown in figure 34.[9]

---

[9] Ibid., pp. 302–303, fig. 2.

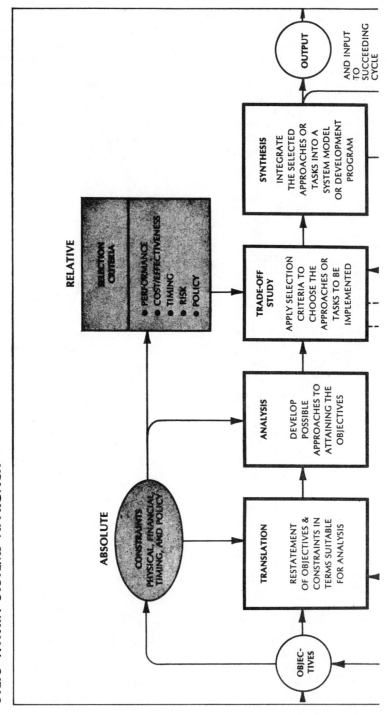

STEPS WITHIN SYSTEMS APPROACH

OBJEC-TIVES

TRANSLATION — RESTATEMENT OF OBJECTIVES & CONSTRAINTS IN TERMS SUITABLE FOR ANALYSIS

ANALYSIS — DEVELOP POSSIBLE APPROACHES TO ATTAINING THE OBJECTIVES

TRADE-OFF STUDY — APPLY SELECTION CRITERIA TO CHOOSE THE APPROACHES OR TASKS TO BE IMPLEMENTED

SYNTHESIS — INTEGRATE THE SELECTED APPROACHES OR TASKS INTO A SYSTEM MODEL OR DEVELOPMENT PROGRAM

OUTPUT — AND INPUT TO SUCCEEDING CYCLE

ABSOLUTE — CONSTRAINTS PHYSICAL, FINANCIAL, TIMING, AND POLICY

RELATIVE — SELECTION CRITERIA
- PERFORMANCE
- COST/EFFECTIVENESS
- TIMING
- RISK
- POLICY

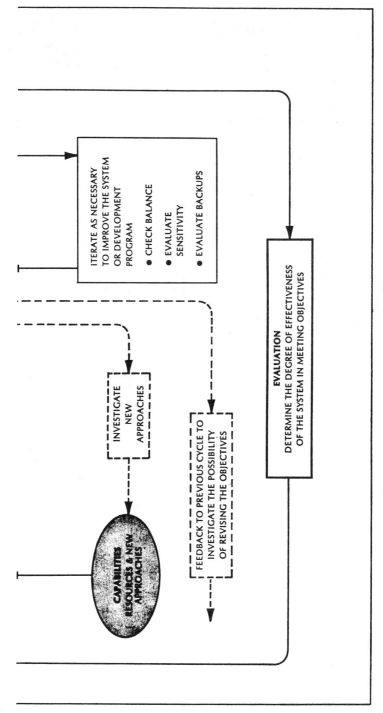

FIG. 34.—STEPS WITHIN SYSTEMS APPROACH

	Nelson (S. 430)	Scott (S. 467)	Kimball (MRI)
Operational program:			
A. Purpose	Mobilize and utilize scientific and engineering manpower of Nation to employ SA and SE to fully employ manpower resources to solve national problems.		Public commission, combining study/action function, to demonstrate feasibility of applying SA to public problems and needs.
B. Administering body	Department of Labor		National Commission
C. Function	1. Make grants to States. 2. Enter into arrangements with universities and other institutions for purpose of causing SA and SE approaches to be applied to national problems.		1. Demonstration projects: Provide for training SA's and relevant public officials. 2. At existing institutions with sufficient pool of manpower and resources. 3. 10 to 15 centers.
Study Program:			
A. Purpose		National Commission on Public Management to study and recommend methods to use SA and management techniques to resolve national and community problems in the nondefense sector.	Same as above
B. Administering body		National Commission	National Commission
C. Function		1. Develop information (on SA). 2. Analyze application to public programs. 3 Categorize national and community problems. 4. Assess proper relationship between government and private. 5. Conduct studies. 6. Schedule seminars. 7. Make recommendations to encourage application.	1. Identify problems to which SA is most applicable. 2. Specify how that approach is best used. 3. Explore problems and pitfalls.
D. Composition		Chairman, Vice Chairman, and all members. Individuals concerned with subject matter and with experience derived from Government, business, labor, teaching, and research. 4 Congressional Members (Morse) and 0 (Scott).	Might be analogous to the Commission on Technology, Automation, and Economic Progress.
E. Reports		1. Preliminary report, 1 year. 2. Final report in 30 months including recommendations and proposals for legislative and administrative action if warranted.	
F. Other			

FIG. 35.—COMPARISON OF LEGISLATION BEFORE SUBCOMMITTEE ON

[Key: SA—Systems analyst; SE—Systems engineering; MRI—Midwest Research Institute; SLFP
OSTI—Organization for

Mushkin (SLFP)	Harr (AIA)	Bryk & Whittenbury (RAC)	Schon (OSTI)
To assist States and their political subdivisions to more adequately meet problems through mobilization of scientific and engineering manpower in SA and SE programs.	To provide for comprehensive long-range and coordinates program to resolve national and State community problems.	To seek improved approaches to solution of national problems through modern management methods including SA and SE.	
Department of Housing and Urban Development.	National Council on Science and Engineering Manpower Resources, Vice President, Chairman.	Designated authority_____	
Demonstration projects: 1 To illustrate implement of PPBS, to provide for employee training and recruitment of staff. 2. Provide State and local training programs.	1. Survey all significant activity. 2. Develop a long-range comprehensive program. 3. Designate and fix responsibility. 4. Insure adequacy of qualified SA, SE, etc. 5. Make appropriate grants to States and other institutions to conduct SA on national and local problems. 6. Review programs annually.	Make grants to States through arrangements with universities and other public or private institutions and organizations to bring SA and other modern management techniques to regional or local problems.	1. Test out solutions to problems in an administrative neutral context (without committing itself to any one department). 2. Establish an interagency pool drawn from HEW, OEO, Labor, HUD, etc.
Commission. To encourage social inventiveness or governmental innovation to meet changing and cumulative needs_____	Same as above_____	Same as above_____	
Not stated_____	Commission on Scientific and Engineering Manpower Resources.	National Commission for Scientific Manpower Utilization.	Place that can provide continuity, prestige auxiliary staff. Possibility: the Brookings Institution.
1. To examine utility of SA, SE, and PPBS. 2. To examine all innovative approaches. 3. To review innovative instruments currently in use. 4. To assess demonstrative aids. 5. Recommend new legislation if required.	1. Review needs of non-defense sector. 2. Review surveys, studies, etc. 3. Review existing programs. 4. Recommend a governmental organization plan.	1. Define relevant national problems. 2. Describe management methods and problem-solving techniques. 3. Suggest appropriate means for bringing management methods to bear on these national problems.	1. Serve as vehicle for stimulating and keeping track of State and regional efforts to take systems approaches to public problems. 2. Hold seminars and conferences in those areas.
-------------------------	20 members appointed by the President, 5 from executive, 4 advisory members from Congress.	Chairman, Vice Chairman, and 11 other members.	Persons from industry, universities, research institutes, State and local government. Duration: 3 to 5 years
-------------------------	-------------------------	-------------------------	
-------------------------	-------------------------	1. Commission supported by studies contracted for by Secretary of Labor. 2. Commission assisted by a panel of experts to perform a "quality control" function.	

THE UTILIZATION OF SCIENTIFIC MANPOWER AND SOLICITED COMPROMISE VERSIONS

—State Local Finances Project; AIA—Aerospace Industries Association; RAC—Research Analysis Corp; Social and Technical Innovation]

An analysis of the proposed legislation, i.e., S. 430 and S. 467, shows that most witnesses favored a merger of the two approaches, thereby allowing a "national commission" to study existing uses of systems technology in correcting social and community problems and then provide financial resources both to State governments and approved private institutions which would be responsible for undertaking pilot projects in selected problem areas.

There was no consensus, however, about the composition nor tenure of such a commission; its placement within the Federal establishment also varied according to the witness. Assignments of grants on a regional (i.e., multi-State) or big city basis received some support. The selection of witnesses resulted in a well-balanced representation of the views of key groups in the public and private sectors, as charted below:

Hearing No.	Federal	State	University	Industry	Association	Foundation	Labor	Miscellaneous
1st (1965)		2	1	5	1		1	2
2d (1966)	5							
3d (1967)	3		3	3	1	2	1	
4th (1967)	1	2	3	2				
Total	9	4	7	10	2	2	2	2

Of the 38 witnesses actually appearing before the special subcommittee, 13 were from the Federal Government (either Congress or the executive branch) and 25 represented private sector groups. In addition to those persons testifying before the subcommittee, other individuals and groups submitted statements regarding the proposed legislation, background information on corporate experience in projects germane to the purpose of the hearings, and ancillary documentation. A matrix comparison of the critiques of the two bills appears below, containing the comments of witnesses Kimball, Harr, and Schon, Senators Nelson and Scott, and contributors—Dr. Selma Mushkin, Dr. Clive Whittenbury and Oliver Bryk—from interested external organizations. This comparison (see fig. 35) includes:

    1. Operational program:
        A: Purpose;
        B: Administering body;
        C: Function; and
    2. Study program:
        A: Purpose;
        B: Administering body;
        C: Function;
        D: Composition;
        E: Reports; and
        F: Other.

The full content of the hearings testimony and supporting material is contained in the 590 pages of the 2-volume documentation entitled "Scientific Manpower Utilization, 1965–66," and "Scientific Manpower Utilization, 1967."

VII. ANALYSIS OF QUESTIONNAIRE RESPONSES ON STATE AND LOCAL USE OF SYSTEMS TECHNOLOGY

A. Introduction With 1968 Subcommittee Letter and Questionnaire

In order to determine the extent to which systems technology was being used at State and local levels the Senate Special Subcommittee on the Utilization of Scientific Manpower determined to obtain certain types of information by sending out two questionnaires, in 1966 and 1968. The intent of these surveys was to supplement the information gathered during the four series of subcommittee hearings in the 89th and 90th Congresses (detailed in sec. VI). Recognizing the necessity of obtaining current information, the special subcommittee, with the assistance of the Science Policy Research Division of the Legislative Reference Service (Library of Congress), formulated and mailed questionnaires to the 50 States, 22 select cities, five regional development organizations, and three U.S. territories. The purpose of this section is to present and discuss the narrative and statistical information contained in the questionnaire responses. Included are brief reviews of other pertinent surveys and related materials.

The intent of the questionnaires was to assist the Nelson subcommittee in assessing the role of systems technology in State and local government. The initial survey, in 1966, was subsequently compiled and included as "additional information" in the volume covering the 1967 hearings.[1] Both the 1966 and 1968 questionnaires reflected several areas of interest to the special subcommittee. The questionnaires identified:

The planning and program areas in which the systems approach was being used.

The type of systems performer (in-house or out-of-house).

Source of funding support.

Function and position within the governmental structure.

Degree of benefit in terms of cost.

Possible approaches for new Federal legislation.

The 1968 questionnaire was expanded slightly to include information on the following:

The Federal agency or program that supported a systems activity.

The inclusion or termination of programs or activities.

Studies, reports or other publications resulting from systems activities.

The advantages and limitations of the use of systems technology.

In the accompanying letter to the questionnaire, the special subcommittee chairman, Senator Gaylord Nelson, expressed the import-

---

[1] Scientific Manpower Utilization, 1967. Op. cit., p. 362.

The author wishes to acknowledge the invaluable assistance of Mrs. Louise Giovane Becker, his research assistant, in preparing the draft manuscript for this section.

ance of understanding the problems facing State and local government in applying successfully systems technology as a new management tool. Senator Nelson's letter describing the nature and scope of the survey and a copy of the subcommittee's "Systems Analysis Questionnaire," 1968 follow:

— — — —

The special Subcommittee on Scientific Manpower Utilization of the Senate Labor and Public Welfare committee has conducted extensive hearings over the past 3 years on ways and means of encouraging the use of the systems approach toward solving problems facing State and local governments.

Late in 1966 we sent you a questionnaire designed to gather comprehensive data on the experience of States and large cities in applying these techniques. The results of the survey were printed in the Senate hearings and a brief summary is enclosed for your information.

In order to bring this information up to date prior to writing a final committee report and drafting legislation, a revised questionnaire is enclosed. The subcommittee is particularly interested in knowing in what instances the systems approach has proven not to be feasible and to learn of new areas of activity.

The range of problems facing State governments includes environmental pollution, transportation, welfare and education services and urban renewal. The subcommittee is attempting to find out to what extent the application of systems technology and automatic data processing is useful, what types of personnel are involved in analyzing the community problems and designing corrective systems, and what these efforts are costing the locality and the Nation.

We hope that someone on your staff will have time to complete the questionnaire and that it can be returned in time for us to prepare our report beginning the week of July 26.

Let me thank you for your further cooperation in this important effort.

Sincerely yours,

GAYLORD NELSON,
Chairman, Special Subcommittee on Scientific Manpower Utilization.

SYSTEMS ANALYSIS QUESTIONNAIRE

1968

I. A. Indicate whether a systems capability exists within your staff
(in-house) or whether you rely on outside consultants (out-of-
house):

| Management Activities | In-House | Out-of-House | | | 1/ |
		Non-Profit	Univ.	Industr.	Fed. Gov.
1. Planning & Policies	----	----	----	----	----
2. Personnel Management	----	----	----	----	----
3. Interagency Activity	----	----	----	----	----
4. Mgt. Standards & Controls	----	----	----	----	----
5. Equipment Selection	----	----	----	----	----
6. Procurement Activity	----	----	----	----	----

Operational Areas

7. Legislatures	----	----	----	----	----
8. Courts	----	----	----	----	----
9. Financial	----	----	----	----	----
10. Taxation	----	----	----	----	----
11. Education	----	----	----	----	----
12. Health and Hospitals	----	----	----	----	----
13. Crime and Corrections	----	----	----	----	----
14. Transportation	----	----	----	----	----
15. Urban Renewal & Growth	----	----	----	----	----
16. Science & Research Promotion	----	----	----	----	----
17. Natural Resources	----	----	----	----	----
18. Pollution Control	----	----	----	----	----
19. Parks & Recreation	----	----	----	----	----
20. Regulation of Commerce, etc.	----	----	----	----	----
21. Labor & Manpower Services	----	----	----	----	----
22. Utilities & Enterprises	----	----	----	----	----
23. Welfare & Anti-poverty	----	----	----	----	----
24. Social Secur. & Veterans	----	----	----	----	----
25. Other	----	----	----	----	----

Note: In answering questions IB, II, and V below, please use the numbers
found in question IA which denote specific management activities or
program areas; for example, #18 is for pollution control. A maximum
of 5 such areas may be indicated for each part of question II and V.

1/ Indicate Federal agency or programs from which support was obtained.
(Use abbreviations, i.e., HUD, DOT, HEW, etc.)

I.B. Have any programs or activities been terminated? (Indicate by number from question I. A..)

Reasons:

I.C. Have any new programs been added since 1966? Are any activities being considered for inclusion in the near future?

II. In which types of activities (as listed above) does your in-house capability have sufficient size and experience to:

A. Conceptualize systems requirements
and a technical approach   ----   ----   ----   ----   ----
B. Monitor the the work of a design
or implementing group   ----   ----   ----   ----   ----

C. Analyze current conditions and
design new approaches   ----   ----   ----   ----   ----
D. Implement new procedures and
techniques   ----   ----   ----   ----   ----
E. Evaluate innovative projects
and equipment use   ----   ----   ----   ----   ----
F. Initiate corrective action as
needed   ----   ----   ----   ----   ----

III. How has your in-house capability been institutionalized?

A. Special advisor for systems analysis or
operations research   ----
B. Line department for systems (or program)
development   ----
C. Assistant for Planning-Programming-Budgeting   ----
D. Computer programming group   ----
E. Automatic data processing facility   ----
F. Other   ----

IV. How much money has been spent for sophisticated systems development? (Indicate Fiscal Year (FY) or Calendar Year (CY).

                                  A. In-house   B. Out-of-house

   ---- 1965
   ---- 1966
   ---- 1967
   ---- 1968
   ---- 1969

V. Referring to the list of activities contained in I. (above), how useful have the systems approach innovations proven to be?

A. Measurable benefit   ---- ---- ---- ---- ---- ---- ----
B. Marginal benefit   ---- ---- ---- ---- ---- ---- ----
C. Too early to assess   ---- ---- ---- ---- ---- ---- ----
D. Negative value   ---- ---- ---- ---- ---- ---- ----
E. Unknown   ---- ---- ---- ---- ---- ---- ----

VI. Federal legislation should provide your government establishment with support through:

A. Consulting services   ----
B. Direct financial subsidy ----
C. Matching funds   ----
D. Training support   ----
E. Enabling legislation   ----
F. Other   ----

VII. Have any reports or studies been published in conjunction with your systems activities? Please give full citations or include copies if possible.

VIII. In your opinion have there been significant achievements or distinct limitations in utilizing the systems approach? Identify such advantages or shortcomings.

Over 75 percent of the jurisdictions who received them replied to Senator Nelson's second questionnaire. The responses received showed a level of activity ranging from high to virtually nothing. Significantly, the general attitude of the various types of groups responding was one of enthusiastic support regardless of the present level of development. There was an expressed desire to enter into greater cooperative efforts and to increase the capability to deal with a wider range of problems. The letters and questionnaire answers reflected a growth and sophistication in the use of systems technology by State and local organizations in the last 2 years. This will be discussed in greater detail in section VII. D.

Although the main intent of the questionnaire was to identify and explore the activity in applying systems technology in specific functional areas, the survey also was helpful in reflecting the changing attitude and current experiences of sub-Federal groups. In addition, the concern and interest of those participants who did not have any activity as well as those who had extensive experience indicated a requirement for a continuing examination of the application of systems analysis and ADP to social and community problems.

The recipients of the questionnaire often had slightly different interpretations of what is meant by "systems analysis," thus making it difficult to achieve complete uniformity in the categorization of responses. The development of the questionnaire analysis hinged on providing an adequate and relevant meaning to the term "systems analysis." Senator Nelson, in giving direction to the respondents in the earlier (1966) survey, had noted that: [2]

* * * Efforts at defining "systems approaches" have not been very successful for the term covers a broad range of related concepts and techniques, from the Defense Department's refinement of performance budgeting concepts to the use of computers for information handling. The subcommittee, however, is particularly interested in techniques for the analysis of problems facing government and the development of alternative policies toward their solution. We are interested in finding out to what extent such analyses now are being carried out at the State and local level, who is doing them, how effective they are in helping with real problems and what experience at the local and State level suggests as the best way to proceed from here.

The difficulty faced by respondents with regard to the terminology was reflected in those that identified their programs in a broad sense with the systems approach and those that applied a restricting definition to their activities (for example, use of ADP only). Therefore, some responses indicated the use of systems analysis only in the context of implementing their automatic data processing activities while others viewed the concept as being fully integrated into the high level decisionmaking process. The handling of the questionnaire responses has required a subjective analysis of these differences so that the opinions expressed can be fully understood and reflected accurately in the statistical aspects of this study.

Two recent studies, discussed below, presented additional information on the significant role that new methods and technology have played in the management and operation of State governments. One of these reflects the increasingly widespread use of ADP and the other highlights the development and implementation of planning, programing, and budgeting systems.

---

[2] Ibid.

Prior to the development of its own questionnaire, the special sub-committee examined an outstanding survey that was conducted jointly by the Council of State Government's Committee on Information Systems and the Public Administration Service.[3] The published report, "Automated Data Processing in State Government," was issued in 1965 and focuses on the historical, legislative, and procedural aspects of ADP in State governments. A follow-up survey, conducted in 1966, concentrated on facts about equipment applications, personnel, and expenditures.[4] This second report, entitled "Automation in State Government 1966–1967," compiled the results from a second questionnaire and will be discussed below.

A companion piece to these two studies is a study on "Automated Data Processing in Municipal Government." This report discusses the role of ADP in the cities,[5] and was based in part on a survey taken by the Public Administration Service in collaboration with the International City Managers, Association. Some of the information in the report was described by Henry Willis in the 1965 "Municipal Year Book of the International City Managers, Association." The study outlined:

> Development of data processing in government and industry.
> ADP applications in the modern municipality.
> Use of computers in cities.
> Conducting feasibility and systems studies.
> Organization for ADP.
> Equipment selection.
> ADP and city personnel.
> Guidelines for installing ADP.

In the 1966 survey, discussed in the second report, it was noted that the 33 States responding to the questionnaire represented the more densely populated areas of the Nation. Generally the larger and more populated States have tended to utilize ADP to a greater extent. There seems to be some evidence that there is a distinct relationship between population density, land area, and State organization complexity and the use of modern management tools and techniques.[6]

The report noted that there has been an increase in the use of outside management consultants. The consultants have greatly aided the States by providing the trained and sophisticated manpower needed in utilizing the new "Third generation" computer systems. In addition, there has been a move on the part of some States to develop better intergovernmental communication; examples of such State-Federal, State-local, or State-to-State activity appear in sections IV and VIII. Along with this trend is the development of Statewide information systems which combine sharing of computer facilities, the pooling of technical capabilities, and the development of intragovernmental coordination mechanisms.[7]

As efforts were made to achieve better utilization of ADP resources, a discussion of the effective use of centralized services was featured

[3] The Council of State Governments and Public Administration Service. Automated data processing in state government. Chicago, Ill., Public Administration Service, 1965. 40 p.
[4] The Council of State Governments and Public Administration Service. Automation in State government 1966–1967: A second report on status and trends. Chicago, Ill., Public Administration Service, 1967. 38 p. Hereafter referred to as automation in State governments, 1966–1967.
[5] Public Administration Service and International City Managers' Association. Automated data processing in municipal government. Chicago, Ill., Public Administration Service, 1966. (34 p.)
[6] "Automation in State government, 1966–1967." (Op. cit., p. 1.)
[7] Ibid., p. 6.

in the report. The obvious considerations that must be given to co-ordination and planning of shared resources were discussed briefly.[8] Not only had the number of computers in use by the States increased noticeably in the period from 1964 to 1966, but there was a related significant increase in the annual outlay for equipment, manpower and contractual services.[9] Some of the facets of systems development identified in the report on ADP in State government were similar to those revealed as a result of the Nelson subcommittee's questionnaires and hearings.

In the preparation of this report for the special subcommittee, a recent survey on the implementation of PPB[10] at the State govern-ment level also was studied. This survey was part of the State-local finances project at The George Washington University.[11] The question-naire, sent out in the fall of 1968, attempted to appraise the scope and extent to which State governments had implemented PPB. States were queried on:

1. The level of activity;
2. The extent that consultants were used; and
3. The use of universities for the training of key State personnel.

The questionnaires were disseminated to known officers of the States and designated grant-in-aid coordinators.

A preliminary report on the survey of PPB implementation in the State governments was provided by Dr. Selma Mushkin, former director of the project. As of February 1969, 26 of the 39 States surveyed replied that they had begun systems implementation while six more noted that they were considering such action. All respondents indicated that orientation or training programs were being considered for their staffs.[12]

1. Number of States considering implementation

Thirty-two of the 39 States replying to the inquiries responded "yes" to the question: Is your State considering the introduction of a PPB system? None of the States reported consideration and rejection of a PPB system. Several States, however, reported a trial beginning effort with the decision on full-scale implementation awaiting results of the trial.

2. Number beginning implementation

Of the 32 States considering the introduction of PPB, 26 or over half of the 50 States indicated that they had taken at least beginning steps toward implementation. Of these 26, two States (Illinois and Kentucky) have indicated that only partial steps are underway. In addition, there are four other States in which a possible future commitment to PPB implementation is reported.

3. Training

Numbers of staff who had had some orientation or training on PPB systems varied widely. Almost all of the 32 States that reported consideration of PPB for State adoption indicated that some staff had participated in an orientation or training program.

---

[8] Ibid., p. 7.
[9] Ibid., p. 17.
[10] See sec. II.E. for a more detailed discussion of PPBS.
[11] Mushkin, Selma J. PPB systems in state, county and city: A report on the 5-5-5 Demonstration. Wash-ington, D.C.: The State-local finances project of The George Washington University, ch. 3 (unpublished).
[12] Ibid.

Respondents to the two Nelson questionnaires, the Council of State Governments' survey of the use of ADP, and the GWU State-Local Finances project survey of PPB activity are listed in figure 36, "A Summary of Recent Surveys of State Governments." The following States indicated some activity in all three categories, i.e., planning-programing-budgeting, systems analysis, and the use of ADP:

California	Oregon
Hawaii	Pennsylvania
Iowa	Vermont
Maryland	Washington
Massachusetts	West Virginia
Missouri	Wisconsin
New York	

The responses to the Nelson questionnaires and the surveys on ADP and PPB revealed the concern and interest at all levels of government in the utilization of modern management tools and techniques. The problems confronting society and the limited resources available have stimulated the State and local governments to examine, experiment with, and apply operationally the new techniques. Significantly, three-fourths of the participants in the Nelson survey expressed the intent to move ahead in making better use of today's new methods.

The survey for the special subcommittee indicated the wide range of systems activity presently being engaged in at the State and local levels. The queried governmental elements expressed the desire to enter into more meaningful dialogue with those who have been successful in using the systems approach. The 1966 and 1968 questionnaires, reviewed in this section, serve to impart a better understanding of the impact of systems technology on State and local governments.

State	State-local finance project preliminary survey on planning-programing-budgeting activity [1]	Nelson subcommittee questionnaires on systems analysis and data processing services		Council of State Governments survey on automatic data processing [3]
		1966	1968 [2]	
Alaska		1966	1968	ADP
Arkasnas	PPB	----	1968	
California	PPB	----	1968	ADP
Colorado				ADP
Connecticut	PPB	1966	1968	
Delaware		1966	----	
Florida	PPB	1966	----	
Georgia	PPB	----	----	ADP
Hawaii	PPB	1966	----	ADP
Idaho		----	1968	
Illinois		1966	1968	ADP
Indiana	PPB	----	----	ADP
Iowa	PPB	1966	1968	ADP
Kansas		----	1968	ADP
Kentucky		1966	----	ADP
Maine		----	1968	ADP
Maryland	PPB	1966	1968	ADP
Massachusetts	PPB	1966	1968	ADP
Michigan	PPB	----	----	ADP
Minnesota				ADP
Missouri	PPB	1966	1968	ADP
Nevada			1968	
New Hampshire				ADP
New Jersey	PPB			
New Mexico	PPB	----	1968	ADP
New York	PPB	1966	1968	ADP
North Carolina	PPB	----	1968	
North Dakota			1968	ADP
Ohio		1966	1968	ADP
Oklahoma		1966		ADP
Oregon	PPB		1968	ADP
Pennsylvania	PPB	1966	1968	ADP
Rhode Island	PPB	1966	1968	
South Carolina			1968	
South Dakota		1966	1968	ADP
Tennessee				ADP
Texas		1966	1968	ADP
Utah	PPB	1966	1968	
Vermont	PPB		1968	ADP
Virginia				ADP
Washington	PPB	1966	1968	ADP
West Virginia	PPB	1966	1968	ADP
Wisconsin	PPB	1966	1968	ADP
Wyoming		1966		ADP

[1] Mushkin, Selma J., project director, "PPB Systems in State, County, and City: A Report on the 5–5–5 Demonstration," Washington, D.C., the State-local finances project of the George Washington University (in process).
[2] Senate Special Subcommittee on the Utilization of Scientific Manpower questionnaires in 1966 and 1968.
[3] Council of State Governments and Public Administration Service. Automated data processing in State government: Status, problems, and prospects, Chicago, Ill., Public Administration Service, 1965, 40 pages.

FIG. 36.—SUMMARY OF RECENT SURVEYS OF STATE GOVERNMENTS

B. Discussion of Responses With Selected Statistics

The responses to the special subcommittee questionnaire reflect the interest of both State and local governments in the use of systems technology. Of the 80 governmental elements surveyed, 58 responded to the 1968 questionnaire; over 75 percent of those responding indicated that they were engaged in some level of systems analysis activity. The range of responses varied from preliminary consideration being given to the implementation of a single ADP application to multiple systems designed to assist in high level decisionmaking. Although the intent of the questionnaire was to determine the extent that systems analysis was being used, the scope was modified due to the inclusion of other ADP services by the respondents. In reviewing the responses the subcommittee recognized the growing trend at all levels of government to make far greater use of modern management methods and tools. The statistics contained herein are presented to develop understanding of the current situation.

In response to question I(a),[13] which requested information on the type of operational or management activity currently being engaged in, the participants indicated the application or program, the type of in-house or out-of-house staffing, and if support was received from not-for-profit, university, industry, or Federal sources. The results of question I are summarized in figure 37, "Indications of Systems Capability by Function (1968)." The chart includes only those participants reporting more than one application or program area of systems activity.

Generally, it was indicated that in-house staffs performed without external support in more than half of the identified activities. Consultants and in-house staffs frequently functioned in combination. Few States and cities relied on outside consultants for their systems effort. For the most part, consultants were used with in-house staff support when only a few applications were being considered. This would seem to indicate that those reporting participants in the early stages of development frequently tend to rely on the use of outside consultants. As more applications are undertaken, in-house capability is often expanded and augmented. In brief, those State and local respondents that indicated many applications or program areas tend to rely on their in-house staffing and the use of consultants in developing their improved systems capabilities.

More than 50 percent of the active respondents indicated that they used systems analysis in more than the six management areas noted in figure 37. The following nonmanagement areas were cited most frequently by the participants as areas of activity:

Financial, Taxation, Education, Health and hospitals, Crime and corrections, Transportation, and Welfare.

---

[13] Questions indicated in this section refer to "Systems Analysis Questionnaire." (See sec. VII.A.)

Column headers:

(1) Planning and policies
(2) Personnel management
(3) Interagency activity
(4) Management standards control
(5) Equipment selection
(6) Procurement activity
(7) Legislatures
(8) Courts
(9) Financial
(10) Taxation
(11) Education
(12) Health and hospitals
(13) Crime and corrections
(14) Transportation
(15) Urban renewal and growth
(16) Science and research promotion
(17) Natural resources
(18) Pollution control
(19) Parks and recreation
(20) Regulation of commerce, etc.
(21) Labor and manpower services
(22) Utilities and enterprises
(23) Welfare and anti-poverty
(24) Social security and veterans
(25) Other

STATES	(1)	(2)	(3)	(4)	(5)	(6)	(7)	(8)	(9)	(10)	(11)	(12)	(13)	(14)	(15)	(16)	(17)	(18)	(19)	(20)	(21)	(22)	(23)	(24)	(25)
Alaska	X	X	X	X	X	X			X	X	Y	Y	Y	Y			Y		Y		Y		Y	X	
Arkansas	Y	X	X	Y	Y	Y			X	X	Y	Y	Y	X				Y	Y	Y	Y	Y	X	Y	X
California	Y	Y	Y	Y	Y	Y			Y	Y	Y	Y	Y	Y	Y	Y	Y	Y	X	X		X	Y		
Colorado	Y	Y	Y	Y	X	X		X	Y	Y	X	X	X	X	X	Y	Y		X	X			X	X	X
Connecticut	X	X	X	X	X	X	X		X	X	X	X	X	Y	X	X	Y	X	X	X	X		X		
Idaho	N	Y	Y	Y	X	X	X		X	X	X	X		Y	X		Y		X		Y		X		
Illinois	X	X	X	X	X	Y	Y		X	X	Y	Y	Y	X	X	X	Y				Y	X	X		
Indiana	X	X	X	X	X				X	X	Y	X	X	X	X	X		X	Y		N	Y	Y	N	
Iowa	Y	Y	Y	Y	X	X	Y		Y	Y	Y	N	N	N	N	N	Y	X	N	N	Y		Y	Y	Y
Kansas	Y	Y	Y	Y	X	Y			Y	Y	N	Y	Y	Y	N	N	Y	Y		X	Y	N	X	Y	
Maine					X	Y	X	N	N	X	Y	Y	Y	Y	Y	N	Y	N	Y	X	X		X	X	
Maryland			N	X	Y		N	Z	X	N	Y	Y	Y	X	Z	Z	Y	X	Y	X	Y	Z	N	Z	
Massachusetts		X	X	N	X		X	X	X	Z	X	Y	Y	Y	Y	Y	Y	Y	X	X	Y		Y	Y	
Missouri	X	X	X	N	X	N	Y	X	X	Y	X	Y	N	X	Y	N	Y	N	Y	X	X	X	X	N	Y
Nebraska	X	X	X	N	N	Z	Z		X	Z	X	Y	X	X	X	X	X	Z	X	X	X	X	X	X	
New Mexico	Y	X	X	Y	X	Y	X	N	X	X	Y	Y	X	X	Y	Y	X	N	Y	X	X	X	X	X	Y
New York	Y	X	X	X	Y	X	X	X	X	Y	X	Y	X	Y	Y	X	X	Y	X	X	X	X	X	X	
North Carolina	X	X	X	X	Y	X	X		X	X	X	X	X	Y	Y	Y	X	Y	X	X	X	X	Y	X	
Ohio	X	X	X	X	X	X	X	X	Y	X	X	X	Y	X	Y	Y	Y	Y	X	X	Y	X	Y	X	
Oregon	X	X	X	Y	X	X	X		X	X	X	X	X	Y	Y	X	Y	Y	X	Y	X	X	Y	X	
Pennsylvania	Y	Y	Y	Y	Y	X	X	Y	X	Y	Y	Y	Y	Y	Y	Y	Y	Y	X	Y	X	Y	Y	X	
Rhode Island	Y	N	Y	Y	X	X	X		X	X	Y	X	X	X	Y	N	X	Y	X	X	Y	Y	Y		X
South Carolina	Y	Y	Y	X	Y	X	Y		X	Y	Y	X	X	Y		Y	N	Y	X	X	Y	Y	Y	N	
South Dakota					X	X			X	Y			X					Y		Y	Y		Y		
Texas	Y	Y	Y	Y	X	X	N	Y	X	N	X	X	Y	X	Y	N	Y	Y	Y	Y	Y	Y	X	Y	
Utah	Y	X	Y	X	X	X	Z	Y	X	Z	X	X	X	X	Y	Z	X	Y	Y		X	Y	X	X	
Vermont	Y	X	Y	Y	X	X	Y		X	Y	X	X	X	X	Y	Y	X	Y	Y		X	Y	X	X	
Washington	X	X	X		X	X	Y	X	X	X	N	N	Y	X	N	N	N	Y		N	N	N	X	X	X
West Virginia	Y	X	X	X	Y	X	Y		X	X	Y	X	Y	X	N	X	X	Y	Y	Z	X	N	X	X	
Wisconsin	X	X	X	X	X	X	X	X	X	X	Y	X	X	Y	X	X	X	X	X	N	X	N	Y	X	X

CITIES																									
Atlanta, Ga	X	X	X	X	X	X	----	X	X	X	----	----	X	X	X	----	X	X	----	X	X	X	----	----	
Buffalo, N.Y	----	----	X	----	----	----	----	----	X	X	----	----	Y	----	X	----	----	----	----	----	----	----	----	----	
Chicago, Ill	Y	Y	X	Y	X	X	----	----	Y	X	Y	Y	Y	Y	Y	----	----	Y	Y	X	Y	X	Y	----	
Cincinnati, Ohio	X	X	X	X	Z	X	X	X	Z	X	X	X	Z	X	X	X	X	X	X	X	X	Y	X	----	
Cleveland, Ohio	Y	Z	Y	----	----	----	----	----	Y	Y	Y	Y	----	Z	----	----	X	----	X	----	Z	X	Z	----	
Denver, Colo	X	X	X	X	X	X	----	X	X	X	----	X	X	X	X	----	----	X	X	----	----	----	X	----	
Detroit, Mich	Y	X	----	Y	----	----	----	----	X	X	----	----	----	----	X	----	----	----	----	----	----	----	----	----	
Houston, Tex	X	----	X	----	----	----	----	X	----	X	----	X	----	----	X	----	----	----	----	----	----	X	----	----	
Kansas City, Mo	----	X	----	----	----	----	----	Z	X	X	----	----	----	----	Y	----	----	----	----	----	----	Y	----	----	
Los Angeles, Calif	X	X	X	Y	X	X	X	----	X	X	----	----	Y	Y[3]	Y	----	----	Y[4]	Y	X	X	X	X	X	Y[5]
New Orleans, La	----	X	----	----	----	----	----	X	X	X	----	----	X	----	X	----	----	----	----	----	X	----	----	----	
New York, N.Y	Y	X	Y	Y	X	X	----	Y	Y	Y	Y	Y	Y	Y	Y	X	----	Y	----	----	----	----	Y	----	Y
Philadelphia, Pa	Y	X	Y	X	X	X	X	X	Y	X	X	X	Y	Y	Y	X	----	Y	X	X	X	X	X	----	
Phoenix, Ariz	X	----	----	X	X	----	----	X	----	----	----	X	----	----	----	----	----	----	----	----	----	X	X	----	
San Diego, Calif	X	X	Y	X	Y	X	----	----	X	----	----	----	X	X	X	----	----	X	X	----	----	X	----	----	X
San Francisco, Calif	X	----	----	----	----	----	----	----	----	----	----	----	----	----	X	----	----	----	----	----	----	----	----	----	
REGIONAL DEVELOPMENT ORGANIZATIONS AND U.S. TERRITORIES																									
Port Authority of New York	Y	Y	X	X	X	X	----	----	X	----	----	----	X	Y	Y	Y		X	----	----	X	----	X	----	Y
Guam	X	Y	----	Y	X	X	X	X	Y	Y	Y	Y	Y	Y	Y	----	Y	Y	----	X	Y	Z	Y	Z	
Puerto Rico	Y	----	----	----	----	----	----	----	----	----	Y	Z	----	Y	Y	----	Y	Y	Y	----	----	Y	----	----	

[1] Motor vehicle, liquor accounting and inventory.
[2] Liquor control board.
[3] Limited application at city level, nonprofit Southern California Transit District.
[4] Limited.
[5] Other: Sanitation, fire prevention, library services.

FIG. 37.—INDICATIONS OF SYSTEMS ANALYSIS CAPABILITY BY FUNCTION, 1968

[X—In-house capability; Y—out-of-house and in-house capability; Z—out-of-house capability]

Responses to question I(a) are also reflected, in part, in figure 38, "Federal Agencies Support of Systems Analysis Activity." More than half of the active participants indicated that support for their programs was obtained from Federal Government sources. While the majority of respondents identified the specific source of the support, six respondents noted that Federal support was given for a distinct function but otherwise failed to designate the specific agency or department. Twenty-two of the active respondents did not state if they received any support from the Federal Government.

The chief Federal agencies identified as rendering support are:

Department of Housing and Urban Development (HUD);
Department of Health, Education, and Welfare (HEW);
Office of Economic Opportunity (OEO);
Department of Transportation (DOT);
Department of Justice (DOJ);
Department of Interior (Int.);
Department of Labor (DOL); and
Department of Commerce (DOC).

In a few instances the National Science Foundation (NSF), Department of Defense (DOD), Veterans' Administration (VA), and the National Institute of Mental Health (NIMH) were designated as also having given support.

State	(1) Planning and policies	(2) Personnel management	(3) Interagency activity	(4) Management standards control	(5) Equipment selection	(6) Procurement activity	(7) Legislatures	(8) Courts	(9) Financial	(10) Taxation	(11) Education	(12) Health and hospitals	(13) Crime and corrections	(14) Transportation	(15) Urban renewal and growth	(16) Science and research promotion	(17) Natural resources	(18) Pollution control	(19) Parks and recreation	(20) Regulation of commerce, etc.	(21) Labor and manpower services	(22) Utilities and enterprises	(23) Welfare and anti-poverty	(24) Social security and veterans	(25) Other
Alaska	HUD			HUD							HEW	HEW	DOJ	DOT			Int.		Int.		DOL		HEW		
Arkansas											HEW	(¹)(²)	(¹)	DOT											
California																									
Colorado																									
Connecticut³																									
Idaho³																									
Illinois³																									
Indiana³																									
Iowa																									
Kansas																									
Maine³																									
Maryland³																									
Massachusetts	HUD		HUD						OEO		HEW	HEW		DOT	(⁶)	(⁶)				(²)	DOL		HEW		
Missouri	(⁵)		HUD								HEW	HEW		DOT	(²)				(²)		DOL	HUD	HEW		
Nebraska					DOL						HEW	HEW													
New Mexico³⁸														DOT											
New York³														DOT											
North Carolina																									
Ohio																									
Oregon			HUD	(³)											(¹⁰)	DOC	(⁶)	Int.		DOC	(²)		HEW		
Pennsylvania³	(²) HUD		OEO						HEW		HEW	HEW	DOJ	DOT	HUD	NSF	HEW	Int.	Int.	DOC	DOL		HEW		
Rhode Island																	Int.				DOL				
South Carolina	HEW HUD						HUD	HEW	HEW	DOJ	HEW HEW	HEW HEW (²)	DOJ DOJ (²)	(¹⁴)	HUD HUD		Int.		Int.	DOC	DOL DOL	DOC	HEW	(¹³)	
South Dakota³													HUD										(²)		
Texas	BOB																								
Utah																									
Vermont																									
Washington³																									
West Virginia³																									
Wisconsin³																									

See footnotes at end of table, p. 174.

CITIES	(1) Planning and policies	(2) Personnel management	(3) Interagency activity	(4) Management standards control	(5) Equipment selection	(6) Procurement activity	(7) Legislatures	(8) Courts	(9) Financial	(10) Taxation	(11) Education	(12) Health and hospitals	(13) Crime and corrections	(14) Transportation	(15) Urban renewal and growth	(16) Science and research promotion	(17) Natural resources	(18) Pollution control	(19) Parks and recreation	(20) Regulation of commerce, etc.	(21) Labor and manpower services	(22) Utilities and enterprises	(23) Welfare and anti-poverty	(24) Social security and veterans	(25) Other
Atlanta[3]																									
Buffalo[3]																									
Cincinnati																									
Chicago	(2)		(2)								(2)	(2)	(2)	DOT (2)	HUD			HEW			(2)		OEO		
Cleveland																									
Dallas[3]																									
Denver[3]																									
Detroit	(16)			(15)																					
Kansas City[3]																									
Los Angeles[3]																									
New Orleans[3]																									
New York City													(10)	(2)	HUD								HEW		
Philadelphia															HUD			HEW							
Phoenix[3]																									
San Diego[3]																									
San Francisco[3]																									

1 NIMH.
2 Did not specify source.
3 Did not indicate Federal support.
4 HEW, OEO, HUD, DOT.
5 HEW, HUD.
6 HEW, NSF, HUD.
7 HUD, Int. (Bureau of Reclamation).
8 HUD will support PPBS.

9 HUD, DOD, DOT, EDA.
10 HUD, DOT.
11 Int. (Federal Water Pollution Control Adm.).
12 HEW, OEO.
13 HEW, V.A.
14 DOT, Interior.
15 HUD, OEO.

FIG. 38.—FEDERAL AGENCIES SUPPORT OF SYSTEM ANALYSIS ACTIVITY (STATES AND CITIES) BY FUNCTION

175

The level of activity varied greatly between the States and cities. Of 46 active respondents, 22 percent of the States and 6 percent of the cities reported that they were applying systems analysis in more than 20 program areas. State governments, for the most part, were able to participate in a wide range of application areas while the cities, because of their jurisdictional limitations, often did not initiate activity in as many applications. Figure 39, "Number of States and Cities Engaged in Systems Analysis Efforts by Level of Activity (1968)," shows the contrast between the two types of governing bodies. The States are concentrated at the high or moderately high level of activity while the majority of the cities are clustered in the charted area of less than 15 applications.

Level of activity	Number of States	Number of cities
High level (20 to more than 25 program areas)	11	3
Moderately high level (15 to 19)	9	4
Moderate level (10 to 14)	7	2
Low level (1 to 9)	3	7

FIG. 39.—NUMBER OF STATES AND CITIES ENGAGED IN SYSTEMS ANALYSIS EFFORTS BY LEVEL OF ACTIVITY, 1968

Since most participants were in the very early stages of implementing the systems approach or ADP services, the responses to question I(b) which requested information on the termination of programs or projects, were for the most part negative. Thirty respondents indicated that they had not terminated any programs. The termination of a program was often the result of a completed program that required no further activity. In a few instances, programs were terminated due to the elimination or withdrawal of funds or support. In two cases, the withdrawal of Federal assistance terminated ongoing programs. For example, the lack of continued support by HUD required the city of Denver to terminate its metro data bank center.

Question I(c) requested participants to indicate whether or not new programs had been added since 1966. Besides those listed in figure 37, the questionnaire participants indicated several additional areas of systems analysis or ADP activity, including—

1. Veterans loans.
2. Vocational rehabilitation.
3. Financial information systems.
4. Educational data systems and educational television systems.
5. Procurement and support services (i.e., mental health area).
6. Pollution control.
7. Urban data centers.

Contemplated for inclusion in the near future were these types of activities: tax commission, natural resources, municipal courts, regional data centers, model cities programs, law enforcement, and natural resources.

The questionnaire also focused on the placement within the governmental structure of the in-house systems staff. This information was useful in comprehending more accurately the type of effort being implemented. Both cities and States, in response to question III, indicated similar in-house organizational structures. A summary of the responses to question III appears in figure 40, "Organization of In-House Capability." A large segment of the respondents indicated that the in-house capability generally was concentrated in "ADP services" groups.

	States	Cities
A. Special adviser for systems analysis or operations research	11	7
B. Line department for systems (or program) development	18	8
C. Assistant for planning-programing-budgeting	14	5
D. Computer programing group	24	13
E. Automatic data processing facility	22	15

[1] Responses to question III of the 1968 Nelson subcommittee questionnaire.

FIG. 40.—ORGANIZATION OF IN-HOUSE CAPABILITY [1]

For the period 1965 to 1969 the survey participants reported spending a total of $82 million for systems analysis and ADP services. Over one-third of this amount was reportedly spent for out-of-house consulting services. In responding to question IV, regarding expenditures, the figures often were "estimated," "approximated," or "projected" budgetary requests; therefore, the reported total of $82 million can only be considered a general indicator of what has been spent on implementing systems technology in State and local government.

Figure 41 prepared from the responses to question IV, lists selected cities' and States' expenditures for the years 1965 to 1969. The respondents represent a wide variation in spending for this 5-year period. In examining the questionnaires, there was a not unexpected increase in the amount of spending as new program areas and applications were added. Few governments reported a decrease in funding or a cessation of individual program support for the period 1965 to 1969. Since some of the respondents' expenditures include support of projects only peripheral to systems analysis, the figures should be viewed as at least representative of the general trend of State and local expenditures for systems analysis and automatic data processing.

State	Number of years funded	Total	Cities	Number of years funded	Total
Alaska	3	$915,000	Chicago	3	$1,650,000
Arkansas	1	60,000	Cincinnati	4	3,370,000
California	2	1,881,000	Cleveland	2	5,200,000
Colorado	4	669,581	Denver	4	212,000
Connecticut	5	1,800,000	Detroit	3	1,141,602
Idaho	3	138,000	Kansas	3	205,000
Kansas	3	1,100,000	Los Angeles	5	4,230,718
Maryland	5	4,544,201	New Orleans		352,000
Massachusetts	5	1,450,000	New York	1	4,500,000
Missouri	3	1,565,000	Philadelphia	3	4,600,000
New York	5	[2] 28,700,000	San Diego	1	225,000
Ohio	5	4,700,000	San Francisco	4	314,000
Rhode Island	4	925,000			
South Dakota	4	[3] 1,530,000			
Texas	4	[2] 3,200,000			
Washington	2	2,830,000			
Wisconsin	4	[4] 677,622			

[1] As reported to subcommittee in question IV of systems analysis questionnaire.
[2] Estimated.
[3] Data processing.
[4] Central systems staff only.

FIG. 41.—TOTAL FUNDS EXPENDED, 1965-69 [1]

Relevant to the special subcommittee's line of investigation is the role that the Federal Government should assume in assisting State and local governments to examine current and potential uses of systems technology. The participating governmental elements were requested, in question VI of the survey, to designate the type of support Federal legislation should provide and in most cases indicated one or more types of support desired. The responses to this key question are summarized in figure 42, "Federal Legislation to Assist in Implementing Systems Indicated by Respondents."

	State	Cities
A. Consulting services	13	6
B. Direct financial subsidy	17	12
C. Matching funds	12	9
D. Training support	19	6
E. Enabling legislation	3	4

FIG. 42.—FEDERAL LEGISLATION TO ASSIST IN IMPLEMENTING SYSTEMS INDICATED BY RESPONDENTS [1]

[1] Responses to question IV of the 1968 special subcommittee questionnaire

The majority of the respondents indicated that direct financial subsidy was desired. Nineteen States indicated that training support would assist them in implementing the systems approach in key problem areas. Many letters accompanying the questionnaire forms placed special stress on the need to train and retrain State and local government personnel. The States also indicated an eagerness to make better use of Federal systems capability by more extensive use of consulting services. There was general agreement that the Federal Government, with its wide range of experience, could be of help both to State and local government in utilizing systems technology.

C. Excerpts From Responses With Summary Comments

It was evident from the responses to the special subcommittee's questionnaires and the accompanying letters that State and local government administrators were actively seeking ways to utilize the new management tools and techniques. The narrative and statistical data showed that systems technology was being used at various levels of government to help solve a wide range of problems. Confidence was expressed that systems analysis could help provide better services to an expanding population. Respondents voiced concern that dramatic steps were needed to find adequate solutions to the social and administrative problems facing our communities. The States and local government, with their limited resources, were anxious to develop the possibility of improving services through the use of systems technology.

Since 1966, when the special subcommittee dispatched its first questionnaire, there has been an increase in the number of applications and problem areas in which systems analysis is being used. The reported success of systems analysis in both the private and public sectors has further stimulated interest in utilizing this new technique. Although some respondents felt that they were using at best a "quasi-systems approach" to solve some of their urgent problems, many expressed a genuine interest in learning more about applying systems analysis.

In examining the respondents' letters it was found that there were several problem areas common to numerous responding governmental elements. One of these problems was the difficulty faced in implementing systems techniques from outmoded organizational structures. Many States and local groups also told of difficulties encountered due to bureaucratic reluctance to accept changes, and organizational policy and procedural patterns that made it cumbersome to work out adequate solutions.

The participants listed manpower problems, organizational inadequacies, and lack of funds as being some of the chief difficulties faced in attempting to implement a systems approach to varied problems. This view is summarized, in part, by the assistant to the city manager of Dallas, Tex., Mr. James R. Favour:

* * * The principal factor which impedes the establishment of a good municipal systems program is not the lack of desire for it but rather the lack of available qualified systems personnel. We think that this can only be overcome by better municipal salaries, and through inservice training program for existing personnel in the use of computers and systems technique.

The need to obtain training for municipal and State employees in new management techniques was reflected in several letters, as well as in the structured responses to question V (also see sec. VII. B.). Many survey participants indicated the need to train systems analysts who then would be essential in the future development of implementing adequate programs at the local or State level.

A number of respondents noted that they were actively seeking to remedy the lack of trained manpower. Mr. Patrick Brady, budget and systems officer for Kansas City, Mo., expressed a desire to find a solution to the problem:

Close analysis of operations in all departments or areas of specialization has repeatedly turned up a common basic underlying problem: the majority of department heads and subordinate supervisors are certainly specialists in their areas, but lack general administrative or management training (covert training in those responsibilities common to virtually all supervisory or management situations.)

Mr. Brady then goes on to elaborate:

Our intended answer to this general administrative problem is based on two proposed levels of concurrent and intensive training and personnel development programs. The first level of training will be designed to stimulate an all-pervading awareness of continuing management responsibilities; that is, planning, coordination, control and so forth. The second level of training will be designated to provide selected staff units and personnel some of the tools of the advanced management sciences necessary for evaluating alternative policies or programs.

Several other responding officials also reiterated the need to promote education and training in order to meet some of the manpower problems. Some localities have made limited advances in meeting educational needs. Mr. Edward J. Martin, director of finance for the city of Philadelphia highlighted some of the efforts being made to train management level personnel in understanding and utilizing the new tools and techniques:

* * * we have made enormous strides in the last two years. We have conducted a series of courses for high level management personnel in such areas as computer sciences and in operation research techniques. The latter course was not intended to make operations research specialists out of management personnel, unless the individual was so motivated. Our major objective here was to acquaint management with the role of operations research in local government.

Attention was brought to a number of key areas that were being considered for future implementation. Many respondents recognized the need to expand and delineate other areas where systems analysis could be used. In New York State, T. N. Hurd, director of the budget, noted that:

* * * virtually every New York State agency has been provided with an internal systems analysis staff as well as computer systems capabilities through either an internal computer system or the State's central computer service. This staff is assisted on certain projects by the State's central management and systems group, which is located in the division of the budget. On other occasions, consultants are employed to provide specialized skills or services not readily available on the State's staff.

The need to incorporate systems methodology and ADP into new areas and revitalize the approaches to already established functions has been one of the major tasks of State and local governments. Many jurisdictions indicated the use of consultants to supplement and expand their present systems analysis capability. Another series of comments indicated that the use of new management techniques served to enhance government's capability to serve the public better.

Although questions regarding planning, programing, and budgeting [14] were not featured in the Nelson subcommittee's questionnaires, it is interesting to note that the inter-relationship between PPB and systems analysis often was discussed by choice by the respondents. For example, Governor Tom McCall's of Oregon letter to Senator Gaylord Nelson commented that PPB was being implemented by the budget division of the Department of Finance. Another State to establish a PPB system has been the State of Arkansas, with support in part by a grant from HUD and other money from private sources. Initial plans provide the Governor with an analytical staff.

In general, respondents reflected a growing trend towards a greater reliance on the use of modern management tools and techniques, including PPBS, operations research, a few basic models, and ADP. The Governor of Guam, Manuel Gerrero, told of the experience of that territory:

We are, however, becoming more knowledgeable in the "systems approach" toward problem-solving and decisionmaking. We have within the Governor's office a bureau of management research whose responsibility is to review from time-to-time the functional requirements and operations of the Government, and to recommend methods for improvements in operation and services. We recently began using automatic data processing equipment for payroll purposes, and we hope to utilize such equipment in other program areas.

Certain discerning State and local authorities have noted the need for management to receive training that will allow the subject group to utilize systems analysis effectively. The importance of management perception of the role of systems technology is summed up by Edward J. Martin, director of finance (Philadelphia, Pa.):

Our general philsophy and policy in the systems area has matured somewhat since 1966. Today, we are often more aware that the systems approach is not only desirable, it is absolutely imperative if local government is to survive in today's crisis situations. We further recognize that this approach can best be implemented through an interdisciplinary team approach which crosses agency lines and makes maximum use of the systems design team concept. The achievements of the present and the future, we feel, are not likely to be the products of

---

14 Planning, Programing and Budgeting (PPB): See sec. II for a detailed description. Also, see app. L, "Program Budgeting in Wisconsin," for a more concise description of an operational system.

any one person's individual insights. We feel that no single science, or discipline, possess [sic] all the techniques necessary to resolve the problems of today.

The excerpts from the letters accompanying the special subcommittee's 1966 and 1968 "Systems Analysis Questionnaire(s)" are selections which highlight selected issues and developments relevant to some of the priority problems confronting State and local governments. The selected excerpts are arranged in broad categories—developments and activities, cooperative measures, problems of definition, and requirements for Federal support—in order to focus on topics of key concern to the participating State and local respondents.

DEVELOPMENTS AND ACTIVITIES

CREATION OF A PLANNING AGENCY

The State's need for a useful input-output system is apparent. When the State Planning Agency was formed, during the latter half of 1966, the position of information systems director was among the first defined. He will have the responsibility of planning, developing, and maintaining a comprehensive State information system which will be the base of systems approaches in the future. It will include the development of a standardized approach to the collection, analysis, and use of demographic, economic, and other data used in State planning and programing. The employee, using this capability, will be able to produce, interpret, and transfer information needed by the Governor, State legislature, State departments and all other decisionmakers in our State governments.

RAYMOND T. OLSEN,
State Planning Director,
Department of Administration, State of Minnesota.

ADP FACILITY

* * * We do have a computer center and a staff of approximately 30 working in the area of data processing. As you are aware, computerization presupposes at least an elementary systems approach. To that extent we have placed on our computers such activities as payrolls, personnel benefits, scale accounting programs and do not involve sophisticated problem-solving techniques. The computerization of these activities has contributed to a basic ground work for the development of performance units and identifiable objectives which will eventually grow into a systems approach for the city. The anticipated growth of our automatic data processing techniques is based on the notion the city is a total system and that, as time passes, we will be able to develop functional areas as subsystems of that total system.

J. D. BRAMAN,
Mayor, Seattle, Wash.

PLANNED SYSTEMS GROWTH

* * * Our State auditor has been given a small systems staff until such time as proper enabling legislation is passed by our State legislature. A bill has been drawn for this purpose which would place the responsibility under our executive branch of government. If this bill is enacted into law our efforts in the data processing area will be substantially enlarged.

WILLIAM H. CORBETT,
Director, Fiscal and Audit, State of Florida.

A STATE COORDINATING MECHANISM

Several years ago, Colorado established a Management Analysis Office to provide in-house capability and overall leadership for State organization, systems analysis, management standards, and controls. Recently, this office has been adapted to the automatic data processing systems, planning and management analysis office. It will continue its former functions but has added the responsibility for overall State planning and coordination for all ADP activity. Although this organization is new, we are looking forward to greater increased ADP benefits at the lowest possible cost.

JOHN A. LOVE,
Governor, State of Colorado.

INITIAL PROGRAM IMPLEMENTATION

In 1965, the Nevada Legislature authorized the creation of a central data processing division in the department of administration. This agency conducts most of the systems analysis on the State level. Projects have been completed or are presently underway in such program areas as education, personnel, purchasing, accounting and welfare.

PAUL LAXALT,
Governor, State of Nevada.

SELECTED PROGRAM ADMINISTRATION

As you know, New York State has been a leader among the States in encouraging systems approaches to our State problems, particularly in the areas of administration of criminal justice, water pollution, transportation planning and statewide economic planning.

T. N. HURD,
Director of the Budget, State of New York.

COOPERATIVE MEASURES

THE 5–5–5 PROJECT

We are presently entering into two other special projects which relate to your subcommittee's interest. One is an intergovernmental project involving five States, five counties, and five municipal governments. This cooperative venture, which is sponsored by the Ford Foundation under the direction of Dr. Selma Mushkin, seeks to refine and advance the program planning and budgeting system of these cooperating jurisdictions along the lines of the PPB system presently being introduced in the Federal Government.

WARREN KNOWLES,
Governor, State of Wisconsin.

VERTICAL PROJECT DEVELOPMENT

At present we see a need for increased effort on the part of Federal and State agencies in the development of training programs in the field of systems analysis and in the development of "vertical projects" which analyze the flow of information among all levels of government in various functional areas such as crime, health, education and transportation. In this regard, we look forward to more extensive cooperation with the Federal Government in joint efforts to develop solutions to some of our major economic and social problems. From our past experience we believe that the use of advanced systems techniques will be essential if this effort is to succeed.

T. N. HURD,
Director of the Budget, State of New York.

INTERSTATE PLANNING

While we do intend to utilize planning-programing-budgeting systems methodology (to evaluate the performance characteristics of various State programs), we do not believe that this methodology can be useful without establishment of a statewide information system. We feel it necessary to establish regional information centers as part of this system and, because at least two of these systems would necessarily bridge State boundaries (in the Kansas City and St. Louis areas), we feel that we must attempt to achieve common design of storage and retrieval programs with adjacent States.

PHILIP V. MAHER,
Director, Office of State and Regional Planning and Community Development, State of Missouri.

PROBLEMS OF DEFINITION

PRELIMINARY PERSPECTIVE

As you mentioned in your letter "system approaches" does not yet seem to be a well defined technique. In completing the questionnaire I have defined it rather broadly to include most of the efforts of our state planning program, which in many respects is similar to the Federal planning, program, budgeting system. In most of the management and program areas outlined in the questionnaire, our efforts are only beginning and the evaluation of their effectiveness must be considered very preliminary.

ROBERT P. HUEFNER,
State Planning Coordinator, State of Utah.

SUBJECTIVE JUDGMENTS

Although the questionnaire has been designed to cover the widest possible range of governmental activities, we have attempted to interpret your requirements and provide the most accurate information possible. Because systems analysis is sometimes difficult to segregate and define within the total scope of management sciences, the questionnaire response is sometimes subject to judgment evaluation.

GEORGE F. GORGOL,
Director of Data Processing, Chicago, Ill.

REQUIREMENTS FOR FEDERAL SUPPORT

PERSONNEL AND FUNDING NEEDS

Advice and assistance can be given by the Federal agencies. We find that we we are competing with the Federal Government for hard-to-get skills, such as operations research people, systems analysts, budget analysts, and so forth. More of these people must be trained. Lastly, the technical people and the computer equipment are expensive. Federal funds should be available for local governments using the "systems approaches."

EDWARD J. MARTIN,
Director of Finance, Philadelphia, Pa.

TRAINING OF MANPOWER

* * * This city is particularly interested in developing a planning-programing-budgeting system similar to that employed by the Department of Defense. Training support in the development and use of this management tool would rank high in our list of recommendations for Federal support.

ILUS W. DAVIS,
Mayor, Kansas City, Mo.

REGIONAL CENTERS

* * * we would urge Federal interest in, and financial support of, interstate efforts to integrate statewide information systems. Secondarily, we would urge Federal support for the design, testing, and continuing operation of major urban regional information centers which, by all accounts, would tax State resources so severely as to jeopardize the development of statewide information systems.

PHILIP V. MAHER,
Director, Office of State and Regional Planning and Community Development, State of Missouri.

INFORMATION EXCHANGE

* * * we would be interested in any programs which would raise the level of our capabilities internally and which would forge a bridge of information with sister States and the Federal Government.

PHILIP H. HOFF,
Governor, State of Vermont.

D. COMPARATIVE ANALYSIS OF THE 1966 AND 1968 QUESTIONNAIRES

The special subcommittee's surveys developed out of a need to better understand the nature and scope of the systems analysis effort at the State and local levels. Both questionnaires were received with interest and enthusiasm. An analysis of the 1966 and 1968 questionnaires reveals that over 90 percent of those queried responded to either questionnaire or both. While a few additional participants responded for the first time to the 1968 survey, there were a few of the 1966 respondents who did not reply to the later questionnaire. Because of this, the total number of respondents to the 1968 questionnaire only equaled the number of 1966 survey participants.

The 1968 questionnaire was expanded slightly (see sec. VII. A.) in order to gain more information about certain aspects of systems development activity in State and local governments. The differences in the questionnaires, although minimal, represent the Nelson subcommittee's desire to obtain more "hard" data and interpretive commentary, and to better understand the extent of Federal support. In addition, the 1968 questionnaire requested respondents to include citations and copies of studies and reports resulting from systems analysis activities. Approximately half of the active respondents included one or more documents that described an actual systems effort. Most of these are listed in the selected references at the conclusion of this report.

Another area of departure between the two questionnaires is that the most recent survey requested the participants to identify those Federal Government departments or agencies involved in supporting a specific systems analysis effort (see fig. 38). Essentially this was an expansion of the 1966 survey which asked only that participants identify if they received Federal support for a specific application.

The 1968 questionnaire asked that expenditures be listed for each year from 1965 through 1969; this was an expansion of the 1966 questionnaire which requested only that a "range of funds" be designated for the year 1966. The 1968 results are summarized in figure 41 and discussed in section VII.B. Figure 43, "Funds Expended by the States and Cities on Systems Analysis in 1966," represents at best a general approximation of expenditures listed by active participants in the 1966 questionnaire. For this one period a grand total of $22½ million reportedly was allocated by the respondents. In the 1968 questionnaire, responding governments indicated that a total of $82 million was spent for the 5-year period (1965–69). Although both figures are averages and only approximate the funds being expended on systems technology activity, they are indicative of the present level of expenditures.

States:		Cities and regional groups:	
1. New York	[1] $4, 250, 000	1. Baltimore	[1] $2, 675, 000
2. Oklahoma	[1] 1, 800, 000	2. Philadelphia	[1] 2, 500, 000
3. Pennsylvania	[1] 1, 125, 000	3. Port Authority of	
4. Wisconsin	[1] 550, 000	New York	[1] 1, 975, 000
5. Connecticut	[1] 375, 000	4. Los Angeles	[1] 1, 975, 000
6. Maryland	[1] 350, 000	5. New York City	[1] 800, 000
7. Massachusetts	[1] 350, 000	6. Chicago	[1] 550, 000
8. Texas	[2] 350, 000	7. Mississippi Re-	
9. West Virginia	[1] 350, 000	search and De-	
10. Alaska	[2] 260, 000	velopment	
11. Rhode Island	[1] 225, 000	Center	[1] 375, 000
12. Kentucky	[1] 225, 000	8. Cincinnati	[2] 310, 000
13. Missouri	[1] 175, 000	9. San Diego	[1] 225, 000
14. Florida	[1] 175, 000	10. Denver	[1] 175, 000
15. Utah	[1] 175, 000	11. Phoenix	[1] 100, 000
16. Washington	[1] 100, 000	12. Atlanta	[1] 100, 000
17. South Dakota	[1] 100, 000	13. Detroit	[1] 100, 000
18. North Dakota	[1] 50, 000	14. Kansas City	[1] 100, 000
19. Wyoming	[1] 50, 000	15. New Orleans	[1] 100, 000
20. Delaware	[1] 50, 000	16. Houston	[2] 66, 000
21. Iowa	[1] 50, 000	17. Cleveland	50, 000
22. Ohio	[2] 7, 500	18. Buffalo	([3])
23. Hawaii	([3])	19. Pittsburgh	([3])
24. Illinois	([3])	20. Green Bay	([3])
25. North Carolina	([3])	21. Seattle	([4])
Subtotal	11, 142, 500	Subtotal	11, 326, 000
		Grand total	22, 468, 500

[1] Estimated average figure.
[2] Stated figure.
[3] Not indicated.
[4] Not available.

Source: Scientific Manpower Utilization, 1967, op. cit., p. 368, table 3

FIG. 43.—FUNDS EXPENDED BY STATES AND CITIES ON SYSTEMS ANALYSIS IN 1966

Federal legislation that would assist State and local governments to use more sophisticated tools and techniques was desired by virtually all the respondents. Recommendations regarding the form and type of Federal support were based primarily on the needs as well as the legal and jurisdictional considerations. Cities, therefore, generally indicated a need for direct financial subsidy while States often emphasized a requirement for training support. In order of preference, Federal legislation most desired by respondents to both questionnaires should provide:

1. Direct financial subsidy.
2. Matching funds.
3. Training support.
4. Consulting services.

At all levels of activity respondents to both questionnaires stated that they found it difficult to obtain qualified systems personnel. A requirement repeatedly expressed was a need for support in the training of existing workers as well as providing special orientation for other affected personnel. In comparison the responses to question VI on Federal legislative action, as shown in the 1966 and 1968 questionnaires there was an increase in 1968 in the number of States desiring training support. The results of the 1968 and 1966 surveys are summarized in figures 37 and 44, respectively. Each lists the participants and the discrete application areas in which systems technology is being used.

[X—In-house capability; Y—out-of-house and in-house capability; Z—out-of-house capability]

STATES	(1) Planning and policies	(2) Personnel management	(3) Interagency activity	(4) Management standards control	(5) Equipment selection	(6) Procurement activity	(7) Legislatures	(8) Courts	(9) Financial	(10) Taxation	(11) Education	(12) Health and hospitals	(13) Crime and corrections	(14) Transportation	(15) Urban renewal and growth	(16) Science and research promotion	(17) Natural resources	(18) Pollution control	(19) Parks and recreation	(20) Regulation of commerce, etc.	(21) Labor and manpower services	(22) Utilities and enterprises	(23) Welfare and anti-poverty	(24) Social security and veterans	(25) Other
Alaska	Y	X	X	X	X	X		X	X	X	X	X	X	X			X		X		X		X	X	
Connecticut	X	X	X	X	X	X	X	X	X	X	X	X	X	X					X		Z	Z	X	Z	Z
Delaware	X		Z	Z	Z	Z	Z	Z	X	X	X	X	X	X	Y	X	Z	Z	Z	Z	X	X	Z	X	X
Florida	X	Z	X		X	X	X		X	Y	Z	X	X			X	Y	X	X		X	Z	X		
Hawaii	Y					X			X	X	X	X	X	X		Y	X	X	X		X		X		
Illinois		X	X	X	X	X			X	X	X	X	X	X			X	X	X		X		X	X	
Iowa	X	X	Y	X	X	X	X	X	X	X	Y	X	X	Y	Y	Z	X	Y	X	X	X	X	X	X	X [1]
Kentucky	X	Y	Y	X	X	X	Z	X	Y	X	Y	X	Y	X	Y	Z	Y	Y	X	Z	Z	X	X	X	
Maryland	X	Y	X	Y	X	X	Z		Y	Z	Y	X	X	Y	X	Z	X	X	X	Z	Y	X	X	Z	
Massachusetts	Y	Z	X		X	X	X		X	X	Y	X	X	Y	Z	Z	X	X	X	X	Z	X	X	Z	
Missouri	X	X	X	X	X	X			X	Y	Y	X	X	Y	X	Z	X	X	Z	X	X	X	X		
New York	X	X	Z		X	X		X	Y	X	Y	X	X	Y	X	X	Y	X	X	X	Y	X	X	Z	X [2]
North Carolina	X	X	X	X	X	X	X	Y	X	X	X	X	X	Y			X	X	Z	X	X	X	X	X	
North Dakota	Z								Y	X	Y	Y	X	Y		Z					Y		Y		
Ohio	Y	X	X		Z		X		Y	X	X	Y	Y	X			X		X	X	Y		Y	Z	
Oklahoma	X								Y	X	X	Y	Y	Y				Y	Y		Y	Y	Y	Z	
Pennsylvania									Y	X	X	X		X									X		
Rhode Island	Z	X	X	Y	X	X	Z		X	X	X	Y		X	Y	Y	Y	Y	X	Y	Y	Y	X	Z	X [3]
South Dakota	Y	X	Y	Z [4]	Y	X	X		X	X	X	Y	Y	Y		Y	Y	Y	X	Y	Y	Y	X	Y	
Texas	X	X	Y	Z	Z	X	X		Y	X	Y	Y	Y	X			X	X	X	X	X	X	X	X	
Utah	Y	X	Y	Y	X	X	X		Y	X	X	Y	X	Y			Y	X	Y	Y	Y	Y	Y	X	X
Washington	X	X	Y	Y	X	X	Z		Y	X	X	X	X	X		Y	X	X	X	X	X	Y	X	X	
West Virginia	X	X	Y	Y	X	X			Y	X	X	Y		X			X			Y	X		X	X	X
Wisconsin	X	X	X	Y	X	X	Y		X	X	X	X	X	Y			X	X	X	X	X	X	X	X	
Wyoming	X	X	X	Y	X	X	X	X	X	X	X	X	X	Y		Z	X		X	X	X	X	X	X [5]	

See footnotes at end of table, p. 186.

FIG. 44.—INDICATIONS OF SYSTEMS ANALYSIS CAPABILITY BY FUNCTION (1966)

[X—In-house capability; Y—out-of-house and in-house capability; Z—out-of-house capability]

CITIES	Planning and policies (1)	Personnel management (2)	Interagency activity (3)	Management standards control (4)	Equipment selection (5)	Procurement activity (6)	Legislatures (7)	Courts (8)	Financial (9)	Taxation (10)	Education (11)	Health and hospitals (12)	Crime and corrections (13)	Transportation (14)	Urban renewal and growth (15)	Science and research promotion (16)	Natural resources (17)	Pollution control (18)	Parks and recreation (19)	Regulation of commerce, etc. (20)	Labor and manpower services (21)	Utilities and enterprises (22)	Welfare and anti-poverty (23)	Social security and veterans (24)	Other (25)
Atlanta, Ga.	X	X	X	X	Y	X	Y	X	X	Y	Y	Y	X	Y	X	X	Y	Y	Y	Y	X	X	Y	Y	
Baltimore, Md.	Y	Y	Y	X	X	Y	Y	Y	Y	X	Y	X	Y		Y			Y	Y	Y	X	X	N		
Buffalo, N.Y.	X	X	X	X	X	X		X	X	X			Y	X	Y			Y	X	N	X	X	Y		X[6]
Chicago, Ill.	X	X		X		X			X	X		X	X		X			N			Y	X			X
Cincinnati, Ohio	N	X	X	X	X	X			X	N		X	X		N		X		X	X	X		X	X	
Cleveland, Ohio	Y[7]		Y[7]			Y[7]	Y[7]		Y[7]	Z	Y[7]	X[8]		Z	Z			Y[7]	X		Y[7]	Y[7]	N		
Denver, Colo.									X	Y	Y[7]	X	X		X	X			X	X	X	X	X[2]		X
Detroit, Mich.	X	X	X	Y	X	X	X	X	X	X	X	X	X	Y	Y	X	X	X	Y	X	X	Y	Y	X	
Green Bay, Wis.	X	X	X	Y	X	X	Y	Y	X	X	X	Y	Y	Y	Y	X	X	X	X	X	X	Y	Y	X	X
Houston, Tex.	Y	X	Y	X	X	X	X	Y	X	X	X	Y	Y	Y	Y	X	Y	X	X	X	X	X	X	X	X
Kansas City, Mo.	X	X	Y	X	X	X	X	X	X	X		Y	Y	Y	Y			X	Y	X	X	Y	X		
Los Angeles, Calif.	X	X	Y	X	X	X		X	X	X															
New York, N.Y.	N	N	N	N	N	N	N	N	X	X	N	X	N	N	X	X	X	X	X	X	X	X	X	X	X
Port Authority of New York	X	Y	Y	X	Y	Y	Y	Z	X	X	X	X	Z	Y	X	X		Y			X	X	X	X	X

REGIONAL DEVELOPMENT ORGANIZATIONS

Mississippi Research and Development

Source: "Scientific Manpower Utilization," 1967, op. cit., pp. 365–366, chart No. 1.

[1] Information systems.
[2] Welfare only.
[3] Highway.
[4] In-house capability being established.
[5] Veterans only.
[6] Building code enforcement.
[7] Green Bay Industrial Authority activity included.
[8] Health only.

FIG. 44.—CONTINUED

The respondents, in each case, indicated the use of in-house or out-of-house support for the application or program listed. Figure 44 provides a summary of State and city functional applications of systems analysis, as learned from the 1966 questionnaire responses. In both surveys, participants indicated that most activities were being accomplished by in-house staff or a combination of in-house staff with support from industry, universities, non-profit groups and/or the Federal Government. Few, if any, relied totally on outside consultants. Both surveys indicated that the number of States and local governments utilizing systems technology remained approximately the same but that the level of activity had, increased.

Figure 45 reflects the increase in the level of activity (in three broad categories) in the States in 1968. There was, however, a noticeable decrease in the number of participants (in 1968) engaged in fewer than nine applications.

A Comparison of the 1966 and 1968 Level of Systems Activity (State Governments)

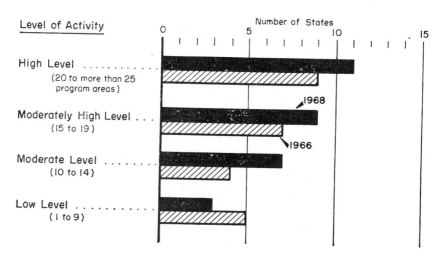

FIG. 45.—A COMPARISON OF THE 1966 AND 1968 LEVEL OF SYSTEMS ACTIVITY (STATE GOVERNMENTS)

On the other hand, figure 46 gives an indication of the types of functional areas where systems devices and procedures are being introduced. The percentage of respondents concerned with management and related activities remained essentially unchanged in the 1966 and 1968 responses. These precentage figures are based on the number of respondents engaged in a special program compared to the total number of active respondents. While management and service functions showed little change, social and economic development and the legal and regulatory categories showed an increase. Although the percentage figure increases in these areas are not dramatic, they do reflect the trend toward an increased interest in applying systems technology to program-oriented areas rather than management applications.

	1966	1968
I. Management and related activities—overall percentage	77	77
Planning and policies.		
Financial.		
Equipment selection.		
Personnel management.		
Taxation.		
Procurement activity.		
Interagency activity.		
Management standards and control.		
II. Services and transportation—overall percentage	60	59
Transportation.		
Utilities and enterprises.		
III. Social and economic development—overall percentage	57	61
Crime and correction.		
Education.		
Welfare and antipoverty.		
Health and hospitals.		
Urban renewal and growth.		
Pollution control.		
Social security and veterans' affairs.		
IV. Utilization of resources—overall percentage	51	51
Parks and recreation.		
Labor and manpower services.		
Natural resources.		
Science and research promotion.		
Overall percentage.		
V. Legal and regulatory—overall percentage	40	45
Legislatures.		
Regulation of commerce.		
Courts.		

FIG. 46.—PERCENTAGE OF ACTIVE RESPONDENTS EMPLOYING THE SYSTEMS APPROACH BY FUNCTION (46 ACTIVE RESPONDENTS), 1966 AND 1968

In reviewing the two surveys, there has been a recognition of certain similarities in the responses. There was a discernible trend toward greater utilization of systems technology with more applications being implemented and others identified for future development. More funds have been expended in developing an in-house capability but yet there is strong support for the continued utilization of outside consultants in conjunction with in-house staff. The responses to both surveys highlighted the wide diversity of activity currently being carried on in the States and cities, and by regional and territorial groups. The sophistication, cost, and funding of these systems efforts reflect an increased awareness by governmental management of the efficacy of utilizing fully the advancements made possible by computer technology and man-machine systems.

E. List of Respondents (1966 and 1968)

In 1966 and 1968 the subcommittee sent out questionnaires to the Governors of the 50 States, the mayors of 22 large cities, and to five regional development organizations. The 1968 questionnaire also was sent to the Governors of Guam, Puerto Rico, and American Samoa. The three listings that follow (figs. 47–49) include all those that responded to the 1966 and 1968 questionnaires. Those respondents that returned the questionnaire are marked with a double asterisk (**) while those that replied in letter form but without completing the questionnaire form are noted with a single asterisk (*).

Respondent	Replied to the questionnaire in—		Respondent	Replied to the questionnaire in—	
	1966	1968		1966	1968
Alabama	*	----------	Nebraska		**
Alaska	**	**	Nevada	*	----------
Arizona		*	New Hampshire		*
Arkansas		**	New Jersey	*	----------
California	*	**	New Mexico		**
Colorado	*	**	New York	**	**
Connecticut	**	**	North Carolina	**	**
Delaware	**		North Dakota	**	----------
Florida	**	----------	Ohio	**	**
Georgia	*	----------	Oklahoma	**	*
Hawaii	**		Oregon	*	**
Idaho		**	Pennsylvania	**	**
Illinois	**	**	Rhode Island	**	**
Indiana	*	**	South Carolina		**
Iowa	**	**	South Dakota	**	**
Kansas		**	Tennessee	*	*
Kentucky	**	----------	Texas	**	**
Maine	*	**	Utah	**	**
Maryland	**	**	Vermont	**	**
Massachusetts	*	**	Washington	*	**
Minnesota	**	----------	West Virginia	**	**
Mississippi		*	Wisconsin	**	**
Missouri	**	**	Wyoming	**	----------

FIG. 47.—STATES RESPONDING TO THE NELSON QUESTIONNAIRE

Respondent	Replied to the questionnaire in—		Respondent	Replied to the questionnaire in—	
	1966	1968		1966	1968
Atlanta, Ga	**	**	Indianapolis, Ind		*
Baltimore, Md	**	----------	Kansas City, Mo	**	**
Buffalo, N.Y	**	**	Los Angeles, Calif	**	**
Chicago, Ill	**	**	New Orleans, La	**	**
Cincinnati, Ohio	**	**	New York, N.Y	**	**
Cleveland, Ohio	**	**	Philadelphia, Pa	**	**
Dallas, Tex		*	Phoenix, Ariz	**	**
Denver, Colo	**	**	Pittsburgh, Pa	**	----------
Detroit, Mich	**	**	San Diego, Calif	**	**
Green Bay, Wis	**	----------	San Francisco, Calif		**
Houston, Tex	**	**	Seattle, Wash	**	----------

FIG. 48.—CITIES RESPONDING TO THE NELSON QUESTIONNAIRE

Respondent	Replied to the questionnaire in—	
	1966	1968
Mississippi Research and Development	*	
Port Authority of New York	**	**
American Samoa		*
Guam		**
Puerto Rico		**

FIG. 49.—REGIONAL DEVELOPMENT ORGANIZATIONS AND U.S. TERRITORIES RESPONDING TO NELSON QUESTIONNAIRE

F. SUMMARY

During the past decade, as the result of the passage of numerous public laws (see sec. V) and the establishment of various Federal assistance programs by executive branch departments and agencies (as discussed in sec. IV), State and local governments have been able to mount more than a token effort to reduce or remove certain social and community problems. Monetary aid and, in some cases, technical assistance, has been available from Federal sources but there has been too little information (or statistical data) interchanged among governments.

The purpose of the series of questionnaires drawn up and disseminated for the Senate Special Subcommittee on the Utilization of Scientific Manpower was to learn more about how State and cities were utilizing systems analysis and automatic data processing in a range of application areas, what steps were being taken to create inhouse systems competence or draw upon external consulting expertise, and where the money was coming from (and in what amounts) to design, test, and implement improved systems.

The graphic depiction of these essential elements of information allows a ready understanding of the scope and nature of regional-State-local acceptance of systems technology and the desire of these governmental groups to adapt for their use many of the tools and techniques originally devised for use in the defense and space environments. In order to broaden the perspective of those who might study the results of an analysis of the 1966 and 1968 special subcommittee questionnaires, two other surveys were examined and their findings presented for purposes of comparison. Both the State-local finances project on PPBS activity and Council of State Governments ADP activity surveys were instrumental in allowing a deeper understanding of the steps being taken in metropolitan areas and by States to avail themselves of the latest management techniques and program-supporting tools.

The questionnaire data also was considered in the light of developmental projects underway in the various States. An example of this was the report prepared by the New York State Business Advisory Committee on Management Improvement which examined several proposals for utilizing systems analysis techniques to develop "comprehensive" solutions to New York State's problems: [15]

Applying a systems approach to State government

The committee is investigating the possibility of developing a systems approach to State government by enlisting the help of technicians from private industry to work with State personnel in looking at one particular management area within a specific State agency. For example, the committee, working through the Secretariat, might arrange for industry specialists to work with a given State agency on a system for measuring productivity within the agency. Initially, the committee is examining the possibility of having one or two State officials meet informally with a group of representatives from various private companies to discuss the kind of technical help with these firms might provide and to relate this to the needs of State agencies.

Center for advanced studies of State and local problems

The committee has prepared a proposal to establish a center for advanced studies—or "think tank"—in New York State.

After preliminary review of the work of the office of planning coordination and the implementation of the planning-programing-budgeting system in New York State, the committee has examined the feasibility of utilizing the services of a center for advanced study to attack problems which:

Involve broad social, economic or technological changes;

Require integrated activities with other States or levels of government;

Demand the skills of an interdisciplinary team; and

Lend themselves to possible solution through the use of scientific management techniques.

[15] New York State. Business Advisory Committee on Management Improvement. Report to Governor Nelson A. Rockefeller (2d annual report), Albany, N.Y., Business Advisory Committee on Management Improvement. May 1968, pp. 5-6.

Another State-oriented activity producing valuable comparative information was that involving the National Governors' Conference Committee on State Planning. This top-level decisionmaking group, supported by the Institute on State Programing for the 1970's, prepared a report entitled "Relevance, Reliance, and Realism." In discussing the role and importance of planning in State governments, the report summarized the many considerations involved in establishing effective planning mechanisms. The Governors were charged by the report to:[16]

Establish your state planning agency as a staff function closely related to your key administrative and management function.

Charge your planning agency to brief you fully on major issues for you, the legislature and the public.

Charge your planning agency to assist you in establishing goals and priorities, and ask it to suggest alternative approaches for achieving these goals.

Install a planning-programing-budgeting system to enhance policy implementation and to increase budgetary control and functional coordination.

Charge your planning agency with the responsibility for establishing a management information system which would make all germane data available and usable to all parts of State government.

Direct your planning agency to make a continuous study of structural and constitutional constraints, and to recommend administrative reorganization and constitutional revision.

Charge your planning agency to stimulate and coordinate adequate multi-State and sub-State planning programs, and to serve as the capstone of this process.

Charge your planning agency with the responsibility to educate your State government—all parts—to the need and benefits of effective planning.

As the analysis of the two special subcommittee questionnaires and their responses proceeded, attention also was given to the means by which States and cities learn of comparable governments' efforts to use systems technology in handling various problems. In certain instances, progress information on projects underway or completed is available from a Federal establishment (e.g., U.S. Department of Commerce Clearinghouse for Scientific and Technical Information). It should be noted, however, that a number of State and cities specifically mentioned the need for better information on (a) Federal sources of information, and (b) developments at their own governmental levels. The communication of State and local activity in utilizing ADP and systems methodology is sporadic at best. For this reason there have been pioneer efforts such as the Management Analysis Newsletter prepared from time to time by the Division of Methods Data Processing and Office Services of the State of Rhode Island. One-time studies, such as the recent "Comprehensive Evaluation of the Urban Crisis in Pennsylvania," also have proven useful to other State and city officials. A private sector publication which furnishes good coverage of automated systems in government is the monthly "Public Automation." Published by the Public Administration Service, this chronicle records systems developments at the regional, State, county, and city levels, and also treats one special topic each month in the "Output" feature.

As a result of scrutinizing the two sets of questionnaire responses, there has been a more complete framework within which to examine the testimony of the witnesses appearing before the special subcommittee (see section VI). Commentary on various uses of computers and PPBS procedures, whether citing the benefits or the shortcomings of these innovative approaches, is more meaningful when considered alongside the broader scope of the questionnaire findings. Also, the requirements

---

[16] National Governors Conference. Committee on State Planning. Relevance, reliance and realism, 1967, p. 17.

for further internal action at the State or local level as well as augmented Federal support become better defined as the result of studying the public legislation, Federal programs, private sector experience, and State-local fledgling efforts.

The major problems identified by the questionnaire respondents merit review, for it is these which must be solved through the improvement of management decisionmaking, information exchange, and problem-oriented program development:

1. Limited funds and lack of initial capital investment.
2. Nonsynchronous budgeting cycles.
3. Lack of management sophistication in understanding how to use systems technology.
4. Inexperience in obtaining available Federal assistance, either monetary or consulting expertise.
5. Tendency to develop fragmented management and program systems.
6. Difficulty in obtaining qualified systems personnel, or retraining on-board staff members.
7. Absence of enabling legislation which could provide the necessary initial impetus for corrective action.
8. Difficulties caused by overlapping jurisdictional control.

VIII. EXAMPLES OF STATE AND LOCAL ACTIVITIES RE-FLECTING THE USE OF SYSTEMS TOOLS AND TECH-NIQUES

A. Introduction

The dilemma facing those responsible for initiating and monitoring corrective action which will reduce or remove the social and community problems discussed in this report centers about the difficulty of finding out what comparable activity has been undertaken elsewhere which may be applicable to the local situation. There is no information exchange system even within governmental circles, although such a network has been proposed in the 90th Congress (S.J. Res. 110, which is discussed in sec. IV. B). Concern about the "reinvention of the wheel" is manifested on every side, as recurring instances of basic efforts attempted in one area when already proven or demonstrated as unsuitable in another area have occurred.

The fragmentation of written recording of projects begun, techniques tested, measurements compiled, and results published has been a major impediment to expeditious improvement of many problem situations. While many Federal-sponsored programs require "progress reports," these often do not find their way into generally accessible repositories—for example, the National Referral Center of the Library of Congress or the Federal Clearinghouse for Scientific and Technical Information—for many months. Other useful documents, such as university research findings or industrial proposals focusing on a key problem area, may never become part of the formal "literature," available to all interested parties. The documentation extant on the use of systems tools and techniques in various program or project areas is limited not only by the relative recentness of the employment of such technology, but by the nature of the reporting practices. Several persons in positions of government or private organizations' leadership have urged that a survey be carried out which would collect, organize, collate, and report upon all such documentation. This approach is featured in S. 467 (and companion bill H.R. 20), as discussed in section V. B. of this report.

Since the exchange of badly needed information on projects in progress or programs completed is largely a matter of happenstance, professional interchange of ideas and findings has been undertaken through the traditional medium of professional meetings. At the Annual Symposium of the Association for Computing Machinery in 1967, for example, the central theme was the "Application of Computers to the Problems of Urban Society." [1] Formal papers were presented and useful discussion was achieved on a cross-section of applications ranging from air pollution and waste disposal to education needs. The proposals and conclusions contained in some of the papers stressed the relationship between data banks and model construction,

---

[1] See Socio-economic planning sciences, vol. 1, note 3, July 1968 which contains a collection of papers presented at the Annual Symposium of the Association for Computing Machinery, New York City, Nov. 10 1967.

discussing these facets of system development (and others) within the context of community conditions, resource availability, and the necessity of maintaining a proper balance between men and machines.

Another meeting of community leaders and technological experts which addressed the interaction of technology with urban problems was sponsored by the National Academy of Engineering and the National Academy of Sciences. This symposium resulted in the preparation of a report entitled "Science, Engineering, and the City," [2] and included contributions on the subjects of urban transportation and construction, education, the role of technology in urban planning, and urban research and development. A series of searching questions were posed for consideration by the participants; included in these were: [3]

1. How useful is the ideal-solution approach to urban problems?

2. Would it be more valuable to use an action-oriented strategy which begins with analysis for reaching an ideal solution, but which then is concerned with currently feasible developments and relates them to systems that already exist in the city?

3. What known science and engineering sophistication is already available?

4. What academic, industrial, and governmental resources are available, and what additional resources are needed?

5. How can each of these resources make the best possible contribution to the whole?

6. How should they be coordinated?

During the course of this meeting, Dr. Donald F. Hornig, Director of the Office of Science and Technology, underscored strongly the responsibility of the scientist in the realm of social and community problems:

We in the scientific community have tended to ignore the magnitude of the urban development task. Responsibility for urban investments and operations are decentralized—tens of thousands of institutions are involved—and each tends to define its own urban mission in parochial terms. This situation has made it difficult to describe adequately major problems in their totality in such areas as housing, education, and transportation and to design workable programs for the implementation of promising solutions.[4]

In this section of the report to the special subcommittee, source material has been obtained as the result of supporting documentation provided by government groups responding to the questionnaire (see sec. VII), material submitted by participants in the hearings before the special subcommittee (sec. VI), and random acquisition from other sponsoring agencies and private sector research elements. In discussing the diverse programs and projects either proposed or undertaken, consideration has been given both to short-term and long-range planning and implementation. Governmental management systems—such as statewide administrative information handling systems or "overview" urban management systems—first will be discussed, followed by commentary on reports and proposals dealing with specific problem applications (i.e., transportation planning, solid waste disposal). Finally, proposals concerning information exchange mechanisms will

[2] Science, Engineering, and the city. Publication 1498. Washington, D.C., National Academy of Sciences, 1967. 142 p.
[3] Ibid., p. vi.
[4] Ibid., p. 61.

be considered within the framework of governmental and private requirements for summary highlight information and supporting narrative and statistical data.

B. Governmental Management Systems

State and local governments today are charged with the responsibility of managing and operating increasingly dynamic and complex environments. The problems harassing public leadership include population mobility, an increased demand for services which often transcends traditional political boundaries, and the need to define anew areas of responsibility for regional, State, county, and local governments. In the period following World War II, a search was undertaken for management personnel capable of directing the revamping of State and metropolitan governments. Concurrently, existing and proposed means of operating these governmental entities were examined. Management reforms were initiated in order to create an orderly process of government capable of accomplishing priority public objectives.

Decisionmaking, traditionally viewed as a subjective process, became the object of considerable controversy as popular discontent with the functioning of local (or State) government was exhibited. The importance of having the right kind of information for the formulation of decisions became a point of focus for the critic of management. In the recent study of New Haven, Conn., the importance of information to the urban manager was expressed in terms of five "problem areas": comprehensiveness, accessibility, relatability, timeliness, and utilization.[5] The exploratory endeavor by the International Business Machines Corp. developed a series of concepts for an urban management system (including files, programs, personnel procedures) against a background of descriptive parameters characterizing the affected urban environment and the problems facing the city's managers. The needs of the leadership to monitor and control what was going on within the city were treated carefully in their relationship to the resources necessary to insure the continued functioning of all city departments. Gradually a structural framework for an urban management information system emerged, expressed in terms of central files, data manipulation capabilities, and system programs. Figure 50 depicts the basic structure of the proposed system.[6]

The importance of a computer-oriented information system appears increasingly in 5-year plans, information system study reports, and ADP implementation project descriptions prepared by State and local authorities. Surveys conducted by groups throughout the country—an example is the listing by state of projects completed, in progress, and requested which is published periodically by the Rhode Island Division of Methods, Data Processing, and Office Services [7]—reflect a rapidly broadening spectrum of applications for management analysis and ADP techniques. In section VII of this report appears an analysis of comments and status data sent to the special subcommittee by dozens of State and city governments.

---

[5] Concepts of an urban management information system, a report to the city of New Haven, Conn. Yorktown Heights, N.Y., "Advanced Systems Development Division," International Business Machines Corp; [no date]. p. 2.

[6] Ibid., p. 25, fig. 1.

[7] Management analysis newsletter. Edited by Division of Methods, Data Processing, and Office Services, State House, Providence, R.I. Issue note 2, October 1968. 12 p.

196

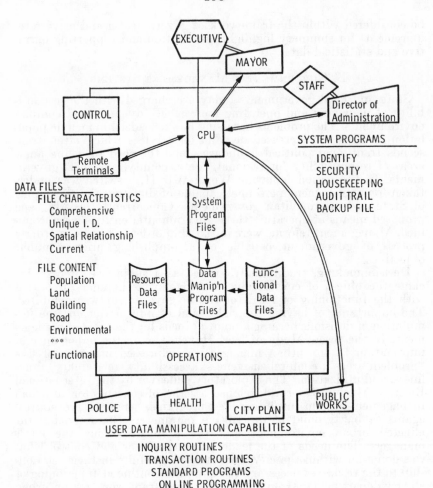

FIG. 50.—THE BASIC STRUCTURE OF AN URBAN MANAGEMENT INFORMATION SYSTEM

The studies conducted by or for State governments, for the most part, have certain common elements:

1. An objective review of the current data processing system, whether manual or machine supported;

2. An enumeration of those departments (e.g., personnel, treasury) having stipulated information requirements and "systems" in being;

3. Graphic representations of information-handling costs and projected system changes; and

4. Alternative paths to improved information handling, usually involving the utilization of ADP equipment.

While some of the studies offer comparisons of centralized versus decentralized ADP-supported systems, the majority of States (and localities) are turning to a single coordinated system. The State of New Hampshire, for example, has sought to expedite its staff's access to key information; figure 51 portrays present versus future availability of information: [8]

FIGURE 51

The General Electric study of the State of New Hampshire's information-handling needs included a thorough system analysis of nearly a score of State agencies, departments, and commissions. The first phase of system development featured an integration of five subsystems: Requisition/procurement, fiscal accounting, payroll, treasury, and retirement. A central computer, as shown in figure 52,[9] would offer both random access and magnetic tape files for storage of the selected information.

[8] State of New Hampshire information system study report. Prepared by Internal Automation Operation, General Electric, May 1966. P. 8, fig. 4.
[9] Ibid., p. 12, fig. 6.

198

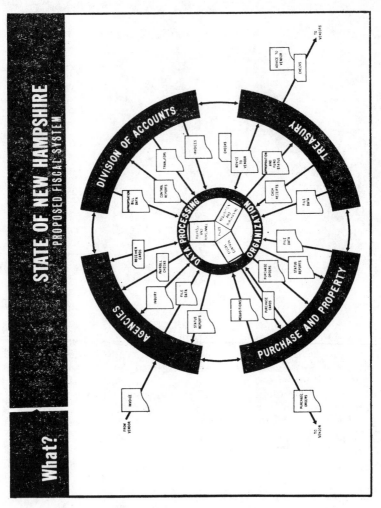

FIG. 52.—CENTRAL COMPUTER PROGRAM

In the case of the State of Alaska, a 5-year implementation plan, based on a Lockheed study, was developed. Analysis of the State's information handling system resulted in a statement of specific objectives, design features of the "Alaska information system," organizational concepts to be used in the improvement of the statewide system, and a detailed plan for developing 13 major system areas (e.g., administrative information system, health and welfare information system). Figure 53 [10] shows the relationship between the major systems within the State and the planned Alaska information system.

In the 1967 "automated data processing planning project" undertaken by Touche, Ross, Bailey & Smart for the State of Washington, five objectives of paramount importance were identified: [11]

1. Promotion of an effective use of ADP in order to economically provide service to State citizens.

2. Recognition of the vesting, legal or otherwise, of citizen service program responsibility at the agency level.

3. Recognition of the present heterogeneous development of ADP utilization.

4. Provision for a natural guided evolution from independent ADP systems to a functioning coordinated statewide system.

5. Provision for an implementation plan that is not dependent on the passage of complex legislation or major realinement of State government organization and responsibilities.

This study developed a conceptual framework for the long-range evolution of ADP systems which could serve the State government components. It is important to note that the use of ADP equipment is not a prerequisite to the development of this capability. Figure 54 graphically portrays the system evolution, with each circle representing ndividual State agencies.[12] A highlight of the report is a carefully developed ADP 5-year plan [13] which denotes the steps to be taken during the 5-year period for each agency. The requirements for data element studies or shared file systems are identified, along with the integration of tab or computer equipment, communications links, and the creation of a common data index.

[10] State of Alaska implementation plan—Alaska information system: July, 1967–July, 1972 [no date]. P. 11, fig. II–3.
[11] Touche, Ross, Bailey & Smart. Automated data processing planning project. Report to the governor of the State of Washington, January 1967. P. 3.
[12] Ibid., p. 68, exhibit 17.
[13] Ibid., pp. 77–81 (especially exhibit 19).

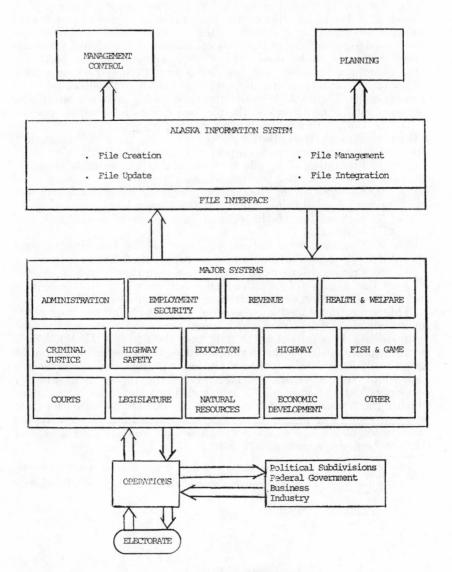

FIG. 53.—ALASKA INFORMATION SYSTEM

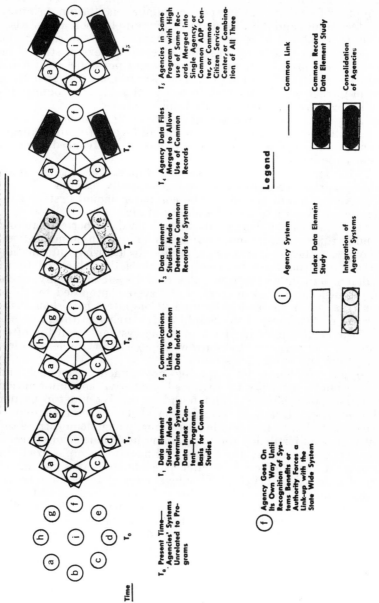

ADP SYSTEMS PLAN — CONCEPTUAL FRAMEWORK FOR
LONG RANGE EVOLUTION OF STATE-WIDE ADP SYSTEMS

Time

T_0 Present Time—
Agencies' Systems
Unrelated to Pro-
grams

T_1 Data Element
Studies Made to
Determine Systems
Data Index Con-
tent—Programs
Basis for Common
Studies

T_2 Communications
Links to Common
Data Index

T_3 Data Element
Studies Made to
Determine Common
Records for System

T_4 Agency Data Files
Merged to Allow
Use of Common
Records

T_5 Agencies in Some
Program with High
use of Same Rec-
ords Merged into
Single Agency, or
Common ADP Cen-
ter, or Common
Citizen Service
Center, or Combina-
tion of All Three

(f) Agency Goes On
Its Own Way Until
Recognition of Sys-
tems Benefits or
Authority Forces a
Link-up with the
State Wide System

Legend

(i) Agency System

Index Data Element
Study

Integration of
Agency Systems

———— Common Link

Common Record
Data Element Study

Consolidation
of Agencies

FIGURE 54

While many of the State and local efforts to improve information-handling systems have concentrated on the computerization of relatively routine functions, an increasing number has begun to apply systems techniques and ADP to more difficult application areas. Also, the need to establish a recognized mechanism for controlling the systems development activity has been stressed in several studies, as has been the requirement to train management personnel in systems methodology and ADP techniques. The 5-year plan prepared by a Wisconsin intragovernmental task force had these four recommendations, all of which were subsequently implemented: [14]

    1. Establish a data processing coordinating committee.

    2. Establish a training program for all levels of State management concerning data processing, systems, and advanced management techniques.

    3. Review present data processing applications to determine what programs are necessary and economically feasible.

    4. Any program placed on the computer should first be substantiated by an adequate feasibility study.

The State and local governments share a need for better planning data, and in this area optimum information handling becomes critical. Not only must these governmental groups decide what information they need and in what form, but they must take action which will result in the design and implementation of a useful system. Finally, there must be knowledgeable evaluation of the system proposed by those who must use it. In his article entitled "Computers in Urban Planning," [15] W. K. Williams, of the Stanford Research Institute, reminds us that "the lack of planning for a planning system" is all too often encountered. The reasons: "too narrow scope, lack of problem understanding, time and money constraints."

One recent technique for exposing government personnel and systems technologists alike to advanced predictive and information manipulation potentialities is that found at the Washington Center for Metropolitan Studies in Washington, D.C. Several computer-supported models have been made operational which demonstrate the feasibility of cross-fertilizing the programing field and social sciences. The project, sponsored by the Office of Education, emphasized the programing of the smallest computer possible so that a maximum number of users could benefit. [16] The developers of these have recognized that many social scientists have become disillusioned with models, finding that many problems could not be quantified readily and that probability techniques then had to be applied. Another key factor in a growing negativism on the part of social scientists was the difficulty of translating the findings of the models into language comprehensible to the government policymakers and program managers.

At the Washington Center for Metropolitan Studies a model has been created called "City I"; this features a simulation activity where teams of decisionmakers representing a community seek to exert either economic or political power in order to influence urban growth

[14] Data processing in Wisconsin State government—a 5-year plan: 1967–72. Department of Administration, Bureau of Systems and Data Processing [no date], p. 3.

[15] Williams, W. K. Computers in urban planning. In Socio-econ. plan. sci., vol. 1, No. 3, July 1968, p. 298.

[16] House, Peter. The simulated city: the use of 2d generation gaming in studying the urban system. Presented at the Annual Symposium of the ACM on "The application of computers to the problems of urban society," Oct. 18, 1968, pp. 1–2.

and development. The structure and use of the City I model is described in appendix C.

Planning for the future, then, is a matter of high-priority common concern to regional, State, and local leadership, and the efficacy of the plans rests with the quality and currency of the information available to the decisionmakers. The governmental information systems discussed in this report are made up of three elements, as shown in figure 55 :[17]

FILE GENERATION	FILE MAINTENANCE	DATA REPORTING
Design of data files	Identification of data	Identification of user
Field surveys	sources:	requirements
Data collection	Internal forms	Design of processing
Coding	External sources	algorithm
Editing	Field surveys	Request procedure
Keypunching	Systems design	Request forms
Programing	Standardization of	Programing
Testing	reporting	Testing
Processing	Design of forms	Processing
Documenting	Programing	Output requirements
	Testing	Tabular
	Keypunching	Graphical
	Processing	Statistical
	Controlling	

FIG. 55.—THREE ELEMENTS OF GOVERNMENTAL INFORMATION SYSTEMS

It has been shown that the responsible leadership of the State or city government is aware of, and has taken steps to improve markedly, the role of information in the management and operation of the full range of public programs. In many areas, strenuous efforts are being made to modernize governmental information handling systems through the use of systems analysis and automatic data processing techniques and equipments. The "Selected References" section of this report contains an illustrative cross-section of project reports, scholarly proposals, and implementation plans. The emphasis will now shift from management systems with an overview orientation to problem-oriented activities where systems technology has played a significant role.

C. Problem-Oriented Programs and Projects

The difficulty of learning about proposals for, or progress on, systems technology utilization in specific problem areas has caused a demand for remedial action. Recent legislation (e.g., S. 467, as discussed in sections IV and V) has stressed the imperative need for a survey of existing programs and projects, by problem area, where systems tools and techniques have been or are being used. One useful listing of these public programs is found in the Catalog of Federal Aids to State and Local Governments.[18] Federal Departments such as Housing and Urban Development, Transportation, Health, Education, and Welfare, and the Office of Economic Opportunity regularly

[17] Williams, W. K., op. cit., p. 299, fig. 1.
[18] U.S. Congress. Senate. Committee on Government Operations. Subcommittee on Intergovernmental Relations. Catalog of Federal aids to State and local governments. (88th Cong., 2d sess., committee print.) Washington, U.S. Govt. Print. Off., 1964. 154 pages.
Supplement, Jan. 4, 1965. (89th Cong., 1st sess., committee print.) Washington, U.S. Govt. Print. Off., 1965. 65 pages.
Second supplement, Jan. 10, 1966. (89th Cong., 2d sess., committee print.) Washington, U.S. Govt. Print. Off., 1966. 257 pages.

finance research and the development of new approaches to old problems.

Authorship of the several dozen research reports, technical memoranda, and implementation plans scrutinized in the preparation of this report, represented industrial (30 percent) and university (25 percent) professionals, with State and local in-house task forces preparing about 30 percent of the documentation. The remaining items were prepared by foundations or regional development groups.

TRANSPORTATION

Transportation planning has been the focus of concern at all levels of government since World War II. In some cases, the emphasis has been placed on regional growth; examples are the Penn-Jersey study and the Bay Area transportation study in California. Other projects have concentrated on more limited aspects of the subject, such as traffic flow in a circumscribed area or the control of a freeway diamond exchange. Air, rail, and highway transportation planning increasingly includes the utilization of computers, simulation models, and the extensive mapping of modular corrective action.

One of the best examples of the use of systems technology in transportation planning is the work performed by the System Development Corp. for the Bay Area Transportation Study Commission (BATSC). In the preparation of a long-range regional ground transportation plan for the nine-county San Francisco area, a large volume of machine-processable information was collected. These data described the economic activities, people, transportation facilities, and land use. The requirements of the analysis team for diversified information which must be acquired, filtered, stored, manipulated, analyzed, and retrieved made the use of ADP a prerequisite to the conduct of the study.

Many data sources must be drawn upon in the conduct of this type of study: public records, household interviews, private agency surveys. In the SDC work for BATSC and on the Penn-Jersey transportation study a set of programs was designed so that a regional growth model could be created. A large scale data management, file processing, graphic display, and statistical analysis system called SPAN (Statistical Processing and Analysis) was devised for the use of planners. Also developed for use in transportation planning were MADAM (Moderately Advanced Data Management), a means for organizing initially unstructured data into accessible files, and DATA-DOX, a series of IBM 1401 routines for producing documentary printouts from reference files. In order to ensure that usable documentation would be available for use by other governmental organizations, the U.S. Department of Housing and Urban Development requested that SDC prepare a report on the BATSC project.[19] The descriptive portion on the basic system activity entitled "BATSC and Its Information System," appears as appendix D.

The analysis and measurement of vehicular traffic at a freeway diamond interchange were featured in a study sponsored jointly by the U.S. Bureau of Public Roads and SDC. On this project, a series of computer models were designed: [20]

---

[19] Kevany, Michael J. An information system for urban transportation planning: the BATSC approach. Technical memorandum (TM–38920/000/01). Santa Monica, Calif., System Development Corporation. 56 pages plus appendix.
[20] The diamond interchange—a mathematical model. In SDC magazine, vol. 11, No. 5, May 1968, pp. 18–19.

1. Model 1—concerned with the merging of on-ramp traffic with freeway traffic;
2. Model 2—consists of a combination of Model 1 and associated upstream freeway and off-ramp traffic;
3. Model 3—simulates the signalized intersections of ramps and the surface road: and
4. Model 4—constitutes the entire interchange.

The modeling of an environment is in no way a mysterious nor incomprehensible process. The step-by-step development, in this instance:

* * * involves the need to measure what actually happens under varying traffic conditions, to reduce these phenonema to mathematical form, to find the mathematic equations that express the relationships among the variable phenomena, and to verify the conditions for which the expression is valid. It is a process of going back and forth between the predicted and observed effect, refining the model until it does really predict what actually occurs.[21]

Simulation of an urban transportation transfer point was studied by the Transportation Research Institute of Carnegie-Mellon University. The transfer point which was modeled linked a "geometrically flexible distributor system and a regional rapid transit system." [22] The objective of this study was to achieve a system balance and design a transfer point featuring smooth flow and minimal delay of passenger movement. The role of simulation in this case interacted directly with the elements of cost of operation and patronage. Figure 56 shows a functional block diagram-transfer point simulation.[23]

In a study performed by the Institute for Research on Land Water Resources at Pennsylvania State University, ADP technology was utilized in examining one aspect of the acquisition of land for highway rights-of-way.[24] When property is condemned and acquired by the State for highway purposes, payment is made for that property condemned and there must be equitable compensation for the "severance effects" suffered by the remaining portion of the property. This research report, prepared with the cooperation of the Pennsylvania Department of Highways and the U.S. Bureau of Public Roads, used a computer in the correlation of highway severance case data. The retrieval program developed has four options which offer a variety of information printout (or card product).

Closely related to transportation network planning is the problem of traffic control. One recent paper—"Digital Computers in Traffic Control—The Experience and the Significance" [25]—discusses the part which signal devices (both coordinated and uncoordinated) play in relieving traffic congestion. The San Jose, Calif., and Wichita Falls, Falls, Tex. systems are supported by IBM 1800 facilities. Information is fed from the traffic detectors to the computer where it is processed and divided into three categories: volume, occupancy, and speed System controllers are able to make decisions either on individual intersectional or group confluence bases. On a test of 32 San Jose heavy traffic intersections, the computer-oriented system resulted in

[21] Ibid., p. 18.
[22] DiCesare, F. and J. C. Strauss. Simulation of an urban transportation transfer point. In Socio-econ. plan. sci., vol. 1, No. 3, July 1968, p. 405.
[23] Ibid., p. 410, fig. 4.
[24] Taylor, Harold H. and Milton C. Hallberg. Use of computers in the storage and retrieval of severance effect information. University Park, Pa., Institute for Science and Engineering and the Institute for Research on Land and Water Resources, The Pennsylvania State University, 1966. 35 pages.
[25] Bermant, O. I. Digital computers in traffic control—the experience and the significance. In Socio-econ. plan. sci., vol. 1, No. 3, July 1968, pp. 415–421.

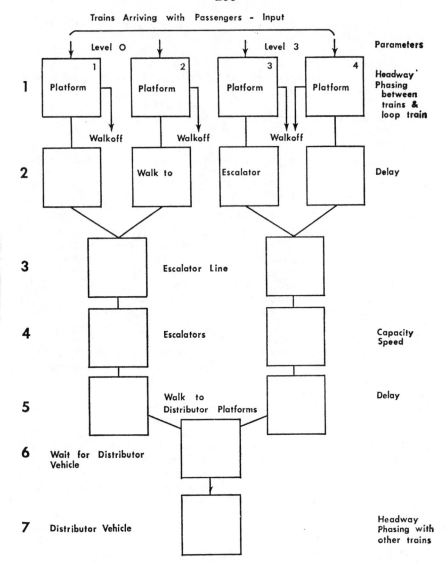

FIG. 56.—FUNCTIONAL BLOCK DIAGRAM—TRANSFER POINT SIMULATION

a 14-percent reduction in average vehicle delay and a 17.8 reduction in the probability of a stop.[26]

ENVIRONMENTAL POLLUTION

Transportation continues to be a major concern to politicians, technologists, and planners alike, but other problems are of equal importance. Environmental pollution today is an accepted part of the metropolitan milieu. The air is contaminated from industrial output, vehicular exhausts, and numerous lesser sources. Rivers and streams no longer can recover through the processes of nature from chemical wastes and raw sewage. Solid waste disposal also is perplexing those responsible for keeping urban areas clean. The planning cycle for

---

[26] Ibid., p. 420.

correcting these conditions—as well as comparable community problems—is placed in a circular graph by Edmund N. Bacon: [27]

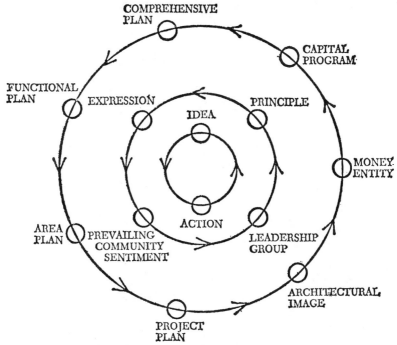

FIG. 57.—PLANNING CYCLE FOR CORRECTING ENVIRONMENTAL POLLUTION

Air pollution abatement increasingly is challenging the systems analyst, who not only must devise feasible corrective action but must keep in mind the realistic constraints imposed by governmental policies and the characteristics of the community involved. Emanuel S. Savas, deputy city administrator in the office of the mayor, the city of New York, believes that there are 10 factors which contribute to the overall pollution picture, and which must be analyzed prior to abatement planning: [28]

1. The activity which produces pollution;
2. Type of emission source;
3. Ownership (public or private) of the emission source;
4. Size (pollution potential) of the emission source;
5. Kind of fuel burned;
6. Kind of combustion equipment employed;
7. Kind of treatment applied to the combustion exhaust products;
8. Specific pollutants emitted;
9. Location of the emission source;
10. Time when the pollution-producing activity occurs.

It should be remembered that correction of one condition in a typical metropolitan environment often compounds the difficulty in coping with allied pollution sources. Figure 58 presents a tabular analysis of pollution factors for abatement planning.[29]

[27] Bacon, Edmund N. Urban process. In Daedalus, Fall 1968: the conscience of the city, p. 1170.
[28] Savas, Emanuel S. Computers in urban air pollution control systems. In Socioecon. plan. sci., vol. 1 1967, p. 160.
[29] Ibid., p. 161, table 3.

Activity (1)	Type of source (2)	Ownership of source (administrative control) (3)	Size (emission potential) (4)	Fuel (5)	Combustion equipment (6)	Exhaust control equipment (7)	Pollutants (8)	Location of source (9)	Time of emission (10)
Space heating	residential commercial -office -store -warehouse theatre auditorium institutional -school -hospital -religious industrial other	private public -federal -state -municipal quasi-public	number of floors 1–4 5–8 9–12 13+ floor space (in square feet) small medium large	Type gas oil -No. 1 -No. 2 -No. 3 -No. 4 -No. 5 -No. 6 coal -soft -hard characteristics chemical composition heating value	gas burner -type A -type B -etc. oil burner -type E -type F -etc. coal burner -hand fired -spreader stoker -underfeed stoker -pulverized coal	type A type B type C etc.	sulfur dioxide particulates -by size distribution carbon monoxide hydrocarbons	Block (all sources located by block via geographic coordinates) for control purposes area sources may be aggregated by: square mile neighborhood zone borough subdivision borough etc. point sources may be treated as individual control points	seasonal -heating -non-heating weekly -workday -weekend/holiday daily -morning (5–9am) -mid-day (9–3pm) -afternoon (3–7pm) -evening (7–11pm) -night (11–5am)
Incineration	(as above)		(as above)	food wastes paper plastics cloth other	type J type K type L etc.	type F type G type H etc.	particulates -by size distribution hydrocarbons sulfur dioxide	Height height above ground where emissions enter the atmosphere; for each source	

Transportation	bus truck automobile -taxi -other aircraft ship railroad	N.A.	gasoline -type A -type B -etc. diesel oil jet fuel etc.	N.A.	gasoline engine -type K -type L -type M -etc. diesel engine -type Q -type R etc.	carbon monoxide nitrogen oxides hydrocarbons metals	may be treated as area and line sources
Power generation	public utility other	small large	(as under 'Space Heating' above)	type P type Q etc.	type U type V type W etc.	(as under 'Space Heating' above)	treated as point sources
Industrial Processing	N.A.	small medium large	(as under 'Space Heating' above, plus process-specific etc. fuels)	type S type T type U	type Y type Z etc.	(as under 'Space Heating' above, plus process-specific pollutants	treated as point sources
Evaporation	architectural coatings industrial solvents	small large	N.A.	N.A.	N.A.	hydrocarbons	N.A.
	gas storage				seals -type 1 -type 2 etc.	may be treated as point sources	N.A.

*Numbers in parentheses at column headings refer to similarly numbered factors in the text.
N.A.= not applicable.

Note: Wind-borne particles produced by the erosion of structures, soil, etc., are not considered in this analysis.

FIG. 58.—ANALYSIS OF POLLUTION FACTORS FOR ABATEMENT PLANNING

In considering the pollution factors and the control policies which might be instituted to achieve a desired level of air quality, the analyst must keep in mind that each alternative can be weighed on the basis of three factors: the political acceptability, the social desirability, and the economic feasibility. In the monograph, "Adapredictive Air Pollution Control for the Los Angeles Basin," [30] it is pointed out that while $100 million has been spent in the past decade in the Los Angeles basin on air pollution control equipment, the desired improvement of the situation is nowhere in sight. The Federal Government estimates losses to the Los Angeles' population of $450 million annually, or 30 times the amount being spent on air pollution control.

Control systems capable of reducing air pollution to an acceptable level may be thought of in terms of five levels: [31]

1. District or county (stationary);
2. Intrastate or interstate (air shed);
3. State (moving);
4. Federal (health and welfare and interstate);
5. National (coordination).

The type of command and control hierarchy required to oversee successfully a regional improvement program is shown in figure 59.[32]

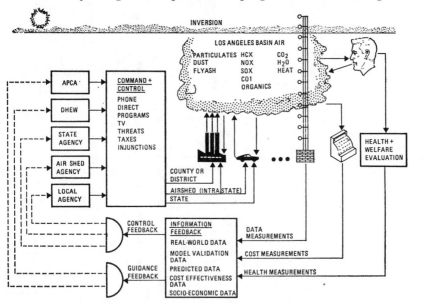

FIG. 59.—COMMAND AND CONTROL HIERARCHY

The elements of a basin distributed model, featuring multiple control zones with computer-supported monitoring elements, are depicted in figure 60.[33]

[30] Ulbrich, E. A. Adapredictive air pollution control for the Los Angeles basin. In Socioecon. plan. sci., vol. 1, No. 3, July 1968, p. 423.
[31] Ibid., p. 424.
[32] Ibid., p. 427, fig. 3.
[33] Ibid., p. 429, fig. 5.

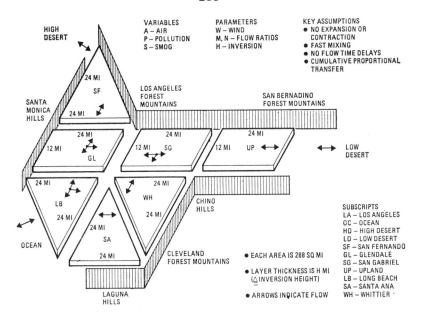

FIG. 60.—LOS ANGELES BASIN DISTRIBUTED MODEL

The purposes to which such an air pollution control system, with its network of sensors, data links, and ADP devices, can be dedicated are several:[34]

1. Provide real-time air-pollution data for the area under control.

2. Provide a validated model for the area under control.

3. Provide a prediction of the future air pollution for the area under control using the fast-time model allowing control before the alert comes.

4. Provide an improved public information tool using the mass media.

5. Provide a scientific base for air-pollution licensing and litigation using the validated model to demonstrate cost-effectiveness of control actions.

6. Provide an improved scientific base for biomedical air-pollution studies.

7. Provide a scientific base for urban planning.

8. Provide a scientific base for cost-effectiveness studies on various control methods.

9. Provide data as required for air-pollution control inspectors and others relative to the emission inventory.

10. Provide improved data as required by independent researchers.

The establishment of an effective water pollution control system would result in several comparable purposes being fulfilled. Congressional elements (see section .IV.B.) have acknowledged the importance of designing and implementing control mechanisms which

---

[34] Ibid., pp. 425, 430.

utilize modern devices and procedures. A recent study of the Delaware estuary was made by a team of men from the General Electric Co. and the U.S. Department of the Interior (FWPCA). In their "Mathematical Simulation of the Estuarine Behavior and Its Applications,"[35] stress was placed on evaluating the characteristics of the watershed; the location, identification, and measurement of the various pollutional sources; and the development of a rational means of estimating the cause and effect relationships "so that accurate forecasts of the results of proposed control schemes can be obtained."[36] A model interrelating the DO (dissolved oxygen) and BOD (biochemical oxygen demand) systems appears in figure 61.[37]

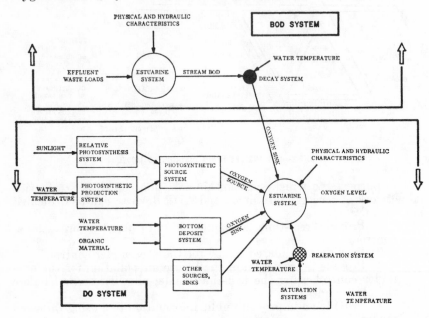

FIG. 61.—DISSOLVED OXYGEN SYSTEM

WATER RESOURCE PLANNING

The application of the systems approach to water resource planning has been the result of interdisciplinary efforts on the part of university and industry groups. Recent Federal legislation—for example, the Water Quality Act (Public Law 89–234) and the Water Resources Planning Act (Public Law 89–80)—reflects the increased attention being paid to systems analysis methodology in selected problem areas. The Technology Planning Center, Inc., has conducted a number of projects which involve the use of PPBS formats and data use, relevance tree diagrams, and simulation and modeling techniques.

One such project for the State of Michigan, under a Federal grant from the Urban Renewal Administration of the Department of

[35] Jeglic, J. M. and G. D. Pence. Mathematical simulation of the estuarine behavior and its applications. In Socioecon. plan. sci., vol. 1, No. 3, July 1968, pp. 363–389.
[36] Ibid., p. 364.
[37] Ibid., p. 367, fig. 2.

Housing and Urban Development, is an excellent example of the use of four different techniques. The PPBS matrix format, shown in figure 62,[38] acts as a three-dimensional link between planning, administrative control, and operations level functions.

The second technique, the relevance tree diagram, possesses the element of forcing policy planners to delineate their program goals as a first step. As figure 63 [39] indicates, water resource planners can view the system alternatives and each subsystem as it relates to a specific goal. Also, relevance tree diagrams help measure the costs of varying combinations of systems.

PPBS FORMAT - WATER RESOURCE CASE

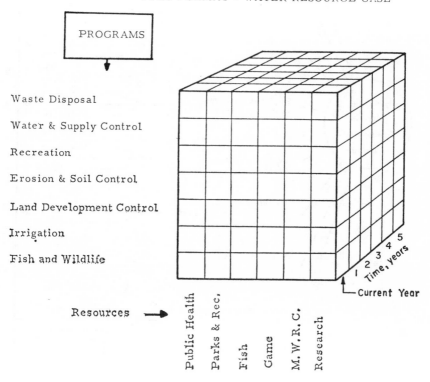

FIG. 62.—PPBS FORMAT—WATER RESOURCE CASE

[38] Systems analysis techniques and concepts for water resource planning and management. Ann Arbor Mich., Technology Planning Center, Inc. [no date]. P. 9, fig. 2.
[39] Ibid., p. 12, fig. 4.

RELEVANCE TREE DIAGRAM - WATER ORIENTED

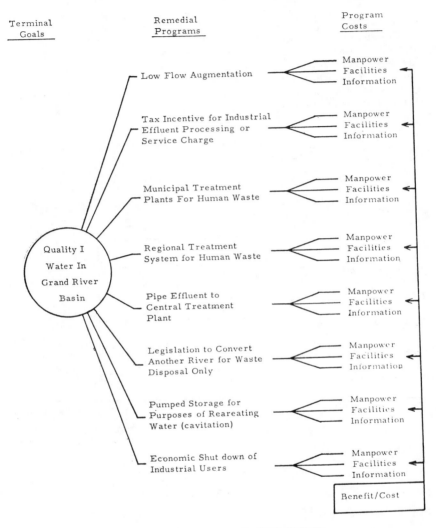

FIG. 63.—RELEVANCE TREE DIAGRAM—WATER ORIENTED

Simulation and modeling comprise the third set of techniques. Three classes of models may be involved: physical models, economic planning and allocation models, and social models. An example of a purely physical problem involving the selection of good water impoundment sites is expressed in the analog model shown in figure 64.[40]

[40] Ibid., p. 20, fig. 6.

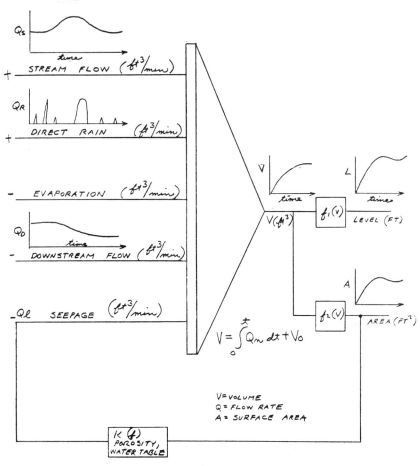

FIG. 64.—ANALOG MODEL OF SELECTION OF GOOD WATER IMPOUNDMENT SITES

The need for a "hybrid model" which combines the social, as well as economic and physical consideration has been explored by the staff of the Technology Planning Center. The discussion of their findings regarding flow-logic-feedback techniques and the use of the hybrid model are contained in appendix E.

HOUSING AND URBAN RENEWAL

Urban renewal is another major problem area which offers challenges and opportunity for the systems technologist. At the Mid-America Conference on Urban Design held in 1966, the focus was placed on renewal and reconstruction of housing and commercial structures alike. Pressures of population—recent projections indicate 14 cities totaling 12 million people by the year 2000 [41]—are forcing government leadership and private enterprise to reexamine methods and materials involved in the creation of masses of new buildings. At the Mid-

[41] Ehrenkrantz, Ezra. Introductory remarks (for section 3, Urban construction). In Science, engineering; and the city, op cit., p. 67. [Quoting Robert C. Wood].

America Conference, the concept of "total downtown renewal" was discussed. The "Pei Plan"[42] (named for the firm responsible for the planning design) features five interlocking elements in a city's downtown district. Buildings are regrouped and more open space is provided. The five ingredients are: a convention center, an entertainment hub, a retail shopping core, a business and financial and government office district, and a downtown residential area.

Implementing such a plan takes money, energy, and the preparation of an imaginative set of procedures and techniques. There are, however, many obstacles to the ready adoption of modern tools and techniques which could speed up the construction of commerical structures, school, and domiciles. Ezra Ehrenkrantz discusses these problems in "Science, Engineering, and the City": [43]

In the building industry today, there are many barriers to the use of new ideas. An architect who perhaps can interest a group of manufacturers to do development work on new products for public work has then to specify these products on the basis of competitive bids with his work drawings and his specifications, describing the results of the research and development work of the one company, so that others can bid on it. Unless a number of bidders can bid to the same set of specifications, the requirements for competititive bidding are not met. Without providing rights to invention, it becomes difficult to get people to put in a great deal of effort on innovation. If one surmounts this initial problem and goes out to bid on the basis of general contractors installing these products in an initial installation the size and scope of usual building programs, one finds that the contingencies increase in cost with respect to future projects. Yet, there is no way of taking advantage of the learning curve and discounting the cost of the initial installation.

One specific project, undertaken by the North Carolina Fund Low Income Housing Demonstration Office, developed because of growing concern that innovative technology in designing and constructing low-income housing was not being used. In the project progress report, "A Computer Based Building Cost Analysis and Design System," [44] emphasis was placed on achieving an optimal combination of geometrical configuration, structure, materials, and siting for a given situation, expressed in terms of cost versus effectiveness.

The study team recognized that the technology of building is growing in complexity far more rapidly than designers can stay abreast of the innovations. Thus, any design tool capable of furnishing selected performance, environmental, and cost data upon demand can be a great asset. The project, begun late in 1965, aimed at the provision of certain categories of key information needed by the designer at the time when he needs them. Grant support for the project came from the Low-Income Housing Demonstration Program (U.S. Department of Housing and Urban Development) and the North Carolina Fund.

The Integrated Building Industry System (IBIS) consisted of a program capable of storing the geometrical description and dimensions of a building plus a frequently updated library of construction costs and specification data. The system possesses the capability of analyzing the design of the building against the matrix of cost and specification data, then producing reports which indicate the probable costs of

[42] Shirk, George H. The Pei Plan—total downtown renewal. In Proceedings of the Mid-America Conference on urban design. Kansas City, Midwest Research Institute [no date], pp. 71-79.
[43] Ehrenkrantz, op. cit., pp. 69-70.
[44] Cogswell, Arthur R., Werner Hausler, and C. David Sides, Jr. A computer based building cost analysis and design system. North Carolina Fund Low Income Housing Demonstration Office, Chapel Hill, North Carolina, March 1967. 29 pages, plus data forms.

construction of the selected design. Time-consuming hand calculations and input procedures are minimized. A few essential elements of information—dimensions and locations of wall, floor, and roof elements relative to a constant point of origin on a global coordinate system, together with listings of material and structural information—are entered into the ADP-supported system using an IBM graphic display system and a light pen (see fig. 65).[45]

FIG. 65.—INFORMATION REGARDING THE GEOMETRICAL FORM AND SIZE OF A BUILDING, TOGETHER WITH MATERIALS AND STRUCTURAL INFORMATION CAN BE INPUT USING THE CATHODE RAY TUBE AND LIGHT PEN. THE DESIGNER INDICATES THE APPROPRIATE TIME ON THE SCOPE WITH THE LIGHT PEN AND USES THE KEYBOARD TO ENTER THE INFORMATION REQUIRED.

---

[45] Ibid., p. 4.

Upon receipt of the key data, the machine is able to compute areas and quantities and issue these to the operator in the form of a series of reports: [46]

    1. Labor—a list of the various categories of labor required for the design, with estimated man-hours in each category.

    2. Materials—a list of the materials required, with quantities and costs.

    3. Equipment—a list of the equipment required, with costs.

Figure 66 shows a graphic display of these three categories of data.[47]

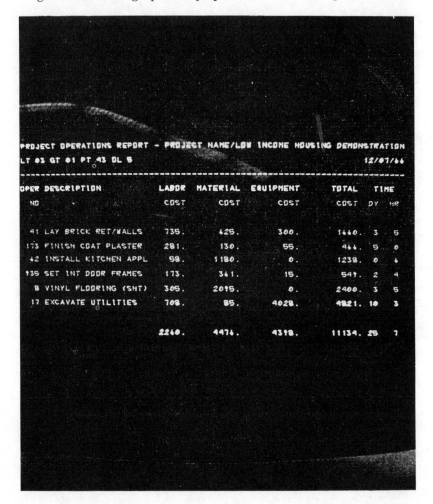

FIG. 66.—DISPLAYED ON THE SCOPE HERE IS AN OUTPUT REPORT INDICATING COSTS OF LABOR, MATERIAL, AND EQUIPMENT BY OPERATION. THE COST OF THE BRICK RETAINING WALLS FOR THE PROJECT UNDER ANALYSIS IS INDICATED, FOR EXAMPLE, AS $1,660 AND THE TIME OF CONSTRUCTION AS 3 DAYS AND 5 HOURS

[46] Ibid., pp. 6–7.
[47] Ibid., p. 18.

Other reports in the series include those concerned with:

4. Operations—a list of the construction operations required by the design under analysis.

5. Suppliers—a list of the material suppliers and shipping distances to the site.

6. Specifications—a printout of outline specifications or, if requested, complete specifications for the project.

7. Details—a list of the construction detail drawings required for the project.

8. Estimate—the estimate of the total construction cost of the project.

This project is representative of the creative thinking which is beginning to produce useful techniques. The problem of using technology in providing, among other things, shelter, will continue to be a subject of intensive study in the future. Such consideration must be made, as noted at the 1965 Engineering Foundation Research Conference on "The Social Consequences of Technology," within a new conceptual context.[48]

One reason that technology has played a no more creative role in the development of better urban areas is, that neither its generators nor its users consider urban requirements in any kind of a systems context—the focus is fragmented among individual products and buying agencies. We have the technology to construct the pieces of the city, but not the whole.

HEALTH SERVICES

The area of health services—administration, scope, resource allocation, measurement of results—has received in recent years much attention by Government managers and private consultants employing systems analysis. The creation of a health services management information system based on a thorough analysis of the elements and interactions integral to this active environment often is a first step. Figure 67 contains the key elements of an existing system together with proposed changes: [49]

[48] Alcott, James. Technology and urban needs. A statement from The Engineering Foundation Research Conference on the Social Consequences of Technology. Kansas City, Midwest Research Institute, 1966, p. 13.

[49] Amstutz, Arnold E. City management—a problem in systems analysis. In Technology review, vol. 71, No. 1, October/November 1968, p. 50.

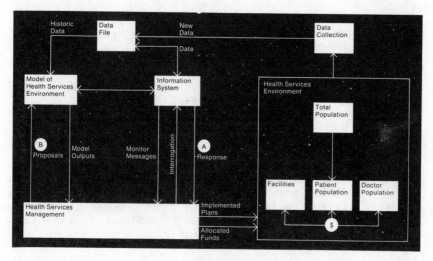

FIGURE 67

FIG. 67.—THE STRUCTURE OF A HEALTH SERVICES MANAGEMENT INFORMATION SYSTEM. MANAGEMENT CAN INTERROGATE THE SYSTEM CONCERNING THE CURRENT STATE OF THE ENVIRONMENT AT THE INTERACTIONS NOTED BY A AND CAN TEST PROPOSED PROGRAMS BY THE INTERACTION SET MARKED B. PROPOSED PLANS ARE COMMUNICATED TO THE INFORMATION SYSTEM, WHICH ESTABLISHES HYPOTHETICAL CONDITIONS FOR THE MODEL. RESULTS OBTAINED FROM THE MODEL ARE THEN TRANSFERRED TO THE INFORMATION SYSTEM, WHICH FORMATS THEM FOR PRESENTATION TO MANAGEMENT. FOLLOWING THIS PROCESS, MANAGEMENT IS ABLE TO EVALUATE THE CONDITIONAL RESULTS OF PROPOSED PROGRAMS USING THE SAME PROCEDURES AND EQUIPMENT EMPLOYED TO ASSESS THE CURRENT STATE OF THE ENVIRONMENT THROUGH INTERROGATION

In an allied area of study, a report was prepared for the Office of Statewide Planning for Vocation Rehabilitation Services in New York State, which stressed the role of systems analysis in this vital problem situation.[50] Significant discussions were held which considered such issues as whether the "systems analysis" philosophy was "out of harmony" with the prevailing philosophy of vocational rehabilitation.[51]

* * * an emphasis upon quantification, interconnected goals, management by objectives and policies for implementation appeared to some in the Division of Vocational Rehabilitation contrary to notions of individual worth and human dignity.

In modeling the Division of Vocational Rehabilitation (DVR) operating processes and goals, Drs. Adams and Balk reviewed the five major resource inputs—law, clients, services, funds, and staff— in terms of a designed processing sequence having five major decision points: diagnosis, selection, planning, service, and cloture. Figure 68 shows DVR as a system which accepts the resource inputs, processes them through several operations, and then produces certain outputs responsive to the identified organizational goals.[52]

Another related service, that of emergency ambulance service, was studied in New York City using the systems approach. Computer simulation was utilized in order to analyze quantitatively projected improvements in ambulance service. Emphasis was placed upon cost

[50] Adams, Harold and Walter Balk. An analysis of the division of vocation rehabilitation. Albany, State University of New York at Albany, 1968. 71 pages.
[51] Ibid., p. 2.
[52] Ibid., p. 16, fig. II.

effectiveness to the end that modifications were recommended in satellite station coverage and the augmenting of the vehicular force. Great care was taken to examine the three alternative plans both in terms of improved service and, if possible, reduced operating costs. Figures 69 and 70 reflect the findings of the systems analysis:[53]

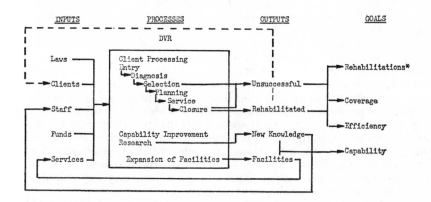

DVR VIEWED
A GOAL-ORIENTED SYSTEM

*Notes:
1. Rehabilitations: Number of cases closed in rehabilitated status during year.
2. Coverage : Number of cases closed during year.
3. Efficiency : Number of rehabilitations per 1 million dollars spent.
4. Capability : Average number of rehabilitations per 100 closed cases during year.

(Items 1 & 2 are product goals - Items 3 & 4 are process goals)

FIG. 68.—DVR VIEWED—A GOAL-ORIENTED SYSTEM

Alternative	Effectivenes (minutes saved)	Cost (dollars per month)	Cost/effective-ness (dollars per minute)	Cost per call
(A) Open a satellite	1.5	$2,343	$0.43	$0.64
(B) Add 3 ambulances	.3	6,286	5.73	1.72
(C) Add satellite and ambulances	2.6	8,629	.91	2.36

FIG. 69.—COST-EFFECTIVENESS OF ALTERNATIVE WAYS TO REDUCE RESPONSE TIMES

Alternative	Effectiveness (percentage points reduced below 20 minutes)	Number of calls per month reduced below 20 minutes	Cost (dollars per month)	Cost/effective-ness (dollars per call reduced)
(A) Open a satellite	3	110	$2,343	$21.30
(B) Add 3 ambulances	2	73	6,286	86.10
(C) Add satellite and ambulances	8	292	8,629	29.50

FIG. 70.—COST-EFFECTIVENESS OF ALTERNATIVE WAYS TO REDUCE EXCESSIVE DELAYS

[53] Emergency ambulance service. City of New York, Office of the mayor, March 8, 1968, p. 61, table 15 and table 16.

EDUCATION

School planning is yet another problem area which has drawn upon the skills of the operations research specialist and the computer technician. With the necessity of predicting the number of classrooms required, the allocation of teaching personnel, and the impact of school populations upon related municipal planning (e.g., transportation, housing), the employment of various new approaches is scarcely surprising. A team from the Wharton School of Finance and Commerce (in Philadelphia) designed a simulation model [54] (see fig. 71) which would be useful in examining the consequences of alternative allocation policies.

The model being discussed, called S.D. Two, contains specific representation of: areas within the district, grouping of students, grades, equipment types, and educational programs. It is sufficiently detailed to permit the "exploration of the effect of alternative educational program mixes or plans both on achievement and on the consumption of resources." [55] Both plans and policies are included as input data for the simulator, which then processes them within certain constraints: operating budgets, capital budgets, limitation on teacher and staff mobility, tenure, and desire for continuity of programs. The simulator then estimates the consequences of these policies in terms of: operating and capital expenditures, programs actually implemented, and changes in student achievement.

Another facet of school planning is considered in "an assignment program to establish school attendance boundaries and forecast construction needs." [56] The approach taken in studying the projected growth in the Bloomfield Hills School District (Oakland County, Mich.)—from 7,500 families in 1967 to an estimated 13,800 families in 1975—has been to develop an assignment program which will provide the following information: [57]

1. Pupil population growth forecasts;

2. Forecast of new school requirements and a basis for evaluation of potential building sites currently owned or available to the district;

3. Ultimate attendance boundaries for the district (the attendance boundaries expected after the district is saturated); and

4. The yearly assignment of geographic areas to the particular schools from the current year until the district is saturated.

The program has three major elements: part I (The Population Generator), part II (Ultimate Assignment of Subcensus Zone to Schools), and part III (Building Program and Yearly Subcensus Zone Assignments). While part I is programed in FORTRAN and accepts punched card census data, parts II and III are programed in SIMSCRIPT and accept punched card output from parts I and II to generate zone assignments which establish attendance boundaries for each school. The logical sequence of events comprising, respectively, parts I, II, and III are set forth in appendix F.

[54] Szekely, Miguel, Martin Stankard, and Roger Sisson. Design of a planning model for an urban school district. In Socioecon. plan. sci., vol. 1, No. 3, July 1968, p. 235, fig. 3.
[55] Ibid., p. 232.
[56] Ploughman, T., W. Darnton, and W. Heuser. An assignment program to establish school attendance boundaries and forecast construction needs. In Socioecon. plan. sci., vol. 1, No. 3, July 1968, pp. 243–258.
[57] Ibid., pp. 244–245.

Design of a Planning Model for an Urban School District

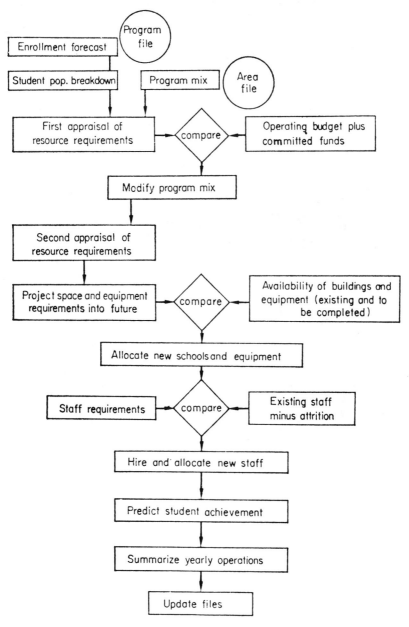

FIG. 71.—SIMULATION MODEL

The problem of racial desegregation of school systems has received much study during the past 15 years. Several plans for assigning pupils on the basis of several criteria in order to achieve a preestablished ethnic balance have been put forward. The utilization of operations research techniques and the computer have provided flexibility in data handling and an enhancement of projection capabilities. In "An Operations Research Approach to Racial Desegregation of School Systems," [58] the focus was placed upon the use of mass transportation to obtain a certain mix by race in each school. The system, called "Minitran," is approached in this way. Given: the distribution, by race, of students in a community; the location and capacity of each school; the ethnic composition desired at each school; and the configuration of mass transportation lines in the community. The objective, then, is to find: a plan of assignment of students to schools which achieves the desired ethnic composition at each school; requires no student to travel more than a specified number of minutes per day; and minimizes the total daily student traveltime. The hypothetical community of "Duckberg" was created as a source of test data; a map of the area appears in figure 72.[59]

[58] Clarke, S. and J. Surkis. An operations research approach to racial desegregation of school systems. In Socioecon. plan. sci., vol. 1, No. 3, July 1968. pp. 259–272.
[59] Ibid.. p. 268, fig. 1 (appendix A).

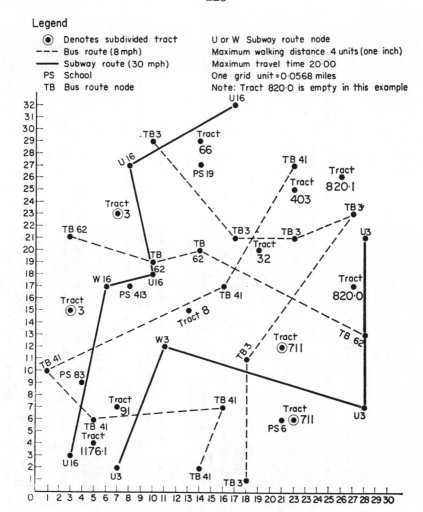

FIG. 72.—MAP OF HYPOTHETICAL COMMUNITY (DUCKBERG)

VIII–47

In developing the minimal transportation times, a set of eight computer programs were written in FORTRAN IV (and called "Minitran"). A typical computer output in the form of a "Table of Assignments" is featured in figure 73.[60]

---

[60] *Ibid.*, p. (appendix C).

APPENDIX C—*Output for hypothetical community ("Duckburg")*

TABLE OF ASSIGNMENTS

```
TRACT   1 (000003    3   15)-
        ASSIGN TO SCHOOL  12 (PS 83    4    9)- RACE A    225, RACE B   794, RACE C   123 - TOTAL   1142.
        RECOMMENDED ROUTE-
        PS  83    9 - TB 41    8 - TB 41    10 - TB 41    1    5    12 - 000003    3    15 -
        ROUND-TRIP DAILY COST $ 0.30 PER STUDENT OR $  342.60 TOTAL, TIME  6.24 MIN PER STUDENT OR  118.77 STUDENT-HRS TOTAL.

TRACT   1 (000003    3   15)- TOTAL STUDENTS  1142, RACE A (PR)   225, RACE B (NEG)   794, RACE C (OTH)   123.

TRACT   2 (1176-1    5   41)-

TRACT   2 (1176-1    5   41)-
        ASSIGN TO SCHOOL  12 (PS 83    4    9)- RACE A      7, RACE B    15, RACE C  1106 - TOTAL   1128.
        RECOMMENDED ROUTE-
        PS  83    9 - TB 41    8 - TB 41    3    5    6 - 1176-1    5    4 -
        ROUND-TRIP DAILY COST $ 0.30 PER STUDENT OR $  338.40 TOTAL, TIME  2.44 MIN PER STUDENT OR   45.87 STUDENT-HRS TOTAL.

TRACT   2 (1176-1    5   41)- TOTAL STUDENTS  1128, RACE A (PR)     7, RACE B (NEG)    15, RACE C (OTH)  1106.

TRACT   3 (000091    7   71)-

TRACT   3 (000091    7   71)-
        ASSIGN TO SCHOOL  12 (PS 83    4    9)- RACE A    116, RACE B   333, RACE C   282 - TOTAL    731.
        RECOMMENDED ROUTE-
        PS  83    4    9 - 000091    7    7 -
        ROUND-TRIP DAILY COST $ 0.00 PER STUDENT OR $    0.00 TOTAL, TIME  0.02 MIN PER STUDENT OR    0.24 STUDENT-HRS TOTAL.

TRACT   3 (000091    7   71)- TOTAL STUDENTS   731, RACE A (PR)   116, RACE B (NEG)   333, RACE C (OTH)   282.

TRACT   4 (000003    7   23)-

TRACT   4 (000003    7   23)-
        ASSIGN TO SCHOOL  13 (PS 413    8   17)- RACE A    225, RACE B   795, RACE C   124 - TOTAL   1144.
        RECOMMENDED ROUTE-
        PS 413    8   17 - TB 62    9   19 - TB 62    6   20 - 000003    7   23 -
        ROUND-TRIP DAILY COST $ 0.30 PER STUDENT OR $  343.20 TOTAL, TIME  2.72 MIN PER STUDENT OR   51.86 STUDENT-HRS TOTAL.

TRACT   4 (000003    7   23)- TOTAL STUDENTS  1144, RACE A (PR)   225, RACE B (NEG)   795, RACE C (OTH)   124.

TRACT   5 (000008   13   15)-

TRACT   5 (000008   13   15)-
        ASSIGN TO SCHOOL  13 (PS 413    8   17)- RACE A     51, RACE B   991, RACE C     8 - TOTAL   1050.
        RECOMMENDED ROUTE-
        PS 413    8   17 - TB 41    9   14 - TB 41   11   13    15 - 000008   13   15 -
        ROUND-TRIP DAILY COST $ 0.30 PER STUDENT OR $  315.00 TOTAL, TIME  3.64 MIN PER STUDENT OR   63.70 STUDENT-HRS TOTAL.

TRACT   5 (000008   13   15)- TOTAL STUDENTS  1050, RACE A (PR)    51, RACE B (NEG)   991, RACE C (OTH)     8.

TRACT   6 (000066   14   29)-

TRACT   6 (000066   14   29)-
        ASSIGN TO SCHOOL  13 (PS 413    8   17)- RACE A    166, RACE B     0, RACE C     0 - TOTAL    166.
        RECOMMENDED ROUTE-
        PS 413    8   17 - U 16   18 - U 16    8   27 - TR 3   10   29 - TB 3   12   27 -
        000066   14   29 -
        ROUND-TRIP DAILY COST $ 0.30 PER STUDENT OR $   49.8C TOTAL, TIME 34.52 MIN PER STUDENT OR   95.51 STUDENT-HRS TOTAL.
```

FIGURE 73

A similar effort was undertaken by Ide Associates, Inc., in mid-1966 for the Philadelphia school district, which sought to create an operational address coding guide and develop systems for computer coding of addresses for the preparation of detailed geographical tabulations of the school population and for the computer transfer of statistical data to maps. One key aspect of this study was the reflection of racial integration and its impact on various planning activities. The systems developed were applied, with the support of automatic data processing programs and equipments, to the coding of addresses, including data collected from teachers and pupils. Problems of standardizing the spelling of street names were coped with using ADP techniques, a block number system was developed, and a computer-generated "map like" street ordered listing was prepared with even and odd addresses presented on a specially formated sheet.

Age and race data were prepared on a school service area basis in order to expedite planning for school, classroom, and bussing. Pupil names were added in the second year of operation, thus allowing the preparation of systemwide directories which are useful in a number of administrative and research functions. Computer over-printing of block data on a city map resulted in many benefits for the users. An example of a map prepared with ADP technology appears as figure 74.[61]

MAP 03B1.01   TOTAL AND NEGRO PUBLIC SCHOOL ENROLLMENT, ALL GRADES, 10-31-67

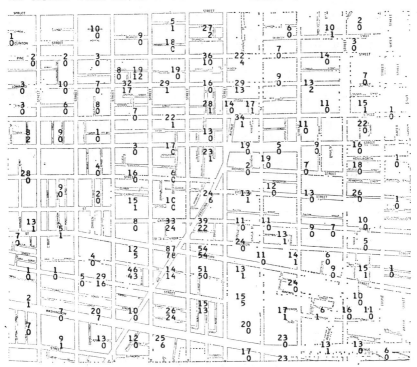

Data preparation and computer recording by Ide Associates, Inc.

PHILADELPHIA CITY PLANNING COMMISSION 1965

FIGURE 74

[61] Ide, Edward A. Address coding to produce age-race data by city block for school planning. Philadelphia, Ide Associates, Inc., 1968. Attachment: map 03B1.01.

In viewing the achievements registered by States and localities in applying systems analysis skills and automatic data processing techniques to various social and community problems, it should be remembered that the tool is always subservient to its controller. The problem areas selected for discussion in this section—transportation planning, environmental pollution, water resource development, urban renewal, health services, and school planning—must be of concern to all of the people. No one is exempt from the impact of a breakdown in planning or the operation of proper services. The responses to the questionnaire prepared by the Senate Special Subcommittee on the Utilization of Scientific Manpower show an increasing commitment on the part of State and city management to the employment of systems technology. Section VII contains a narrative and graphic presentation of the developments throughout the country on a problem-by-problem basis.

As developmental projects are initiated at the various levels of government, the need for communication between elements within a State or region becomes more and more important. The focus of comment now will rest upon programs and projects involving information exchange mechanisms and their role in upgrading regional or multiple community planning and functioning.

D. Information Exchange Mechanisms

One requirement which is preempting the attention and resources of numerous Federal departments and State and local governments is to develop programs and procedures which are responsive to the continually changing conditions which govern our existence. The Department of Housing and Urban Development, which views community planning as a continuous process, has sought consultant counsel in examining urban and regional information management systems. The System Development Corporation was commissioned by HUD to undertake an exhaustive survey of this key area, and commenced by posing a series of questions which should be considered by any group seeking to cope with this difficult problem: [62]

1. What equipment best suits each application?
2. What benefits are offered by new developments such as general-purpose programs, natural language processing, time-sharing?
3. What techniques might enhance regional and national cooperation by making each organization's data available to and usable by others?
4. How should systems be designed to provide means of perceiving trends and evaluating alternatives?
5. What are the best ways to gather significant information and keep it up to date?
6. What cost and manpower requirements are involved?
7. How can information privacy be maintained?

In essence, these very questions were addressed by the Lockheed study team which undertook the State of California information system analysis in 1965. (See section VI for further discussion.)

---

[62] Lanham, Richard, "People, problems, and planning." In SDC magazine, vol. 10, No. 10, October 1967. p. 7.

VARIOUS TYPES OF EXCHANGE MECHANISMS

The information exchange mechanisms being considered or implemented are of several types. First, there is the national network designed to store, manipulate, and communicate information of a special type; for example, the National Crime Information Center (NCIC) of the Federal Bureau of Investigation. This system links local, State, and Federal agencies through a computer-supported capability, with various types of communications links being utilized. Records on file include stolen vehicles, vehicles used in the perpetration of felonies, stolen engines and transmissions, stolen or missing license plates when all plates issued for a specific vehicle are missing, stolen guns, other items of stolen property which are serially identifiable, and wanted persons. The system is operational 7 days a week, 22 hours a day; transactions with the NCIC computer—entries and inquiries—number in excess of 10,000 per day. The projected nationwide network is depicted in figure 75.[63]

A second type of information system is exemplified by the recommended Southwestern Regional Information Dissemination Network [64] proposed to the U.S. Department of Commerce by a group of universities in the State of Texas. The aim of this system is to make selected scientific and technical information, and documents containing this type of information, available to private sector elements in the region being served.

HIGH SPEED CIRCUITS
LOW SPEED CIRCUITS
TERMINAL INTERCHANGES
TERMINALS

Courtesy Washington Evening Star.

A preliminary concept of a completed nationwide network.

FIGURE 75

[63] A national crime information center. Reprinted from the FBI law enforcement bulletin, May 1966. p. 3. Also see NCIC progress report. Reprinted from the FBI law enforcement bulletin, September 1967. 7 pages.
[64] A southwestern regional information dissemination network. A special merit proposal to the U.S. Department of Commerce technical services program by the Coordinating Board, Texas College and University System [no date]. 98 pages plus appendixes.

A third type of information sharing (or exchange) mechanism might be the coordinated metropolitan data system suggested by the Public Affairs Research Institute of San Diego State College for the metropolitan San Diego area. Emphasis is placed on the establishment of realistic criteria for such a (coordinated) system: [65]

1. The system should be flexible and lend itself either to expansion or redesign in the light of increased experience and changing requirements.

2. The system should be related to the regular operating processes of the system participants.

3. The system must be user-oriented.

4. The system must return benefits to data suppliers.

5. The system must provide adequate safeguards to protect the confidentiality of data and to insure proper authorization for use of data in the system.

6. The system must not exceed the manpower, equipment, or financial resources of the participating agencies.

7. The system must enjoy the full support of heads of the participating agencies.

8. The development of any system must bear in mind other statewide and regional information systems and changes in Bureau of Census procedures.

The importance of establishing an awareness within the Congress, certain Federal agencies, and State and local governments of the existence of these systems, or proposals for their creation, cannot be overemphasized. Such information exchange and sharing mechanisms rely heavily on automatic data processing and advanced communications equipment and techniques. To establish and maintain a multifaceted system such as the ones described is costly and takes a long time. Here the role of the systems analyst, interacting closely with the responsible city manager, State planning section, or Federal overseer is critical. A full appreciation of the resources necessary to implement even a fairly straightforward system must be gained by all cognizant decisionmakers, as noted above.

Regional information sharing has engaged the efforts of government, university, and industrial experts alike. A discussion of a few major studies published during the past 5 years follows, with the selected projects being chosen within the context of three major types of information exchange (or sharing) mechanisms. The first features several components, each with its own "system," which may include ADP equipment; exchange is performed without central coordination, and may take the form of hard copy, teletype, telephone, or tape (magnetic or paper). A simple graphic of this mechanism is shown in figure 76. It should be noted that equal interchange of information between all components is not necessarily the case.

The second type of mechanism has several components, each with its individual information handling capability (including ADP), but with central coordination and administration at one location. Direction of the exchange of information stems from the "command and control" point, but no master index to holdings nor reproduction of materials within the system takes place in this type of system. Figure 77 shows this system network.

[65] Haak, Harold H. and W. Richard Bigger. A coordinated data system for Metropolitan San Diego. San Diego, Public Affairs Research Institute, San Diego State College, Sept. 1, 1966. p. v.

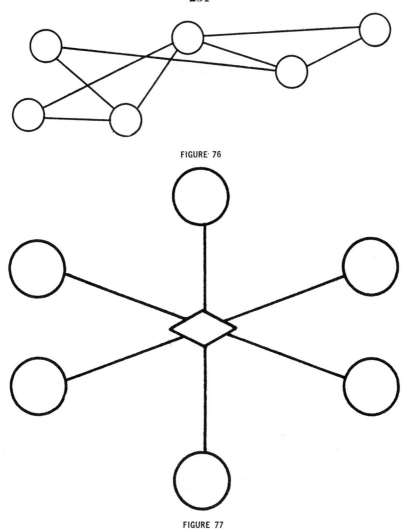

FIGURE 76

FIGURE 77

The third type of system for exchanging and sharing information combines central direction and coordination with the maintenance of a current index of holdings of the system members. In addition, it may offer a central stores capability; that is, multiple copies of reports, etc., of probable broad interest will be stored for expected distribution. Also, photoreproduction services may be offered. These additional benefits, needless to say, are costly to establish and maintain, but may be essential to the functioning of a particular exchange arrangement. In the early days of the IBM Corp. Selective Dissemination of Information (SDI) system, copies of any system holdings (or their reproductions) had to be obtained from the originating company element. Today, the sophisticated ITIRC (IBM Technical Information Retrieval Center) system provides all of the advantages enumerated above. Figure 77 depicts this third type of mechanism.

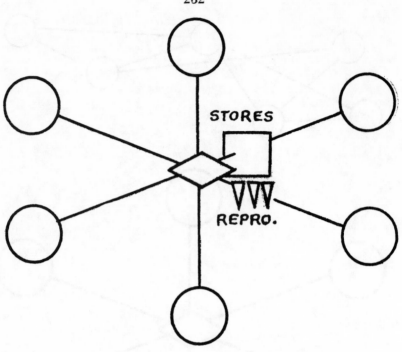

FIG. 78.—EXAMPLES OF REGIONAL AND MUNICIPAL SYSTEMS

EXAMPLES OF REGIONAL AND MUNICIPAL SYSTEMS

Discussion of a few typical information exchange mechanisms follows, based on systems proposed or implemented since 1964. In the study jointly prepared by the University of Connecticut and the Travelers Research Center, Inc., entitled "A Regional Municipal Information Handling Service for the Capitol Region," [66] the increasing cost of data handling was noted:

> The cost of collecting, processing, and reporting data is increasing at an average annual rate of 10 percent and this trend will continue for many years to come, unless this function is automated through the utilization of computer based systems.

Awareness also was registered that the principal cost in using computer technology is not in the "hardware" but in the effective application of that equipment through appropriate systems design and programs. The analyst team concluded that the most efficient— as expressed in terms of cost versus quality—means of applying ADP technology to the information needs of a region and its components is through the organization of a " 'unified' regional municipal information handling service." Prior to a detailed analysis of the

[66] University of Connecticut and the Travelers Research Center, Inc. A regional municipal information handling service for the capitol region. Report of research supported by the Connecticut Research Commission, vol. 1, October 1967. p. 1.

service per se, an examination was made of the relationship between "operating data" and "management information." These same considerations are being faced by governmental units at every level, and will often combine to force action on modernizing the overall information handling system. Figure 79 indicates graphically the relationship between operating data and management information.[67]

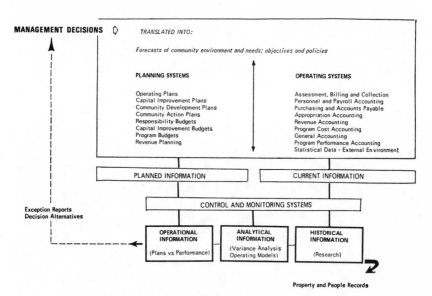

FIG. 79.—THE RELATIONSHIP OF OPERATING DATA AND MANAGEMENT INFORMATION

This study also discusses at length the savings to be realized through the adoption of an intergated ADP-based system. For example, the cost comparison based on an hour of computer time is: $75 for a regional computer, $200–$225 for a commercial computer, and $150 to $1,000 for an individual municipality's computer (contingent upon the size of the town).[68] Useful cost figures, for selected townships, are included, and definite savings are indicated when the centralized concept is followed.

An interurban computer network embodying some similar characteristics and facing some similar problems has been designed for San Gabriel Managers Association by the Herbert H. Isaacs Research and Consulting firm. The area affected is shown in figure 80.[69]

[67] Ibid., chart 1 (between pp. 13 and 14).
[68] Ibid., p. 27.
[69] Interurban automation. In public automation, vol. 3, No. 2, February 1967, p. 1.

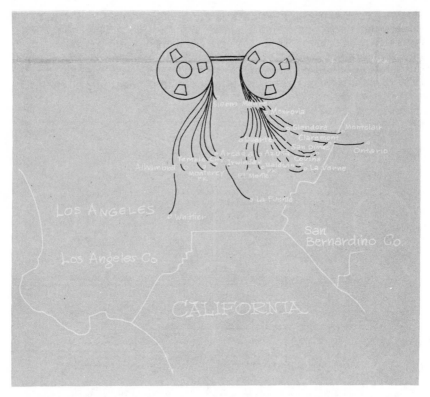

FIG. 80.—AN INTERURBAN COMPUTER NETWORK UNDER CONSTRUCTION IN CALIFORNIA WOULD LINK 20 CITIES AND TOWNS TO A SINGLE, LARGE COMPUTER CENTER CAPABLE OF HANDLING THEIR DATA PROCESSING NEEDS ON A TIME-SHARING BASIS.

In analyzing the needs of the 11 cities involved, the analysis team first examined the extent of information sharing within each city on the basis of selected departmental activity. Only manual or semiautomatic data processing was in progress. In the manual approach, all data collection, filing, and reproduction was done by hand, with arithmetic computations and report preparation being carried out with the use of basic office devices. In the semiautomatic approach, all data collection and the bulk of filing was done by hand, but computational work and report preparation were performed using an IBM 6400, or upon occasion by an external service bureau organization. The extent to which information was exchanged traditionally was constrained by the cost of restructuring a given file so that it might be useful to one or more other departments within a city, or to users in adjoining communities. This inhibition exists throughout all governmental elements—Federal, State, and local—and points up the need for standardizing the formating and filing of information, and the importance of having machine systems which can exchange data without expensive buffering operations or the actual physical transfer through rekeypunching or some comparable action.

The San Gabriel study selectively scrutinized certain areas of prime concern to the cities' management: Appropriation accounting, utility billing, police statistical data, and city planning information. The basic requirements of a multiple user, ADP-oriented system were discussed, with emphasis on such questions as: why is a simultaneous user, time sharing system preferred to a sequential user batch processing system? Also, the trade offs between generalized and specific program languages were reviewed with reference to the local situation. The analysis resulted in a series of recommendations for a centralized network, as reflected in figure 81.[70]

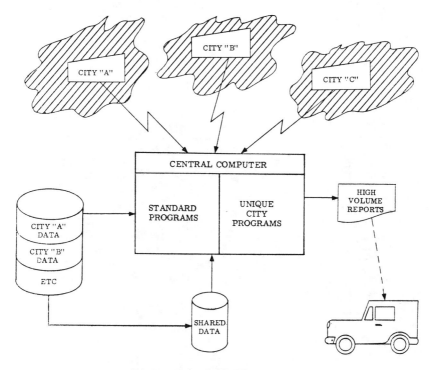

FIG. 81.—VALLEY COMPUTER SYSTEM

The Cincinnati-Hamilton County Regional Computer Center concept originated as the result of a study by Ernst & Ernst consultants in 1965. Operation CINSY (Cincinnati Information System) evolved, involving 25 city agencies. An inhouse staff of analysts, programers, and operators was formed, and a machine capability established. Initial emphasis was placed upon providing priority information to police officials. The consultant study indicated that certain factors dictated the need for a regional information center which could support urban and rural law enforcement groups, regardless of size. Project CLEAR (County Law Enforcement Applied Regionally) was created to serve the 38 police departments in the Cincinnati metropolitan area early in 1967. While the functioning of a police information system has been given initial priority, Project CLEAR also will include: economic information systems (e.g., accounting), environment information systems (water distribution), land information systems (county auditor and county recorder

[70] Isaacs, Herbert H. and Michael McCune. The feasibility of a central computer system for San Gabriel Valley cities. Los Angeles, Herbert H. Isaacs Research and Consulting, Inc., Jan. 31, 1967, p. 69, fig. 12.

activities), health information systems (public health and general hospital services), and people information systems (income tax). Task force studies have been undertaken in the specific areas noted parenthetically. Care has been taken to link the Cincinnati-Hamilton County effort to the State highway patrol office in Columbus, Ohio, and to the National Crime Information Center (discussed earlier in this section) in Washington, D.C. Two RCA Spectra 70/45 computers are the heart of the machine support. Figure 82 shows how the CLEAR Communications Center services the multiple information needs of the region.[71]

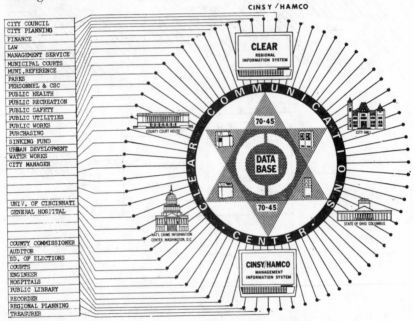

FIG. 82.—MANAGEMENT INFORMATION SYSTEM

In Detroit, Mich., the initiative for developing an overall plan for a regional ADP system was taken by the Metropolitan Fund, Inc., Data Processing Policy Committee. This nonprofit corporation concentrates on developing research and action on metropolitan problems. In presenting the findings of its systems analysis study, the recommendation was made that:[72]

Local units of government in southeastern Michigan should centralize data processing using one large centrally located computer system.

(a) The computer system would be a "utility" that would be available to all government-associated functions and schools, supplying computer power in the same manner that a powerplant supplies electrical power.

(b) Users would gain access to the computer power via telephone lines and computer input and output equipment located in their own offices. The users would be called remote users since their input and output equipment is physically remote in terms of the central system.

(c) The complex, consisting of the centrally located systems and the multitude of remote users, would constitute a Total Information Processing System (TIPS).

---

[71] Cincinnati-Hamilton County Regional Computer Center. Annual report 1967. P. 39.
[72] Stitelman, Leonard, "Automation in Government: A Computer Survey of the Detroit Metropolitan Region." Detroit, Metropolitan Fund, Inc., November 1967, pp. xi–xii.

(d) The TIPS would be capable of performing all of the different varieties of information processing peculiar to governments, including the information processing performed by counties, municipalities, cities, grade schools through universities, city and county hospitals, police departments, departments of public works, sanitation departments, traffic controllers, taxing agencies, unemployment administration, water districts, et cetera. Such information processing applications as payroll, tax billing, utility billing, school and class scheduling, personnel administration, et cetera, could all be performed using the same central system.

(e) Central files built up by the various users could be made available to remote users in the State and Federal Governments (who are also undertaking programs such as the program here described).

(f) The central system would operate on an "online," "real-time," "time-shared" basis. The users are "online" in that they have direct access to the central system whether their remote input and output equipment is in Livonia, Garden City, or Detroit; the users operate on a "real-time" basis since the response to their input is virtually instantaneous (especially in terms of existing response or turn-around times); the users operate on a "time-shared" basis since many users can be connected to the central system, running different jobs, all at the same time.

The study effort stressed the importance of achieving a mutually accessible data base and ADP capability—one which would service the more than 400 Government jurisdictions in the metropolitan area—through the more efficient and economical use of available resources. Existing machine and personnel capabilities were inventoried and assessed, in order to arrive at the recommendations listed above. Figure 83 shows the number of computer installations available to serve the six-county region,[73] and figure 84 indicates the data processing personnel on hand, to furnish expert analytical and programing support, plus machine operation.[74]

The Total Information Processing System visualized by the design team epitomizes the type of effort being undertaken across the Nation by multijurisdictional groups. The sharing of files, the provision of a quick-time access to centrally stored data, the establishment of compatible formats for data and machine programing—all of these factors are instrumental in the successful design and operation of a coordinated regional information handling system.

Jurisdiction	In operation	On order[1]	Total
County	(4)	(1)	(5)
Oakland	1		1
Wayne	3	1	4
Municipality	(4)	(1)	(5)
Detroit[2]	3		3
Pontiac		1	1
Wyandotte	1		1
School district	(6)	(4)	(10)
Macomb County Intermediate	1		1
Oakland County Intermediate	1		1
Washtenaw County Intermediate	1		1
Detroit[3]	1		1
Dearborn		1	1
Hazel Park		1	1
Livonia	1		1
Port Huron area		1	1
Waterford	1		1
Wayne Community	1		1
Total	14	6	20

FIG. 83.—COMPUTER INSTALLATIONS IN DETROIT SIX-COUNTY AREA LOCAL GOVERNMENTS BY JURISDICTION

[1] Does not include replacements for current equipment.
[2] Two computers under controller counted as one installation.
[3] Two computers counted as one installatioi.

[73] Ibid., p. 4, table 1.1.
[74] Ibid., p. 9, table 2.1.

Jurisdiction	Programers and systems analysts	Total EDP personnel
County:		
Oakland	3	13
Wayne	3	24
Municipality:		
Detroit	19	82
Wyandotte	1	5
School district:		
Macomb intermediate	4	7
Oakland Intermediate	5	19
Detroit	6	28
Livonia [1]	2	2
Waterford	4	11
Wayne Community	10	3
Total	48	194

[1] Data preparation at schools.

FIG. 84.—DATA PROCESSING PERSONNEL IN DETROIT SIX-COUNTY AREA COMPUTER INSTALLATIONS

EXAMPLE OF A NATIONAL NETWORK

The final example of an information exchange mechanism concerns a requirements survey, network design, and organizational planning study prepared for the Department of Housing and Urban Development by the Midwest Research Institute. The report, published late in 1968, is entitled "A National Clearinghouse Network for Urban Information." [75] The network discussed would be broader than any of the others described in this section. It would encompass information centers inside of HUD, a few Federal agencies, and yet other public and private entities. Specific information investment goals for such an operation would have to be delineated and a group of organizations and individuals whose impact on urban affairs would qualify them as bona fide users listed; for the purposes of the study it was assumed that this user list would number less than 5,000. From nine candidate HUD programs preselected for study, five programs were chosen as "windows" through which the urban information community could be viewed. These five programs thus became a focal point for the study:

1. Rent Supplement Program (Federal Housing Administration).

2. Neighborhood Facility Grants (Renewal and Housing Assistance).

3. Urban Renewal Project Grants (Renewal and Housing Assistance).

4. Model City Grants (Model Cities).

5. Urban Planning Assistance Grants (Metropolitan Development).

In developing an overall concept for a clearinghouse, the designers studied the potential users, their goals, participation in HUD pro-

[75] A national clearinghouse network for urban information. Requirements survey, network design and organizational planning. Progress report No. 1 for Department of Housing and Urban Development, Sept. 24, 1968. Kansas City, Midwest Research Institute. 30 pages.

grams, information sources and categories, and flow projections. The role of people in such an information system is portrayed in figure 85.[76]

INFORMATION FLOW PROJECTIONS

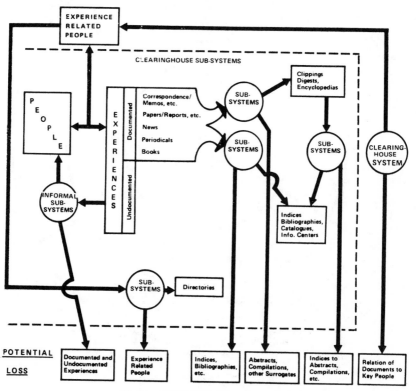

FIG. 85.—INFORMATION FLOW PROJECTION

A number of information categories emerged as the result of creating a conceptual "map" which characterized the various information environments in which urban "actors" move. Figure 86 [77] centers on HUD's internal management information environment, and then expands outward to include HUD in-house information activities, Federal information activities outside HUD, and private sector information elements. Similar charts could prove useful to the decision-makers responsible for shaping solutions to related social and community problems.

[76] Ibid., p. 23, fig. 4.
[77] Ibid., p. 25, fig. 5.

240

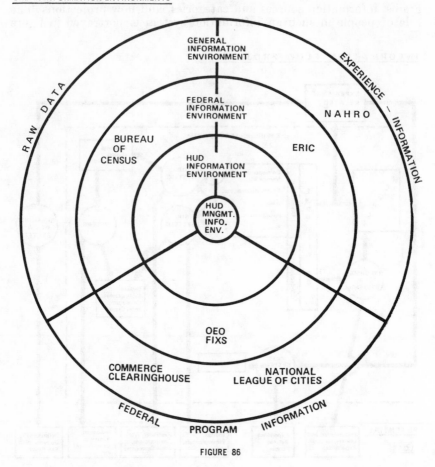

FIGURE 86

E. Summary

As the proposals submitted by university and industrial researchers have been reviewed and discussed, and the results of actual experimentation on various projects have been revealed, the need for a careful compilation of these factual and interpretive data appears to be very real. So fragmented is the information on the far-flung State, local, and sometimes regional projects, that the government manager who is seeking comparable program information may never discover anything of value to him. Legislative suggestions, such as the Scott-Morse and Kennedy proposals (see section IV. B.), take cognizance of the need for information exchange and the surveying of program results for storage in a file which may be accessed by public and private organizations.

The development and application of special computer programs, multidimensional models, and sophisticated man-machine techniques have been noted through a selection of examples. Many of the technological innovations matured in the aerospace community, which now strives to adapt its capabilities to the civil sector. As the result of encouragement from the top levels of government and a desire to diversify, several of the large companies are dedicating resources to the new problem areas. Five major companies reportedly have established internal structures in order to acquire new business in such areas as urban redevelopment and waste management.[78] Six other firms have formed a consortium called Urban Systems Associates, Inc., which will counsel State and municipal customers on how to qualify for Federal money which will allow them to undertake corrective programs for their localities.[79]

The extent to which private industry can be expected to invest its own funds in order to develop problem solving approaches and hardware suitable to alleviate the social and community problems increasingly is being debated. A conference held at Airlie House entitled "The Urban Challenge: the Management and Institutional Response" brought together leaders from industry, analysts of a wide range of special problems from universities, and selected government representatives. One recommendation resulting from this meeting was to form an Urban Management Institute, which could be academically situated and offer a point of cohesion for developing orientation and training seminars, an information system for handling relevant socioeconomic data, and the establishment of a full-time staff. Gen. Bernard A. Schriever, the conference chairman, also identified four major steps which could be taken in order to position the Nation to better overcome urban problems:

1. Creation of a national commission;
2. Creation of a regional planning authority for each urban area;
3. Creation of city system management offices; and
4. Development of long-term programs for cities' rehabilitation.[80]

The involvement of certain Federal agencies in sponsoring research and development in these new areas has been indicated. In particular, the Department of Housing and Urban Development, the Department of Commerce, the Office of Economic Opportunity, and the Department of Health, Education, and Welfare have been active. The U.S. Arms Control and Disarmament Agency commissioned two studies dealing with the potential for applying systems technology to civil sector problems: "Technological Innovation in Civilian Public Areas," prepared by Analytic Services, Inc.,[81] and "Defense Systems Resources in the Civil Sector," a University of Denver Research Institute effort.[82]

---

[78] Gregory, William H. Aerospace and social challenge, part 1: Industry probes socioeconomic markets. In Aviation Week & Space Technology, vol. 88, No. 24, June 10, 1968, p. 39. This series is continued in the June 17, 1968 (part 2) and July 1, 1968 (part 3) issues.

[79] Down Washington corridors. In Scientific-Engineering-Technical Manpower Comments, vol. 5, No. 11, December 1968, p. 3.

[80] The urban challenge: the management and institutional response. A conference sponsored by the George Washington University, North American Rockwell Corp., and The System Development Corp., Airlie House, Warrenton, Va., June 19–21, 1968. [Remarks of Gen. Bernard A. Schriever].

[81] Black, Ronald P. and Charles W. Foreman. Technological innovation in civilian problem areas. Falls Church, Va., Analytic Services, Inc., September 1967, 95 pages plus appendixes.

[82] Gilmore, John S., John J. Ryan and William S. Gould. Defense systems resources in the civil sector: an evolving approach, an uncertain market. Denver, University of Denver Research Institute, July 1967, 201 pages.

Through the media of studies, symposia, and seminars, the policy-makers in government are commencing to comprehend the jargon and the approaches of the systems community. Conversely, the "think" groups, with their staffs of experienced technologists, are beginning to view some of their job challenges in terms of the world about them. These confrontations, either on a group or man-to-man basis, can spell the difference between success and failure as resources are marshaled to overcome the Nation's problems.

APPENDIX A

Glossary of Terms

The purpose of this glossary is to provide a basis of definition for selected terms in the areas of systems analysis and design, operations research, and automatic data processing. These definitions originally were contained in the "Datamation ADP Glossary." [1] Excluded from this selection are terms with esoteric meanings, terms reflecting complexities or details of an operation or mechanisms, and others whose usage is so common as to render their repetition here unnecessary.

Access, random

1. Pertaining to the process of obtaining information from or placing information into storage where the time required for such access is independent of the location of the information most recently obtained or placed in storage.

2. Pertaining to a device in which random access, as defined in definition 1, can be achieved without effective penalty in time.

Address

1. An identification, represented by a name, label, or number, for a register or location in storage. Addresses are also a part of an instruction word along with commands, tags, and other symbols.

2. The part of an instruction which specifies and operand for the instruction.

Alphabetic-numeric

The characters which include letters of the alphabet, numerals, and other symbols such as punctuation or mathematical symbols.

Alphanumeric

A contraction of alphabetic-numeric.

Analysis, numerical

The study of methods of obtaining useful quantitative solutions to mathematical problems, regardless of whether an analytic solution exists or not, and the study of the errors and bounds on errors in obtaining such solutions.

Analysis, Systems

The examination of an activity, procedure, method, technique, or a business to determine what must be accomplished and how the necessary operations may best be accomplished.

---

[1] Datamation ADP glossary. Published by F. D. Thompson Publications, Inc., Los Angeles, Calif. [no date]. 62 pp.

Analyst

A person skilled in the definition of and the development of techniques for the solving of a problem; especially those techniques for solutions on a computer.

Application

The system or problem to which a computer is applied. Reference is often made to an application as being either of the computational type, wherein arithmetic computations predominate, or of the data processing type, wherein data handling operations predominate.

Assemble

1. To integrate subroutines that are supplied, selected, or generated into the main routine, by means of preset parameters, by adapting, or changing relative and symbolic addresses to absolute form, or by placing them in storage.

2. To operate, or perform the functions of an assembler.

Auto-Abstract

1. A collection of words selected from a document, arranged in a meaningful order, commonly by an automatic or machine method.

2. To select an assemblage of key words from a document, commonly by an automatic or machine method.

Automation

1. The implementation of processes by automatic means.

2. The theory, art, or technique of making a process more automatic.

3. The investigation, design, development, and application of methods of rendering processes automatic, self-moving, or self-controlling.

Automation, source data

The many methods of recording information in coded forms on paper tapes, punched cards, or tags that can be used over and over again to produce many other records without rewriting. Synonymous with SDA.

Binary

A characteristic, property, or condition in which there are but two possible alternatives; for example, the binary number system using two as its base and using only the digits zero and one. Related to (decimal, binary coded) and clarified by (systems, number).

Bionics

The application of knowledge gained from the analysis of living systems to the creation of hardware that will perform functions in a manner analogous to the more sophisticated functions of the living system.

Bit

1. An abbreviation of binary digit.

2. A single character in a binary number.

3. A single pulse in a group of pulses.

4. A unit of information capacity of a storage device. The capacity in bits is the logarithm to the base two of the number of possible states of the device. Related to (capacity, storage).

Buffer

1. An internal portion of a data processing system serving as intermediary storage between two storage or data handling systems with different access times or formats; usually to connect an input or output device with the main or internal high-speed storage. Clarified by (storage, buffer (4)).

2. A logical OR circuit.

3. An isolating component designed to eliminate the reaction of a driven circuit on the circuits driving it; e.g., a buffer amplifier.

4. A diode.

Calculator

1. A device that performs primarily arithmetic operations based upon data and instructions inserted manually or contained on punch cards. It is sometimes used interchangeably with computer.

2. A computer.

Card, edge notched

A card of any size provided with a series of holes on one or more edges for use in coding information for a simple mechanical search technique. Each hole position may be coded to represent an item of information by notching away the edge of the card into the hole. Cards containing desired information may then be mechanically selected from a deck by inserting a long needle in a hole position and lifting the deck to allow the notched cards to fall from the needle. Unwanted cards remain in the deck.

Card, eighty (8C) Column

A punch card with 80 vertical columns representing 80 characters. Each column is divided into two sections, one with character positions labeled zero through nine, and the other labeled eleven (11) and twelve (12). The 11 and 12 positions are also referred to as the X and Y zone punches, respectively. Related to (card, punch) and (card, ninety column).

Card, Punch

A heavy stiff paper of constant size and shape, suitable for punching in a pattern that has meaning, and for being handled mechanically. The punched holes are sensed electrically by wire brushes, mechanically by metal fingers, or photoelectrically by photocells. Related to (card, 80 column) and (card, 90 column).

Cell

1. The storage for one unit of information, usually one character or one word.

2. A location specified by whole or part of the address and possessed of the faculty of store. Specific terms such as column, field, location, and block, are preferable when appropriate.

Center, Data Processing

A computer installation providing data processing service for others, sometimes called customers, on a reimbursable or nonreimbursable basis.

Character

1. One symbol of a set of elementary symbols such as those corresponding to the keys on a typewriter. The symbols usually include the

decimal digits 0 through 9, the letters A through Z, punctuation marks, operation symbols, and any other single symbols which a computer may read, store, or write.

2. The electrical, magnetic, or mechanical profile used to represent a character in a computer, and its various storage and peripheral devices. A character may be represented by a group of other elementary marks, such as bits or pulses.

Chart, flow

A graphic representation of the major steps of work in process. The illustrative symbols may represent documents, machines, or actions taken during the process. The area of concentration is on where or who does what rather than how it is to be done. Synonymous with (process chart) and (flow diagram).

Circuit

1. A system of conductors and related electrical elements through which electrical current flows.

2. A communications link between two or more points.

Code, instruction

The list of symbols, names, and definitions of the instructions which are intelligible to a given computer or computing system.

Coding

The ordered list in computer code or pseudo code, of the successive computer instructions representing successive computer operations for solving a specific problem.

Collate

To merge two or more ordered sets of data or cards, in order to produce one or more ordered sets which still reflect the original ordering relations. The collation process is the merging of two sequences of cards, each ordered on some mutual key, into a single sequence ordered on the mutual key.

Collator

A device used to collate or merge sets or decks of cards or other units into a sequence. A typical example of a card collator has two input feeds, so that two ordered sets may enter into the process, and four output stackers, so that four ordered sets can be generated by the process. Three comparison stations are used to route the cards to one stacker or the other on the basis of comparison of criteria as specified by plugboard wiring.

Command

1. An electronic pulse, signal, or set of signals to start, stop, or continue some operation. It is incorrect to use command as a synonym for instruction.

2. The portion of an instruction word which specifies the operation to be performed.

Comparator

1. A device for comparing two different transcriptions of the same information to verify the accuracy of transcription, storage, arithmetic operation or other processes, in which a signal is given dependent

upon some relation between two items; i.e., one item is larger than, smaller than, or equal to the other.

2. A form of verifier.

Compatibility, equipment

The characteristic of computers by which one computer may accept and process data prepared by another computer without conversion or code modification.

Compile

To produce a machine language routine from a routine written in source language by selecting appropriate subroutines from a subroutine library, as directed by the instructions or other symbols of the original routine, supplying the linkage which combines the subroutines into a workable routine and machine language. The compiled routine is then ready to be loaded into storage and run; i.e., the compiler does not usually run the routine it produces.

Computer

A device capable of accepting information, applying prescribed processes to the information, and supplying the results of these processes. It usually consists of input and output devices, storage, arithmetic, and logical units, and a control unit.

Computer, analog

A computer which represents variables by physical analogies. Thus any computer which solves problems by translating physical conditions such as flow, temperature, pressure, angular position, or voltage into related mechanical or electrical quantities and uses mechanical or electrical equivalent circuits as an analog for the physical phenomenon being investigated. In general it is a computer which uses an analog for each variable and produces analogs as output. Thus an analog computer measures continuously whereas a digital computer counts discretely. Related to (machine, data processing).

Computer, automatic

A computer which performs long sequences of operations without human intervention.

Computer, digital

A computer which processes information represented by combinations of discrete or discontinous data as compared with an analog computer for continuous data. More specifically, it is a device for performing sequences of arithmetic and logical operations, not only on data but its own program. Still more specifically it is a stored program digital computer capable of performing sequences of internally stored instructions, as opposed to calculators, such as card programed calculators, on which the sequence is impressed manually. Related to (machine, data processing).

Computer, general purpose

A computer designed to solve a large variety of problems; for example, a stored program computer which may be adapted to any of a very large class of applications.

Control

1. The part of a digital computer or processor which determines the execution and interpretation of instructions in proper sequence, in-

cluding the decoding of each instruction and the application of the proper signals to the arithmetic unit and other registers in accordance with the decoded information.

2. Frequently, it is one or more of the components in any mechanism responsible for interpreting and carrying out manually initiated directions. Sometimes it is called manual control.

3. In some business applications, a mathematical check.

4. In programing, instructions which determine conditional jumps are often referred to as control instructions, and the time sequence of execution of instructions is called the flow of control.

Control, process

Descriptive of systems in which computers, most frequently analog computers, are used for the automatic regulation of operations or processes. Typical are operations in the production of chemicals wherein the operation control is applied continuously and adjustments to regulate the operation are directed by the computer to keep the value of a controlled variable constant. Contrasted with (control, numerical).

Conversion

1. The process of changing information from one form of representation to another; such as, from the language of one type of machine to that of another or from magnetic tape to the printed page. Synonymous with (conversion, data).

2. The process of changing from one data processing method to another, or from one type of equipment to another; for example, conversion from punch card equipment to magnetic tape equipment.

Cybernetics

The field of technology involved in the comparative study of the control and intracommunication of information-handling machines and nervous systems of animals and man in order to understand and improve communication.

Data

A general term used to denote any or all facts, numbers, letters, and symbols, or facts that refer to or describe an object, idea, condition, situation, or other factors. It connotes basic elements of information which can be processed or produced by a computer. Sometimes data is considered to be expressible only in numerical form, but information is not so limited. Related to (information).

Data reduction

The process of transforming masses of raw test or experimentally obtained data, usually gathered by automatic recording equipment, into useful, condensed, or simplified intelligence.

Decision

The computer operation of determining if a certain relationship exists between words in storage or registers, and taking alternative courses of action. This is effected by conditional jumps or equivalent techniques. Use of this term has given rise to the misnomer "magic brain"; actually the process consists of making comparisons by use of arithmetic to determine the relationship of two terms (numeric, alphabetic, or a combination of both); for example, equal, greater than, or less than.

Definition, problem

The art of compiling logic in the form of general flow charts and logic diagrams which clearly explain and present the problem to the programer in such a way that all requirements involved in the run are presented.

Descriptor

An elementary term, word, or simple phrase used to identify a subject, concept, or idea.

Device, film optical sensing

A piece of equipment capable of reading the contents of a film by optical methods; that is, a system consisting of a light source, lenses, photocells, and a film moving mechanism. The output of the device is digitized and transferred directly to an electronic computer. An example of such a device is the FOSDIC system developed jointly by the Bureau of Census and the National Bureau of Standards.

Dictionary, automatic

The component of a language translating machine which will provide a word-for-word substitution from one language to another. In automatic searching systems, the automatic dictionary is the component which substitutes codes for words or phrases during the encoding operation. Related to (translation, machine).

Disk, magnetic

A storage device on which information is recorded on the magnetizable surface of a rotating disk. A magnetic disk storage system is an array of such devices, with associated reading and writing heads which are mounted on movable arms. Related to (storage, disk).

Document

1. A form, voucher, or written evidence of a transaction.
2. To instruct, as by citation of references.
3. To substantiate, as by listing of authorities.

Documentation

The group of techniques necessary for the orderly presentation, organization, and communication of recorded specialized knowledge, in order to maintain a complete record of reasons for changes in variables. Documentation is necessary not so much to give maximum utility as to give an unquestionable historical reference record.

Drops, false

The documents spuriously identified as pertinent by an information retrieval system, but which do not satisfy the search requirements, due to causes such a improper coding, punching spurious or wrong combinations of holes, or improper use of terminology. Related to (noise).

Drum, magnetic

A cylinder having a surface coating of magnetic material, which stores binary information by the orientation of magnetic dipoles near or on its surface. Since the drum is rotated at a uniform rate, the information stored is available periodically as a given portion of the surface moves past one or more flux detecting devices called heads located near the surface of the drum.

Edit

To rearrange data or information. Editing may involve the deletion of unwanted data, the selection of pertinent data, the application of format techniques, the insertion of symbols such as page numbers and typewriter characters, the application of standard processes such as zero suppression, and the testing of data for reasonableness and proper range. Editing may sometimes be distinguished between input edit (rearrangement of source data) and output edit (preparation of table formats).

Equipment, automatic data processing

1. A machine, or group of interconnected machines, consisting of input, storage, computing, control, and output devices, which uses electronic circuitry in the main computing element to perform arithmetic and/or logical operations automatically by means of internally stored or externally controlled programed instructions. Synonymous with (equipment, electronic data processing).

2. The data processing equipment which directly supports or services the central computer operation. Clarified by (equipment, peripheral).

Equipment, input

1. The equipment used for transferring data and instructions into an automatic data processing system.

2. The equipment by which an operator transcribes original data and instructions to a medium that may be used in an automatic data processing system.

Equipment, off-line

The peripheral equipment or devices not in direct communication with the central processing unit or a computer. Synonymous with auxiliary equipment.

Equipment, on-line

Descriptive of a system and of the peripheral equipment or devices in a system in which the operation of such equipment is under control of the central processing unit, and in which information reflecting current activity is introduced into the data processing system as soon as it occurs. Thus, directly in-line with the main flow of transaction processing. Synonymous with in-line processing, and on-line processing.

Equipment, output

The equipment used for transferring information out of a computer.

Equipment, peripheral

The auxiliary machines which may be placed under the control of the central computer. Examples of this are card readers, card punches, magnetic tape feeds and high-speed printers. Peripheral equipment may be used on-line or off-line depending upon computer design, job requirements and economics. Clarified by equipment, automatic data processing and by equipment off-line.

Equipment, tabulating

The machines and equipment using punch cards. The group of equipment is called tabulating equipment because the main function of installations of punch card machines for some 20 years before the first automatic digital computer was to produce tabulations of information resulting from sorting, listing, selecting, and totaling data on punch cards. This class of equipment is commonly called PCM or tab equipment. Similar to (machine, electrical accounting), clarified by (tabulator).

Evaluation, performance

The analysis in terms of initial objectives and estimates, and usually made on-site, of accomplishments using an automatic data processing system, to provide information on operating experience and to identify corrective actions required if any.

Feedback

The part of a closed loop system which automatically brings back information about the condition under control.

Field, fixed

A given field on punch cards or a given number of holes along the edge of an edge punched card, set aside for the recording of a given type or classification of information.

Field, free

A property of information processing recording media which permit recording of information without regard to a preassigned or fixed field; for example, in information retrieval devices information may be dispersed in the record in any sequence or location.

File

An organized collection of information directed toward some purpose. The records in a file may or may not be sequenced according to a key contained in each record.

File, detail

A file of information which is relatively transient. This is contrasted with a master file which contains relatively more permanent information; for example, in the case of weekly payroll for hourly employees, the detail file will contain employee number, regular time, and overtime, the hours such employee has worked in a given week, and other information changing weekly. The master file will contain the employee's name, number, department, rate of pay, deduction specifications, and other information which regularly stays the same from week to week.

File, master

A file containing relatively permanent information.

Flag

1. A bit of information attached to a character or word to indicate the boundary of a field.

2. An indicator used frequently to tell some later part of a program that some condition occurred earlier.

3. An indicator used to identify the members of several sets which are intermixed. Synonymous with (sentinel).

Flip-flop

1. A bistable device; that is, a device capable of assuming two stable states.

2. A bistable device which may assume a given stable state depending upon the pulse of history of one or more input points and having one or more output points. The device is capable of storing a bit of information.

3. A control device for opening or closing gates; that is, a toggle. Synonymous with (Eccles-Jordan circuit) and (Eccles-Jordan trigger).

Format

The predetermined arrangement of characters, fields, lines, page numbers, and punctuation marks, usually on a single sheet or in a file. This refers to input, output, and files.

Fortran

A programing language designed for problems which can be expressed in algebraic notation, allowing for exponentiation and up to three subscripts. The Fortran compiler is a routine for a given machine which accepts a program written in Fortran source language and produces a machine language routine object program. Fortran II added considerably to the power of the original language by giving it the ability to define and use almost unlimited hierarchies of subroutines, all sharing a common storage region if desired. Later improvements have added the ability to use Boolean expressions, and some capabilities for inserting symbolic machine language sequences within a source program.

FOSDIC

*F*ilm *O*ptical *S*ensing *D*evice for *I*nput to *C*omputers, same as (device, film optical sensing).

Frame, main

1. The central processor of the computer system. It contains the main storage, arithmetic unit and special register groups. Synonymous with (CPU) and (central processing unit).

2. All that portion of a computer exclusive of the input, output, peripheral and in some instances, storage units.

Gap

1. An interval of space or time used as an automatic sentinel to indicate the end of a word, record, or file of data on a tape; e.g., a word gap at the end of a word, a record or item gap at the end of a group of words, and a file gap at the end of a group of records or items.

2. The absence of information for a specified length of time or space on a recording medium, as contrasted with marks and sentinels which are the presence of specific information to achieve a similar purpose. Marks are used primarily internally in variable word length machines. Sentinels achieve similar purposes either internally or externally; however, sentinels are programed rather than inherent in the hardware. Related to (gap, file) and (symbol, terminating).

3. The space between the reading or recording head and the recording medium, such as tape, drum, or disk. Related to (gap, head).

Hardware

The physical equipment or devices forming a computer and peripheral equipment. Contrasted with (software).

Head

A device which reads, records or reases information in a storage medium, usually a small electromagnet used to read, write or erase information on a magnetic drum or tape or the set of perforating or reading fingers and block assembly for punching or reading holes in paper tape or cards.

Heuristic

Pertaining to trial and error methods of obtaining solutions to problems.

Hollerith

A widely used system of encoding alphanumeric information onto cards, hence Hollerith cards is synonymous with punchcards. Such cards were first used in 1890 for the U.S. Census and were named after Herman Hollerith, their originator.

Index, word

An index based on the selection of words as used in a document, without giving thought to synonyms and more generic concepts related to the term selected.

Indexing, coordinate

An indexing scheme by which descriptors may be correlated or combined to show any interrelationships desired for purposes or more precise information retrieval.

Indicators

The devices which register conditions, such as high or equal conditions resulting from a comparison of plus or minus conditions resulting from a computation. A sequence of operations within a procedure may be varied according to the position of an indicator.

Information

A collection of facts or other data especially as derived from the processing of data. Related to (data).

Input

1. Information or data transferred or to be transferred from an external storage medium into the internal storage of the computer.
2. Describing the routines which direct input as defined in (1) or the devices from which such information is available to the computer.
3. The device or collective set of devices necessary for input as defined in (1).

Input-output

A general term for the equipment used to communicate with a computer and the data involved in the communication. Synonymous with (I/O).

Instruction

1. A set of characters which defines an operation together with one or more addresses, or no address, and which as a unit, causes the computer to perform the operation on the indicated quantities. The term instruction is preferable to the terms command and order; command is reserved for a specific portion of the instruction word; that is, the part which specifies the operation which is to be performed,

order is reserved for the ordering of the characters, implying sequence, or the order of the interpolation, or the order of the differential equation. Related to (code (1)).

2. The operation or command to be executed by a computer, together with associated addresses, tags, and indices.

Intelligence, artificial

The study of computer and related techniques to supplement the intellectual capabilities of man. As man has invented and used tools to increase his physical powers, he now is beginning to use artificial intelligence to increase his mental powers. In a more restricted sense, the study of techniques for more effective use of digital computers by improved programing techniques.

Interface

A common boundary between automatic data processing systems or parts of a single system.

Interlock

To arrange the control of machines or devices so that their operation is interdependent in order to assure their proper coordination.

Interpret

1. To print on a punchcard the information punched in that card.
2. To translate nonmachine language into machine language instructions.

Interpreter

1. A punchcard machine which will take a punchcard with no printing on it, read the information in the punched holes, and print a translation in characters in specified rows and columns on the card.
2. An executive routine which, as the computation progresses, translates a stored program expressed in some machine like pseudocode into machine code and performs the indicated operations, by means of subroutines, as they are translated. An interpreter is essentially a closed subroutine which operates successively on an indefinitely long sequence of program parameters, the pseudoinstructions and operands. It may usually be entered as a closed subroutine and left by a pseudocode exit instruction.

Interrupt

To temporarily disrupt the normal operation of a routine by a special signal from the computer. Usually the normal operation can be resumed from that point at a later time.

I/O

The abbreviation for input/output. Synonymous with (input-output).

Key

1. A group of characters which identifies or is part of a record or item, thus any entry in a record or item can be used as a key for collating or sorting purposes.
2. A marked lever manually operated for copying a character; for example, a typewriter, paper tape perforator, card punch, manual keyboard, digitizer or manual word generator.
3. A lever or switch on a computer console for the purpose of manually altering computer action.

Keypunch

1. A special device to record information in cards or tape by punching holes in the cards or tape to represent letters, digits, and special characters.

2. To operate a device for punching holes in cards or tape.

Key-verify

To use the punchcard machine known as a verifier, which has a keyboard, to make sure that the information supposed to be punched in a punchcard has actually been properly punched. The machine signals when the punched hole and the depressed key disagree.

Language

A system for representing and communicating information or data between people, or between people and machines. Such a system consists of a carefully defined set of characters and rules for combining them into larger units, such as words or expressions, and rules for word arrangement or usage to achieve specific meanings.

Language, Algorithmic

An arithmetic language by which numerical procedures may be precisely presented to a computer in a standard form. The language is intended not only as a means of directly presenting any numerical procedure to any suitable computer for which a compiler exists, but also as a means of communicating numerical procedures among individuals. The language itself is the result of international cooperation to obtain a standardized algorithmic language. The International Algebraic Language is the forerunner of ALGOL. Synonymous with (ALGOL) and clarified by (language, international algebraic).

Language, Artificial

A language specifically designed for ease of communication in a particular area of endeavor, but one that is not yet natural to that area. This is contrasted with a natural language which has evolved through long usage.

Language, Common Business Oriented

A specific language by which business data processing procedures may be precisely described in a standard form. The language is intended not only as a means for directly presenting any business program to any suitable computer, for which a compiler exists, but also as a means of communicating such procedures among individuals. Synonymous with (COBOL).

Language, machine oriented

1. A language designed for interpretation and use by a machine without translation.

2. A system for expressing information which is intelligible to a specific machine; for example, a computer or class of computers. Such a language may include instructions which define and direct machine operations, and information to be recorded by or acted upon by these machine operations.

3. The set of instructions expressed in the number system basic to a computer, together with symbolic operation codes with absolute addresses, relative addresses, or symbolic addresses. Synonymous with (language, machine); clarified by (language); related to (language, object); and contrasted with (language, problem oriented).

Language, problem oriented

1. A language designed for convenience of program specification in a general problem area rather than for easy conversion to machine instruction code. The components of such a language may bear little resemblance to machine instructions.

2. A machine independent language where one needs only to state the problem, not the how of solution. Related to (generators, program) and contrasted with (language, procedure oriented).

Language, procedure oriented

A machine independent language which describes how the process of solving the problem is to be carried out; for example, FORTRAN. Contrasted with (language, problem oriented).

Library

1. A collection of information available to a computer, usually on magnetic tapes.

2. A file of magnetic tapes.

Location

A storage position in the main internal storage which can store one computer word and which is usually identified by an address.

Logic

1. The science dealing with the criteria or formal principles of reasoning and thought.

2. The systematic scheme which defines the interactions of signals in the design of an automatic data processing system.

3. The basic principles and application of truth tables and interconnection between logical elements required for arithmetic computation in an automatic data processing system. Related to (logic, sumbolic).

Logic, symbolic

1. The study of formal logic and mathematics by means of a special written language which seeks to avoid the ambiguity and inadequacy of ordinary language.

2. The mathematical concepts, techniques and languages as used in 1, whatever their particular application or context. Synonymous with (mathematical logic) and related to (logic).

Machine, electrical accounting

The set of conventional punchcard equipment including sorters, collators and tabulators. Synonymous with (EAM) and clarified by (equipment, tabulating).

Masking

1. The process of extracting a nonword group or a field of characters from a word or a string of words.

2. The process of setting internal program controls to prevent transfers which otherwise would occur upon setting of internal machine latches.

Match

A data processing operation similar to a merge, except that instead of producing a sequence of items made up from the input, sequences are matched against each other on the basis of some key. The following is a schematic of a two-item match:

Sequence A		Sequence B
1	_____	1
2	_____	3
3	_____	3
4	_____	4
5	_____	5
6	_____	6
7	_____	7
8	_____	11
9	_____	11
10	_____	13
11	_____	
12	_____	
13	_____	

Matrix

1. An array of quantities in a prescribed form; in mathematics, usually capable of being subject to a mathematical operation by means of an operator or another matrix according to prescribed rules.

2. An array of coupled circuit elements; for example, diodes, wires, magnetic cores, and relays, which are capable of performing a specific function; such as the conversion from one numerical system to another. The elements are usually arranged in rows and columns. Thus a matrix is a particular type of encoder or decoder. Clarified by (encoder) and (decorder).

Merge

To combine items into one sequenced file from two or more similarly sequenced files without changing the order of the items.

Method, Monte Carlo

A trial and error method of repeated calculations to discover the best solution of a problem. Often used when a great number of variables are present, with interrelationships so extremely complex as to forestall straight forward analytical handling.

Mistake

A human failing; for example, fault arithmetic, use of incorrect formula, or incorrect instructions. Mistakes are sometimes called gross errors to distinguish from rounding and truncation errors. Thus computers malfunction and humans make mistakes. Computers do not make mistakes and humans do not malfunction, in the strict sense of the word. Contrasted with (error (2)).

Mnemonic

Pertaining to the assisting, or intending to assist, human memory; thus a mnemonic term, usually an abbreviation, that is easy to remember; for example, mpy for multiply and acc for accumulator.

Model, mathematical

The general characterization of a process, object, or concept, in terms of mathematics, which enables the relatively simple manipulation of variables to be accomplished in order to determine how the process, object, or concept would behave in different situations.

Modify

1. To alter a portion of an instruction so its interpretation and execution will be other than normal. The modification may permanently change the instruction or leave it unchanged and affect only the current execution. The most frequent modification is that of the effective address through use of index registers.

2. To alter a subroutine according to a defined parameter.

Monitor

To supervise and verify the correct operation of a program during its execution, usually by means of a diagnostic routine used from time to time to answer questions about the program.

Multiplex

The process of transferring data from several storage devices operating at relatively low-transfer rates to one storage device operating at a high-transfer rate in such a manner that the high-speed device is not obliged to wait for the low-speed devices.

Multiplexing

1. The transmission of a number of different messages simultaneously over a single circuit.

2. Utilizing a single device for several similar purposes or using several devices for the same purpose; for example, a duplexed communications channel carrying two messages simultaneously.

Multiprocessor

A machine with multiple arithmetic and logic units for simultaneous use.

Multiprograming

A technique for handling numerous routines or programs simultaneously by means of an interweaving process.

Notation

1. The act, process, or method of representing facts or quantities by a system or set of marks, signs, figures, or characters.

2. A system of such symbols or abbreviations used to express technical facts or quantities; as mathematical notation.

3. An annotation; note.

Number

1. The, or a total, aggregate, or amount of units.

2. A figure or word, or a group of figures or words, representing graphically an arithmetical sum; a numeral, as the number 45. Clarified by (systems, number).

3. A numeral by which a thing is designated in a series; as a pulse number.

4. A single member of a series designated by consecutive numerals; as, a part number.

5. A character, or a group of characters, uniquely identifying or describing an article, process, condition, document, or class; as, a 6SN7 tube.

6. To count; enumerate.

7. To distinguish by a number.

Number, binary

A number, usually consisting of more than one figure, representing a sum, in which the individual quantity represented by each figure is based on a radix of two. The figures used are 0 and 1.

Number, binary coded decimal

A number usually consisting of successive groups of figures, in which each group of four figures is a binary number that represents but does not necessarily equal arithmetically, a particular figure in an associated decimal number; e.g., if the three right most figures of a decimal number are 262, the three right most figures groups of the binary coded decimal number might be 0010, 0110, and 0010.

Number, coded decimal

A number consisting of successive characters or a group of characters in which each character or group of characters usually represents a specific figure in an associated decimal number; e.g., if the figures of a decimal number are 45, the coded decimal number might be represented as GQ, or LIZZ, or 1101 or 0110.

Number, decimal

A number, usually or more than one figure, representing a sum, in which the quantity represented by each figure is based on the radix of 10. The figures used are 0, 1, 2, 3, 4, 5, 6, 7, 8, and 9.

Octal

Pertaining to eight; usually describing a number system of base or radix eight; e.g., in octal notation, octal 214 is 2 times 64, plus 1 times 8, plus 4 times 1, and equals decimal 140. Octal 214 in binary-coded-octal is represented as 010, 001, 100; octal 214, as a straight binary number is written 10001100. Note that binary coded octal and straight binary differ only in the use of commas; in the example shown, the initial zero in the straight binary is dropped. Clarified by (number, octal).

Off-line

Descriptive of a system and of the peripheral equipment or devices in a system in which the operation of peripheral equipment is not under the control of the central processing unit. Clarified by (equipment, off-line).

On-line

Descriptive of a system and of the peripheral equipment or devices in a system in which the operation of such equipment is under the control of the central processing unit, and in which information reflecting current activity is introduced into the data processing system as soon as it occurs. Thus, directly in-line with the main flow of transaction processing. Clarified by (equipment, on-line); synonymous with (in-line processing), and (on-line processing).

Open-ended

The quality by which the additionof new terms, subject headings, or classifications does not disturb the preexisting system.

Operation

A defined action. The action specified by a single computer instruction or pseudo instruction.

Operation, parallel

The performance of several actions, usually of a similar nature, simultaneously through provision of individual similar or identical devices for each such action. Particularly flow or processing of information. Parallel operation is performed to save time over serial operation Parallel operation usually requires more equipment. Contrasted with (operation, serial).

Operation, real time

The use of the computer as an element of a processing system in which the times of occurrence of data transmission are controlled by other portions of the system, or by physical events outside the system, and cannot be modified for convenience in computer programing. Such an operation either proceeds at the same speed as the events being simulated or at a sufficient speed to analyze or control external events happening concurrently.

Operation, scheduled

The periods of time during which the user plans to use specified equipment. Such a designation must be made a given number of hours in advance, provided however, that such scheduled hours of the operation may be modified after that time in the event of an emergency or in the event that equipment failure creates a need for such rescheduling. Usually the foregoing is further modified in that during the performance period the hours rescheduled as a result of equipment failure usually are not considered as scheduled hours of operation in computing equipment effectiveness.

Operation, Serial

The flow of information through a computer in time sequence using only one digit, word, line or channel at a time. Contrasted with (operation, parallel).

Operations Research

The use of analytic methods adopted from mathematics for solving operational problems. The objective is to provide management with a more logical basis for making sound predictions and decisions. Among the common scientific techniques used in operations research are the following: linear programing, probability theory, information theory, game theory, monte carlo method, and queuing theory. Synonymous with (O.R.).

Optimize

To rearrange the instructions or data in storage so that a minimum number of time consuming jumps or transfers are required in the running of a program.

Output

1. The information transferred from the internal storage of a computer to secondary or external storage, or to any device outside of the computer.
2. The routines which direct 1.
3. The device or collective set of devices necessary for 1.
4. To transfer from internal storage on to external media.

Panel, control

1. An interconnection device, usually removable, which employs removable wires to control the operation of computing equipment. It is used on punch card machines, to carry out functions which are under control of the user. On computers it is used primarily to control input and output functions.

2. A device or component of some data processing machines, which permits the expression of instructions in a semifixed computer program by the insertion of pins, plugs, or wires into sockets, or hubs in the device, in a pattern to represent instructions, and thus making electrical interconnections which may be sensed by the data processing machine. Synonymous with (plugboard) and related to (pinboard.)

Panel, graphic

A master control panel which, pictorially and usually colorfully, traces the relationship of control equipment and the process operation. It permits an operator at a glance, to check on the operation of a far-flung control system by noting dials, valves, scales, and lights.

Parallel

1. To handle simultaneously in separate facilities.

2. To operate on two or more parts of a word or item simultaneously. Contrasted with (serial).

Parameter

1. A quantity in a subroutine, whose value specifies or partly specifies the process to be performed. It may be given different values when the subroutine is used in different main routines or in different parts of one main routine, but which usually remains unchanged throughout any one such use. Related to (parameter, program).

2. A quantity used in a generator to specify machine configuration, designate the subroutines to be included, or otherwise to describe the desired routine to be generated.

3. A constant or a variable in mathematics, which remains constant during some calculation.

4. A definable characteristic of an item, device, or system.

Phone, data

A generic term to describe a family of devices available to facilitate data communication.

Planning-programing-budgeting systems (PPBS)[2]

PPBS is a system aimed at helping management make better decisions on the allocation of resources among alternative ways to attain government objectives. Its essence is the development and presentation of relevant information as to the full implications—the costs and benefits—of the major alternative course of action.

Plotter

A visual display or board in which a dependent variable is graphed by an automatically controlled pen or pencil as a function of one or more variables.

---

[2] Mushkin, Selma J. Planning, Programing, Budgeting for City, State, County Objectives. What is PPB? State-local finances project, George Washington University. 1967. p. 1.

Precision
 1. The degree of exactness with which a quantity is stated.
 2. The degree of discrimination or amount of detail; for example, a 3 decimal digit quantity discriminates among 1,000 possible quantities. A result may have more precision than it has accuracy; for example, the true value of pi to six significant digits is 3.14159; the value 3.14162 is precise to six figures, given to six figures, but is accurate only to about five.

Printer, high-speed
 A printer which operates at a speed more compatible with the speed of computation and data processing so that it may operate on-line. At the present time a printer operating at a speed of 250 lines per minute, 100 characters per line is considered high-speed. Synonymous with HSP.

Printer, line
 A device capable of printing one line of characters across a page; that is, 100 or more characters simultaneously as continuous paper advances line by line in one direction past type bars or a type cylinder that contains all characters in all positions.

Printer, xerographic
 A device for printing an optical image on paper in which dark and light areas of the original are represented by electrostatically charged and uncharged areas on the paper. The paper is dusted with particles of finely powdered dry ink and the particles adhere only to the electrically charged areas. The paper with ink particles is then heated, causing the ink to melt and become permanently fixed to the paper.

Process
 A general term covering such terms as assemble, compile, generate, interpret, and compute.

Processing, automatic data
 Data processing performed by a system of electronic or electrical machines so interconnected and interacting as to reduce to a minimum the need for human assistance or intervention. Synonymous with (ADP) and related to (system, automatic data processing).

Processing, batch
 A technique by which items to be processed must be coded and collected into groups prior to processing.

Processing, centralized data
 Data processing performed at a single, central location on data obtained from several geographical locations or managerial levels. Decentralized data processing involves processing at various managerial levels or geographical points throughout the organization.

Processing, data
 1. The preparation of source media which contain data or basic elements of information, and the handling of such data according to precise rules of procedure to accomplish such operations as classifying sorting, calculating, summarizing, and recording.
 2. The production of records and reports. Synonymous with (data handling).

Processing, electronic data

Data processing performed largely by electronic equipment. Synonymous with (EDP) and related to (processing, automatic data).

Processing, information

A less restrictive term than data processing, encompassing the totality of scientific and business operations performed by a computer.

Processing, real time

The processing of information or data in a sufficiently rapid manner so that the results of the processing are available in time to influence the process being monitored or controlled. Synonymous with (real time system).

Processor

1. A generic term which includes assembly, compiling, and generation.

2. A shorter term for automatic data processor or arithmetic unit.

Program

1. The complete plan for the solution of a problem, more specifically the complete sequence of machine instructions and routines necessary to solve a problem.

2. To plan the procedures for solving a problem. This may involve, among other things, the analysis of the problem, preparation of a flow diagram, preparing details, testing, and developing subroutines, allocation of storage locations, specification of input and output formats, and the incorporation of a computer run into a complete data processing system. Related to (routine).

Program, internally stored

A sequence of instructions, stored inside the computer in the same storage facilities as the computer data, as opposed to external storage on punched paper tape and pin boards.

Programer

A person who prepares problem solving procedures and flow charts and who may also write and debug routines.

Programing, automatic

The method or technique whereby the computer itself is used to transform or translate programing from a language or form that is easy for a human being to produce, into a language that is efficient for the computer to carry out. Examples of automatic programing are compiling, assembling, and interpretive routines.

Programing, interpretive

The writing of programs in a pseudo-machine language, which is precisely converted by the computer into actual machine language instructions before being performed by the computer.

Programing, linear

A technique of mathematics and operations research for solving certain kinds of problems involving many variables where a best value or set of best values is to be found. This technique is not to be confused with computer programing, although problems using the technique may be programed on a computer. Linear programing is most likely to be feasible when the quantity to be optimized, sometimes called the objective function, can be stated as a mathematical expression in terms of the various activities within the system, and when this expression is simply proportional to the measure of the activities; i.e., is linear, and when all the restrictions are also linear.

Protection file

A device or method which prevents accidental erasure of operative data on magnetic tape reels.

Punch card

A machine which punches cards in designated locations to store data which can be conveyed to other machines or devices by reading or sensing the holes. Synonymous with (card punch unit).

Range

1. All the values which a function or word may have.
2. The difference between the highest and lowest of these values.

Read

1. To sense information contained in some source.
2. The sensing of information contained in some source.

Reader, character

A specialized device which can convert data represented in one of the type fonts or scripts read by human beings directly into machine language. Such a reader may operate optically; or if the characters are printed in magnetic ink, the device may operate magnetically or optically.

Record

1. A group of related facts or fields of information treated as a unit, thus a listing of information, usually in printed or printable form.
2. To put data into a storage device.

Record unit

1. A separate record that is similar in form and content to other records; e.g., a summary of a particular employee's earnings to date.
2. Sometimes refers to a piece of nontape auxiliary equipment; e.g., card reader, printer or console typewriter.

Requirements, information

The actual or anticipated questions which may be posed to an information system.

Retrieval, information

The recovering of desired information or data from a collection of documents or other graphic records.

Routine

A set of coded instructions arranged in proper sequence to direct the computer to perform a desired operation or sequence of operations. A subdivision of a program consisting of two or more instructions that are functionally related; therefore, a program. Clarified by (subroutine) and related to (program).

Routine, heuristic

A routine by which the computer attacks a problem not by a direct algorithmic procedure, but by a trial and error approach frequently involving the act of learning. Synonymous with (heuristic program).

Run

The performance of one program on a computer, thus the performance of one routine, or several routines linked so that they form an automatic operating unit, during which manual manipulations by the computer operator are zero, or at least minimal.

Scan

To examine every reference or every entry in a file routinely as a par of a retrieval scheme; occasionally, to collate.

Selective Dissemination of Information (SDI) [3]

SDI is that service within an organization which concerns itself with channeling of new items of information, from whatever source, to those points within the organization where the probability of usefulness, in connection with current work or interests, is high.

Search

To examine a series of items for any that have a desired property or properties.

Sense

1. To examine, particularly relative to a criterion.
2. To determine the present arrangement of some element of hardware, especially a manually set switch.
3. To read punched holes or other marks.

Sensing, Mark

A technique for detecting special pencil marks entered in special places on a punchcard and automatically translating the marks into punched hole.

Sequence

1. To put a set of symbols into an arbitrarily defined order; i.e., to select A if A is greater than or equal to B, or select B if A is less than B.
2. An arbitrarily defined order of a set of symbols; that is, an orderly progression of items of information or of operations in accordance with some rule.

Serial

1. The handling of one after the other in a single facility, such as transfer or store in a digit-by-digit time sequence, or to process a sequence of instructions one at a time; that is, sequentially.
2. The time sequence transmission of storage of, or logical operations on the parts of a word, with the same facilities for successive parts. Related to (operations, serial) and contrasted with (parallel (2)).

Shop, closed

The operation of a computer facility where programing service to the user is the responsibility of a group of specialists, thereby effec-

[3] Luhn, H.P. Selective Dissemination of New Scientific Information with the Aid of Electronic Processing Equipment. American Documentation, April 1961: 131-138.

tively separating the phase of task formulation from that of computer implementation. The programers are not allowed in the computer room to run or oversee the running of their programs. Contrasted with (shop, open).

Shop, open

The operation of a computer facility where computer programing, coding, and operating can be performed by any qualified employee of the personnel of the computing center itself and where the programer may assist in or oversee the running of his program on the computer. Contrasted with (shop, closed).

Simulation

1. The representation of physical systems and phenomena by computers, models, or other equipment; e.g., an imitative type of data processing in which an automatic computer is used as a model of some entity; for example, a chemical process. Information enters the computer to represent the factors entering the real process, the computer produces information that represents the results of the process, and the processing done by the computer represents the process itself.

2. In computer programing, the technique. of setting up a routine for one computer to make it operate as nearly as possible like some other computer.

Simulator

1. A computer or model which represents a system or phenomenon and which mirrors or maps the effects of various changes in the original, enabling the original to be studied, analyzed, and understood by means of the behavior of the model.

2. A program or routine corresponding to a mathematical model or representing a physical model.

3. A routine which is executed by one computer but which imitates the operations of another computer.

Software

The totality of programs and routines used to extend the capabilities of computers, such as compilers, assemblers, narrators, routines, and subroutines. Contrasted with (hardware).

Solid state

The electronic components that convey or control electrons within solid materials; for example, transistors, germanium diodes, and magnetic cores. Thus, vacuum and gas tubes are not included.

Sort

To arrange items of information according to rules dependent upon a key or field contained in the items or records; for example, to digital sort is to sort first the keys on the least significant digit, and to resort on each higher order digit until the items are sorted on the most significant digit.

Sort, merge

To produce a single sequence of items, ordered according to some rule, from two or more previously unordered sequences, without changing the items in size, structure, or total number, although more than one pass may be required for a complete sort, items are selected during each pass on the basis of the entire key.

Sorter

A machine which puts items of information into a particular order; for example, it will determine whether A is greater than, equal to or less than B and sort or order accordingly. Synonymous with (sequencer).

Spot, flying

A small, rapidly moving, spot of light, usually generated by a cathode ray tube and used to illuminate successive spots of a surface containing dark and light areas. The varying amount of light reflected is detected by a phototube and used to produce a time succession of electronic signals which effectively describe the surface.

Station, inquiry

The remote terminal device from which an inquiry into computing or data processing equipment is made.

Storage

1. The term preferred to memory.
2. Pertaining to a device in which data can be stored and from which it can be obtained at a later time. The means of storing data may be chemical, electrical, or mechanical.
3. A device consisting of electronic, electrostatic, electrical, hardware, or other elements into which data may be entered, and from which data may be obtained as desired.
4. The erasable storage in any given computer. Synonymous with (memory).

Storage, auxiliary

A storage device in addition to the main storage of a computer; for example, magnetic tape, disk or magnetic drum. Auxiliary storage usually holds much larger amounts of information than the main storage, and the information is accessible less rapidly. Contrasted with (storage, main).

Storage, buffer

1. A synchronizing element between two different forms of storage, usually between internal and external.
2. An input device in which information is assembled from external or secondary storage and stored ready for transfer to internal storage.
3. An output device into which information is copied from internal storage and held for transfer to secondary or external storage. Computation continues while transfers between buffer storage and secondary or internal storage or vice versa take place.
4. Any device which stores information temporarily during data transfers. Clarified by (buffer).

Storage, external

1. The storage of data on a device which is not an integral part of a computer, but in a form prescribed for use by the computer.
2. A facility or device, not an integral part of a computer, on which data usable by a computer is stored such as, off-line magnetic tape units, or punchcard devices.

Storage, internal

1. The storage of data on a device which is an integral part of a computer.

2. The storage facilities forming an integral physical part of the computer and directly controlled by the computer. In such facilities all data are automatically accessible to the computer; for example, magnetic core, and magnetic tape on-line. Synonymous with (internal memory) and contrasted with (storage, external).

Storage, magnetic disk

A storage device or system consisting of magnetically coated disks, on the surface of which information is stored in the form of magnetic spots arranged in a manner to represent binary data. These data are arranged in circular tracks around the disks and are accessible to reading and writing heads on an arm which can be moved mechanically to the desired disk and then to the desired track on that disk. Data from a given track are read or written sequentially as the disk rotates. Related to (storage, disk).

Storage, magnetic core

A storage device in which binary data is represented by the direction of magnetization in each unit of an array of magnetic material, usually in the shape of toroidal rings, but also in other forms such as wraps on bobbins. Synonymous with (core storage).

Storage, magnetic drum

The storage of data on the surface of magnetic drums. Related to (drum, magnetic).

Storage, magnetic tape

A storage device in which data is stored in the form of magnetic spots on metal or coated plastic tape. Binary data are stored as small magnetized spots arranged in column form across the width of the tape. A read-write head is usually associated with each row of magnetized spots so that one column can be read or written at a time as the tape traverses the head.

Storage, main

Usually the fastest storage device of a computer and the one from which instructions are executed. Contrasted with (storage, auxiliary).

Storage, random access

A storage technique in which the time required to obtain information is independent of the location of the information most recently obtained. This strict definition must be qualified by the observation that we usually mean relatively random. Thus, magnetic drums are relatively nonrandom access when compared to magnetic cores for main storage, but are relatively random access when compared to magnetic tapes for file storage. Synonymous with (random access memory) and contrasted with (storage, sequential access).

Store

1. To transfer an element of information to a device from which the unaltered information can be obtained at a later time.

2. To retain data in a device from which it can be obtained at a later time.

Study, application

The detailed process of determining a system or set of procedures for using a computer for definite functions or operations, and establishing specifications to be used as a base for the selection of equipment suitable to the specific needs.

Subroutine

1. The set of instructions necessary to direct the computer to carry out a well-defined mathematical or logical operation.

2. A subunit of a routine. A subroutine is often written in relative or symbolic coding even when the routine to which it belongs is not.

3. A portion of a routine that causes a computer to carry out a well-defined mathematical or logical operation.

4. A routine which is arranged so that control may be transferred to it from a master routine and so that, at the conclusion of the subroutine, control reverts to the master routine. Such a subroutine is usually called a closed subroutine.

5. A single routine may simultaneously be both a subroutine with respect to another routine and a master routine with respect to a third. Usually control is transferred to a single subroutine from more than one place in the master routine and the reason for using the subroutine is to avoid having to repeat the same sequence of instructions in different places in the master routine. Clarified by (routine).

Symbol

A substitute of representation of characteristics, relationships, or transformations of ideas or things.

Synthesis

The combining of parts in order to form a whole; for example, to arrive at a circuit or a computer or program, starting from performance requirements. This can be contrasted with analysis, which arrives at performance, given the circuit or program.

System

An assembly of procedures, processes, methods, routines or techniques united by some form of regulated interaction to form an organized whole.

System, automatic data processing

The term descriptive of an interacting assembly of procedures, processes, methods, personnel, and automatic data processing equipment to perform a complex series of data processing operations.

System, data processing machine

An assembly of data processing machines united by some form of regulated interaction to form an organized whole.

System, information

The network of all communication methods within an organization. Information may be derived from many sources other than a data processing unit, such as by telephone, by contact with other people, or by studying an operation.

System, information retrieval

A system for locating and selecting, on demand, certain documents, or other graphic records relevant to a given information requirement from a file of such material. Examples of information retrieval systems are classification, indexing, and machine searching systems.

System, management information

A communications process in which data are recorded and processed for operational purposes. The problems are isolated for higher level decisionmaking and information is fed back to top management to reflect the progress or lack of progress made in achieving major objectives.

System, peek-a-boo

An information retrieval system which uses peek-a-boo cards; that is, cards into which small holes are drilled at the intersections of coordinates (column and row designations) to represent document numbers. Synonymous with (batten system) and (cordonnier system) and related to (card, aspect).

System, uniterm

An information retrieval system which uses uniterm cards. Cards representing words of interest in a search are selected and compared visually. If identical numbers are found to appear on the uniterm card undergoing comparison these numbers represent documents to be examined in connection with the search. Related to (card, aspect) and (indexing uniterm).

Table

A collection of data in a form suitable for ready reference, frequently as stored in sequenced machine locations or written in the form of an array of rows and columns for easy entry and in which an intersection of labeled rows and columns serves to locate a specific piece of data or information.

Tabulator

A machine which reads information from one medium; for example, cards, paper tape, and magnetic tape and produces lists, tables, and totals on separate forms or continuous paper. Synonymous with (machine, accounting), and clarified by (equipment, tabulating).

Tag

A unit of information, whose composition differs from that of other members of the set so that it can be used as a marker or label. A tag bit is an instruction word that is also called a sentinel.

Tape

A strip of material, which may be punched, coated, or impregnated with magnetic or optically sensitive substances, and used for data input, storage or output. The data are stored serially in several channels across the tape transversely to the reading or writing motion.

Tape, magnetic

A tape or ribbon of any material impregnated or coated with magnetic or other material on which information may be placed in the form of magnetically polarized spots.

Tape, paper

A strip of paper capable of storing or recording information. Storage may be in the form of punched holes, partially punched holes, carbonization or chemical change of impregnated material, or by imprinting. Some paper tapes, such as punched paper tapes, are capable of being read by the input device of a computer or a transmitting device by sensing the pattern of holes which represent coded information.

Tape, program

A tape which contains the sequence of instructions required for solving a problem and which is read into a computer prior to running a program.

Telemetering

The transmission of a measurement over long distances, usually by electromagnetic means.

Test, systems

1. The running of the whole system against test data.
2. A complete simulation of the actual running system for purposes of testing out the adequacy of the system.
3. A test of an entire interconnected set of components for the purpose of determining proper functioning and interconnection.

Theory, game

A mathematical process of selecting an optimum strategy in the face of an opponent who has a strategy of this own.

Theory, information

The mathematical theory concerned with information rate, channels, channel width, noise and other factors, affecting information transmission. Initially developed for electrical communications, it is now applied to business systems, and other phenomena which deal with information units and flow of information in networks.

Theory, queuing

A form of probability theory useful in studying delays or line-ups at servicing points.

Theory, probability

A measure of likelihood of occurrency of a chance event, used to predict behavior of a group, not of a single item in the group.

Time, access

1. The time it takes a computer to locate data or an instruction word in its storage section and transfer it to its arithmetic unit where the required computations are performed.
2. The time it takes to transfer information which has been operated on from the arithmetic unit to the location in storage where the information is to be stored, Synonymous with (read time); (real time) and related to (time, write) and (time word (2)).

Time-sharing

The use of a device for two or more purposes during the same overall time interval, accomplished by interspersing component actions in time.

Transfer

1. The·conveyance of control from one mode to another by means of instructions or signals.

2. The conveyance of data from one place to another.

3. An instruction for transfer.

4. To copy, exchange, read, record, store, transmit, transport, or write data.

5. An instruction which provides the ability to break the normal sequential flow of control. Synonymous with (jump), and (control transfer).

Tube, cathode ray

1. An electronic vacuum tube containing a screen on which information may be stored by means of a multigrid modulated beam of electrons from the thermionic emitter storage effected by means of charged or uncharged spots,

2. A storage tube.

3. An oscilloscope tube.

4. A picture tube.

Unit, arithmetic

The portion of the hardware of a computer in which arithmetic and logical operations are performed. The arithmetic unit generally consists of an accumulator, some special registers for the storage of operands and results supplemented by shifting and sequencing circuitry for implementing multiplication, division, and other desired operations. Synonymous with ALU.

Unit, assembly

1. A device which performs the function of associating and joining several parts or piecing together a program.

2. A portion of a program which is capable of being assembled into a larger whole program.

Uniterm

A word, symbol, or number used as a descriptor for retrieval of information from a collection; especially, such a descriptor used in a coordinate indexing system. Related to (card, aspect); (descriptor); (indexing, coordinate); (docuterm).

Verify

To check a transcribing operation, by a compare operation. It usually applies to transcriptions which can be read mechanically or electrically.

Vocabulary

A list of operating codes or instructions available to the programer for writing the program for a given problem for a specific computer.

Vocabulary, sophisticated

An advanced and elaborate set of instructions. Some computers can perform only the more common mathematical calculations such as addition, multiplication, and subtraction. A computer with a sophisticated vocabulary can go beyond this and perform operations such as linearize, extract square root, and select highest number.

Word

An ordered set of characters which occupies one storage location and is treated by the computer circuits as a unit and transferred by the computer circuits as a unit and transferred as such. Ordinarily a word is treated by the control unit as an instruction, and by the arithmetic unit as a quantity. Work lengths may be fixed or variable depending on the particular computer.

Write

1. To transfer information, usually from main storage, to an output device;

2. To record data in a register, location, or other storage device or medium.

Xerography

A dry copying process involving the photo electric discharge of an electrostatically charged plate. The copy is made by tumbling a resinous powder over the plate, the remaining electrostatic charge discharged and the resin transferred to paper or an offset printing master.

APPENDIX B

List of Witnesses and Contributors

Throughout the course of the four series of hearings conducted by the Senate Special Subcommittee on the Utilization of Scientific Manpower, numerous witnesses from universities, industry, and all echelons of government presented testimony. In addition, other individuals submitted prepared statements or ancillary material at the invitation of the subcommittee. The hearings are published in two volumes:

Volume I, "Scientific Manpower Utilization, 1965–66," describes the hearings held in California (November 19, 1965) and Washington, D.C. (May 17 and 18, 1966).

Volume II, "Scientific Manpower Utilization, 1967," contains the testimony given in Washington, D.C. (January 24–26, 27, and March 29–30, 1967), and augmenting documentation.

While the first volume addresses only S. 2662, the "Scientific Manpower Utilization Act," the second volume is concerned with S. 430 (the 90th Congress version of S. 2662) and S. 467, which calls for the creation of a National Commission on Public Management.

For the purposes of this listing, "witnesses" are those persons who appeared in person before the special subcommittee. "Contributors" are those who submitted prepared statements or other supporting documentation. The alphabetical listing contains the name of each participant, followed by an indication of the volume of hearings and appropriate page numbers where his testimony or other contribution appears.

Daniel Alpert, Dean, The Graduate College, University of Illinois, Urbana, Ill.: (II, 206, witness).

Kathleen Archibald, Assistant Director, Public Policy, Research Organization, University of California, Irvine, Calif.: (II, 178, witness).

Patricia Arnold, Economic Committee of Women for Peace: (I, 142, witness).

Arthur W. Barber, Deputy Assistant Secretary (Arms and Trade Control), International Security Affairs, Department of Defense: (II, 193, witness).

Kurt W. Bauer, executive director, Southeastern Wisconsin Regional Planning Commission: (II, 274, contributor).

Louise Becker, research assistant, Science Policy Research Division, Legislative Reference Service, Library of Congress: (II, 362, contributor).

Dennis W. Brezina, research assistant, Science Policy Research Division, Legislative Reference Service, Library of Congress: (II, 362, contributor).

Hon. Edmund G. Brown, Governor of the State of California: (I, 6, witness).

Robert L. Chartrand, information sciences specialist, Science Policy Research Division, Legislative Reference Service, Library of Congress: (II, 345, 325, contributor).

Dwight B. Culver, M.S., manager, life support systems division, Aero-jet-General Corp., Azusa, Calif.: (I, 57, witness).

Richard M. Cyert, dean, graduate school of administration, Carnegie Institute of Technology, Pittsburgh, Pa.: (II, 266, witness).

Richard P. Daly, president, Aries Corp., McLean, Va.: (II, 280, contributor).

Ward Dennis, Northup Corp.: (II, 43, witness).

Richard E. Engler, Jr., Human Sciences Research, Inc., McLean, Va.: (II, 281, contributor).

Alain C. Enthoven, Assistant Secretary for Systems Analysis U.S. Department of Defense: (I, 146, witness).

John C. Geyer, professor, Johns Hopkins University, Baltimore, Md.: (II, 171, witness).

Don A. Godall, legislative action, general manager, Chamber of Commerce of the United States.: (II, 289, contributor).

William Gorham, Assistant Secretary for Program Coordination, U.S. Department of Health, Education, and Welfare: (I, 177, witness).

Paul Grogan, Director, Office of State Technical Services, Department of Commerce: (II, 128, witness).

Karl G. Harr, Jr., president, Aerospace Industries Association: (II, 34, witness).

James Hodgson, consultant, operation research, Systems Engineering: (II, 369, contributor).

Walter G. Hollander, State of Wisconsin: (II, 260, contributor).

Jack Jones, assistant to the president, North American Aviation, Inc., Los Angeles, Calif.: (I, 93, witness).

Joseph A. Kershaw, Assistant Director, U.S. Office of Economic Opportunity: (I, 194, witness).

Charles N. Kimball, president, Midwest Research Institute, Kansas City, Mo.: (II, 97, witness).

Robert W. Krueger, president, Planning Research Corp.: Los Angeles, Calif., Washington, D.C.: (II, 289, contributor).

John Kuhn, project manager, Space-General Corp., El Monte, Calif.: (I, 123, witness).

Kenneth T. Larkin, Director of Special Programs, Lockheed Missiles and Space Co.: (I, 29, witness).

Louis J. Lauler, project leader, special programs, Lockheed Missiles and Space Co.: (I, 34, witness).

Frank Lehan, president, Space-General Corp., El Monte, Calif.: (I, 122, witness).

Robert Lekachman, department of economics, State University of New York, Stony Brook, Long Island, New York, N.Y.: (II, 90, witness).

Robert A. Mang, Friends Committee on Legislation of California: (I, 140, witness).

Michael Michaelis, Washington office, Arthur D. Little, Inc., Washington, D.C.: (II, 217, witness).

Vincent J. Moore, assistant director, office of planning coordination, State of New York, Albany, N.Y.: (II, 232, witness).

Hon. F. Bradford Morse, Representative in Congress from the State of Massachusetts: (II, 14, 17, 26, witness).

Leonard Woodcock, vice president, United Automobile, Aerospace & Agricultural Implement Workers of America, AFL–CIO: (II, 187, contributor).

John V. Zuckerman, associate professor of management, University of Southern California: (I, 48, witness).

APPENDIX C

Description of City Model I

The Washington Center for Metropolitan Studies in Washington, D.C., has developed several systemic models for use by a wide range of public officials, social scientists, and selected citizens interested in community interactions determined and affected by political and economic power and holdings. One of these is called "City I," originally mentioned in section VIII. B. of this report. The structure, scope, and use of this model are described below.

The *City I* [1] model is played with nine teams, representing the elite decisionmakers in the community. As in region II these decisionmakers have two principal overt means of making their influence felt: economic power and political power. The interrelationships between the economic and the political sectors and their joint influence on urban growth and development are among the interrelationships emphasized in the game.

The model is played on a board of 625 squares, each representing 1 square mile of land. Because we have the physical facility to operate models of this type at the Washington center, we use a 6-foot square plexiglass board. In one corner of the land area represented on the board is an urban community or city with a population of approximately 300,000 people. The rest of the board is divided into three towns. These four governmental jurisdictions are within a single county, and all of the political decisions are made, in effect, by the county as a whole. All three towns and the city are connected by a series of roadways which have different levels of traffic capacity.

The model incorporates nine types of private land use and two types of governmental land uses; a school sector, which provides elementary, high school, and college facilities for an area; and a municipal service sector, which provides police, fire, hospital, and other services. Two other governmental land uses are parkland and highway rights-of-way.

The governmental part of the model provides the medium for role-playing. Each team plays a separate role which may change each round. One team is elected chairman, and a councilman is elected from each of the four political jurisdictions. The chairman appoints teams to be public works and safety department, highway department, school department, finance department, and planning and zoning department. Teams which are not otherwise designated assume the role of citizen and mass media teams and act as critics of the other teams.

---

[1] Contained in: House, Peter, the Simulated City: the Use of Second Generation Gaming in Studying the Urban System. Presented at the Annual Symposium, the Application of Computers to the Problems of Urban Society. Oct. 18, 1968.

Play progresses in a series of rounds with each round divided into three general portions, although there are no time limits imposed by the director. The model requires only that voting be done at a certain time and that all actions be completed by the end of a round. During the first portion of a round, players elect and appoint their officials and, as entrepreneurs, make their economic plans for the round.

The operator then distributes these economic decision forms to the teams. These plans are then used as a basis for the bureaucracy's estimate of necessary public actions in the round. For example, the highway department requests funds for upgrading and maintaining highways and terminals. The school superintendent asks for money to construct and operate schools. The finance department produces a budget on the basis of its interpretation of departmental requests and expected revenues. This budget is passed to the chairman, who reviews and modifies it as he sees fit, then submits it to the council for approval.

After the budget is approved, funds are allocated to the different departments to provide public facilities for the urban area. During this period, teams carry out their economic development and provide governmental services where needed. The success of the teams' performance in their roles, in conjunction with the opinions of the citizens and mass media teams, determines the chances of a team's holding political office in the next round.

City I is a rather complicated model and yet compared to reality, the game is still quite simple. As noted earlier, it is limited by a number of factors: By the size of the computer available, by the amount of available information to be fed into the model, and finally, by our own lack of experience.

Computer breakthroughs

Much of the original research design for *City I* had to be restricted or eliminated altogether because of computer limitations, but in several notable instances the computer programing abilities of our staff allowed the research design to go beyond original expectations. The most notable example of this was the optimizer function.

This programing breakthrough in the gaming area simulates a market economy and allow some of the more routine and burdensome economic functions of the game to be performed by the computer. The optimizer assumes that buyers purchase from sellers who offer the best net price. The net price is the sales price minus transportation costs of the buyer to go from home to store. The optimizer functions in the model assign workers to jobs, buyers of goods and services to establishments selling goods and services, students to school units, and residences to municipal service units. In the cases where only transportation costs are involved between user and supplier; that is, for schools, municipal services, and terminals, the total transportation cost is minimized. In the case of employment assignments, net wages (wages minus transportation costs to work) are maximized. In the case of goods and service establishments, net price (store price minus transportation costs to the store) is minimized. Thus, with the optimizer, teams are not bothered with the detailed problems of finding the best (cheapest) assignments for work, et cetera, and goods and service suppliers get immediate results from a change in prices charged.

In our earlier urban models if a new commercial sector was built or changed its prices, the team owning the sector had to contract separately with each customer. With the optimizer in *City I* a new sector or a sector with a new price is automatically allocated customers on the basis of a local market system.

For example, the optimizer chooses for a particular residential sector the 10 cheapest (including transportation costs) personal service sectors in the area. The optimizer then investigates the personal service sector which would yield that residence the greatest gain. Since each commercial sector has a limited capacity, the cheapest personal service sector may be unavailable. In that case, the optimizer investigates the next most economical transaction, and so on. When the most economical transaction is determined, the residential sector is assigned to that commercial sector. Then the remaining capacity of that personal services establishment is decreased appropriately. The entire optimization (i.e., all necessary assignments) for each developed parcel occurs at the end of each round.

As another example of the kind of problems solved by the computer, let me cite the employment optimizer. It is the job of the employment optimizer to assign employees to employers at the highest possible salary. It compares employee lists of net income (net of transportation expenses) for 10 employers and then assigns employees to employers on the basis that the greatest net incomes are assigned first. When all the obvious solutions have been derived, the computer then begins assigning employees with equal net incomes. These assignments are not strictly rational and are subject to some random variation.

The employment optimizer is carried out four separate times with four socioeconomic classes of workers. The first class optimized is high income. If these jobs are filled in such a fashion that the number of high-income positions equals the number of high-income residents, they are matched. If there are too few high-income residents in the system, the employees are imported from outside the local economy at twice the average salary level. If there are too many high-income residents, then the surplus gets the first choice of middle-income jobs (and gets paid middle-income salaries). This sequence of job assignment forces all unemployment down to the slum residents as the optimizers iterate through the four income levels.

After the assignments have been made by the computer, a subroutine then goes back over them and notes travel to work over various road sections. These journeys are taken into consideration when calculating amount of congestion and road depreciation. These figures—for congestion and depreciation—are used to calculate travel costs for the whole system.
tion of business services.

Program interrelations

All parts of the *City I* program are integrally related. Let me give a few examples of the program interrelationships:

(1) The chances of a parcel being struck by natural disaster are related to the value ratio (condition) of the development. If a residence unit is struck, both the degree of damage and the number of people killed are affected by the value ratio of the municipal service unit serving that residence. The destruction of a place of work or the death

of residents has obvious economic, social, and political consequences in the model.

(2) The minimum cost to travel on a road is a constant, depending on the type of road. However, the actual cost to travel varies with the amount of traffic on the road, the condition (value ratio) of the road, and the amount spent by the government to maintain the road. The more traffic on a road, the faster it deteriorates. A road of greater design capacity can accommodate more traffic at less cost to the traveler, but the construction of such a road costs the government more than that of a smaller road.

(3) The amount of money the government has to spend depends on both the government income and the appropriations approved by the council. Government income is affected by, among other things, team income and the value of property. There is a business cycle which affects the gross income of basic industry and the interest rate on loans.

(4) Each round the value of each parcel of land is reassessed. All owned land depreciates by 3 percent. Then each parcel is assessed in relation to the value of the surrounding parcels. This procedure causes a rippling effect in land values which is reflected in the taxes on each parcel.

(5) The tax rate and the Government appropriations are voted by the council. At the beginning of a game, the three suburban councilmen are played by the computer. Each of the four councilmen has a different amount of voting power, based on the size of his constituency as calculated by the computer. The degree of preference, positive or negative, with a mean and a standard deviation, of each computer councilman is fed into the computer. For example, the councilman in the wealthy estate area opposes all Government expenditures which might lead to industrial development in his town. The preliminary votes of the four councilmen are then taken. Then they bargain over issues, trading votes on issues of lesser importance for others' votes on issues of greater importance. An issue crucial to one may be of little importance to another. The final vote is taken and the approved budget printed and sent to Government departments.

Each action in the public or private areas has several ramifications which are shown in the detailed output for a team in each of its two functions. The total output at the end of each round is about 75 computer pages, increasing with the rounds played.

APPENDIX D

Bay Area Transportation Study Commission and Its Information System

The problems encountered in developing a metropolitan transportation system are no better identified, nor progressive steps to a solution scheduled, than in the Bay area transportation study. The creation of an information system to be used in the modular establishment of a complex transportation plan is described below:[1]

To meet the rising need for comprehensive, coordinated planning of all transportation facilities in the nine counties and 7,000 square miles of the San Francisco Bay area, the Bay Area Transportation Study Commission (BATSC) was created by the California State Legislature in 1963 and instructed to prepare a master regional transportation plan, including recommendations for implementing the plan, provisions for a continuing transportation planning process, and coordination among separate planning activities at the regional level in the Bay area.

BATSC has had a close working relationship with the Association of Bay Area Governments (ABAG) and relies on ABAG as the regional planning agency for the area. With ABAG, BATSC has conducted a joint data release operation. BATSC also maintains close relationships with local agencies in the region through the commission itself, through a series of committees and through the exchange of information.

The study is being carried out by the commission staff consisting of (at the time of this report) about 25 transportation planners, economists, and related specialists supported by administrative and clerical presonnel. The staff is now in the process of preparing a final report to the legislature.

The BATSC data base

BATSC has an extensive and growing data base—physically stored on some 1,100 reels of magnetic tape—consisting of data acquired from a variety of sources, including newly collected data from field surveys and existing data from public (Federal, State, and local) and private agencies. The BATSC information system was specifically designed to permit flexibility in file organization, and the computer software provides a capability for readily organizing files as most appropriate for specific analysis or data processing tasks. BATSC conceived of a dynamic data base that would continue to grow; the information system was designed so that newly acquired data, as well as new data arising from analyses made on existing data, could be readily incorporated into the data base. Thus, in principle, any data generated by an analyst as a result of a data processing task would itself become part of the data base, would be described and documented as part of the data base management system, and would be available to all other users of the data base.

BATSC has acquired data from a variety of sources, and through a variety of collection techniques: aerial photography, roadside interviews, etc. The major data acquisition activity was the home interview survey which has resulted in about 10 million pieces of information at a cost of about $1,500,000. The home interview survey—an extension of the origin-and-destination type survey common to transportation planning—was conducted for a 5-percent sample of the Bay area and included an extensive set of data items, such as data on household characteristics (rent, income, etc.), individual members of household (education, employment, etc.), as well as data on each trip taken by individuals in the household (origin, destination, mode, purpose, etc.). Field survey personnel completed elaborate questionnaires as a result of home interviews, and the questionnaires were then converted to punched cards. The home interview data utilized about 1.5 million punched cards, which were then converted to about 16 reels of magnetic tape. About 1½ hours of IBM 7094 time is required to reprocess the complete home interview file.

---

[1] Kevany, Michael J. An information system for urban transportation planning: the BQTSC approach, op. cit. pp. 11–13.

The data base in addition contains data from existing demographic, economic, land use and traffic studies. Also included are data from such common sources as the U.S. Census data and State employment data. The major portion of the data base has resulted from a series of field inventories of land use, highway and transit facilities, parking, and employment.

Data collection, and the home interview survey in particular, took place concurrently with the development of the information system. Though BATSC particularly stressed the importance of early, "Off-the-shelf" information system support for all phases of the study, it was necessary at BATSC to proceed with various data processing tasks even before the information system was fully operational.

In regional planning it is customary to analyze the region in terms of smaller areas or zones. BATSC used several systems of areal units, ranging from the block to the county levels. These zone systems are reflected in the organization of the BATSC data base. Information available for these geographic entities is operated on through use of various planning models, including residential and employment location models, demographic and economic projection models, the Control Data Corporation's Tran Plan package, the network evaluation procedures and various other methods.

The study information system

The BATSC information systems was designed to meet at least three primary objectives: (1) to permit a closer interaction between an analyst and his data, and provide him with capabilities for processing and analyzing data without the assistance of a computer programer; (2) to provide for the systematic management of a growing data base containing data from a variety of sources in a variety of formats; and (3) to facilitate the release of data to other agencies in the area. In addition, there was a need to have the information system operational very early in the study.

As a matter of policy, implemented in practice, BATSC has placed the analyst (nonprogramer) in closer interaction with his data. User-oriented systems such as the SPAN system described below permit the user, not necessarily acquainted with computer programing, to call for data processing tasks through simple English-like directives. The data processing system operates as an open shop, available equally to programers and analysts. Through the use of user-oriented general-purpose software, and as a result of deliberate study policy and procedures, the analysts—transportation planners, economists, and social scientists—have enjoyed more direct access to the data processing system capabilities and the data base than has been the case in other transportation studies generally.

During most of the study, BATSC has had a small in-house computer, first a five-tape IBM 1401 with 8,000 characters of core, replaced in December 1966 by a 16K core, four-tape Honeywell 200/120. The in-house 1401 computer was made available on a full-time basis to BATSC personnel for use in simple file manipulation, recordkeeping, and preparation of magnetic tapes for use on the larger scale computers also used by the study. A significant proportion of the usage of the Honeywell configuration during the 7 months that the machine was installed at BATSC was devoted to reprograming 1401 programs. Utilized through service bureaus were an IBM 7094, primarily for the processing of large files and for statistical and other data analysis; and a CDC 3800 for transportation and land-use modeling. By August 1967, the study ceased maintaining an in-house computer facility and began purchasing 1401 computer time through a local service bureau. The original plan for on-premises hardware called for the installation of a Calcomp magnetic tape plotter; budgetary cutbacks forced the study to settle for plotter time purchased from a service bureau instead.

BATSC personnel have used a variety of programing languages (e.g., Autocoder, Fortran) and software packages (e.g., SPAN, MADAM). Rather than prepare specific computer programs for each operation on a file, BATSC has employed general-purpose systems that can perform a wide range of processing activities with a minimum of simplified instructions. The information system was based primarily on the availability of the SPAN system, a user-oriented general-purpose software package developed specifically to support urban and regional planning.

SPAN, a large-scale data management file processing and statistical analysis system operating on the IBM 7094 computer, consists of a set of modules that provide a variety of capabilities and are designed so that one module can accept the results from operations performed by another module. SPAN capabilities include file manipulation, statistical and mathematical analysis, report generation, and graphic display. Development of SPAN was initiated at the Penn-Jersey transportation study and was carried on at SDC. Certain capabilities—the Midas processor, a graphic display module, and capabilities to handle spatial data— were developed for and under contract to BATSC.

The extent of SPAN's user orientation and the benefits derived therefrom have been expressed by two BATSC workers as follows:

SPAN functions as a user-oriented file processing system which can be used by nonprogramers to conceive and implement file processing and analysis jobs at a striking level of sophistication. The analysis staff, some 20 people at BATSC, have been trained in SPAN in a series of short lectures combined with practical working [experience] * * *. This extent of user participation has allowed BATSC to use only four programers for major file processing * * * [that might require] an intensive background in programing and data processing. Once a file is prepared, the analysts are given full rein to carry out their analysis without being dependent upon the programing section.

APPENDIX E

Utilization of a Hybrid Model

The value of a dynamic, computer-supported model to Federal, State, and local government management must be understood in terms of the construction, orientation, characteristics, and employment of that model. A so-called hybrid model combines the essential ingredients of social, economic, and physical models. It embodies the flow-logic-feedback technique so useful to the analyst. Reproduced below from the report on "Systems Analysis Techniques and Concepts for Water Resource Planning and Management," prepared by the Technology Planning Center, Inc., is detailed material on the nature and uses of the hybrid model.[1]

The need for a hybrid model

In an effort to integrate social, as well as economic and physical considerations in a dynamic model for water planning and management, the model must meet the following "systems" specifications:

(*a*) The model should portray all activities and events which occur between a specific area of concern and the ultimate terminal goals or objectives sought.

(*b*) The model should include all elements, components, and variables, inherent in the problem in their proper relationship to one another; i.e., economic, physical, and social.

(*c*) A means must be provided for isolating all cost, benefit, and value variables in the model.

(*d*) The logical nature of the decision process must be portrayed, specifically isolating classes of decisions which are made by or for people.

(*e*) A clarification of decision points, choices, and alternatives should be implicit in the model.

(*f*) The relationships between antecedent data, information, and intelligence related to decisionmaking should be evident.

(*g*) The model should be in a form suitable for development into a simulator with a minimum of mathematical transition.

(*h*) The model should be clear and easy to understand.

In light of the preceding specifications, a comprehensive river basin model borrows heavily from the individual physical, economic, and social models. Moreover, and in fact, the comprehensive model is actually a "hybrid," incorporating all three models as interacting components. Such a model is illustrated in figure 8, "Hybrid River Basin Model."

[1] "Systems Analysis Techniques and Concepts for Water Resource Planning and Management." Prepared by the Staff of Technology Planning Center, Inc., Ann Arbor, Mich., pp. 23-32.

Social Models

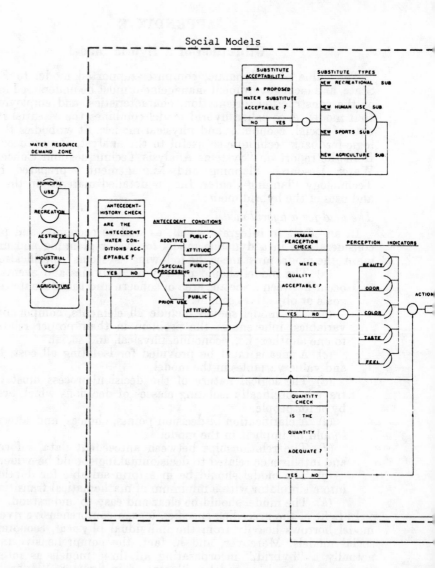

FLOW LOGIC – FEEDBACK MODEL

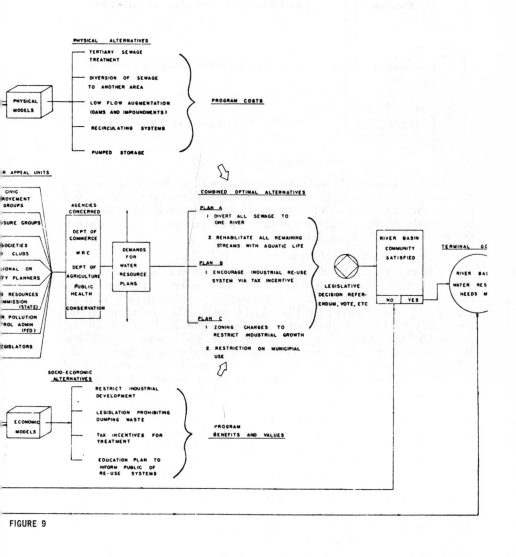

PHYSICAL ALTERNATIVES
- TERTIARY SEWAGE TREATMENT
- DIVERSION OF SEWAGE TO ANOTHER AREA
- LOW FLOW AUGMENTATION (DAMS AND IMPOUNDMENTS)
- RECIRCULATING SYSTEMS
- PUMPED STORAGE

PHYSICAL MODELS

} PROGRAM COSTS

R APPEAL UNITS

- CIVIC IMPROVEMENT GROUPS
- SURE GROUPS
- SOCIETIES
- CLUBS
- IONAL OR TY PLANNERS
- RESOURCES MMISSION (STATE)
- R POLLUTION TROL ADMIN (FED)
- EGISLATORS

AGENCIES CONCERNED

DEPT. OF COMMERCE

W R C

DEPT. OF AGRICULTURE

PUBLIC HEALTH

CONSERVATION

DEMANDS FOR WATER RESOURCE PLANS

COMBINED OPTIMAL ALTERNATIVES

PLAN A
1. DIVERT ALL SEWAGE TO ONE RIVER
2. REHABILITATE ALL REMAINING STREAMS WITH AQUATIC LIFE

PLAN B
1. ENCOURAGE INDUSTRIAL RE-USE SYSTEM VIA TAX INCENTIVE

PLAN C
1. ZONING CHANGES TO RESTRICT INDUSTRIAL GROWTH
2. RESTRICTION ON MUNICIPIAL USE

LEGISLATIVE DECISION REFERENDUM, VOTE, ETC

RIVER BASIN COMMUNITY SATISFIED

| NO | YES |

TERMINAL GC

RIVER BAS WATER RES NEEDS M

SOCIO-ECONOMIC ALTERNATIVES
- RESTRICT INDUSTRIAL DEVELOPMENT
- LEGISLATION PROHIBITING DUMPING WASTE
- TAX INCENTIVES FOR TREATMENT
- EDUCATION PLAN TO INFORM PUBLIC OF RE-USE SYSTEMS

ECONOMIC MODELS

} PROGRAM BENEFITS AND VALUES

FIGURE 9

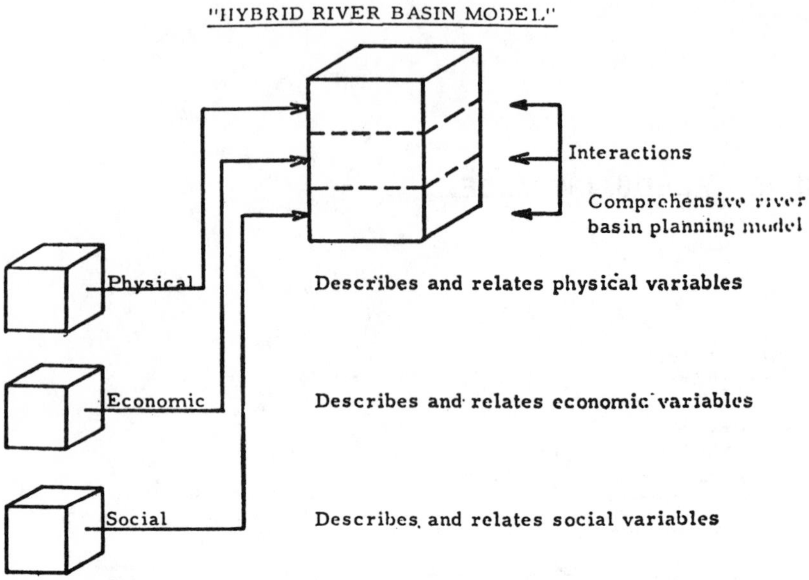

"HYBRID RIVER BASIN MODEL."

FIG. 8.—HYBRID RIVER BASIN MODEL

Although a comprehensive quantitative hybrid model of this type has not yet been developed, the Technology Planning Center recommends that research should presently be started in this direction, particularly on isolating the relevant social factors as they relate to the economic and physical variables.

The flow-logic-feedback technique

One promising approach to developing a hybrid model adaptable to computer simulation which has met with success in other public service problem areas employs a specialized modeling technique developed by the Technology Planning Center. The technique known commonly as a flow-logic-feedback model, lends itself especially well to the analysis of combined social, physical, and economic problems. We have stated earlier that modeling is simply the setting up of a scheme in miniature, as an abstraction of important situations. With the flow-logic-feedback model, one can set up hypothetical problem situations and then determine logical solutions to these problems.

In an exemplary spirit, the Technology Planning Center constructed a very simplified hybrid model using this flow-logic-feedback technique. Illustrated in figure 9, "Water Management Flow-Logic-Feedback Model," and described below, the intention is to communicate the flavor of the technique and give the reader an understanding of its application and an appreciation for its utility.

Before proceeding with an explanation of the model, a clarification of several terms is in order. *Flow* indicates that an event moves along a time sequence which, in turn, is associated with cost to the system. *Logic* indicates that there is a causal relationship between events that occur in time. Feedback stands for the various paths which may be taken by an individual as the result of decisions made by him or for him. By means of the flow-logic-feedback model, we may predict future activity, its scale, and its cost to society. Elements of chance can be built into such a diagram, and personal, as well as administrative decision points are clearly indicated and analyzed in order to predict possible future occurrences.

The model, then, outlines the processes by which water related problems arise, formulates alternative solutions and, ultimately, meets the community's water needs. Although limited in substantive detail, the model depicted on the following page cogently lays out and ties together all the planning components of a river basin.

Description of the illustrated model

The terminal goal of the river basin model illustrated is to satisfy the water resource needs of a community. The model begins by enumerating the various demand segments on water resources: recreation, industrial, municipal consumption, et cetera. From this point, it is necessary to determine whether the antecedent water conditions are acceptable: that is, has the water in any way been tampered with so that public attitude is affected? If the conditions are not acceptable, the unacceptable antecedent conditions must be determined and corrected.

Programing further, we note that the affirmative or negative response to whether a proposed water substitute is acceptable will trigger different courses of action on the route to meeting the water resources needs of the river basin. Assuming that a water substitute is not acceptable, we proceed directly to the next decision box; namely, "Is the water quality acceptable?" Passing various decision points, the planner is repeatedly triggered to action based on the decisions made along the way. Alternative paths or solutions are set out for him in logical sequence.

If the water quantity is inadequate, one or more of the enumerated "action" groups—civic improvement groups, legislators, et cetera—in vocalizing their demands will bring the problem to the attention of the public agency or agencies concerned with the water problem. It is at this juncture that the necessity for modeling and simulation is most apparent. Through the combined development of physical, social, and economic models as previously discussed, water resource planning agencies possess the ability to weigh the implications of pursuing alternative solutions. After water agencies designate one alternative, the final decision as to its acceptability rests with the community. If the solution is not adequate, the water resource planners must reevaluate the problem and search for new solutions. If adequate, the terminal goal of meeting community water resource needs will have been accomplished.

Experience has indicated that serendipity in this approach to modeling is profuse and offers a very high probability of developing insights into many unexpected relationships. It is very likely that new requirements for improved information needs can be identified and that forecasted changes in program operations will illustrate the power of an integrated "systems" approach to problem solving.

Through the development of a flow-logic-feedback model format, such as the one previously illustrated in figure 9, we thus have available an analytic "guiding star" with three points of utility, all related to use under the PPBS concept.

IMMEDIATE USES OF THE HYBRID MODEL

Use I—A policy planning benefit/cost predictor

Since future decisionmaking is directly related to the problem of presenting and transferring information to the decisionmaker, the flow-logic-feedback model has an important use as a benefit/cost predictor.

It is a well-known axiom in educational circles that optimal information transfer occurs through concrete, direct, personal experience. In planning, however, the decisionmaker must always project himself into the future, which is almost always nebulous and uncertain. To counteract this, however, a good technological aid is available through the design and development of "contrived experience" or simulation devices. In the past, these devices have taken on the forms mentioned earlier; that is, flight trainers, wind tunnels, marine model basins, atomic power reactors simulators, et cetera.

In the public services area, it is entirely possible to design a simulator in which benefit/cost alternatives might be displayed and programed on a highly "time compressed" simulation basis. In this way, a future situation can be projected in increments of 1 year, and the consequences of earlier decisions can be observed in, let us say, the fifth year of the projection. Mistakes made earlier will undoubtedly "catch up," and the simulator will thus provide useful predictive information on the consequences of an earlier decision. Based on experience to date, it is entirely conceivable that a 5-year set of planning decisions can be compressed in a single day of simulator operation.

A legislator, public official, land developer, planner, voter, or leader of a public opinion group, might thus become a water resource planner for 1 day, a health planner the next day, a recreation planner the next, and so on. This is all possible and technically feasible through the development of a family of "contrived experience" simulators. Technology has thus produced a means by which complex planning decisions can be assisted through an automated device.

By using "contrived experience" simulation, based on a social flow-logic-feedback model, water resource planners can examine, for example, the effects of alternative sets of programs on a particular river basin and the area's inhabitants. The water management information system proposed in part VI of this report will have the capacity and flexibility to provide "initial condition" data to such a simulator.

The relationships between a water management information system, river basin models, and the river basin simulators are illustrated in figure 10, "Simulation process," on the following page.

The combined flow-logic-feedback "contrived experience" simulation approach offers many other advantages to natural resource planners and researchers. First, the technique forces the water resource planners to make a detailed and thorough assessment of the social, as well as physical and economic implications of water programs under study. The inherent relationships of these phenomena must be understood. Coupled with that understanding, the development of new concepts which can be conveniently tested for benefit-cost feasibility by simulation is possible. Secondly, a realistic analysis can be made of various program characteristics based on interpretations of the behavior exhibited by the computer model. The water management planner can actually "see," by observing an electronic recorder or plotter, the

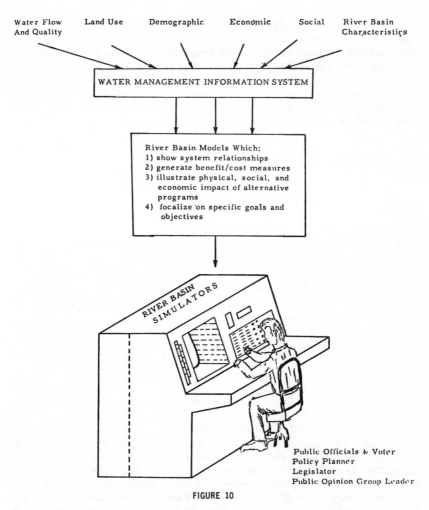

Water Flow   Land Use   Demographic   Economic   Social   River Basin
And Quality   Characteristics

WATER MANAGEMENT INFORMATION SYSTEM

River Basin Models Which:
1) show system relationships
2) generate benefit/cost measures
3) illustrate physical, social, and
   economic impact of alternative
   programs
4) focalize on specific goals and
   objectives

RIVER BASIN SIMULATORS

Public Officials & Voter
Policy Planner
Legislator
Public Opinion Group Leader

FIGURE 10

intermediate benefit and cost at any point in the model. Since the model potentially can include all relative characteristics, the system or program is available for exploratory study as an integrated whole.

The payoff of this planning approach, of course, is to generate data to policy planners for use in making tradeoff analyses between competing programs within the restraints of a fixed total budget.

Use II—Flow-logic-feedback diagrams in public education programs

Because of the inherent nature of the flow-logic-feedback model, it can present a clear picture of physical, economic, and social variables as they relate to the general public. A water demand model, for example, would clearly portray all the factors (economic, physical, and social), which would bear upon the optimal allocation of a fixed quantity of water in a river basin. The variables and their relationship to each other are clearly presented and provide a means of portraying the solution tradeoffs to the public at large.

A river basin simulator would, also, have educational utility in that an interested citizen could, through simulation see what the repercussions of a proposed change would be on his personal situation.

The accompanying figure 11, "A cross-media information transfer spectrum," presents a cross-media information transfer spectrum which can serve as an aid in considering this vital task. It can readily be seen that the use of media most closely replicating direct experience produces the best and most lasting form of information transfer. Speeches, on the other hand, are the most abstract form of information transfer unless augmented with various other combinations of audio-visual devices.

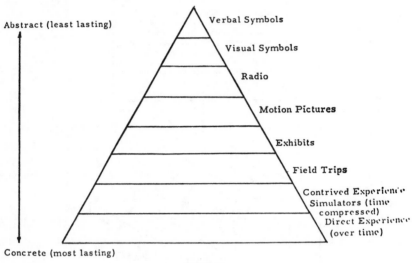

FIGURE 11

Use III—Determining information needs

The last and possibly the most important immediate use of the flow-logic-feedback model, as related to water resource planning, is its utility as a point of reference for determining specific individual and statistical information needs of various water agencies. By scanning vertically through figure 9, "Water management flow-logic-feedback model," the types of specific information required can be differentiated according to the needs of operating agencies, researchers, and policy planners. In other words, the diagram automatically organizes and catalogs data and information differences according to ultimate user needs. Using the model in this fashion, it serves as a point of reference for relating studies involved with data acquisition, organization, codification, storage, and retrieval.

This approach also places data acquisition surveys into proper perspective. Through such a model, the relevance and importance of various attitudinal and social factors can be assessed.

These are but a few of the areas in which questions arise with regard to information which should be collected, those items that should be retained, those items that should be deleted, and those items which are not currently collected but which should be included in a total water resource management information system.

APPENDIX F

Computer Programs for School Planning

The development and utilization of computer programs for application to social and community problems are of high interest to those who must ensure the smooth functioning of current activities and plan for expanded and often significantly modified programs for the future. In this example, featuring the creation of a series of three computer programs, the importance of being able to establish realistic predictions of pupil loads and the related problem of providing enough classrooms has been stressed. The presentation of the logical sequence of events comprising three distinct computer programs was excerpted from appendixes A, B, and C of "An Assignment Program to Establish School Attendance Boundaries and Forecast Construction Needs." [1]

[1] Ploughman, T. W., Darnton, and W. Heuser. An assignment program to establish school attendance boundaries and forecast construction needs. In Socidecon. plan. sci., v. 1, n. 3, July 1968, pp. 250–258.

Part I. FORTRAN Computer program

The logical sequence of events comprising Part I. (The Population Generator) computer programs are as follows:

Specify number of zones, smoothing factor, the pupils/family ratio at maturation, number of years from saturation to maturation, the pupils/family ratio for new families, and subcensus zone data.

For each zone calculate number of new families to saturation, number of years growth takes place, and number of families from year 1–25.

For each zone calculate an *adjusted* pupil/family ratio as a function of the previous years ratio and the ratio at maturation up to the year of saturation.

For each zone calculate number of pupils for years from saturation to maturation via straight line to reach maturation ratio. Then hold number of pupils constant to year 25.

For each zone calculate the yearly pupils/family ratio as a result of previous forecasts.

For each of 315 zones print-out yearly data—year, families, pupils, and ratio.

Print-out the sum of yearly forecasts for zones to each school based on year 1 assignments.

Same for each of 33 census tracts.

Print-out yearly forecasts for total district—year, families, pupils, and ratio.

The computer print-out is illustrated by the tables reproduced in Figs. A-1, A-2, and A-3.

An Assignment Program to Establish School Attendance Boundaries

Subcensus zone statistics—ELEM Subcensus zone 64. Smoothing X = 0·80

Year	Families	Pupils	Ratio	Adj Ratio
1	6·00	3·00	0·50	0·50
2	6·00	3·00	0·50	0·50
3	6·00	3·00	0·50	0·50
4	6·00	3·00	0·50	0·50
5	6·00	3·00	0·50	0·50
6	13·50	9·60	0·71	0·50
7	21·00	15·63	0·74	0·67
8	28·50	21·20	0·74	0·70
9	36·00	26·41	0·73	0·70
10	36·00	25·01	0·69	0·69
11	36·00	23·61	0·66	0·66
12	36·00	22·21	0·62	0·62
13	36·00	20·80	0·58	0·58
14	36·00	19·40	0·54	0·54
15	36·00	18·00	0·50	0·50
16	36·00	18·00	0·50	0·50
17	36·00	18·00	0·50	0·50
18	36·00	18·00	0·50	0·50
19	36·00	18·00	0·50	0·50
20	36·00	18·00	0·50	0·50
21	36·00	18·00	0·50	0·50
22	36·00	18·00	0·50	0·50
23	36·00	18·00	0·50	0·50
24	36·00	18·00	0·50	0·50
25	36·00	18·00	0·50	0·50

FIG. A–1. Subcensus zone yearly table, families and pupils, elementary.

ELEM summary table for census zones 1–4. Smoothing factor X = 0·80

	Zone–1			Zone–2			Zone–3			Zone–4		
Year	Families	Pupils	Ratio	Families	Pupils	Ratio	Families	Pupils	Ratio	Families	Pupils	Ratio
1	124	62	0·50	150	82	0·55	198	129	0·65	226	185	0·82
2	148	83	0·56	167	96	0·57	210	134	0·64	244	187	0·76
3	164	95	0·58	181	106	0·58	222	139	0·62	254	184	0·72
4	172	100	0·58	189	110	0·58	234	143	0·61	264	181	0·69
5	176	101	0·57	192	109	0·57	241	144	0·60	274	179	0·65
6	176	98	0·56	195	109	0·56	243	141	0·58	284	177	0·62
7	176	95	0·54	199	109	0·55	244	137	0·56	295	174	0·59
8	176	92	0·52	202	108	0·54	244	133	0·54	301	169	0·56
9	176	90	0·51	205	108	0·53	245	130	0·53	305	170	0·56
10	176	88	0·50	209	109	0·52	245	125	0·51	310	170	0·55
11	176	88	0·50	212	111	0·52	245	124	0·50	311	168	0·54
12	176	88	0·50	212	110	0·52	245	123	0·50	311	164	0·53
13	176	88	0·50	212	109	0·51	245	123	0·50	311	161	0·52
14	176	88	0·50	212	108	0·51	245	123	0·50	311	159	0·51
15	176	88	0·50	212	108	0·51	245	122	0·50	311	157	0·51
16	176	88	0·50	212	107	0·50	245	122	0·50	311	156	0·50
17	176	88	0·50	212	106	0·50	245	122	0·50	311	155	0·50

FIG. A–2. Summary Tables for Census Tracts 1–4.

Summary table for 315 subcensus zones. Smoothing factor $X = 0.80$

Year	Families	Pupils	Ratio
1	6438·00	3995·00	0·62
2	8028·64	5248·77	0·65
3	8931·29	5821·99	0·65
4	10066·90	6558·22	0·65
5	11113·84	7153·08	0·64
6	12145·37	7683·38	0·63
7	12648·69	7723·68	0·61
8	13101·94	7711·70	0·59
9	13537·57	7795·99	0·58
10	13830·83	7779·31	0·56
11	14113·31	7775·06	0·55
12	14151·05	7571·78	0·54
13	14188·78	7463·22	0·53
14	14192·19	7343·16	0·52
15	14195·59	7248·10	0·51
16	14199·00	7184·51	0·51
17	14199·00	7121·05	0·50
18	14203·75	7116·03	0·50
19	14208·50	7110·64	0·50
20	14213·25	7113·13	0·50
21	14218·00	7115·38	0·50
22	14218·00	7113·45	0·50
23	14218·00	7112·55	0·50
24	14218·00	7111·66	0·50
25	14218·00	7110·77	0·50

FIG. A–3. Summary table for 315 subcensus zones, elementary pupils.

APPENDIX B

Part II. Simscript computer program

The logical sequence of events comprising Part II ("The Ultimate Assignment of Subcensus Zones to Schools") computer programs are as follows:

Assign zones having only one potential school.

Assign zones to the closest school having a class average below the district average. Zones having the farthest maximum distance school are assigned first.

Assign zones to their closest school having a class average below the upper limit.

Assign unassigned zones to a potential school where the resulting class average is the lowest. Order of selection of zones is by number of students.

A subsection of the computer program is reproduced in Fig. B-1. Examples of program output are reproduced in Figs. B-2 and B-3.

An Assignment Program to Establish School Attendance Boundaries

```
C

C       ***THIS SECTION ASSIGNES ZONES WITH SINGLE POTENTIAL SCHOOLS

C

        DO TO 11, FOR EACH ZONE I

        LET NPSKL = 0

        DO TO 9, FOR EACH SKOOL J, WITH(PSKLS(I,J))EQ(I)

        LET NPSKL = NPSKL + 1

    9   LOOP

        IF (NPSKL)NE(0), GO TO 10

        CALL GOOF(I)

   10   IF (NPSKL)GR(I), GO TO 11

        FIND FIRST, FOR EACH SKOOL J,WITH(PSKLS(I,J))EQ(I),WHERE JX

        LET SSKL(I) = JX

        IF(CLSD(JX))EQ(I),LET FORC(I) = I

        LET PUPLS(JX) = PUPLS(JX) + STDTS(I)

        IF(PUPLS(JX)/SIZE(J))GE(CLSAV),LET CLSD(JX) = I

   11   LOOP

C

C       ***THIS SECTION ASSIGNES ZONES TO THE CLOSEST SCHOOL HAVING

C           A CLASS AVERAGE BELOW THE DISTRICT AVERAGE.  ZONES HAVING

C           THE FARTHEST MAXIMUM DISTANCE SCHOOL ARE ASSIGNED FIRST

C

   12   DO TO 30,FOR EACH ZONE I,WITH(SSKL(I))EQ(0)
```

FIG. B–1. A portion of SIMSCRIPT program.

Report of zones

Zone No.	No. of students	School assigned	I If forced	Distance to school
1	29·0	3	0	1·45
2	18·0	3	0	1·18
3	21·0	3	0	0·70
4	6·0	3	0	0·83
5	4·0	3	0	0·64
6	4·0	3	0	0·88
7	4·0	3	0	1·15
8	4·0	3	0	0·88
9	4·0	3	0	1·14
10	—	3	0	0·95
11	—	6	0	0·62
12	—	3	0	0·65
13	—	3	0	0·74
14	21·0	6	0	0·41
15	17·0	3	0	0·44
16	29·0	3	0	0·20
17	17·0	3	0	0·24
18	24·0	6	0	0·53
19	1·0	6	0	0·74
20	27·0	3	0	0·44
21	30·0	3	0	0·27
22	9·0	3	0	0·67
23	23·0	3	0	0·38
24	20·0	3	0	0·62
25	11·0	3	0	1·39
26	9·0	3	0	1·11
27	28·0	3	0	1·17
28	24·0	3	0	0·68
29	48·0	3	0	1·02
30	53·0	3	0	1·11
31	4·0	3	0	1·74
32	11·0	3	0	1·97
33	—	6	0	2·12
34	62·0	6	0	1·74
35	87·0	6	0	1·91
36	6·0	6	0	1·83
37	24·0	6	0	1·71
38	30·0	6	0	1·50
39	25·0	0	0	0·99
40	113·0	6	0	1·05
41	—	6	0	0·94
42	—	6	0	0·79
43	—	6	0	0·64
44	65·0	6	0	0·53
45	—	6	0	1·00

FIG. B–2. Bloomfield Hills school district report of zone assignment results elementary schools—1988.

An Assignment Program to Establish School Attendance Boundaries

School 5. Size of school -21. ROOMS, No. of pupils -660, class average -31·4.

Zones assigned are
183-
184-
185-
186-
187-
189-
190-
191-
192-
193-
194-
195-
196-.
197-
229-
230-
231-
232-
235-
236-
237-
257-
258-
259-
260-
261-
262-
263-
264-
266-
267-
268-
275-
276-
277-
278-
279-
280-
281-
282-
283-
284-
285-
286-

FIG. B–3. Summary of subcensus zones assigned to school No. 5.

Part III. Simscript computer program

The logical sequence of events comprising Part III (Building Program and Yearly Subcensus Zone Assignments) computer programs are as follows:

DETERMINE NEW SCHOOL BUILDING SCHEDULE. WHENEVER THE DISTRICT CLASS AVERAGE EXCEEDS 28, ENOUGH NEW SCHOOLS MUST BE ADDED TO BRING THAT AVERAGE BELOW 28. THE PARTICULAR SCHOOL (S) SELECTED IN ANY. YEAR WHOSE ULTIMATE ATTENDANCE ZONES ARE CLOSEST TO THEIR PEAK PUPIL POPULATION. (THE NEW SCHOOL SCHEDULE CAN BE ENTERED EXTERNALLY IN PART OR IN TOTAL)

START IN TIMING ROUTINE—ADVANCE TO NEXT SCHOOL YEAR.

BRING IN ANY NEW SCHOOLS CALLED FOR IN BUILDING SCHEDULE.

CLASSIFY ALL OLD SCHOOLS AS ABOVE, IN RANGE OR BELOW THE ACCEPTABLE RANGE ($\pm 2 \cdot 5$) AROUND THE DISTRICT AVERAGE FOR THAT YEAR, BASED ON THE ATTENDANCE AREA OF THE LAST SCHOOL YEAR.

FOR EACH SCHOOL THAT IS ABOVE THE ACCEPTABLE RANGE, ATTEMPT TO MOVE ANY OF THE ZONES CURRENTLY ASSIGNED TO IT THAT ULTIMATELY ATTEND THE NEW SCHOOL(S) BUILT THIS YEAR. USE THE DECISION MATRIX (TABLE I-C) TO SEE IF THE MOVE CAN BE MADE. THE ORDER OF ZONES SELECTED IS BY THE LONGEST TIME SINCE REASSIGNMENT.

RE-EXAMINE THE SCHOOLS THAT WERE ABOVE THE ACCEPTABLE RANGE TO SEE IF THEY ARE STILL ABOVE.

FOR EACH SCHOOL THAT IS STILL ABOVE THE ACCEPTABLE RANGE, SEE IF TEMPORARY CLASSROOMS CAN BE USED TO SOLVE THE PROBLEM. THE FOLLOWING CONSTRAINTS MUST BE MET:

    (1) THEY MUST BE REMOVED IN 3 YR OR LESS.

    (2) ONLY 3 CAN BE ADDED TO ANY SCHOOL.

FOR EACH SCHOOL THAT IS STILL ABOVE THE ACCEPTABLE LIMITS, UNLOAD ALL ITS ZONES, THEN REASSIGN ZONES (THE CLOSEST ZONES THAT ULTIMATELY ATTEND THIS SCHOOL) UNTIL THE SCHOOL IS AT THE LOWER LIMIT OF THE ACCEPTABLE RANGE OF THE DISTRICT CLASS AVERAGE. THOSE ZONES LEFT OVER ARE PLACED IN A SET OF ZONES TO BE EVALUATED FOR REASSIGNMENT.

FOR EACH ZONE TO BE EVALUATED FOR REASSIGNMENT, ATTEMPT TO REASSIGN THE ZONE TO EITHER ITS ULTIMATE SCHOOL (IF IT IS NOW ASSIGNED TO IT) OR THE CLOSEST SCHOOL THAT IS IN OR BELOW THE

An Assignment Program to Establish School Attendance Boundaries

ACCEPTABLE RANGE OF THE DISTRICT CLASS AVERAGE—THE ORDER OF ZONES SELECTED IS BY THOSE ZONES WHOSE REASSIGNMENT HAS THE LEAST EFFECT ON POSSIBLE REASSIGNMENTS OF THE REMAINING UN-ASSIGNED ZONES—THE APPROPRIATE DECISION MATRIX (TABLE 1-C OR 2-C) IS USED IN ATTEMPTING TO MAKE THESE REASSIGNMENTS.

RE-EXAMINE THE SCHOOLS THAT ORIGINALLY WERE BELOW THE ACCEPTABLE RANGE OF THE DISTRICT CLASS AVERAGE. USE THE APPROPRIATE DECISION MATRIX TO TRY TO MOVE ZONES TO THIS SCHOOL THAT EITHER (1) ATTEND THIS SCHOOL ULTIMATELY OR (2) ARE THE CLOSEST ZONES OF THOSE THAT DO NOT ATTEND THE SCHOOL ULTIMATELY.

REPORT THE ZONE ASSIGNMENTS FOR THIS YEAR.

RETURN TO THE TIMING ROUTINE AND ADVANCE TO THE NEXT SCHOOL YEAR.

Variance of school ave. from district	1	2	Years since zone moved 3	4	5	6
27	X					
26	X					
25	X					
24	X					
23	X					
22	X					
21	X					
20	X					
19	X					
18	X					
17	X					
16	X			X: Move is allowed		
15	X			$\longrightarrow$		
14	X					
13	X					
12	X					
11	X					
10	X	X				
9		X				
8		X	X			
7			X			
6			X	X		
5				X	X	
4					X	X
3						X
2						X
1	Move not					X
0	allowed					X
	$\longleftarrow$					

TABLE 1-C. ELEMENTARY SCHOOLS. MOVING A ZONE INTO ITS "FINAL" SCHOOL

Variance of school ave. from district	Years Since Zone Moved					
	1	2	3	4	5	6
27	X					
26	X					
25	X					
24	X					
23	X					
22	X					
21	X					
20	X					
19	X					
18	X					
17	X					
16	X					
15	X		X: Move is allowed $\longrightarrow$			
14	X					
13	X					
12	X	X				
11		X				
10		X	X			
9			X	X		
8				X		
7				X	X	
6					X	X
5						X
4		Move not allowed				X
3		$\longleftarrow$				X
2						X
1						X
0						X

TABLE 2-C. ELEMENTARY SCHOOLS. MOVING A ZONE INTO A SCHOOL OTHER THAN THE FINAL

APPENDIX G

California Regional Land Use Information System—A TRW Civil Systems Project

(By George K. Chacko, Ph. D., Senior Staff Scientist, TRW Washington Operations)

TRW project objective: Utilize land information

In his testimony on January 25, 1967, before the U.S. Senate Special Subcommittee on the Utilization of Scientific Manpower, chaired by Senator Gaylord Nelson, Dr. Simon Ramo, vice chairman of the board, TRW Inc., referred to the California Regional Land Use Information System as a project in which systems technology was being applied to the social and community problems. The project was performed by one of the five operating groups of TRW Inc., viz., TRW Systems Group, under contract P/218–PD–CON–1 during 1966–67. The specific objective was to develop an improved land use information method so that the various departments within the Government jurisdiction, as well as different agencies of the Government, might more effectively share and utilize land use information.

Central American exhibition

The U.S. Information Agency selected this project to be part of the U.S. National Exhibition in El Salvador, San Salvador, in November 1968. The San Salvador Fair emphasized for the first time the role of private enterprise in the Alliance for Progress. The 500,000 viewers, including the dignitaries from government and private sectors, were exposed briefly to the parallels between Central America and California as Allies in Progress in better land utilization through systems approach. In Figure 1 the parallels in population and acreage are shown. The theme of the TRW exhibit in the USIA exhibition was the applicability of systems approach to the development of an effective land use information system.

TRW systems approach

The application of the systems approach by TRW to the California Regional Land Use Information System is highlighted in figure 2 where the emphasis is on the user and his needs, and the development and exchange of the needed information.

1.0 Operational objectives

The project, undertaken for the State of California, had, as its operational objectives:

Review the land use data *requirements* of the major users of such data in California.

Obtain a *consensus* among representatives of major users regarding a compatible system of data classification and coding.

Establish *policy and standards* for data interchange.

Design a land use information system whitin the conceptual framework of the Statewide Federated Information System.

Demonstrate elements of the proposed system using Santa Clara County as the demonstration region.

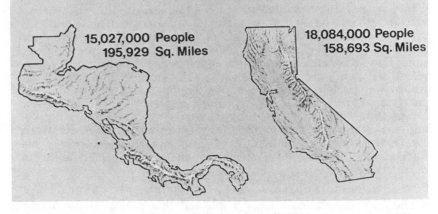

BETTER LAND UTILIZATION THROUGH SYSTEMS APPROACH

15,027,000 People
195,929 Sq. Miles

18,084,000 People
158,693 Sq. Miles

FIG. 1.—BETTER LAND UTILIZATION

TRW SYSTEMS APPROACH

● FOCUS ON THE USER, IDENTIFYING HIS NEEDS

● LOCATE THE NEEDED INFORMATION

● FACILITATE THE EXCHANGE OF THIS INFORMATION

FIG. 2.—TRW SYSTEMS APPROACH

2.1 Data collection—Interim survey

To accomplish the operational objectives, a two-step approach was taken. First, 143 interviews were conducted with organizations in the State which had a wide spectrum of responsibilities, but all of which had the common concern about land-related information. In table I the agencies that were contacted in the interim survey are listed. As many as 262 individual interviews were held at this stage, which included people at eight different levels—Federal, State, county, city, district, regional, private, and miscellaneous. The individual interviewees were generally enthusiastic about the project objectives, and gave freely of their time and experience in outlining their own concern with land data. It was discovered that the nature and variety of land-related data were even broader than had been initially anticipated.

2.2 Data collection—Questionnaire survey

The second of the two-step approach was the questionnaire survey. A questionnaire was designed and coordinated with a number of groups working in related areas in the State. The questionnaire provided a list of data elements identified in the initial survey and supplemented from the literature. Respondents were asked to indicate the source, storage media, local unit of collection, and other attributes about each data element. They were also to indicate present or future data element needs. In figure 3 the information questionnaire is shown with the blowup of column 4—Source of data. It is remarkable that the questionnaire which has as many as 412 elements was consciously filled by as high as 65.6 percent of those to whom the questionnaires were sent. The 554 replies to the 844 questionnaires represented some 35,000 records. In terms of computer terminology, the returns represented 10 million computer "words" as shown in figure 4.

TABLE I.—*Interviewees in the interim survey*

FEDERAL

Department of Agriculture:
    Economic Research Service
    Federal Extension Service
    Forest Service
    Program Evaluation and Programing Area
    Rural Community Development Service
    Soil Conservation Service
    Statistical Reporting Service
Department of Commerce:
    Office of Statistical Standards
    Bureau of the Census
    U.S. Coast and Geodetic Survey
    Office of Regional Economic Development
Department of Interior:
    Bureau of Outdoor Recreation
    National Park Service
    Geological Survey
    Federal Water Pollution Control Administration
    Bureau of Land Management
    Bureau of Reclamation
Department of Housing and Urban Development:
    Office of Metropolitan Development
    Urban Planning Commission
    Office of Planning Standards and Coordination
Department of Housing and Community Development:
    Office for Plans and Programs
    Demonstration Branch
Office of the President: Bureau of the Budget
Department of Defense: Corps of Engineers
Federal Home Loan Bank
Advisory Commission on Intergovernmental Relations

TABLE I.—*Interviewees in the interim survey*—Continued

Department of Agriculture:
    Division of Agriculture Economics
    Bureau of Agricultural Statistics
Department of Conservation:
    Division of Forestry
    Division of Soil Conservation
Department of Public Works:
    Division of Highways
    Division of Aeronautics
    Transportation Agency
Department of Employment:
    Standards and Methods Section
    Research and Statistics Section
Resources Agency:
    Division of Soil Conservation
    Resources Planning
Department of Parks and Recreation:
    Statewide Recreation Planning
    Division of Beaches and Parks
Department of Water Resources:
    Planning Branch
    San Francisco Bay District
Department of Housing and Community Development:
Department of Fish and Game
Department of General Services:
    Statewide Information Systems Project
    Property Acquisition Services
    Systems Analysis Office
Department of Finance:
    State Lands Division
    State Office of Planning
Franchise Tax Board
State Controller
Department of Investment:
    Division of Real Estate
    Real Estate Commissioner
State Board of Equalization
Department of Industrial Relations: Division of Labor Statistics and Research
Department of Public Health: California Health Information for Planning Service (CHIPS) Project
Geological Hazards Advisory Committee
Senate Fact Finding Committee on National Resources
Steering Committee for California Land Use Information System Project

County Supervisors Association of California: Planning and Research Division
Santa Clara County:
    Department of Agriculture
    General Services Agency
    Data Processing Department
    Department of Public Works
    Planning Department
    Assessors Office
Los Angeles County: Assessors Office
Orange County:
    Planning Department
    Data Services Department
San Diego County:
    Planning Department
    Electronic Data Processing Services Department

TABLE I.—*Interviewees in the interim survey*—Continued

CITIES

Los Angeles:
    Department of City Planning
    Data Services Bureau
Palo Alto:
    City Controller
    Planning Department
    Data Processing Department
San Diego: Administrative Management Department
Redondo Beach: Development Agency
League of California Cities

DISTRICTS

Santa Clara County Flood Control and Water District
Metropolitan Water District of Southern California
East Bay Municipal Utility District

PRIVATE

Southern California Edison Company: Area Development Department
Southern Pacific Company: Industrial Development Department
Pacific Gas and Electric Company:
    Management Information Systems
    Commercial Department
    Area Development
Kern County Land Company:
    Real Estate
    Oil Exploration
Caldwell, Banker and Company
Title Insurance and Trust Company:
    System and Research Division
    Corporate Business and Development
John B. Joynt and Associates, Inc.
R. L. Polk and Company: Urban Statistical Division
Anthony Medin and Associates
Land Research Company

REGIONAL

Association of Bay Area Governments: Bay Area Automated Information Systems Coordinating Committee
Bay Area Transportation Study: ABAG–BATSC Data Service
Bay Area Council

MISCELLANEOUS

California Real Estate Association:
    Legislative Committee
    Governmental Relation
Resources for the Future, Inc.
University of California: Institute of Urban and Regional Development
Council of Planners: Subcommittee on Data Processing and Information Systems
American Institute of Planners

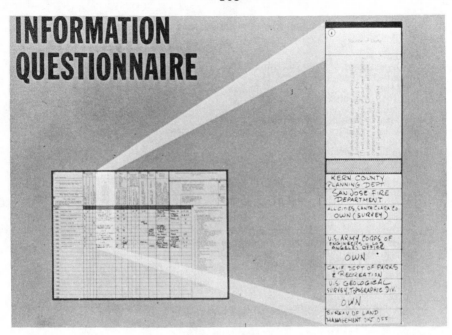

FIG. 3.—INFORMATION QUESTIONNAIRE

FIG. 4.—TRW CALIFORNIA LAND USE INFORMATION SURVEY

3.1   User requirements analysis

What do the data indicate as the principal user needs that are met by the available data? What are the major deficiencies?

The prime users of land use data are found to be:

assessor
fire marshal
sealer of weights and measures
agricultural department
health department
planning department
flood control and controller
land development review committee
registrar of votes
recorder
welfare department

The preliminary analysis to identify users and user needs involved a survey of the property subsystem organization, the present status of operational land use data, current and projected requirements, data collection, data maintenance, and data compatibility procedures. A survey team composed of representatives of each department and TRW staff members studied the status of the present data system to decide on a basis for the design and implementation of the property subsystem. The property subsystem was one of 10 subsystems which were found essential to the California Regional Land Use Information System. Pertinent data files of interest were examined; a data inventory form was prepared and filled out for each data item. A summary flow chart of the information flow regarding property was developed from the intensive interviews and study of existing data collection and data using procedures as shown in figure 5.

3.2   User data needs unfulfilled at present

Figure 5 shows the lack of information blow in Santa Clara County between the departments generating data and the departments requiring additional data for effective operations. Though not as inten-

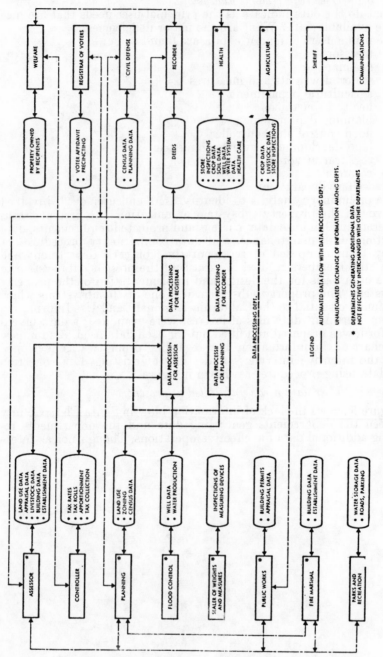

FIG. 5.—PRESENT FLOW OF PROPERTY INFORMATION IN SANTA CLARA COUNTY

sively studied as at the county level, a survey was made of the user data needs unfulfilled at present at the State level. During the questionnaire survey discussed in section 2.2, the 844 recipients of questionnaires were asked to indicate the data items which were "not used but required for their current or future operations." It is interesting to note that every single one of the 412 data elements was indicated as needed by at least one agency. The data elements mentioned most often were:

Water resource data
Land resource data
Road and street data
Utility and service data
Land and structure use data
Urban use intensity data.

In table II are listed the actual number of responses by Federal, State, county, city, district, and private sector groups. It is seen that at the Federal level the unmet data needs represent 5.3 percent of the total data count; at the State level, 20.2 percent; and at the private sector level, 25.4 percent.

Type of data	Federal Unmet data needs	Federal Total data count	State Unmet data needs	State Total data count	Counties Unmet data needs	Counties Total data count	Cities Unmet data needs	Cities Total data count	District Unmet data needs	District Total data count	Private companies Unmet data needs	Private companies Total data count
Environment:												
Climatological	2	52	13	117	21	261	11	128		12	6	30
Topographic	1	29	2	53	8	111	1	92	1	14		14
Geological	4	60	2	127	48	233	24	162		18	9	31
Soil	1	42	25	131	53	288	24	207		30	7	31
Water resource	2	109	46	193	122	504	32	290	1	32	27	67
Land resource	1	18	27	80	30	96	12	67		6	6	14
Vegetation	1	35	19	86	37	161	6	46		6	10	17
Wildlife	2	26	21	70	23	53	3	4		1	7	7
Other environmental hazard	2	7	22	44	40	126	11	41	1	4	6	10
Improvement:												
Road and street	5	95	61	355	169	850	58	1,290		28	28	58
Utility and service (facilities, characteristics, routes, etc.)	9	67	73	294	116	590	43	503		50	23	72
Parcel		41	34	219	68	478	32	547		31	14	67
Nonstructural improvements	4	57	44	174	51	540	27	564	1	27	20	71
Structures (all types)	4	137	128	666		1,506	129	1,390		163	28	248
Valuation:												
Tax	3	26	29	110	41	496	13	232		40	18	73
Market value	2	52	27	158	51	298	18	101	1	24	16	54
Income	2	14	8	42	15	76	5	13		7		9
Insurance		2		9	7	35		6		14	5	9
Tenure:												
Property ownership	6	64	16	320	90	878	45	381	1	49	26	83
Easement	1	16	7	92		176	4	200		13	3	17
Lien	1	11	2	35	12	110		15		6	5	12
Permits and licenses	1	10	10	80	32	240		200		3	5	14
Activity:												
Classification (in terms of current or future activities)	1	11	15	72	8	39	2	109		5	6	23
Zoning	3	16	13	57	18	182	5	202		5	7	28
Land and structure use	1	37	44	175	69	289	30	267		14	12	48
Water use	1	27	14	94	41	178	7	110	1	10	9	128
Legal constraints or obligations (affecting use)	1	2		16	6	51	3	81		4	2	3

Intensity:												
Water use intensity		42	29	157	79	219	14	152	1	10	15	43
Fire control	1	9	10	30	21	102	4	85		8	5	5
Agricultural use	1	28	16	67	16	192		1		1	3	7
Forest and range use intensity	1	9	12	27	9	16				2	1	2
Extractive use intensity	1	7	12	46	22	76	6	19				8
Commercial fishing		2	5	14	11	14	1	5		5		
Recreational use	1	19	20	72	41	109	2	48				4
Urban use intensity		36	45	161	71	340	21	320		36	1	67
Commercial use intensity			4	10	6	13	2	11				3
Industrial use intensity			6	21	18	29	8	30		1		7
Projected intensity		39	59	180	95	296	41	273		5	11	58
Total	66	1,248	940	4,654	1,565	10,251	644	8,192	8	684	341	1,342

TABLE II.—USER DATA NEEDS UNFULFILLED AT PRESENT

4.1 Information system—LOGIC (Local Government Information Control)

Turning from the California Regional Land Use Information System to its application to Santa Clara County (already introduced in section 3.1), the board of supervisors of the county unanimously adopted in April 1965 an information system called local government information control (LOGIC), which was aimed at as comprehensive an information system as possible, fulfilling the many hitherto unmet needs at the county level. Of the 10 subsystems constituting the information system, the Property Subsystem (referred to in section 3.1) comprising all data describing the land of Santa Clara County and its uses was asked to be developed by TRW systems in January 1967. On February 6, 1967, the TRW Systems team started on the initial operations of the study.

LOGIC is not only a computer-based system, but also a system of cooperation between the departments which will contribute to greater functional efficiency. However, long before the data can be handled on the computer, vital problems such as incompatibility standardization, quality control, maintenance, privacy, and cross indexes must be resolved.

The first step in systematizing the property subsystem was to determine the primary files of the system and the information flow between them. All files would be under the control of a master program which would schedule the orderly processing, maintenance, storage, retrieval, recording, summarizing, and display of both qualitative and quantitative as represented in figure 6.

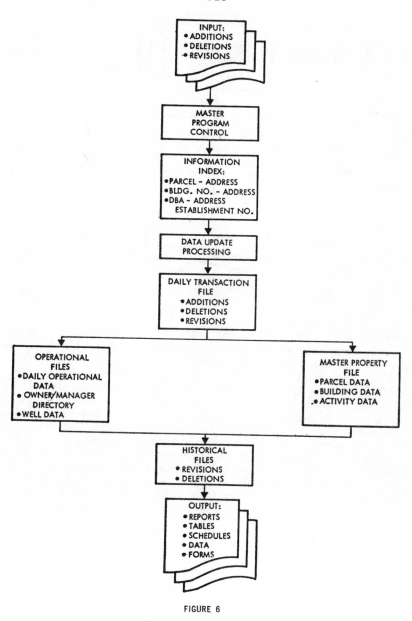

FIGURE 6

The basic elements of a typical user-oriented information processing system is presented in figure 7. The user describes on a set of control cards, the processes and tasks he wishes performed on his data. He may submit data interspersed with control specifications, or require the system to operate on prestored data from a library magnetic tape file. A variety of processing and analyzing programs are available on call from the system library tape. The results from a user application may be either the addition of data files to the tape library or the display of printed reports or tables.

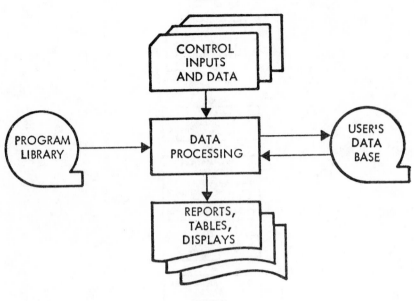

FIGURE 7

Central American application—An illustration

The implementation of the California Regional Land Use Information System is a future development to which planned progress can be made as a result of the TRW project study and recommendations. However, the systems approach which emphasizes the user and his needs, the location of the needed information, and the facilitation of the exchange of appropriate information between the originator and the user of different types of data, can well be applied to the marketing and management of agriculture, say in Central America.

In figure 8, a basic problem facing agriculture is presented. Four principle products indigenous to the region are: sugarcane, coffee, cattle, and bananas. In the illustration, equal amount of land is devoted to each product. The question facing the user of the agricultural land is: What should be the optimum allocation of land to the four different products? In a highly simplified manner, systems approach would emphasize that the optimum allocation is not a question of cultivation alone but also of marketing, as shown in figure 9. To insure that a proper balance is struck between production and distribution, management is indispensable (fig. 10). The crucial element that systems approach brings to the aid of management is scientific analysis in which both machines and minds work together (fig. 11) to arrive at a rational allocation of land among the four different products in such a way that there is balance between with market demand on the one hand, and coordinated production on the other (fig. 12).

CALIFORNIAN SYSTEMS APPROACH
TO LATIN AMERICAN AGRICULTURAL PROBLEMS

FIG. 8.—CALIFORNIAN SYSTEMS APPROACH TO LATIN AMERICAN AGRICULTURAL PROBLEMS

FIG. 9.—NOT A QUESTION OF CULTIVATION ALONE BUT ALSO OF MARKETING

...AND OF MANAGEMENT

FIG. 10.—... AND OF MANAGEMENT

...AIDED BY SCIENTIFIC ANALYSIS (SYSTEMS APPROACH)

FIG. 11.—... AIDED BY SCIENTIFIC ANALYSIS (SYSTEMS APPROACH)

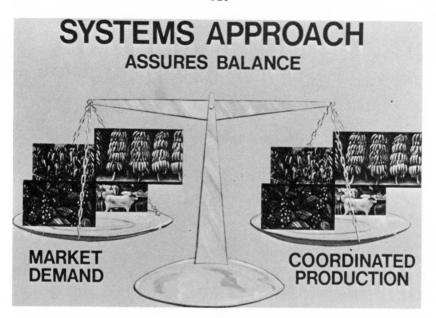

FIG. 12.—SYSTEMS APPROACH ASSURES BALANCE

APPENDIX H

Program Budgeting in Wisconsin

The State of Wisconsin established, in 1959, the first planning-programing-budgeting system in the United States. This occurred during the administration of Governor Gaylord Nelson, later to serve as Senator from that State. The significance of comprehensive planning, whether at the Federal, State, or local level, was recognized by the Senate Special Subcommittee on the Utilization of Scientific Manpower (with Senator Nelson acting as chairman). The full text of the article entitled "Program budgeting in Wisconsin" by John W. Reynolds and Walter G. Hollander provides a useful insight into the steps by which a modern planning-programing-budgeting system was established, and the political as well as technological aspects of that process.[1]

PROGRAM BUDGETING IN WISCONSIN

(By John W. Reynolds and Walter G. Hollander)

The enactment last June of a short, simple bill signaled the green light for Wisconsin to take a giant stride toward achieving a major improvement in state budgeting.

The measure was entitled simply *A bill relating to the development of program budgeting.* To a casual reader, it effected what might be considered a few minor technical changes in the state budget law. Its key section merely amended the law specifying the format of budget bills and appropriation requests by adding the words ". . . or other meaningful classifications."

But the significance of this small amendment is large. It means that budget requests and appropriations no longer will have to be divided into object lines, i.e., personal services, materials and expense and capital—the things that are purchased. Rather, they will be broken down into the services that are to be performed and the programs to be carried out.

This does not imply we are going to eliminate all use of detailed items of expenditure. It does mean, however, that we will have the flexibility necessary to separate those items of cost detail which are important to administrators in evaluating and controlling the efficiency of their operations from those cost groupings necessary for program policy decisions which determine the nature and scope of state services.

Passage of the bill formalized the legislature's endorsement of program budgeting and propelled the state budget office (the Bureau of Management in the Department of Administration) into the final stage of its previously quiet, but determined, drive to implement this modern budgeting concept. Plans have now been completed to recast the 1965–67 biennial budget into a totally new format—with both agency requests and legislative appropriations to be structured along well defined program lines. The entire project is being done without any outside consulting assistance—a unique situation among the few states that have introduced this modern budgeting concept.

The success of these efforts, we believe, is a remarkable achievement in executive-legislative cooperation, especially because it developed and took place during a period of partisian division between the two branches.

The objectives of program budgeting—to assist policy-makers more easily to weigh the alternatives available—are in no way related to partisan ideologies or to the relative balance of power between executive and legislature. But nearly all legislative measures stand the chance of being caught up in these conflicts. Fortunately, in Wisconsin, the adoption of program budgeting did not.

[1] Originally published in Scientific manpower utilization, 1967, op: cit., pp. 260–265.

How did it happen? How will Wisconsin's budget look in program form? Why do we think program budgeting is important for Wisconsin and where do we think it will lead the state? Those are the questions we hope to answer in this article. The answers should provoke thoughtful discussion and consideration of this new budgeting method, not only among government administrators and in the academic community, but also by elected policy makers in other states, and in local governments

Before seeking to answer the questions, however, it is pertinent to look briefly at the concept of program budgeting and its objectives. The term itself is somewhat forbidding to the uninitiated. What we understand it to mean is simply the casting of budget requests and appropriations into groupings on the basis of the services performed and for whom they are carried out, i.e., what is done and for whom it is done. Thus, in the Department of Public Welfare we have, as separate programs, mental health services, correctional services and family services. These describe what is done and for whom it is done. A more detailed description of the program definitions will be given later. The important thing, though, is that they describe what programs the Department of Public Welfare carries on in those areas. The dollars requested or spent in each of these areas can be weighed against the results acheived. This is certainly far more significant than determining whether an agency should be authorized another clerk, or a new typewriter, or a trip to a professional conference by a staff member.

We are not sure who originated the idea of structuring the budget and appropriations according to the programs that are to be carried out. We do know, however, that it was the highly respected Hoover Commission which, on the federal level at least, first used the term and defined its use. Its Task Force Report on Fiscal, Budgeting and Accounting Activities said that "a program or performance budget should be substituted for the present budget, thus presenting . . . expenditure requirements in terms of services, activities and work projects rather than in terms of the things bought. Such a budget would not detract from congressional responsibility and should greatly improve and expedite committee consideration."

Some confusion has apparently occurred because of the introduction of the term "performance" or "program and performance" in connection with program budgeting. They are often used interchangeably, as seen above in the Hoover Commission report. Here in Wisconsin we refer simply to program budgeting. However, we plan to introduce new techniques of performance measurement which will help administrators, the Governor and the legislature to evaluate programs better and to make more meaningful budget policy decisions. These performance measurement techniques, in our view, are not an integral part of program budgeting, but they do evolve logically out of it and enhance the entire budgeting system.

As all experienced political practitioners are aware, major revision of a state's budgeting system does not just happen. Resistance to change in the way of doing things is one of the great stumbling blocks to any kind of major improvement in the management of government services. This fear of change exists among politicians as well as government administrators and employees. Indeed, the various interest groups which often exert strong influence on policies and practices frequently collaborate to maintain the status quo.

Breaking down this resistance to change, dissolving the fear that program budgeting would mean loss of control over expenditures, and educating the various interests to the meaning and advantages of this new system did not occur overnight. In fact, this process began in Wisconsin nearly six years ago.

Initial steps

The first step was in the conversion to program budgeting of one major department—Conservation—in 1959. Financed through a segregated fund, and receiving less strict scrutiny than general fund agency budgets, this proved to be an acceptable department with which to experiment in using the new format. The Conservation Department cooperated willingly, and the interest expressed by legislators was sincere, if not enthusiastic.

This was followed in 1961 by the conversion of the Motor Vehicle Department to a program budget—again, an agency financed through a segregated fund. But that year the Board of Health was also switched over to a program budget— the first agency funded by general purpose revenues to come under the new system.

In the meantime, in 1960, the Governor had invited a group of business executives to study the state's administrative practices and make recommendations for improvement. This group split into several task forces. To each task force was assigned an administrative analyst from the state budget office. The task force on budgeting, which had an opportunity to 'compare the old budgeting methods with the new formats that were being developed for the Conservation and Motor Vehicle Departments and the Board of Health, recommended strongly that the state convert all agencies to program budgets.

In 1962 a legislative interim committee, charged with finding ways for improving efficiency and economy in government, recommended that we convert to program budgeting on a gradual basis. Early in 1964, a citizens' committee appointed by the legislature to examine expenditures and administrative practices enthusiastically endorsed the work that had been done toward conversion to program budgeting and urged speedy adoption of this new method.

The endorsement of program budgeting by these groups and the support they lent were major mileposts in Wisconsin's journey toward full implementation of program budgeting.

Underlying considerations

All during this same period, there was a growing awareness among individual legislators, especially the members of the Legislature's Joint Committee on Finance, that the old methods of budget review were no longer workable. While some of them understandably harbored reservations about the idea of a complete revision of Wisconsin's budget practices, many became convinced that such a step would soon be imperative. They had applauded the work of the budget office, which over the several previous biennia had considerably improved the budget document and its explanatory material. They were also impressed by the support shown for the idea by impartial citizen and taxpayer groups.

Still another factor was developing during this same period which strongly influenced legislative thinking. Over the past decade, Wisconsin's budget, like those in nearly all other states, expanded considerably, due almost entirely to rising school enrollments, increasing demands on higher education and more expensive health and welfare treatment and rehabilitation programs. Citizens expected and were being provided more state services than ever before. The trend began shortly after World War II, but was intensified as the effect of the postwar baby boom made itself apparent in the middle 1950's. As a result of these developments, state budgets grew, not only in the level of expenditures, but also in the volume and complexity of the detailed material needing review.

Assignment to Department of Administration

Because of the frustration of trying to comprehend the budget and make meaningful policy decisions in the face of this growth and complexity, the Finance Committee members decided that something had to be done to lighten their burden and make their task more purposeful. At this point, in the fall of 1963, both the Governor and the Joint Finance Committee asked that the Department of Administration take the necessary steps to proceed with full implementation of program budgeting for the 1965–67 biennium. In his directive to the Department of Administration, the Governor said:

"The recent budget session pointed out the potential for improved methods of evaluating the costs and alternatives of State government services. Many legislators have indicated a desire for a budget which clearly identifies state services and the costs, and which promotes selection of alternative program service possibilities. In addition, interested civic groups, newspaper reporters and editorial writers, and many private citizens seek the policy implicit in our present budge figures, but frequently seek in vain. What is needed is fiscal data which explains, by program, the services provided by state government."

The Joint Finance Committee in a letter to the department pointed out that ". . . When an agency presents its budget to the committee in terms of its proposed contribution to the State of Wisconsin, the Joint Committee is then able to make the most meaningful policy recommendation in terms of the wishes of the elected representatives of the people. Therefore, . . . we urge the Department of Administration to take the necessary steps to implement program planning and budgeting for all state agencies."

Fruition in 1964

Armed with these directives, the Department of Administration proceeded with an all-out effort to develop a meaningful budget format that would achieve the desired objective—to define agency programs in a way that would clearly spell out the services provided and which would facilitate more meaningful budget policy decisions by administrators, the Governor and the legislature.

To accomplish this, and to help all parties involved to understand the new concept better, agency administrators were asked to prepare program formats (with necessary subdivisions) for their entire operations. When these were agreed upon with the Department of Administration, the existing appropriations (for 1963–65) were recast into the new format. A complete prototype program budget was developed for seven separate departments. Another document was prepared, showing the program breakdown for all state agencies. These were presented to the Governor and Joint Finance Committee. For both the Governor and the committee this was the first substantive description of what had been talked about for a long time. Both endorsed the proposals enthusiastically. A short time later the program budget bill was passed.

We are now engaged in the final phase of the conversion. New forms have been developed and new instructions prepared. Agencies are currently in the process of developing their 1965–67 budget requests on a program basis.

Key factors for success

The most significant things which contributed to the success of this effort to convert to program budgeting in Wisconsin were these: (1) the budget office established a channel of open communication with the legislative appropriation committee, based on mutual respect and trust; (2) the Governor showed genuine interest and gave vital support to the project; (3) the support of outside groups was gained by factual, straightforward explanations of the methods, objectives and benefits of program budgeting; (4) a group of highly trained, competent budget personnel was able to translate the professional jargon of program budgeting into terms which could be understood by the decision makers—the Governor and the legislature; (5) awareness grew that budget methods established in another era no longer were sufficient for today's needs.

THE PROGRAM BUDGET FORMAT

What will Wisconsin's program budget look like, and how will it improve the budget process?

Briefly the budget will simply divide each agency's total operations into major components of services—programs. This is the level on which appropriations will be made. In the budget document, and in supporting material, each of these programs will be broken down into sub-programs and activities. Following the "what for whom" rationale noted earlier, which is the basis for the major program, each sub-program will describe this relationship more specifically.

For example, using again the Department of Public Welfare, following are the subprograms that fall under the major program of family services:

A. Family Services Administration
B. Child Welfare Services
C. Services to the Blind
D. Services to the Aged
E. Aid to the Disabled
F. Special Aids to Local Units of Government
G. Aids to Individuals

These sub-programs describe what aspect of family services are to be provided to each group being served.

The remaining breakdown—by activity—answers the question "how?": i.e., how, or by what administrative techniques will the group be served? Thus, under Child Welfare Services, we have the following activities:

1. Child Center (institution for emotionally disturbed, dependent and neglected children.
2. Boarding Home care for Foster Children
3. Special Projects
4. Aid to Dependent Children
5. Licensing and Direct Services
6. Community and County Services

In the budget document, each of these programs, sub-programs and activities will be described in narrative form, and data will be presented showing the level of service and results of the component parts. The requested budget will be portrayed in terms of what change, if any, is proposed in the level of accomplishment.

Another important feature of the new budgeting method is the "total funds" concept. Under it the programs will be structured not according to the source of funds which finance them, but according to the services to be provided. This will eliminate the great fragmentation which has characterized Wisconsin's budgets up to now. It will enable the Governor and legislature to review programs for their relative worth and accomplishment, irrespective of whether they are financed by general purpose revenues, agency receipts, segregated revenues or federal funds.

A new form has been developed which will serve as the primary tool in analyzing why and in what ways a proposed program will change in cost from one fiscal period to the next. Essentially, this form will categorize each of the factors which contribute to a change, up or down in the proposed level of expenditure. It will also help to identify the change in the number of people served or in the kind and quality of services provided. Use of this technique will largely eliminate excessive reliance on objects of expenditure, which could only vaguely and often inaccurately be related to proposed accomplishment.

EXPECTED RESULTS

We believe this new method of constructing budgets will produce several very important results. First of all, it will help agency administrators to carry out their budget preparation on the basis of what they hope to accomplish. Thus it will serve both as a tool for reviewing the efficiency of existing operations and their results, and also as a basis for planning future services.

A second important result, we believe, will be a substantial upgrading of the budget decision-making process. In the budget process, the Governor proposes and the legislature establishes state policy. Previously, there has been no good way for these policy decisions to be formulated. The Governor and legislature were asked to determine, simply, *what is to be bought*—in terms of personnel, services and goods. They were forced to transcribe these data into their own concepts of service level. Now they will be asked to decide first *what is to be done*—what services are to be provided to whom and at what level, and the data will be uniform for both. These significant policy decisions can only be made if the Governor and legislature are given information on a basis which encourages such decisions.

A third benefit will be a substantial improvement in the ease with which the public can comprehend, and therefore appraise, responsible government. This is very important, for only if interested citizens are able to know and understand what government is doing can they intelligently judge its value and effectiveness.

We believe that the introduction of program budgeting will have an important effect on the quality of budget preparation and planning, and will improve the basis for sound decision making by elected policy makers. Because the programs will be structured according to the groups they serve, and will encompass the funds derived from all sources, a much more comprehensive review of the budget requests will be possible and program policy alternatives will be more readily discernible.

Where do we go from here? In a directive to the department heads, describing how the state would proceed with the implementation of program budgeting, the Governor outlined a three-part program to improve financial administration.

1. Budgets stated in terms of services to the people in order to present fiscal policies in the context of services to be accomplished.
2. Criteria to measure the cost of state services and the performance of state agencies in providing these services.
3. Organization for long-range fiscal planning to establish state goals and institute plans to achieve those goals.

The first part of the three, of course, is the implementation of program budgeting. The other two, we believe, are made possible by program budgeting and follow logically from it. Development of unit costs and yardsticks to measure the performance of state agency operations is feasible only if these operations are divided into meaningful units. The programs, with their respective subdivisions, fit very neatly into this requirements. Thus, unit cost data and performance measurement will both evolve out of and also build on the essential foundation—program budgeting.

Long-range fiscal planning and the establishment of goals for future programs is a second logical outgrowth of program budgeting. The natural extension of this, in turn, is the development of better departmental program plans and the synthesis of these plans into broad functional plans. These plans, because they are based on the programs, will further strengthen and unify the budget process.

When these things are accomplished, state government operations will not only be much more comprehensible; there will be a meaningful basis and systematic techniques for planning, reviewing, modifying and carrying out state programs. This, we think, will help both the Governor and the legislature to weigh the value and effectiveness of state government, and to balance and decide upon the essential policy alternatives available to them. It will also help interested citizens to understand and evaluate the services their state government provides.

Program budgeting is not a panacea. Neither are performance measurement and long-range fiscal and program planning. They will not, by themselves, hold down budget levels or provide better service. But if used properly they will promote better planning and better decision making.

Such a program will not only make state government more efficient, but also more responsible to the needs and wishes of the people its serves—a prime objective to all those who are interested in democratic government.

As elected representatives of opposing political beliefs, and serving two separate branches of state government, we are proud of the way in which our forces united to accomplish this significant improvement for Wisconsin.

APPENDIX I

List of Congressional Committees and Subcommittees

As the numbers and types of nondefense, nonspace public problems have multiplied, the two Chambers of the Congress have taken steps to create new legislative entities, or expand the responsibilities of established groups, so that an overview capability could be insured. In some instances, situations involving overlapping jurisdictions were a result, but adjustments were undertaken to clarify committee and subcommittee responsibility. The following listing sets forth those standing, special, select, and joint committees (and their elements) which have jurisdiction over one or more of the national problems treated in this report.

SENATE COMMITTEES, U.S. CONGRESS

Appropriations, Senate Committee on

To consist of 24 Senators, to which committee shall be referred all proposed legislation, messages, petitions, memorials, and other matters relating to appropriation of the revenue for the support of the Government.

> Department of the Interior and Related Agencies, Subcommittee on Departments of Labor and Health, Education, and Welfare and Related Agencies, Subcommittee on
> Department of Transportation, Subcommittee on

Banking and Currency, Senate Committee on

To consist of 15 Senators: (1) Banking and currency generally, (2) Financial aid to commerce and industry, other than matters relating to such aid which are specifically assigned to other committees under this rule, (3) Deposit insurance, (4) Public and private housing, (5) Federal Reserve System (6) Gold and silver, including the coinage thereof, (7) Issuance of notes and redemption thereof, (8) Valuation and revaluation of the dollar, (9) Control of prices of commodities, rents, or services.

> Housing and Urban Affairs, Subcommittee on

Commerce, Senate Committee on

To consist of 19 Senators: (1) Interstate and foreign commerce generally, (2) Regulation of interstate railroads, buses, trucks, and pipelines, (3) Communication by telephone, telegraph, radio, and television, (4) Civil aeronautics except aeronautical and space activities of the National Aeronautics and Space Administration, (5) Merchant marine generally, (6) Registering and licensing of vessels and small boats, (7) Navigation and the laws relating thereto, including pilotage, (8) Rules and international arrangements to prevent collisions at sea, (9) Merchant marine officers and seamen, (10) Measures relating to the regulation of common carriers by water and to the inspection of merchant marine vessels, lights and signals, lifesaving equipment, and fire protection on such vessels, (11) Coast and Geodetic Survey, (12) The Coast Guard, including lifesaving service, lighthouse, lightships, and ocean derelicts, (13) The U.S. Coast Guard and Merchant Marine Academies, (14) Weather Bureau, (15) Except as provided in paragraph (c), the Panama Canal and interoceanic canals generally, (16) Inland waterways, (17) Fisheries and wildlife, including research, restoration, refuges, and conservation, (18) Bureau of Standards, including standardization of weights and measures and the metric system.

> Aviation, Subcommittee on
> Communications, Subcommittee on
> Surface Transportation, Subcommittee on

Government Operations, Senate Committee on

To consist of 15 Senators; (a) Budget and accounting measures, other than appropriations, (b) reorganizations in the executive branch of the Government. Such committee shall have the duty of (a) receiving and examining reports of the Comptroller General of the United States and of submitting such recommendations to the Senate as it deems necessary or desirable in connection with the subject matter of such reports; (b) studying the operations of Government activities at all levels with a view to determining its economy and efficiency; (c) evaluating the effects of laws enacted to reorganize the legislative and executive branches of the Government; (d) studying intergovernmental relationships between the United States and municipalities, and between the United States and international organizations of which the United States is a member.

> Executive reorganization, Subcommittee on
> Government Research, Subcommittee on
> Intergovernmental Relations, Subcommittee on

Interior and Insular Affairs, Senate Committee on

To consist of 17 Senators: (1) Public lands generally, including entry, easements, and grazing thereon, (2) Mineral resources of the public lands, (3) Forfeiture of land grants and alien ownership, including alien ownership of mineral lands, (4) Forest reserves and national parks created from the public domain, (5) Military parks and battlefields, and national cemeteries, (6) Preservation of prehistoric ruins and objects of interest on the public domain, (7) Measures relating generally to the insular possessions of the United States except those affecting their revenue and appropriations, (8) Irrigation and reclamation, including water supply for reclamation projects, and easements of public lands for irrigation projects, (9) Interstate compacts relating to apportionment of waters for irrigation purposes, (10) Mining interests general, (11) Mineral land laws and claims and entries thereunder, (12) Geological survey, (13) Mining schools and experimental stations, (14) Petroleum conservation and conservation of the radium supply in the United States, (15) Relations of the United States with the Indians and the Indian tribes, (16) Measures relating to the care, education, and management of Indians, including the care and allotment of Indian lands and general and special measures relating to claims which are paid out of Indian funds.

Parks and Recreation, Subcommittee on
Public Lands, Subcommittee on
Water and Power Resources, Subcommittee on

Judiciary, Senate Committee on

To consist of 17 Senators: (1) Judicial proceedings, civil and criminal, generally, (2) Constitutional amendments, (3) Federal court and judges, (4) Local courts in the territories and possessions, (5) Revision and codification of the statutes of the United States, (6) National penitentiaries, (7) Protection of trade and commerce against unlawful restraints and monopolies, (8) Holidays and celebrations, (9) Bankruptcy, mutiny, espionage, and counterfeiting, (10) State and territorial boundary lines, (11) Meetings of Congress, attendance of Members and their acceptance of incompatible offices, (15) Immigration and naturalization, (16) Apportionment of Representatives, (17) Measures relating to claims against the United States, (18) Interstate compacts generally.

Improvements in Judicial Machinery, Subcommittee on

Labor and Public Welfare, Senate Committee on

To consist of 17 Senators: (1) Measures relating to education, labor, or public welfare generally, (2) mediation and arbitration of labor disputes, (3) wages and hours of labor, (4) convict labor and the entry of goods made by convicts into interstate commerce, (5) regulation or prevention of importation of foreign laborers under contract, (6) child labor, (7) labor statistics, (8) labor standards, (9) school-lunch program, (10) vocational rehabilitation, (11) railroad labor and and railroad retirement and unemployment, except revenue measures relating thereto, (12) U.S. Employees' Compensation Commission, (13) Columbia Institution for the Deaf, Dumb, and Blind; Howard University; Freedmen's Hospital; and St. Elizabeths Hospital, (14) public health and quarantine, (15) welfare of miners, (16) vocational rehabilitation and education of veterans, (17) veterans' hospitals, medical care and treatment of veterans, (18) soldiers' and sailors' civil relief, (19) readjustment of servicemen to civil life.

Education, Subcommittee on
Employment, Manpower and Poverty, Subcommittee on
Health, Subcommittee on
Veterans' Affairs, Subcommittee
National Science Foundation, Special Subcommittee on
Evaluation and Planning of Social Programs, Special Subcommittee on

Post Office and Civil Service, Senate Committee on

To consist of 12 Senators: (1) The Federal civil service generally, (2) the status of officers and employees of the United States, including their compensation, classification, and retirement, (3) the postal service generally, including the railway mail service, and measures relating to ocean mail and pneumatic-tube service; but excluding post roads, (4) postal-savings banks, (5) census and the collection of statistics generally, (6) The National Archives.

Civil Service, Subcommittee on
Health Benefits and Life Insurance, Subcommittee on

Public Works, Senate Committee on

To consist of 15 Senators: (1) Flood control and improvement of rivers and harbors, (2) public works for the benefit of navigation and bridges and dams (other than international bridges and dams), (3) water power, (4) oil and other pollution of navigable waters, (5) public buildings and occupied or improved grounds of the United States generally, (6) measures relating to the purchase of sites and construction of post offices, customhouses, Federal courthouses, and Government buildings within the District of Columbia, (7) measures relating to the Capitol Building and the Senate and House Office Buildings, (8) measures relating to the construction or reconstruction, maintenance, and care of the buildings, and grounds of the Botanic Gardens, the Library of Congress, and the Smithsonian Institution, (9) public reservations and parks within the District of Columbia including Rock Creek Park and the Zoological Park, (10) measures relating to construction or maintenance of roads and post roads.

    Air and Water Pollution, Subcommittee on
    Flood Control—Rivers and Harbors, Subcommittee on
    Roads, Subcommittee on
    Economic Development, Special Subcommittee on

Aging, Special Senate Committee on

To consist of 20 Members of the Senate. Studies and investigates the problems of the aging and reports findings and makes recommendations to the U.S. Senate, but does not report legislation.

    Consumer Interests of the Elderly, Subcommittee on
    Employment and Retirement Incomes of the Elderly, Subcommittee on
    Federal, State, and Community Services for the Elderly, Subcommittee on
    Health of the Elderly, Subcommittee on
    Housing for the Elderly, Subcommittee on
    Long-Term Care, Subcommittee on

Small Business, Select Senate Committee on

To consist of 17 Senators. "* * * It shall be the duty of such committee to study and survey by means of research and investigation all problems of American small-business enterprises, and to obtain all facts possible in relation thereto which not only would be of public interest, but which would aid the Congress in enacting remedial legislation and to report to the Senate from time to time the results of such studies and surveys. No proposed legislation shall be referred to such committee and such committee shall not have power to report by bill or otherwise have legislative jurisdiction.

    Science and Technology, Subcommittee on

HOUSE COMMITTEES, U.S. CONGRESS

Agriculture, House Committee on

To consist of 33 members: (a) Adulteration of seeds, insect pests, and protection of birds and animals in forest reserves, (b) agriculture generally, (c) agricultural and industrial chemistry, (d) agricultural colleges and experiment stations, (e) agricultural economics and research, (f) agricultural education extension services, (g) agricultural production and marketing and stabilization of prices of agricultural products, (h) animal industry and disease of animals, (i) crop insurance and soil conservation, (j) dairy industry, (k) entomology and plant quarantine, (l) extension of farm credit and farm security, (m) forestry in general, and forest reserves other than those created from the public domain, (n) human nutrition and home economics, (o) inspection of livestock and meat products, (p) plant industry, soils, and agricultural engineering, (q) rural electrification.

Forests, Subcommittee on
Conservation and Credit, Special Subcommittee on
Rural Development, Special Subcommittee on

Appropriations, House Committee on

To consist of 51 members: (a) Appropriation of the revenue for the support of the Government.

Department of Interior and Related Agencies, Subcommittee on
Departments of Labor and Health, Education, and Welfare and Related Agencies, Subcommittee on
Independent Offices and Department of Housing and Urban Development, Subcommittee on
Public Works, Subcommittee on

Banking and Currency, House Committee on

To consist of 35 members: (a) Banking and currency generally, (b) control of price of commodities, rents, or services, (c) deposit insurance, (d) Federal Reserve System, (e) financial aid to commerce and industry, other than matters relating to such aid which are specifically assigned to other committees under this rule, (f) gold and silver including the coinage thereof, (g) issuance of notes and redemption thereof, (h) public and private housing, (i) valuation and revaluation of the dollar.

Housing, Subcommittee on

Education and Labor, House Committee on

To consist of 35 members: (a) Measures relating to education or labor generally, (b) child labor, (c) Columbia Institution for the Deaf, Dumb, and Blind; Howard University; Freedmen's Hospital; and Saint Elizabeths Hospital, (d) convict labor and the entry of goods made by convicts into interstate commerce, (e) labor standards, (f) labor statistics, (g) mediation and arbitration of labor disputes, (h) regulation or prevention of importation of foreign laborers under contract, (i) school-lunch program, (j) U.S. Employees' Compensation Commission, (k) vocational rehabilitation, (l) wages and hours of labor, (m) welfare of miners.

Education, General Subcommittee on
Education, Select Subcommittee on
Education, Special Subcommittee
Labor, General Subcommittee on
Labor, Select Subcommittee on
Labor, Special Subcommittee on

Government Operations, House Committee on

To consist of 35 members: (a) Budget and accounting measures, other than appropriations, (b) reorganization in the executive branch of the Government. Such committee shall have the duty of: (1) Receiving and examining reports of the Comptroller General of the United States and of submitting such recommendations to the House as it deems necessary or desirable in connection with the subject matter of such reports; (2) studying the operation of Government activities at all levels with a view to determining its economy and efficiency; (3) evaluating the effects of laws enacted to reorganize the legislative and executive branches of the Government; (4) studying intergovernmental relationships between the United States and the States and municipalities, and between the United States and international organizations of which the United States is a member.

Government Activities, Subcommittee on
Intergovernmental Relations, Subcommittee on
Natural Resources and Power, Subcommittee on
Research and Technical Programs, Subcommittee on

Interior and Insular Affairs, House Committee on

To consist of 34 members: (*a*) Forest reserves and national parks created from the public domain, (*b*) forfeiture of land grants and alien ownership, including alien ownership of mineral lands, (*c*) geological survey, (*d*) interstate compacts relating to apportionment of waters for irrigation purposes, (*e*) irrigation and reclamation, including water supply for reclamation projects, and easements of public lands for irrigation projects, and acquisition of private lands when necessary to complete irrigation projects, (*f*) measures relating to the care, education, and management of Indians, including the care and allotment of Indian lands and general and special measures relating to claims which are paid out of Indian funds, (*g*) measures relating generally to Hawaii, Alaska, and the insular possessions of the United States, except those affecting the revenue and appropriations, (*h*) military parks and battlefields, and national cemeteries, (*i*) mineral land laws and claims and entries thereunder, (*j*) mineral resources of the public lands, (*k*) mining interests generally, (*l*) mining schools and experimental stations, (*m*) petroleum conservation on the public lands and conservation of the radium supply in the United States, (*n*) preservation of prehistoric ruins and objects of interest on the public domain, (*o*) public lands generally, including entry, easements, and grazing thereon, (*p*) relations of the United States with the Indians and the Indian tribes.

Irrigation and Reclamation, Subcommittee on
Mines and Mining, Subcommittee on
National Parks and Recreation, Subcommittee on
Public Lands, Subcommittee on

Interstate and Foreign Commerce, House Committee on

To consist of 37 members: (1) Interstate and foreign commerce generally, (2) regulation of interstate and foreign transportation, except transportation by water not subject to the jurisdiction of the Interstate Commerce Commission, (3) regulation of interstate and foreign communications, (4) civil aeronautics, (5) weather bureau, (6) interstate oil compacts; and petroleum and natural gas, except on the public lands, (7) securities and exchanges, (8) regulation of interstate transmission of power, except the installation of connections between Government water projects, (9) railroad labor and railroad retirement and unemployment, except revenue measures relating thereto, (10) public health and quarantine, (11) inland waterways.

  Communications and Power, Subcommittee on
  Public Health and Welfare, Subcommittee on
  Transportation and Aeronautics, Subcommittee on

Merchant Marine and Fisheries, House Committee on

To consist of 37 members: (1) Merchant marine generally, (2) registering and licensing of vessels and small boats, (3) navigation and the laws relating thereto, including pilotage, (4) rules and international arrangements to prevent collisions at sea, (5) merchant marine officers and seamen, (6) measures relating to the regulation of common carriers by water (except matters subject to the jurisdiction of the Interstate Commerce Commission), and to the inspection of merchant marine vessels, lights and signals, lifesaving equipment, and fire protection on such vessels, (7) the Coast Guard, including lifesaving service, lighthouses, lightships, and ocean derelicts, (8) U.S. Coast Guard and Merchant Marine Academies, (9) coast and geodetic surveys, (10) the Panama Canal and the maintenance and operation of the Panama Canal, including the administration, sanitation, and government of the Canal Zone, and interoceanic canals generally, (11) fisheries and wildlife, including research, restoration, refuges, and conservation.

  Fisheries and Wildlife Conservation, Subcommittee on

Post Office and Civil Service, House Committee on

To consist of 26 members: (1) The Federal civil service generally, (2) the status of officers and employees of the United States, including their compensation, classification, and retirement, (3) the postal service generally, including the railway mail service and measures relating to ocean mail and pneumatic-tube service; but excluding post roads, (4) postal savings banks, (5) census and the collection of statistics generally, (6) the National Archives.

  Census and Statistics, Subcommittee on
  Manpower and Civil Service, Subcommittee on

Public Works, House Committee on

To consist of 34 members: (1) Flood control and improvement of rivers and harbors, (2) public works for the benefit of navigation, including bridges and dams (other than international bridges and dams), (3) water power, (4) oil and other pollution of navigable waters, (5) public buildings and occupied or improved grounds of the United States generally, (6) measures relating to the Capitol Building and the Senate and House Office Buildings, (8) relating to the construction or reconstruction, maintenance, and care of the buildings and grounds of the Botanic Gardens, the Library of Congress, and the Smithsonian Institution, (9) public reservations and parks within the District of Columbia, including Rock Creek Park and the Zoological Park, (10) measures relating to the construction or maintenance of roads and post roads, other than appropriations therefor; but it shall not be in order for any bill providing general legislation in relation to roads to contain any provision for any specific road, nor for any bill in relation to a specific road to embrace a provision in relation to any other specific road.

Flood Control, Subcommittee on
Rivers and Harbors, Subcommittee on
Roads, Subcommittee on
Watershed Development, Subcommittee on
Federal-Aid Highway Program, Special Subcommittee on
Appalachia, Ad Hoc Subcommittee

Rules, House Committee on

To consist of 15 members: (1) The rules, joint rules, and order of business of the House, (2) recesses and final adjournment of Congress, (3) to sit and act whether or not the House is in session.

Science and Astronautics, House Committee on

To consist of 32 members: (*a*) Astronautical research and development, including resources, personnel equipment, and facilities, (*b*) Bureau of Standards, standardization of weights and measures, and the metric system, (*c*) National Aeronautics and Space Administration, (*d*) National Aeronautics and Space Council, (*e*) National Science Foundation, (*f*) outer space, including exploration and control thereof, (*g*) science scholarships, and (*h*) scientific research and development.

Advanced Research and Technology, Subcommittee on
Science, Research, and Development, Subcommittee on

Veterans' Affairs, House Committee on

To consist of 25 members: (1) Veterans' measures generally, (2) pensions of all the wars of the United States, general and special, (3) life insurance issued by the Government on account of service in the Armed Forces, (4) compensation, vocational rehabilitation, and education of veterans, (5) veterans' hospitals, medical care, and treatment of veterans, (6) soldiers' and sailors' civil relief, (7) readjustment of servicemen to civil life.

Education and Training, Subcommittee on
Hospitals, Subcommittee on
Housing, Subcommittee on

Small Business, Select House Committee to Conduct a Study and Investigation of the Problems of

To consist of 15 members: Studies and investigates problems of small business and reports findings and makes recommendations to the House, but cannot report legislation.

Small Business Problems in Urban Areas, Subcommittee on

JOINT COMMITTEES, U.S. CONGRESS

Economic, Joint Committee on

"Composed of ten Members of the Senate * * * and ten Members of the House of Representatives, * * *. In each case, the majority party shall be prepresented by six members and the minority party shall be represented by four members. (b) It shall be the function of the joint committee—(1) To make a continuing study of matters relating to the Economic Report: (2) to study means of coordinating programs in order to further the policy of this Act; and (3) as a guide to the several committees of the Congress dealing with legislation relating to the Economic Report, no later than March 1, of each year (beginning with the year (1947) to file a report with the Senate and the House of Representatives containing its findings and recommendations with respect to each of the main recommendations made by the President in the Economic Report, and from time to time to make other reports and recommendations to the Senate and House of Representatives as it deems advisable."

 Economic Progress, Subcommittee on

 Economy in Government, Subcommittee on

 Urban Affairs, Subcommittee on

Reduction of Nonessential Federal Expenditures, Joint Committee

Composed of: (1) six Members of the Senate, appointed by the President of the Senate, three from the Committee on Finance and three from the Committee on Appropriations, (2) six Members of the House, appointed by the Speaker, three from the Committee on Ways and Means and three from the Committee on Appropriations, and (3) the Secretary of the Treasury and the Director of the Bureau of the Budget. The Committee studies all expenditures of the Federal Government with a view toward recommending the elimination of those it considers nonessential. At the conclusion of its studies, the committee shall report to the President and the Congress concerning its conclusions and recommendations.

APPENDIX J

List of Commissions, Councils, Task Forces, and Committees

Several types of advisory or monitoring mechanisms have been created by the Congress, the President, or a responsible officer in the executive branch. These bodies are variously called commissions, councils, task forces, or committees. Their areas of purview are numerous, but the selection which follows reflects those advisory entities which exist to scrutinize certain public problems ranging from urban problems to regional economic development to transportation facilitation. Each group is described in terms of the authority which caused the body to be created, its responsibility, and whether or not reports will be prepared in the course of its activity.[1]

Summary listing of advisory bodies:

Advisory Commission on Intergovernmental Relations [2]
Advisory Council on Developing Institution
Advisory Council on Education of Disadvantaged Children
Advisory Council on Health Insurance for the Disabled
Advisory Council on Insured Loans to Students
Advisory Council on Insured Loans to Vocational Students
Advisory Council on Medical Assistance
Advisory Council on Quality Teacher Preparation
Appalachian Regional Commission
Citizens' Advisory Committee on Recreation and Natural Beauty
Coastal Plains Regional Commission
Federal Advisory Council on Regional Economic Development
Four Corners Regional Commission
Great Lakes Basin Commission
Health Insurance Benefits Advisory Council
Kansas-Oklahoma Arkansas River Commission
National Advisory Commission on Health Facilities
National Advisory Commission on Health Manpower
National Advisory Commission on Rural Poverty
National Advisory Committee on Adult Basic Education
National Advisory Council on Extension and Continuing Education
National Advisory Council on Regional Medical Problems
National Advisory Council on Supplementary Centers and Services
National Commission on Product Safety
National Commission on the Causes and Prevention of Violence
National Commission on Urban Problems
New England Regional Commission
New England River Basins Commission
Ozarks Regional Commission

---

[1] Abstracted from Gayle T. Harris, Committees, Commissions, Boards, Councils, and Task Forces Created to Advise the President, the Congress or Executive Agencies Since 1965. Washington, D.C., Library of Congress, Legislative Reference Service, Aug. 1968, 218 p.
[2] See Section IV. I for descriptive information on duties and reports.

President's Commission on Law Enforcement and the Administration
of Justice
President's Committee on Health Manpower
President's Committee on Urban Housing
Quetico-Superior Committee
Souris-Red-Rainy River Basins Commission
Task Force on Handicapped Children and Child Development
Transportation Facilitation Committee
Upper Great Lakes Regional Commission
Water Resources Council

Advisory Council on Developing Institutions

Authority.—Public Law 89–329, November 8, 1965.

Responsibility.—To advise the Commissioner of Education with respect to policy matters arising in the administration of the Developing Institutions program, and to assist him in identifying those institutions through which the purposes of the program may best be achieved, and in establishing priorities for use in approving applications for participation in the program.

Reports.—No reports have been issued nor are any required by law.

Advisory Council on Education of Disadvantaged Children

Authority.—Public Law 89–10, April 11, 1965.

Responsibility.—To review the effectiveness and administration of title I of the Elementary and Secondary Education Act of 1965 (Financial Assistance to Local Educational Agencies for the Education of Children of Low-Income Families), and to make recommendations for improvements in administration and effectiveness.

Reports.—First Annual Report, March 31, 1966; Summer Education for Children of Poverty, November 25, 1966; Report, January 31, 1967; Special Report on the Teacher Corps, August 20, 1967.

Advisory Council on Health Insurance for the Disabled

Authority.—Public Law 90–248, January 2, 1968.

Responsibility.—To advise the Secretary of Health, Education, and Welfare on the unmet needs of the disabled for health insurance, the costs involved in providing the disabled with insurance protection to cover the cost of hospital and medical services, and on ways of financing this insurance.

Report.—Final report required to be submitted to the Secretary prior to January 1, 1969, for transmission to Congress and to the Boards of Trustees established by the Social Security Act.

Advisory Council on Insured Loans to Students

Authority.—Public Law 89–329, November 8, 1965.

Responsibility.—To advise the Commissioner of Education on policy matters arising in the administration of the Federal, State, and private programs of low interest insured loans to students, including policies and procedures governing the making of advances for reserve funds, and the Federal payments to reduce student interest costs.

Reports.—None required by law.

Advisory Council on Insured Loans to Vocational Students

Authority.—Public Law 89–287, October 22, 1965.

Responsibility.—To advise the Commissioner of Education in the preparation of general regulations and on policy matters arising in the administration of programs authorized by the National Vocational Student Loan Insurance Act, including policies and procedures governing the making of advances for reserve funds, the Federal payment to reduce student interest rates, and the making of direct loans.

Reports.—No reports required by law.

Advisory Council on Medical Assistance

Authority.—Public Law 90–248, January 2, 1968.

Responsibility.—To advise the Secretary of Health, Education, and Welfare on the administration of the Social Security Amendments of 1967 and to make suggestions on the improvement of such administration.

Reports.—No reports are required by the legislation.

Advisory Council on Quality Teacher Preparation

Authority.—Public Law 89–329, November 8, 1965.

Responsibility.—To review the administration of the various teacher education programs created by title V of the act, and to recommend improvement of those programs.

Reports.—Reports were never filed.

Appalachian Regional Commission

Authority.—Public Law 89–4, March 9, 1965.

Responsibility.—To plan, develop, and coordinate a 6-year effort to raise the economic potential of the Appalachian area.

Reports.—The Appalachian Regional Commission Annual Report, 1966; Second Annual Report of the Appalachian Regional Commission, 1967. (S. Doc. 294, 90th Cong. 2d sess.)

Citizens' Advisory Committee on Recreation and Natural Beauty

Authority.—Executive Order 11278, May 4, 1966.

Responsibility.—To advise the President and the President's Council on Recreation and Natural Beauty on matters concerning "(1) outdoor recreation and the beautification of our Nation's cities and countryside, (2) the correlation of natural beauty and outdoor recreation activities by Federal agencies and bureaus, and (3) local, State, and private outdoor recreation and natural beauty activities." (Sec. 202, Executive Order 11278.) The Committee is also to advise the Council in evaluating the progress of the Council.

Reports.—First Annual Report, June 29, 1967. The Committee anticipates publication of a second annual report and a manual for the assistance of local citizen organizations in organizing and carrying out local beautification and civic improvement projects and programs.

Coastal Plains Regional Commission

Authority.—Public Law 89–136, August 26, 1965.

Responsibility.—To encourage the economic development of the coastal plains region through Federal and State government and private enterprise efforts.

Reports.—A Report on the Initial Action Planning Program of the Coastal Plains Regional Commission, 1967. (The first annual report is anticipated during fiscal year 1969.)

Federal Advisory Council on Regional Economic Development

Authority.—Executive Order 11386, December 28, 1967.

Responsibility.—To aid the Secretary of Commerce by reviewing proposed long-range development plans prepared by regional commissions serving economically depressed areas; to recommend desirable development objectives and programs for such regions and for Alaska; to review proposed designations of additional economic development regions under title V of the Public Works and Economic Development Act of 1965; and to review Federal programs relating to regional economic development, to develop basic policies and priorities with respect to such programs, and to recommend administrative or legislative action needed to stimulate and further regional economic development.

Reports.—No reports have been issued.

Four Corners Regional Commission

Authority.—Public Law 89-136, August 26, 1965.

Responsibility.—To encourage development of, at least to a nationwide average, the economic development in the area of the adjoining corners of Utah, Colorado, New Mexico, and Arizona.

Report.—An annual report is required after the commission has functioned for a full year. The first such report is anticipated prior to the close of fiscal year 1969.

Great Lakes Basin Commission

Authority.—Executive Order 11345, April 20, 1967 (pursuant to authority granted the President by Public Law 89-80, July 22, 1965).

Responsibility.—To conduct water and related land resource planning within those portions of the eight Great Lake States drained by the St. Lawrence River system. To prepare and keep current a comprehensive and coordinated plan for water resource development in the region. To recommend long-range priorities for investigation, planning, and construction of projects.

Report.—First annual report anticipated during summer 1968.

Health Insurance Benefits Advisory Council

Authority.—Public Law 89-97, July 30, 1965.

Responsibility.—To advise the Secretary of Health, Education, and Welfare on the administration of and formulation of regulations under title XVIII of the Social Security Act, and to study utilization of hospital and other medical care and services under which payment may be made under the title, and to make recommendations for improvements if necessary.

Report.—Report to the President of the United States from the Industry-Government Special Task Force on Travel, February 1968.

Note.—On March 6, 1968, the President requested Ambassador McKinney to remain on duty with his task force to coordinate efforts to implement the recommendations contained in the report. This statement, as nearly as we have been able to ascertain—or, indeed, as nearly as the staff itself can ascertain—also renamed the Task Force the "Presidential Commission on Travel."

Kansas-Oklahoma Arkansas River Commission

Authority.—Public Law 89–789, November 7, 1966.

Responsibility.—As stated in Article I of the Arkansas River Basin Compact, Kansas-Oklahoma, the purposes are: "A. To promote interstate comity between the States of Kansas and Oklahoma; B. To divide and apportion equitably between the States of Kansas and Oklahoma the waters of the Arkansas River Basin and to promote the orderly development thereof; C. To provide an agency for administering the water apportionment agreed to herein; D. To encourage the maintenance of an active pollution-abatement program in each of the two States and to seek the further reduction of both natural and manmade pollution in the waters of the Arkansas River Basin."

Reports.—No reports have been issued to date, although annual reports are anticipated. The first annual meeting is scheduled for July 1968.

National Advisory Commission on Health Facilities

Authority.—Appointed by the President. Announced in his message to Congress on education and health, February 28, 1967.

Responsibility.—To study national needs for total system of health facilities, including hospitals, extended care facilities, nursing homes, long term care institutions, and clinics. To consider the future of the Hill-Burton program, and to make recommendations for financing the construction and modernization of health facilities.

Reports.—No reports have been issued. The President has requested a report by approximately October 1968.

National Advisory Commission on Health Manpower

Authority.—Executive Order 11279, May 7, 1966.

Responsibility.—To examine and evaluate the current and prospective national requirements for, and the availability of, manpower to meet the national health needs. To examine Government and private programs in training health manpower. To recommend new methods for training of additional doctors and highly trained health personnel.

Report.—Report of the National Advisory Commission on Health Manpower, November 1967 (in 2 volumes).

National Advisory Commission on Rural Poverty

Authority.—Executive Order 11306, September 27, 1966.

Responsibility.—To make a study and appraisal of rural income, labor, employment, food, housing, health, and cultural opportunities. To make recommendations, including the administration of State, local, and Federal Government programs, for eliminating poverty in rural America.

Report.—The People Left Behind: A Report by the President's National Advisory Commission on Rural Poverty. September 1967.

National Advisory Committee on Adult Basic Education

Authority.—Public Law 89–750, November 3, 1966.

Responsibility.—To advise the Commissioner of Education in the preparation of general regulations with respect to policy matters arising in the administration of adult basic education programs, including policies and procedures governing the approval of State plans, policies to eliminate duplication and to coordinate programs created by the act with other programs offering adult education activities and services. To make annual reports to the President and to Congress of findings and recommendations for improvement in Federal laws relating to adult education activities and services.

Reports.—First annual report anticipated by June 1968.

National Advisory Council on Extension and Continuing Education

Authority.—Public Law 89–329, November 8, 1965.

Responsibility.—To advise the Commissioner of Education in the preparation of general regulations and with respect to policy matters arising in the administration of the law, including policies and procedures governing the approval of State plans under its provisions and other programs offering extension or continuing education activities and services. To review the administration and effectiveness of all federally supported extension and continuing education programs, including community service programs and to make recommendations with respect thereto.

Reports.—First Annual Report of the National Advisory Council on Extension and Continuing Education: Presented to the President of the United States and the Secretary of Health, Education, and Welfare, March 31, 1967.

National Advisory Council on Regional Medical Problems

Authority.—Public Law 89–239, October 6, 1965.

Responsibility.—To encourage and assist, through grants, the establishment of regional cooperative arrangements among medical schools, research institutions, and hospitals for research and training in the fields of heart disease, cancer, stroke, and related diseases; through such arrangements to allow doctors and medical institutions to provide treatment based on the latest available information; by these means, to improve the health manpower and facilities available to the Nation.

Report.—Report on Regional Medical Programs to the President and the Congress, submitted by William H. Stewart, U.S. Department of Health, Education, and Welfare, June 1967.

National Advisory Council on Supplementary Centers and Services

Authority.—Public Law 90–247, January 2, 1968

Responsibility.—To review and evaluate the administration of the program of extended educational methods provided for in the act, including its effectiveness. To report to the President and to Congress recommendations for the improvement in the operation and administration of title III.

Report.—Report on Regional Medical Programs to the President and the Congress, submitted by William H. Stewart, U.S. Department of Health, Education, and Welfare, June 1967.

National Commission on Product Safety

Authority.—Public Law 90–146, November 20, 1967.

Responsibility.—To make a comprehensive study of the scope and adequacy of measures now employed to protect consumers against unreasonable injuries caused by hazardous household products. To report to the President and Congress findings and recommendations for such legislation as it feels may be necessary.

Report.—None required until the final report is due, as stipulated by law, on or before November 20, 1969.

National Commission on the Causes and Prevention of Violence

Authority.—Executive Order 11412, June 10, 1968.

Responsibility.—To study the causes, occurrence, and control of violence, "from assassination that is motivated by prejudice and by ideology, and by politics and by insanity; to violence in our city streets and even in our homes."

Report.—The President has requested a report by December 1968.

National Commission on Urban Problems

Authority.—Public Law 89–117, August 10, 1965 (sec. 301).

Responsibility.—To make recommendations to the President and the Congress on zoning and land use, building codes, housing codes, property taxation, and State and Federal taxes affecting housing and the growth of cities, and development standards, with special emphasis on achieving an adequate supply of housing for low-income families.

Reports.—Progress report, March 15, 1967; The Challenge of America's Metropolitan Population Outlook—1960 to 1985, June 1968.

New England Regional Commission

Authority.—Public Law 89–136, August 26, 1965.

Responsibility.—To examine the New England region economy, to identify its strengths and weaknesses, and, based on those findings, to recommend and initiate programs to foster its overall economic development.

Report.—New England Regional Commission; first annual report, fiscal year, 1967.

New England River Basins Commission

Authority.—Executive Order 11371, September 6, 1967 (under authority of Public Law 89–90, July 22, 1965).

Responsibility.—To study needs for water and related land resources in New England. To recommend plans and programs for meeting these needs, including programs for water supply, flood control, water quality, power, recreation, fish and wildlife.

Reports.—The first annual report is required by the end of the present fiscal year.

Ozarks Regional Commission

Authority.—Public Law 89–136, August 26, 1965.

Responsibility.—To develop long-range overall economic development programs for the Ozarks area, to improve the economic base of the area, to bring the per capita income in the area closer to the national average, and to encourage cooperation between Federal, State and local government agencies in obtaining these ends.

Reports.—Ozarks Region—An Opportunity for Growth, June 1967; Activities of the Ozarks Regional Commission: First Annual Report. (S. Doc. 73, 90th Cong., 2d sess., Apr. 1, 1968.)

President's Commission on Law Enforcement and the Administration of Justice

Authority.—Executive Order 11236, July 23, 1965.

Responsibility.—To "inquire into the causes of crime and delinquency, measures for their prevention, the adequacy of law enforcement and administration of justice, and the factors encouraging respect or disrespect for law, at the national, State, and local levels * * * (to) develop standards and make recommendations for actions which can be taken by the Federal, State, and local governments, and by private persons and organizations, to prevent, reduce, and control crime and increase respect for law enforcement and related activities, improvements in techniques, organization, and administration of law enforcement activities, improvements in correction and rehabilitation of convicted offenders and juvenile delinquents, promotion of better understanding between law enforcement officials and other members of the community, and promotion of greater respect for law throughout the community." (Sec. 2 of Executive Order 11236.)

Reports.—The Challenge of Crime in a Free Society, February 1967; Supporting documents: Task Force Report: The Police; Task Force Report: The Courts; Task Force Report: Corrections; Task Force Report: Juvenile Delinquency and Youth Crime; Task Force Report: Organized Crime; Task Force Report: Science and Technology; Task Force Report: Assessment of Crime; Task Force Report: Narcotics and Drugs; Task Force Report: Drunkenness; Various Research Studies; and Selected Consultants' Papers.

President's Committee on Health Manpower

Authority.—Executive Order 11279, May 7, 1966.

Responsibility.—To study national requirements for manpower to meet the civilian and military health needs of the Nation; to evaluate present programs for providing health manpower and to recommend improvements; to develop policies for Government and private institutions for providing better training for those in the health services.

Report.—The President's Committee may have contributed to the report of the National Advisory Commission on Health Manpower, but it did not, in any case, publish a separate report of its own.

President's Committee on Urban Housing

Authority.—Presidential directive, June 3, 1967. (The President's intention to appoint such a committee was announced in his message to Congress on urban and rural property, Mar. 14, 1967.)

Responsibility.—To study the means by which the private sector of the economy may be encouraged to participate in the rebuilding of urban slums.

Report.—A final report was expected to be submitted to the President in June 1968.

Quetico-Superior Committee

Authority.—Executive Order 11342, April 12, 1967. (Re-establishing Committee; see Section 6 of the Order for previous authorizations.)

Responsibility.—To promote the protection of the primitive wilderness quality of the Quetico-Superior Country lying between the Rainy and Pigeon River drainages in Ontario and the State of Minnesota. To advise and consult with concerned executive departments and agencies of the United States Government and the State of Minnesota in promoting this ideal.

Reports.—Report to the President, June 1934; Report to the President, December 1953; Report to the President, 1961; Statement of Charles S. Kelly, May 22, 1964.

Souris-Red-Rainy River Basins Commission

Authority.—Executive Order 11359, June 29, 1967.

Responsibility.—To conduct water and land resource planning in the area of the Souris, Red, and Rainy River basins, as authorized by the Water Resources Act of 1966, and to recommend long range priorities for the investigation, planning, and construction of water resource projects.

Reports.—No formal reports indicated.

Task Force on Handicapped Children and Child Development

Authority.—Established by the Secretary of Health, Education, and Welfare at the direction of the President, July 4, 1966.

Responsibility.—To review all existing Federal programs for the handicapped, including prevention, research, training, and manpower, services, construction, and income maintenance.

Report.—Draft Report of the Task Force on Handicapped Children and Child Development, U.S. Department of Health, Education, and Welfare, Office of the Secretary, January 1967.

Transportation Facilitation Committee

Authority.—Public Law 89–670, October 15, 1966. (Appointed by the Secretary of Transportation to carry out the requisitions of Section 2(b) (1) of the Act.)

Responsibility.—To study and to recommend changes in transportation procedures in order to simplify the movement of people and goods.

Report.—No formal reports as yet. At the time the committee was formed, the department issued an explanatory manual on its organization, entitled "Transportation Facilitation Program."

Upper Great Lakes Regional Commission

Authority.—Public Law 89–136, August 26, 1965.

Responsibility.—To identify the potential of the region for economic growth, and to recommend programs and projects for improvement of the region through cooperation between Federal and State Governments.

Reports.—Upper Great Lakes Regional Commission: A New Approach to Economic Development (brochure) ; A Progress Report: Upper Great Lakes Regional Commission, October 1967.

Water Resources Council

Authority.—Public Law 89–80, July 21, 1965.

Responsibility.—To maintain a continuing study of the adequacy of water supplies necessary to meet the needs of the Nation, to appraise the adequacy of existing and proposed policies and programs in the coordination of water conservation efforts of the Federal Government and water control areas, and to make recommendations to the President with respect to Federal policies relating to water conservation. Grants: Victor Mockus.

Reports.—The Council's first formal report is scheduled for publication in September 1968.

APPENDIX K

List of Regional Councils

The role of regional councils has expanded markedly during the years since World War II, and has led to the establishment in 1967 of the National Service to Regional Councils, which is cosponsored by the National League of Cities and the National Association of Counties. Partial support is derived by a matching grant from the Ford Foundation. A list of the various types of regional councils, by State, was published originally in a "1968 Directory of Regional Councils," and is included in this report as a reference tool for governmental and private sector users.[1] A prefacing section describes the purpose and characteristics of the several categories of regional organizations.

REGIONAL COUNCILS

The basic characteristics of regional councils as defined by the National Service to Regional Councils are:

They are multijurisdictional organizations which involve more than one local government and encompass a portion of a State or portions of contiguous States;

They have as their prime purpose increasing cooperation among the local governments of the area. This purpose may relate to land use or transportation planning or may include numerous mutual challenges and problems of the local units.

They generally have been legally created under State law, interlocal agreement or as nonprofit corporations. In their creation, they required in some form, agreement of the local governments involved; and these governments frequently are represented by their elected officials. In most cases, local governments also provide a portion of funding. Types of regional councils include councils of governments, regional planning commissions, economic development districts, local development districts, transportation study groups, and others.

While the above terms are commonly used to identify types of regional councils, specific organizations may not have these terms stated as such in their names. An organization may be a combination of more than one type of regional council, and there may be more than one regional council in a given area. Definitions contained herein are those developed and used by National Service.

(For the purpose of the following definitions of types of regional councils, the term "citizen" is used to describe a person who is not an elected official.)

Councils of governments (councils of elected officials) are associations of local governments represented by their elected officials. Major purposes are: to provide a forum for discussion of issues and challenges

---

[1] The National Service to Regional Council's "1969 Directory of Regional Councils" is in preparation. The new directory will contain additional listings and an updating of all addresses.

commonly shared by the member governments; to determine policy and priorities on these issues; to implement decisions through the member governments; and to coordinate Federal, State, and local programs with regional impact. While comprehensive planning is a prime concern of these councils, their interests are oriented to any and all areawide concerns of the local governments.

In many regions, a council of governments is formed under specific State enabling legislation or general State interlocal agreement legislation. In other areas, a nonprofit corporation is the basis for legal organization.

Representation on the governing body is predominantly of local government elected officials.

Regional planning commissions are organizations with the prime responsibility for comprehensive planning, traditionally with emphasis on land use planning or coordination of local plans for more than one governmental jurisdiction. A city-county planning commission is a regional planning commission if it has multijurisdictional representation and responsibilities. Some county planning commissions may also qualify as regional planning commissions because of the nature of their responsibilities.

Many regional planning commissions are officially agencies of the State, formed by specific State act or general enabling legislation.

Representation on the governing body usually is predominantly citizens appointed by the State or local governments involved.

Economic development districts are generally nonprofit corporations which coordinate public and private efforts within a multicounty area to promote economic progress and development. A district can be created only if the area in which it would be located needs specified Federal Government criteria such as high unemployment or low average income. Districts are designated by the Economic Development Administration of the U.S. Department of Commerce after consultation with the State and counties in which a district is located. The primary responsibility of these districts is to develop an overall economic development program and implement it, with the objective of sustaining growth. The main tools used by them are providing of public works, community facilities, business loans, technical assistance, and research and information.

Most district governing bodies are made up predominantly of citizens who represent the major economic and social interests and the local governments of the area.

Local development districts are similar in concept and purpose to economic development districts. Primary aim also is economic and social development, with emphasis on action and projects. These districts are the outgrowth of the Appalachian Regional Development Act of 1965, and they are formed by and work through the States in the Appalachian region. In a few cases, the States have allowed local development districts enough autonomy to directly apply and qualify for a wider range of Federal programs than those specified for Appalachian redevelopment. Where either a local development district or an economic development district exists, it will generally be honored as the official district for both programs.

The governing body is composed predominantly of citizens representing planning commissions, industry and other employers, the State and the local governments involved.

Transportation and study groups are organizations directly responsible for the planning of highways and mass transportation for more than one jurisdiction. Many transportation planning groups have been set up to comply with the Highway Act of 1962. Most of these are not regional councils, by National Service's definition, as they are extensions of state highway department planning efforts. In other cases, these groups have become projects of existing regional councils.

The Transportation Study Groups that are defined by National Service as regional councils are those that have become more broadly oriented in planning efforts and more representative of citizen and local government interests. An example of this type of regional council is the Tri-State Transportation Commission which covers the consolidated New York metropolitan area in parts of Connecticut, New Jersey and New York.

The governing bodies of such groups are similar to those of regional planning commissions, usually with emphasis on state and transportation-oriented persons.

OTHER REGIONAL COUNCILS

In addition to the type of regional groups listed above, there are a few unique experiments in regional cooperation which we classify as regional councils because of their multi-jurisdictional nature and emphasis on planning and coordination in the public sector. Examples of unique regional councils are the Metropolitan Council in the Twin Cities area and the Hudson River Valley Commission.

The multi-state regional commissions administered by the Department of Commerce and created by Congress in 1965 are not regional councils by our definition. Each includes several states and has state and federal representation on its governing body. There is a minimum of local government involvement. These multi-state groups are: The Appalachian Regional Commission, Ozarks Regional Commission, Upper Great Lakes Regional Commission, New England Regional Commission, Four Corners Regional Commission and Coastal Plains Regional Commission.

ALABAMA

Calhoun County Council of Governments, City Hall, Anniston, Alabama 36201, phone: (205) 236–3421, Ext. 81. Acting Secretary: J. William Mallory; Chairman: Alvis A. Hamric. 1968.
Lee County Council of Local Governments, City Hall, Auburn, Alabama. Secretary: H. R. Thornton; Chairman: Judge Ira Weissinger.
Birmingham-Jefferson County Regional Planning Commission, 2121 Building, Room 1324, Birmingham, Alabama 35203. Phone: (205) 251–8139. Director: Robert J. Juster; Chairman: C. D. Price. 1963.
Birmingham (In the process of forming a council), Mr. Conrad M. Fowler, Sr., Probate Judge, Shelby County Courthouse, Columbiana, Alabama 35051.
North Central Alabama Regional Planning and Development Commission, P.O. Box 2132, Decatur, Alabama 35601, Phone: (205) 355–4515. Acting Executive Director: Gary L. Voketz; Chairman: Robert McGukin.
Huntsville (In the process of forming a council), Mr. Robert Gunn, Intergovernmental Relations Director, Madison County Courthouse, Huntsville, Alabama.
South Alabama Regional Planning Commission, 155 St. Joseph Street, Mobile, Alabama 36602, Phone: (205) 433–6542. Acting Director: John H. Friend; Chairman: Norman J. Walton, 1964.
Montgomery-Montgomery County Regional Planning Commission, Suite 210, Washington Building, Montgomery, Alabama 36104, Phone: (205) 265–7854. Director: Donald L. Horton; Chairman: Tandy D. Little, Jr. 1967.

Muscle Shoals Council of Local Governments, P.O. Box 2475, Muscle Shoals, Alabama 35660, Phone: (205) 383–3861. Executive Director: Stanley E. Munsey; Chairman: David W. Pruett. 1967.

Talladega (In the process of forming a council), Mayor J. L. Hardwick, City Hall, Talladega, Alabama.

Tuscaloosa Area Council of Local Governments, P.O. Box 86, Tuscaloosa, Alabama 35401, Phone: (205) 345–5545. Director: Lewis E. McCray; Chairman: Judge David M. Cochrane. 1966.

West Alabama Regional Development Council, P.O. Box 3199, First Federal Building, Tuscaloosa, Alabama 35401, Phone: (205) 553–6200. Acting Director: Mr. R. Vance Miles, Jr.; President. Judge Robert H. Kirksey.

Valley Council of Local Governments (See Georgia).

ARIZONA

Maricopa Association of Governments, 3800 North Central Avenue, Phoenix, Arizona 85012, Phone: (602) 277–5458. Secretary: John J. DeBolske; Chairman: B. L. Tims.

Tuscon Urban Area Regional Reviewing Committee, c/o Pima County Planning Commission, 411 Trans-America Building, Tuscon, Arizona 85701, Phone: (602) 792–8361. Director: John S. Tsaguris; Chairman: Thomas S. Jay. 1967.

ARKANSAS

North Central Arkansas Economic Development District, Inc., P.O. Box 796, Batesville, Arkansas 72501, Phone: (501) RI3–9396. Director: Max C. McElmurry; Chairman: Charles B. Delgado.

Arkhoma Regional Planning Commission, 104 North 16th Street, Fort Smith, Arkansas 72901, Phone: (501) 785–2651. Executive Director: Lon Hardin; President: J. Fred Patton. 1967.

Western Arkansas Economic Development District, Inc., 104 North 16th Street, Fort Smith, Arkansas 72901, Phone: (501) 785–2651. Executive Director: Lon Hardin; President: Judge Ben A. Geren. 1966.

West Central Arkansas Economic Development District, Inc., Municipal Building, Hot Springs, Arkansas 71901, Phone: (501) NA4–2508. Director: Ray L. Taylor; President: Lon Warneke. 1967.

Metropolitan Area Planning Commission, 216 Pulaski County Courthouse, Little Rock, Arkansas 72201, Phone: (501) FR4–7531. Executive Director: Jason Rouby; Chairman: James S. Binder. 1955.

Southeast Arkansas Economic Development District, Inc., 1306 Cherry Street, P.O. Box 6806, Pine Bluff, Arkansas 71601, Phone: (501) 536–1971. Executive Director: Paul D. Bates; President: Donald H. Smith.

Northwest Arkansas Regional Planning Commission, Box 402, Springdale, Arkansas 72765, Phone: (501) PL1—7125. Director: Kenneth D. Riley; Chairman: Dr. C. Garland Melton, Jr.; 1966.

Ark-Tex Council of Governments (See Texarkana, Texas).

Mississippi-Arkansas-Tennessee Council of Governments (see Memphis, Tennessee).

CALIFORNIA

Kern County Planning Commission, 1103 Golden State Highway, Bakersfield, California 93301, Phone: (805) 327–2111, Ext. 2615. Director: Jack L. Dalton; Chairman: Kenneth Aitken. 1967.

Association of Bay Area Governments, Hotel Claremont, Berkeley, California 94705, Phone: (415) 841–9730. Executive Director: Warren Schmid; President: James P. Kenny. 1961.

Council of Fresno County Governments, Room 300, Hall of Records, Fresno, California 93721, Phone: (209) 268–6011, Ext. 445. Secretary: Mel G. Wingett; Chairman: Mayor Edward T. O'Neill. 1967.

Kings County Regional Planning Agency, 400 N. Douty Street, Hanford, California 93230, Phone: (209) 582–2511, Ext. 34. Secretary: Vincent G. Peterson; Chairman: Evan Cody. 1967.

Southern California Association of Governments, Suite 801, 606 South Hill Street, Los Angeles, California 90014, Phone: (213) 627–8681. Executive Director: Wilber E. Smith; President: Thomas Bradley. 1965.

Merced County Association of Governments, Merced County Court Building, Merced, California 95340, Phone: (209) 722–7411, Ext. 263. Planning Director: Hal Colwell; Chairman: Emory O'Banion. 1967.

Stanislaus Area Advisory Planning Association, 403 County Courthouse, Modesto, California 95354. Chairman: Ronald Eneboe. 1967.

Shasta County-Cities Area Planning Commission, 1855 Placer Street, Redding, California 96001, Phone: (916) 241-2315. Director: James J. Herbert; Chairman: Frank Doherty. 1967.

Sacramento Regional Area Planning Commission, 926 "J" Building, Suite 1100, Sacramento, California 95814, Phone: (916) 443-6805. Executive Director: James A. Barnes; Chairman: Albert J. Talkin.

Monterey Peninsula Area Planning Commission, P.O. Box 1208, Salinas, California 93901, Phone: (408) 424-8611. Director: E. W. DeMars; Chairman: Dr. Edward Marcucci. 1958.

San Diego County Comprehensive Planning Organization, 207 County Administration Center, 1600 Pacific Highway, San Diego, California 92101, Phone: (714) 239-7711. Study Coordinator: Daniel C. Cherrier; Secretary: D. K. Speer. 1964.

Santa Barbara County-Cities Area Planning Council, 105 East Ampamu Street, Santa Barbara, California 93104. Phone: (805) 966-1611. Executive Secretary: Ace R. Southergill; Chairman: Robert D. MacClure. 1967.

Santa Cruz County Regional Planning Agency, 701 Ocean Street, Santa Cruz, California 95060, Phone: (408) 425-2197. Secretary: Louis B. Muhly; Chairman: William P. Murphy. 1967.

California Regional Planning Agency, Box 1212, South Lake Tahoe, California 95705. Chairman: J. Allen Bray.

Tahoe Regional Planning Commission, Tribune Building, Box 3475, South Lake Tahoe, California 95705, Phone: (916) 544-5294. Chairman: W. S. Meneley.

San Joaquin Areawide Planning and Review Agency, 222 East Weber Avenue, Stockton, California, Phone: (209) 944-2661. Secretary: James K. Mahoney; Chairman: Loren Powell. 1967.

COLORADO

Pikes Peak Area Council of Governments, c/o George H. Fellows, City Manager, City Hall, Colorado Springs, Colorado 80901. Chairman: Stan Johnson.

Inter-County Regional Planning Commission, 2475 West 26th Avenue, Denver, Colorado 80211, Phone: (303) 433-7391. Executive Director: Justis K. Smith; Chairman: Mayor Robert Knecht. 1955.

Pueblo Regional Planning Commission, 314 East Seventh Street, Pueblo, Colorado 81003, Phone: (303) 543-6006. Director: C. Allan Blomquist; Chairman: E.H. Pemberton. 1958.

Southern Colorado Economic Development District, 419 Arthur Street, Pueblo, Colorado 81005, Phone: (303) 545-8680. Executive Director: T. Surla, Jr.; President: Thomas Healy.

CONNECTICUT

Valley Regional Planning Agency, 366 Main Street, Ansonia, Connecticut 06401, Phone: (203) 735-8689. Planning Director: Victor Claman; Chairman: John T. Hoye.

Regional Planning Agency of South Central Connecticut, 60 Connolly Parkway, Hamden, Connecticut 06514, Phone: (203) 777-4795. Executive Director: Norris C. Andrews; Chairman: George Brower Cash. 1948.

Capital Region Council of Elected Officials, c/o Greater Hartford Chamber of Commerce, 250 Constitution Plaza, Hartford, Connecticut 06103. Phone: (203) 525-4451. Director: Dana Hanson; Chairman: Lewis B. Rome. 1966.

Capitol Region Planning Agency, 15 Lewis Street, Hartford, Connecticut 06103, Phone: (203) 522-6143. Director: Robert D. Brown; Chairman: Seymore E. Lavitt.

Midstate Regional Planning Agency, P.O. Box 139, Middletown, Connecticut 06457, Phone: (203) 347-7214. Planning Director: Irwin M. Kaplan; Chairman: George M. Eames.

New Haven, Connecticut (In the process of forming a council), Greater New Haven Chamber of Commerce, 152 Temple Street, New Haven, Connecticut 06506.

Southwestern Regional Planning Agency, 83 East Avenue, Norwalk, Connecticut 06851, Phone: (203) 866-5543. Planning Director: Richard Carpenter; Chairman: Stearns Woodman.

Southeastern Connecticut Regional Planning Agency, 139 Boswell Avenue, Norwich, Connecticut 06360, Phone: (203) 889-6757. Director: Richard B. Erickson; Chairman: Frank Leigner, Jr. 1961.

Central Connecticut Regional Planning Agency, 49 West Main Street, Plainville, Connecticut 06062, Phone: (203) 747–5776. Director: Melvin Schneidermeyer; Chairman pro Tempore: Frank M. D'Addabbo. 1966.

Greater Bridgeport Regional Planning Agency, 301 Professional Building, White Plains Road, Trumbull, Connecticut 06611, Phone: (203) 268–0014. Planning Director: Mrs. Mary B. Sowchuk; Chairman: Charles H. Parks. 1960.

Central Naugatuck Valley Regional Planning Agency, 20 East Main Street, Waterbury, Connecticut 06702, Phone: (203) 753–1548. Planning Director: Duncan M. Graham; Chairman: Walter L. Hunt. 1960.

Windham Regional Planning Agency, 33 Church Street, Willimatic, Connecticut 06226, Phone: (203) 423–8611. Planning Director: Robert C. Young; Chairman: Ralph R. Crosthwaite.

Tri-State Transportation Committee (Covers consolidated SMSA), (see New York City).

DELAWARE

New Castle County Land Use and Transportation Planning Program, 4613 Robert Kirkwood Highway, Wilmington, Delaware 19808, Phone: (302) 998–0156. Program Director: H. R. Carlson.

Regional Council of Elected Officials (Includes Wilmington) (See Philadelphia, Pennsylvania).

DISTRICT OF COLUMBIA

Metropolitan Washington Council of Governments, 1225 Connecticut Avenue, NW., Washington, D.C. 20036. Executive Director: Walter A. Scheiber; President: Mayor Walter E. Washington (D.C.); Chairman: Supervisor Frederick A. Babson (Fairfax County, Va.). 1957.

FLORIDA

Northwest Florida Development Council, County Courthouse, Bonifay, Florida 32425, Phone: (904) 547–3665. Executive Director: Barry A. Boswell; President: N. Devane Williams. 1966.

Broward County Area Planning Board, 363 Broward County Courthouse, Fort Lauderdale, Florida 33301, Phone: (305) 525–1641. Planning Director: Valentyne E. Brennan; Chairman: Earle R. Kraft.

Jacksonville-Duval Area Planning Board, 830 American Heritage Life Building, Jacksonville, Florida 32202, Phone: (904) 355–1578. Executive Director: Marvin C. Hill; Chairman: William K. Jackson.

Metropolitan Dade County Planning Commission, 1351 N.W. Twelfth Street, Miami, Florida 33125, Phone: (305) 377–7921. Planning Director: Reginald R. Walters; Chairman: Mayor Chuck Hall (Dade County).

Escambia-Santa Rosa Regional Planning Council, P.O. Box 486, Pensacola, Florida 32502, Phone: (904) 434–1027. Executive Director: Daniel F. Krumel; Chairman: Edward E. Harper. 1967.

Tampa Bay Regional Planning Council, 3151 Third Avenue North, Suite 535, St. Petersburg, Florida 33713, Phone: (813) 898–0891. Executive Director: Don K. King; Chairman: D. William Overton. 1962.

Tallahassee-Leon County Planning Commission, P.O. Box 726, Tallahassee, Florida 32301, Phone: (904) 224–9181. Director: Mr. Phillip W. Pitts; Chairman: Jack Yaeger, Jr.

East Central Florida Regional Planning Council, 2323 South Washington Avenue, Titusville, Florida 32780, Phone: (305) 267–5453. Director: Gordon D. Wagner; Chairman: Claude H. Wolfe. 1962.

Palm Beach County Area Planning Board, Box 1548, West Palm Beach, Florida 33402, Phone: (305) 832–1671. Director: Donald O. Morgan; Chairman: John H. Flancher. 1966.

GEORGIA

Northeast Georgia Area Planning and Development Commission, P.O. Box 1724, Athens, Georgia 30601, Phone: (404) 548–3141. Executive Director: Burton Sparer; Chairman: Donald I. Bloemer. 1961.

Atlanta Region Metropolitan Planning Commission, 900 Glenn Building, Atlanta, Georgia 30303, Phone: (404) 522–7577. Executive Director: Glenn E. Bennett; Chairman: Nelson Severinghous. 1960.

Metropolitan Atlanta Council of Local Governments, 900 Glenn Building, Atlanta, Georgia 30303, Phone: (404) 522–7577. Coordinator: Wayne Moore, Jr.; Chairman: Mayor L. Marion Nolan (College Park). 1964.

353

Augusta SMSA Council of Local Governments, Room 704, Municipal Building, Augusta, Georgia 30902, Phone: (803) 648–8301. Secretary: Lawrence D. Connor, Jr.; Chairman: Mayor H. O. Weeks (Aiken, S.C.)

Augusta-Richmond County Planning Commission, Room 704, Municipal Building, Augusta, Georgia 30902, Phone: (404) 724–4391, Ext. 237. Executive Director: Lawrence D. Connor, Jr.; Chairman: Hugh O. Busbia.

Central Savannah River Area Planning and Development Commission, 630 Ellis Street, Augusta, Georgia 30902, Phone: (404) 722–7521. Executive Director: Tim F. Maund; Chairman: Walter Harrison.

Altamaha Area Planning and Development Commission, P.O. Box 328, Baxley, Georgia 31513, Phone: (912) 367–3637. Executive Director: Jerry O. Bange; Chairman: Randall M. Walker. 1963.

Coastal Area Planning and Development Commission, 102 Old City Hall, Brunswick, Georgia 31520, Phone: (912) 264–3121. Executive Director: David S. Maney; Chairman: Anthony A. Alaimo.

Albany-Dougherty County Planning Commission, P.O. Box 346, Camilla, Georgia 31730, Phone: (912) 336–5616. Planning Director: George Hudson; Executive Director: Carroll Underwood; Chairman: Bernard Reeves.

Southwest Georgia Planning and Development Commission, P.O. Box 346, Camilla, Georgia 31730, Phone: (912) 336–5616. Executive Director: Carroll C. Underwood, Chairman: L. C. Boddiford, Jr. 1963.

Lower Chattahoochee Valley Area Planning and Development Commission, P.O. Box 1908, Columbus, Georgia 31902, Phone: (404) 324–4221. Executive Director: Richard K. Allen; Chairman: Alton H. Fendley. 1962.

Valley Council of Local Governments, Box 1060, Columbus, Georgia 31902, Phone: (404) 322–6748. Secretary: Edward D. Baker, Chairman: Don Bailey. 1967.

Heart of Georgia Planning and Development Commission, P.O. Box 484, Eastman, Georgia 31023, Phone: (912) 374–4245. Executive Director: Kenneth A. Sibal; Chairman: Jack E. McGinty.

West Central Georgia Area Planning and Development Commission, P.O. Box 6, Ellaville, Georgia 31806, Phone: (912) 937–2241. Executive Director: Frank Moore; Chairman: Joseph S. Eason. 1965.

Georgia Mountains Planning and Development Commission, P.O. Box 1294, Gainesville, Georgia 30501, Phone: (404) 532–6541. Executive Director: Oliver T. Terriburry; Chairman: Jef Walraven. 1962.

Chattahoochee-Flint Area Planning and Development Commission, P.O. Box 1363, La Grange, Georgia 30240, Phone: (404) 882–2575. Executive Director: William Lundberg; Chairman: John J. Hood. 1964.

Middle Georgia Area Planning Commission, P.O. Box 4586, Macon, Georgia 31208, Phone: (912) 743–5862. Director: William W. Hibbert III; Chairman: Sam A. Nunn, Jr. 1965.

Oconee Area Planning and Development Commission, P.O. Box 707, Milledgeville, Georgia 31061, Phone: (912) 452–2238. Executive Director: J. McDonald Wray; Chairman: Ben E. Gooch.

Coosa Valley Area Planning and Development Commission, P.O. Box 1424, Rome, Georgia 30161, Phone: (404) 234–8507. Executive Director: Richard H. Orton; Chairman: J. M. Tutton.

Chatham County-Savannah Metropolitan Planning Commission, Box 1038, Savannah, Georgia 31402, Phone: (912) 236–9523. Executive Director: Eugene B. Culpepper; Chairman: Albert Lufburrow. 1955.

Chatham County Municipal Association, City Hall, Savannah, Georgia 31402, Phone: (912) 233–9321. Secretary: Picot B. Floyd; Chairman: Rufus Bazemore. 1963.

Georgia Southern Area Planning and Development Commission, 26 Siebald Street, Statesboro, Georgia 30458, Phone: (912) 764–6241. Executive Director: William T. Greer; Chairman: H. C. Hearn, Jr. 1965.

Coastal Plain Area Planning and Development Commission, P.O. Box 1223, Valdosta, Georgia 31602, Phone: (912) 244–2048. Executive Director: Hal A. Davis; Chairman: F. K. Reyher.

Slash Pine Area Planning and Development Commission, P.O. Box 1276, Waycross, Georgia 31501, Phone: (912) 283–3831. Executive Director: Max W. Harral; Chairman: Pete J. Gibson. 1963.

Chattanooga Area Regional Council of Governments, (see Chattanooga, Tennessee).

HAWAII

Honolulu Planning Department, 629 Pokukaina Street, Honolulu, Hawaii 96813, Phone: Operator–58061. Planning Director: Frank B. Skrivanek; Mayor: Neal S. Blaisdell (Honolulu).

IDAHO

Ada Development Council, Ada County Courthouse, Boise, Idaho 83702, Phone: (208) 343–4605. Secretary: Lynn Rogers; Chairman: Leon Fairbanks. 1968.

ILLINOIS

McLean County Regional Planning Commission, 208 Hillside Lane, Bloomington, Illinois 61701, Phone: (309) 967–6735. Chairman: Larry LeFebvre. 1967.

Greater Egypt Regional Planning and Development Commission, 211½ West Main Street, Carbondale, Illinois 62901, Phone: (618) 549-3306. Executive Director: Franklin H. Moreno; Chairman: Dr. Allen Y. Baker. 1961.

Council of Governments of Cook County, Center for Research in Urban Affairs, Loyola University, 820 North Michigan Avenue, Chicago, Illinois 60611, Phone: (312) 944–0800, Ext 572. Director: Dr. James M. Banovetz; Chairman: Jack D. Pahl. 1967.

Northeastern Illinois Regional Planning Commission, 400 West Madison Street, Chicago, Illinois 60606, Phone: (312) 263–1266. Executive Director: Matthew L. Rockwell; President: John W. Baird. 1957.

Southwestern Illinois Metropolitan Area Planning Commission, 121A West Main Street, Collinsville, Illinois 62234. Executive Director: Theadore Mikesell; President: Wetzel Harness.

Macon County Regional Planning Commission, Municipal Center, Decatur, Illinois 62523, Phone: (217) 423–7541, Ext. 221. Director: Charles Reed; Chairman: Rex Brown. 1966.

East-West Gateway Coordinating Council, 234 Collinsville Avenue, East St. Louis, Illinois 62201, Phone: (618) 274–2750. Executive Director: Eugene G. Moody; Chairman: Mayor Alvin G. Fields, Sr. (East St. Louis). 1965.

Tri-County Regional Planning Commission, 304 Peoria County Courthouse, Peoria, Illinois 61602, Phone: (309) 673–5611. Executive Director: Donald J. Irving; Chairman: Edwin Mitchell. 1958.

Rockford-Winnebago Regional Planning Commission, 425 East State Street, Rockford, Illinois 61104, Phone: (815) 965–4711, Director: Michael J. Meehan; Chairman: Mark Sommer, Jr. 1955.

Bi-State Metropolitan Planning Commission, 1504 Third Avenue, Rock Island, Illinois 61201, Phone: (309) 788–6338. Executive Director: William S. Luhman; Chairman: Richard LeBuhn. 1966.

Metropolitan Springfield Council of Governments, Municipal Building, Springfield, Illinois 62701, Phone: (217) 544–5731, Ext. 324. Secretary: Bradley B. Taylor; Chairman: Willard Tobin. 1964.

Sangamon County Regional Planning Commission, 311 Municipal Building, Springfield, Illinois 62701, Phone: (217) 544–5731. Executive Director: Bradley B. Taylor; Chairman: Robert Summers.

Champaign County Regional Planning Commission, 104 South Bennett Street, Urbana, Illinois 61801, Phone: (217) 365–3313. Executive Director: Richard Maltby; Chairman: Henry I. Green.

INDIANA

Lake-Porter Regional Transportation and Planning Commission, Lake County Courthouse, Crown Point, Indiana 46307, Phone: (219) 663–5606. Director: Norman Tufford; Chairman: Steve W. Manich. 1966.

Evansville-Vanderburgh Metropolitan Planning Commission, Courthouse Annex, Third Floor, Evansville, Indiana 47708, Phone: (812) 425–8188. Planning Director: William D. Jones; President: Dr. B. F. Shepp. 1963.

Metropolitan Planning Commission of Marion County, Metropolitan Planning Department, 2041 City-County Building, Indianapolis, Indiana 46204, Phone: (317) 633–3699. Executive Director: F. Ross Vogelgesang; Chairman: Charles L. Whistler.

Tippecanoe County Area Plan Commission, Courthouse, Lafayette, Indiana 47901, (317) 742–6827. Executive Director: E. H. Worley; Chairman: William K. Bennett. 1959.

Delaware-Muncie Metropolitan Planning Commission, 105 East Washington Street, Muncie, Indiana 47305, Phone: (317) 284–9915. Director of Planning: Donald E. Best; Chairman: James Fraizer. 1965.

Area Planning Commission of St. Joseph County, 129 West Colfax Avenue, South Bend, Indiana 46601, Phone: (219) 233–2955. Executive Director: John K. Wilson; Chairman: J. Frank Miles. 1965.

Area Planning Department for Vigo County, 17 Harding Avenue, Terre Haute, Indiana 47801, Phone: (812) 232–0974. Director: Lee Robert Mann. 1967.

O–K–I Regional Transportation and Development Study Committee (see Cincinnati, Ohio).

Falls of The Ohio Metropolitan Council of Governments (see Louisville, Kentucky).

IOWA

Rathbun Regional Planning Commission, 201 South Main Street, Albia, Iowa 52531, Phone: (515) 932–7126. Chairman: Henry Kolling. 1967.

Northwest Iowa Regional Planning Commission, P.O. Box 98, Ashton, Iowa 51232, Phone: (712) 754–2111. Director: Edward Schippman; Chairman: Bernard L. Braband.

Linn County Regional Planning Commission, Courthouse, Cedar Rapids, Iowa 52401, Phone: (319) 364–8121. Director: Don B. Salyer; Chairman: Robert M. L. Johnson. 1964.

Council Bluffs Metropolitan Area Planning Commission, 209 Pearl Street, Council Bluffs, Iowa 51501, Phone: (712) 322–4061 (City Hall). Director: James D. Richter; Chairman: Philip Willson. 1965.

Omaha-Council Bluffs Metropolitan Area Planning Agency (see Omaha, Nebraska).

Bi-State Metropolitan Planning Commission (Covers Davenport Area) (see Rock Island, Illinois).

Central Iowa Regional Planning Commission, 820 Locust Street, Des Moines, Iowa 50309, Phone: (515) 244–3257. Director: Robert W. Mickle; Chairman: Harold Parnum. 1965.

Mid-Iowa Association of Local Governments, 1133 66th Street, Des Moines, Iowa 50311, Phone: (515) 284–6706. Chairman: Mayor Clarence Millsap (Windsor Heights); Vice Chairman: Mayor Thomas Urban (Des Moines). 1968.

Dubuque County Metropolitan Area Planning Commission, 411 Fischer Building, Dubuque, Iowa 52001, Phone: (319) 588–3684. Director of Planning: Donald D. Gilson; Chairman: Clifford Knippel. 1965.

Siouxland Interstate Metropolitan Planning Commission, P.O. Box 447, Sioux City, Iowa 51103. Phone: (712) 277-2121, Ext. 222. Director: Donald M. Meisner; Chairman: Ernest L. Albertsen. 1965.

Black Hawk County Metropolitan Planning Commission, P.O. Box 689, Waterloo, Iowa 50704, Phone: (319) 233–5112. Director of Planning: Hugh J. Copeland; Chairman: D. F. Mirrielees. 1964.

KANSAS

Metropolitan Planning Commission (see Kansas City, Missouri).

Mid-America Council of Governments (see Kansas City, Missouri).

Greaer Topeka Intergovernmental Council, City Hall, Topeka, Kansas 66603, Phone: (913) CE 5–9261 (City Hall). Chairman: Mayor Charles W. Wright Topeka).

Topeka-Shawnee County Metropolitan Planning Agency, Shawnee County Courthouse, Room 209, Topeka, Kansas 66603, Phone: (913) 235–9261. Director: James H. Schlegel; Chairman: Conant Wait.

South Central Kansas Regional Council of Governments, Center for Urban Studies, Wichita State University, Wichita, Kansas 67208, Phone: (316) 683-7561. Director: Dr. Hugo Wall; Chairman: Mayor Anthony Snyder (Winfield). 1966.

Wichita-Sedgwick County Metropolitan Planning Commission, 104 South Main Street, Wichita, Kansas 67202, Phone: (316) 262–8211. Director of Planning: C. Bickley Foster; Chairman: W. Harold Mooney, 1958.

KENTUCKY

City-County Planning Commission, 227 North Upper Street, Lexington, Kentucky 40507, Phone: (606) 252–8808. Executive Director: William H. Qualls; Chairman: David C. Lagrew. 1928.

Falls of The Ohio Metropolitan Council of Governments, 901 Fiscal Court Building, Louisville, Kentucky 40202, Phone: (502) 504–3572. Director: Wilbert F. Watkins; Chairman: Kenneth A. Schmied, City Hall, Louisville.

Northern Kentucky Area Planning Commission, P.O. Box F, Newport, Kentucky 41072, Phone: (606) 431–2580. Executive Director: Herbert P. Moore; Chairman: George H. Neack.

O-K-I Regional Transportation and Development Study Committee, see Cincinnati, Ohio).

Kyova Interstate Planning Commission, (see Huntington, West Virginia).

LOUISIANA

Capital Regional Planning Commission, 101 St. Ferdinand, Suite 205, Baton Rouge, Louisiana 70801, Phone: (504) 342–6018. Executive Director: Chester H. Jordan; Chairman: Edward E. Evans. 1967.

Evangeline Economic Development District Council, Courthouse, Lafayette, Louisiana 70501, Phone: (318) 233–3215. Executive Director: Patrick M. Killeen; President: Lloyd A. Girouard.

Lafayette Planning Commission, P.O. Box 2154, Lafayette, Louisiana 70501, Phone: (318) 234–4245. Director: Charles Bonnette; Chairman: B. Roy Domingue.

Greater Lake Charles-Sulphur-Westlake Metropolitan Planning Commission, 308 Iris Street, Lake Charles, Louisiana 70601, Phone: (318) 436–5243. Director: Edward J. Strenk.

Regional Planning Commission for Jefferson, Orleans and Saint Bernard Parishes, 333 St. Charles Avenue, New Orleans, Louisiana 70130, Phone: (504) 523–1382. Director: Charles F. O'Doniel, Jr.; Chairman: Emile Prattini.

Caddo-Bossier Council of Local Governments, Room 304, City Hall, Shreveport, Louisiana 71102. Phone: (318)424–4171 (City Hall). Executive Director: John Gallagher; Chairman: Mayor Clyde E. Fant (Shreveport). 1967.

MAINE

York County Regional Planning Commission, York County Court House, Alfred, Maine 04002, Phone: (207) 324–2952. Executive Director: Arthur T. Lougee; Chairman: Arthur Moulton.

Eastern Maine Economic Development District, PRIDE, Inc., City Hall, Brewer, Maine 04412, Phone: (207) 942–6774. Executive Director: Eden Elwell; Chairman: Raynor I. Crosman.

Portland, Maine (In the process of forming a council), Mr. John Salisbury, Executive Secretary, Maine Municipal Association, 89 Water Street, Hallowell, Maine 04347.

Androscoggin Valley Regional Planning Commission, 181 Russel Street, Lewiston, Maine 04240, Phone: (207) 783–6562. Director: James O. Nesbitt; Chairman: Albion B. Richer. 1962.

Greater Portland Regional Planning Commission, 562 Congress Street, Portland, Maine 04101, Phone: (207) 774–0601. Chairman: Floyd Rutherford.

MARYLAND

Regional Planning Council, 701 St. Paul Street, Baltimore, Maryland 21202, Phone: (301) 837–9000. Executive Director: Robert N. Young; Chairman: J. Jefferson Miller. 1964.

Delmarva Advisory Council, 132 East Maine Street, Salisbury, Maryland 21801, Phone: (301) 742–9271. Executive Director: Worthington J. Thompson; President: David A. Clements, Jr.

Tri-County Council for Southern Maryland, Waldorf, Maryland 20601, Phone: (301) 645–2693. Executive Director: John H. Mills; Chairman: John T. Parran, Jr. 1965.

Metropolitan Washington Council of Governments (see District of Columbia).

MASSACHUSETTS

Metropolitan Area Planning Council, 44 School Street, Boston, Massachusetts 02108, Phone: (617) 523–2454. Director: Robert G. Davidson; Chairman: Rev. Seavey Joyce, S.J.

Dukes County Planning Commission, County Commission Courthouse, Edgartown, Massachusetts 02539, Phone: (617) 627–4668. Chairman: Dean Swift.

Southeastern Massachusetts Regional Planning Commission, 123 North Main Street, Fall River, Massachusetts 02720, Phone: (617) 678–3991. Executive Director: William Barbour; Chairman: Leon W. Dean. 1955.

Franklin Regional Planning Commission, Franklin County Department of Planning, Courthouse, Greenfield, Massachusetts 01301, Phone: (413) 773–3003. Director: Donald Caven; Chairman: Thomas Herlihy.

Cape Cod Planning and Economic Development Commission, 365 Main Street, Box 23, Hyannis, Massachusetts 02601, Phone: (617) 775–3532. Director: E. Fletcher Davis; Chairman: Fred Lawrence.

Central Merrimack Valley Regional Planning District, 477 Essex Street, Lawrence, Massachusetts 01840, Phone: (617) 686–0361. Director Richard H. Young; Chairman: James Bannan. 1959.

Greater Lowell Area Regional Planning Commission, 73 East Merrimac Street, Lowell, Massachusetts 01852, Phone: (617) 454–4101. Director: Harold Kramer; Chairman: Arthur E. Hammar.

Berkshire Regional Planning Commission, Courthouse, Pittsfield, Massachusetts 01201, Phone: (413) 443–8740. Executive Director: Karl Hekler; Chairman: Charles Driscoll.

Old Colony Planning Council, 27 Belmont Street, West Bridgewater, Massachusetts 02401, Phone: (617) 583–1833. Director: Alvin Jack Sims; Chairman: Merton H. Ouderkirk. 1967.

Lower Pioneer Valley Regional Planning Commission, 1499 Memorial Avenue, West Springfield, Massachusetts 01089, Phone: (413) 739–5383. Planning Director: Krzysztof M. Munnich; Chairman: Edward C. Peck, Jr. 1962.

Central Massachusetts Regional Planning Commission, 70 Elm Street, Worcester, Massachusetts 01609, Phone: (617) 756–7717. Planning Director: David H. Kellogg; Chairman: Ralph W. Hager. 1963.

Committee on Urban-Suburban Cooperation, 70 Elm Street, Worcester, Massachusetts 01609. Chairman: Sherman S. Ludden.

MICHIGAN

Bay Regional Planning Commission, 702 County Building, Bay City, Michigan 48706, Phone: (517) 892–6011. Director: William A. Lynch; Chairman: David Karbowski.

Southeast Michigan Council of Governments, 211 W. Fort Street, Detroit, Michigan 48226, Phone: (313) 961–4266. Staff Director: E. Robert Turner; Chairman: Larry Mainland. 1967.

Genessee County Metropolitan Planning Commission, 930 Beach Street, Flint, Michigan 48502, Phone: (313) 232–7186. Director-Coordinator: Thomas H. Haga; Chairman: Kenneth C. Mac Gillivray.

Kent-Ottawa Regional Planning Commission, Kent County Courthouse, Grand Rapids, Michigan 49502, Phone: (616) 456–3731. Director: Joseph A. Fendt; Chairman: Clifford Briggs. 1966.

Association of Grand Rapids Area Governments, 3063 Wilson Avenue, Grandville, Michigan 49418, Phone: (616) 456–3166. Chairman: Mayor William A. Trimmer (Grandville). 1967.

Jackson Metropolitan Area Regional Planning Commission, 312 South Jackson, Room 514, Jackson, Michigan 49201, Phone: (517) 782–8131. Director: Frederick L. Barcley; Chairman: Walter R. Boris.

Kalamazoo Metropolitan Planning Commission, 227 West Michigan Avenue, Kalamazoo, Michigan 49005, Phone (616) 343–1201, Ext. 51. Director: Bruce A. Watts; Chairman: Julius T. Wendzel. 1960.

Tri-County Regional Planning Commission, 535 North Clippert Street, Lansing, Michigan 48912, Phone: (517) 372–1810. Executive Director: William C. Roman; Chairman: Myles G. Boylan. 1956.

Muskegon County Metropolitan Planning Commission, County Building, Muskegon, Michigan 49444, Phone: (616) 726–4711. Director: R. T. Dittmer; Chairman: Robert K. Hunter. 1964.

Saginaw County Metropolitan Planning Commission, Courthouse, Saginaw, Michigan 48605, Phone: (517) 793–9100. Planning Director: Martin R. Cramton; Chairman: Ron Heimlein.

Northwest Michigan Economic Development District Commission, Courthouse, Traverse City, Michigan 49684, Phone: (616) 946–5922. Executive Director: Donald E. Goostrey; Chairman: A. R. Jacobs. 1967.

MINNESOTA

Arrowhead Economic Development District, 800 Longsdale Building, 302 W. Superior St., Duluth, Minnesota 55802, Phone: (218) 722–5545. Executive Director: Rudy R. Esala; President: Floyd R. Anderson.

Duluth-Superior Metropolitan Planning and Transportation Study, 402 City Hall, Duluth, Minnesota 55802, Phone: (218) 727–4522, Ext. 281. Director: Donn R. Wiski.

Head of The Lakes Council of Governments, % 409 City Hall, Duluth, Minnesota 55802, Phone: (218) 727–4522, Ext. 281. Staff Director: Donn R. Wiski: President: Mayor Charles Deneweth (Superior, Wis.).

Iron Range Planning Board, 2004 4th Avenue East, Hibbing, Minnesota 55746, Phone: (218) 263–3306. Chairman: Mayor Joe Taveggia (Hibbing).

Metropolitan Council, Suite 101, Capitol Square, Tenth and Ceder Streets, St. Paul, Minnesota 55101, Phone: (612) 645–9194. Executive Director: Robert T. Jorvig; Chairman: James L. Hetland, Jr. 1967.

MISSISSIPPI

Gulf Regional Planning Commission, P.O. Box 268, Bay St. Louis, Mississippi 39520, Phone: (601) 467–5462. Executive Director: Jack Different; Chairman: R. Gordon Williams. 1965.

Southern Mississippi Economic Development District, Inc., P.O. Box 2057, 719 West Scooba Street, Hattiesburg, Mississippi 39401, Phone: (601) 582–9104. Executive Director: Henry W. Pyne; President: J. O. Cagle.

Southwest Mississippi Economic Development District, Inc., P.O. Box 686, McComb, Mississippi 39648, Phone: (601) 684–7250. Executive Director: G. Frank Oakes; President: J. O. Emmerich.

Mississippi-Arkansas-Tennessee Council of Governments (see Memphis, Tennessee).

MISSOURI

Metropolitan Planning Commission, 127 West Tenth Street, Suite 366, Kansas City, Missouri 64105, Phone: (816) 221–7133. Executive Director: Stuart Eurman; Chairman: Frank Corbett.

Mid-America Council of Governments, 127 West Tenth Street, Suite 366, Kansas City, Missouri 64105, Phone: (816) 221–7133. Secretary: Kent E. Crippin; Chairman: Charles E. Curry. 1967.

Green Hills Regional Planning Commission, c/o City Hall, Kingston, Missouri 64650. Chairman: Mayor Stephen J. Millett.

Ozark Foothills Regional Planning Commission, 118 North Broadway, Poplar Bluff, Missouri 63901, Phone: (314) 785–7980. Executive Director: Howard M. Rubin; Chairman: Gene Parrent. 1967.

Metropolitan Planning Commission of Greater St. Joseph and Buchanan County, City Hall, St. Joseph, Missouri 64501, Phone: (816) 233–0275. Director: Lorin A. Dunham; Chairman: James Sollars.

East West Gateway Coordinating Council (Covers St. Louis Area) (see East St. Louis, Illinois).

Boot Heel Regional Planning Commission, Sikeston, Missouri 63801. Phone: (314) 471–2173. Chairman: Pat Lea.

Missouri Valley Planning Commission, City Hall, Slater, Missouri 65349. Chairman: Arthur Preston.

Springfield Metropolitan Area Planning Agency, 301 City Hall, Springfield, Missouri 65802, Phone: (417) 865–1611. Director of Planning: Harold S. Hass; Chairman: Harry Boswell. 1961.

Show-Me Regional Planning Commission, Johnson County Courthouse, Warrensburg, Missouri 64903. Phone: (816) 747–9117 (City Hall). Chairman: Robert Rackett.

South Central Ozark Regional Planning Commission, Williams Building, West Plains, Missouri 65775. Phone: (417) 256-6100. Executive Director: Arthur G. Gutfahr; Chairman: Kenneth B. Ross. 1967.

MONTANA

Billings-Yellowstone City-County Planning Board, 303 Courthouse, Billings, Montana 59101. Phone: (406) 252–5181. Director: Orien R. Gossett; Chairman: Darrell Booth. 1948.

Great Falls City-County Planning Board, Room B-4, Civic Center Building, Great Falls, Montana 59401, Phone: (406) 452-8561, Ext. 61. Director: Robert P. Roberts; Chairman: William J. Hess. 1957.

NEBRASKA

Lincoln-Lancaster County Planning Commission, City Hall, Lincoln, Nebraska 68508, Phone: (402) 435-2981. Planning Director: Douglas E. Brogden; Chairman: Edwin Perry. 1958.
Northeastern Nebraska Joint Planning Commission, Oakland, Nebraska 68045. Chairman: Robert Bogue.
Omaha-Council Bluffs Metropolitan Area Planning Agency, 1910 Harney Street, Omaha, Nebraska 68102, Phone: (402) 341-5899. Director: O. A. Kinney, Jr.; President: Lynn Landgren; Chairman: Fred Jacobberger. 1967.
North Platte Valley Joint Planning Commission, Scottsbluff, Nebraska 69631. Interim Chairman: Rubin Sitzman.
Siouxland Interstate Metropolitan Planning Council (see Sioux City, Iowa).

NEVADA

Clark County Regional Planning Council, 400 Stewart Avenue, Las Vegas, Nevada 89101. Phone: (702) 385-1221, Ext. 240. Director: Don J. Saylor; Chairman: Mayor Oran K. Grayson (Las Vegas). 1966.
Regional Planning Commission of Reno, Sparks and Washoe County, P.O. Box 1286, Reno, Nevada 89504, Phone: (702) 323-0701 Ext. 245. Director: Richard J. Allen; Chairman: W. W. Maynard. 1947.
Tahoe Regional Planning Commission (see South Lake Tahoe, California).

NEW HAMPSHIRE

New Hampshire-Vermont Development Council, Inc., 10 Allen Street, Hanover, New Hampshire 03577, Phone: (603) 643-2844. Executive Director: Vincent R. Dahlfred; President: Mrs. Jean L. Hennessey.
Southern New Hampshire Planning Commission, 908 Elm Street, Manchester, New Hampshire 03101, Phone: (603) 669-4664. Executive Director: James E. Minnoch; Chairman: E. Led Kanteres. 1966.

NEW JERSEY

Atlantic County Planning Board, 2322 Pacific Avenue, Atlantic City, New Jersey 08401, Phone: (609) 348-4361. Director: John R. Gideonse; Chairman: Carl Valore, Jr.
Cumberland County Planning Board, Cumberland County Courthouse, Bridgeton, New Jersey. Director: John Holland.
Delaware Valley Regional Planning Commission (Includes Camden area) (see Philadelphia, Pennsylvania).
Regional Conference of Elected Officials (Includes Camden Area) (see Philadelphia, Pennsylvania).
Salem County Planning Board, County Courthouse, Salem, New Jersey 08079, Phone: (609) 935-4477. Director: Gerald L. Walker; Chairman: Joseph A. Hassler, Jr. 1961.
Lake Hopatcong Regional Planning Board (Route 10, Succassunna, New Jersey 07876, Phone: (201) 584-7400 (City Clerk). Chairman: David L. Reifer.
Tocks Island Regional Advisory Council (see Stroudsburg, Pennsylvania).
Tri-State Transportation Committee (Covers consolidated SMSA) (see New York City).

NEW MEXICO

Middle Rio Grande Council of Governments of New Mexico, 505 Marquette Avenue, N.W., Suite 1320, Albuquerque, New Mexico 87102, Phone: (505) 243-8661. Director: Stephen George, Jr.; Chairman: Edward V. Balcomb. 1967.
North Central New Mexico Economic Development District, P.O. Box 4248, Santa Fe, New Mexico 87501, Phone: (505) 827-2014. Executive Director: Leo T. Murphy; President: Nick L. Salazar. 1967.

NEW YORK

Capitol District Regional Planning Commission, 220 Terminal Building, Albany County Airport, Albany, New York 12211, Phone: (518) 869-0178. Executive Director: Louis Lex, Jr.; Chairman: Mayor Erastus Corning (Albany). 1967.

Lake Champlain-Lake George Regional Planning Board, New York State Office of Planning Coordination, 488 Broadway, Albany, New York 12202, Phone: (518) 474–2994. Chairman: Charles E. Hawley.

Southern Tier East Regional Planning Board, 307 County Courthouse, Binghamton, New York 13901, Phone: (607) 772–2114. Director: Joseph M. Missavage; Chairman: William B. Graham.

Black River-St. Lawrence Economic Development Commission, Inc., St. Lawrence University, Canton, New York 13617, Phone: (315) 386–4551, Ext. 264. Executive Director: Arthur C. Mengel; President: Joseph A. Romola.

Erie and Niagara Counties Regional Planning Board, 2880 Grand Island Boulevard, Grand Island, New York 14072, Phone: (716) 693–2727. Director: Leo J. Nowak, Jr.; Chairman: Lester S. Miller. 1966.

Nassau-Suffolk Regional Planning Board, Veterans Memorial Highway, Hauppauge, New York 11787, Phone: (510) 724–1919. Executive Director: Lee E. Koppelman; Chairman: Leonard W. Hall. 1965.

Mohawk Valley Economic Development District, Inc., 19 West Main Street, Mohawk, New York 13407, Phone: (315) 866–4671. Executive Director: John M. Ladd; President: Ara T. Dildilian. 1966.

Metropolitan Regional Council, 155 East 71st Street, New York, New York 10021, Phone: (212) 628–6803. Executive Director: Robert P. Slocum; Chairman: Mayor John V. Lindsay (New York). 1966.

Tri-State Transportation Commission, 100 Church Street, New York, New York 10007, Phone: (212) 433–4200. Director: J. Douglas Carroll, Jr.; Chairman: David J. Goldberg. 1966.

Eastern Adirondack Economic Development Commission, Box K, Port Henry, New York 12974, Phone: (518) 546–7267, Executive Director: Karl L. Hoffman; President: Charles R. Clark.

Genesee Area Regional Planning Board, 301 County Office Building, Rochester, New York 14614; Chairman: John W. Baybutt.

Monroe County Planning Council, 301 County Office Building, Rochester, New York 14614, Phone: (716) 454–7200. Planning Director: William E. Uptegrove; Chairman: John W. Baybutt.

Central New York Planning and Development Board, 321 East Water Street, Syracuse, New York 13202, Phone: (315) 422–8276. Executive Director: Robert C. Morris; Chairman: Rhea M. Eckel. 1966.

Hudson River Valley Commission, 105 White Plains Road, Tarrytown, New York 10591, Phone: (914) 631–8800. Executive Director: Alexander Aldrich; Chairman: Frank Wells McCabe. 1966.

Herkimer-Oneida Joint Planning Board, Oneida County Courthouse, Utica, New York 13501, Phone: (315) 735–3371. Director: James Barwick; Chairman: Harold Myers. 1963.

NORTH CAROLINA

Metropolitan Planning Board of Ashville-Buncombe Counties, P.O. Box 7148, Ashville, North Carolina 28807, Phone: (704) 253–3611. Executive Director: Dean Y. Matthews; Chairman: William A. V. Cecil.

Western North Carolina Regional Planning Commission. Box 7148, Asheville, North Carolina 28807, Phone: (704) 253–3611. Executive Director: Dean Y. Matthews; Chairman: Weldon Weir. 1960.

Charlotte-Mecklenburg Planning Commission, City Hall Annex, Charlotte, North Carolina 28202, Phone: (704) 376–0731. Planning Director: William E. McIntyre; Chairman: Walter D. Toy, Jr. 1954.

Charlotte, North Carolina (In the process of forming a Council), Dr. James G. Martin, Chairman, Mecklenburg County Board of Commissioners, P.O. Box 697, Davidson, North Carolina 28036.

Mideast Economic Development Commission, Farmville,. North Carolina 27828, Phone: (919) 753–3049. Chairman: Carl V. Venters, Jr.

Cumberland County Joint Planning Board, Room 206, City Hall, Fayetteville, North Carolina 38301, Phone: (919) 483–2605. Director: John L. Booth; Chairman: Karl L. Sloan. 1967.

Guilford County Planning Board, P.O. Box 3427, Greensboro, North Carolina 27401, Phone: (919) 273–3611. Planning Director: Lindsay W. Cox; Chairman: J. Phal Hodgin.

Neuse River Economic Development Commission, Hookerton, North Carolina 28538, Phone: (919) 747–3304. Chairman: A. C. Edwards.

Southeastern Economic Development Commission, P.O. Box 96, Raeford, North Carolina 28376, Phone: (919) 875–3452. Chairman: Col. James R. Fout, Jr.

Research Triangle Regional Planning Commission, P.O. Box 12255, Research Triangle Park, North Carolina 27709, Phone: (919) 549–8184. Executive Director: Pearson H. Stewart; Chairman: Oscar R. Ewing. 1959.

Cleveland Association of Governmental Officials, 400 West Marion Street, Shelby, North Carolina 28150, Phone: (704) 487–6367. Secretary: Malcolm E. Brown; Chairman: Lamar Young. 1964.

Lower Cape Fear Council Of Local Governments, % City Hall, Wilmington, North Carolina 28401, Phone: (919) 762–4323. Secretary: E. C. Brandon, Jr.; Chairman: R. W. Cheers. 1968.

Forsyth Council Of Governments, Government Center, Southwest 3rd Street, Winston-Salem, North Carolina 27101, Phone: (919) 724–5511. Chairman: George Chandler. 1967.

Winston-Salem-Forsyth County Planning Board, Room 11, City Hall, Winston-Salem, North Carolina 27101, Phone: (919) 722–4141. Director of Planning: J. Ben Rouzie, Jr.; Chairman: F. Gaither Jenkins.

NORTH DAKOTA

Fargo Metropolitan Area Council Of Governments, City Hall, Fargo, North Dakota 58102, Phone: (701) 232–0232. Chairman: Mayor Herschel Lashkowitz (Fargo).

OHIO

Tri-County Regional Planning Commission, 415 South Portage Path, Akron, Ohio 4432, Phone: (216) 535–2644. Executive Director: James E. Farmer; Chairman: John F. Seiberling, Jr. 1957.

Wood County Planning Commission, 541 West Wooster Street, Bowling Green, Ohio 43402, Phone: (419) 354–5822. Director: Milton A. Bengston; Chairman: Gerald Avery.

Stark County Regional Planning Commission, 624 County Office Building, Canton, Ohio 44702. Phone: (216) 454–5651, Ext. 301. Director: J. Dale Cawthorne; Chairman: Dale Corbett. 1957.

O-K-I Regional Transportation And Development Study Committee, 309 Vine Street, Cincinnati, Ohio 45202, Phone: (513) 621-7060. Director: Fred R. Rauch; President: Frank Ferris II.

Cleveland Seven County Transportation Land Use Study, 439 The Arcade, Cleveland, Ohio 4414, Phone: (216) 241-2414. Executive Director: William B. Henry; Chairman: Wayne Parsons. 1965.

Regional Planning Commission, 415 The Arcade, Cleveland, Ohio 44114, Phone: (216) 861-6805. Director: William B. Henry; President: Robert Rawson. 1947.

Cleveland, Ohio (In the process of forming a Council), Mayor George J. Urban, President, Cuyahoga County, Mayors and City Managers Association, City Hall, South Euclid, Ohio 44121.

Franklin County Regional Planning Commission (In the process of forming a Council), 514 South High Street, Columbus, Ohio 43215, Phone: (614) 228-2663. Director: Harmon T. Merwin; Chairman: Will Hellerman.

Miami Valley Council of Governments, c/o Department of Political Science, University of Dayton, Dayton, Ohio 45409. Acting Executive Director: Lyndon E. Abbott; Chairman: Robert A. Haverstick. 1968.

Miami Valley Regional Planning Commission, 44 South Ludlow Street, Dayton, Ohio 45402, Phone: (513) 224–0303. Executive Director: Dale F. Bertsch; Chairman: Charles Harbottle. 1964.

Tuscarawas Valley Regional Advisory Committee, Reeves National Bank Building, Dover, Ohio 44622, Phone: (614) 453–0666 (Pres.). Field Coordinator: Morris C. Shawkey; President: Gene L. MacDonald. 1968.

Lima-Allen County Regional Planning Commission, 204–205 Dominion Building, Lima, Ohio 45801, Phone: (419) 224–6726. Director: Phillip N. Boyle; President: Charles Friedman. 1965.

Richland County Regional Planning Commission, 40 South Park Street, Mansfield, Ohio 44902, Phone: (216) 522–9454. Director: Edward T. Meehan; President: Harold Marshall. 1959.

Area Cooperation Committee (In the process of forming a Council), City of Maumee, City Hall, Maumee, Ohio 43537, Phone: (419) 893–8751 (City Hall). Executive Secretary: S. E. Klewer.

Ohio Valley Regional Development Commission, c/o Sciuto County Commissioners Office, Courthouse, Portsmouth, Ohio 45662, Phone: (614) 353–3969. President: Carl E. Mauntel. 1967.

Clark County-Springfield Regional Planning Commission, City Building, Springfield, Ohio 45501, Phone: (513) 323-9731. Planning Director: H. Clarke Mahannah; Chairman: Edward W. Rheins.
Clark County-Springfield Transportation Study, City Building, Springfield, Ohio 45501, Phone: (513) 323-9731 (City Building). Director: Walter A. Szcesney; Chairman: J. C. Shouvlin.
Brooke, Hancock, Jefferson Counties Metropolitan Planning Committee, P.O. Box 383, Steubenville, Ohio 43952, Phone: (614) 283-4111.
Jefferson County Metropolitan Planning Commission, P.O. Box 383, Steubenville, Ohio 43952, Phone: (614) 283-4111. Director: Nicholas Kaschak; President: C. W. Daily.
Steubenville-Weirton Area Transportation Study (see Weirton, West Virginia).
Toledo-Lucas County Planning Commission, 445 Huron Street, Toledo, Ohio 43604, Phone: (419) 255-1500. Director: Lawrence F. Murray; Chairman: Arthur R. Kline.
Trumbull County Planning Commission, County Administration Building, Warren, Ohio 44481, Phone: (216) 399-8811. Planning Director: Edward L. Kutevac; Chairman: R. H. Taylor.
Mahoning County Planning Commission, 21 West Boardmen Street, Youngstown, Ohio 44503, Phone: (216) 747-2092. Planning Director: James C. Ryan; Chairman: Thomas J. Roche.
Mahoning-Trumbull Counties Comprehensive Transportation and Development Study, 1616 Covington Street, Youngstown, Ohio 44510, Phone: (216) 746-4665. Director: Bruce G. Cunningham; Chairman: John Palermo. 1963.
Belmont, Ohio and Marshall Metropolitan Planning Committee, (see Wheeling, West Virginia). Kyova Interstate Planning Commission, (see Huntington, West Virginia).

OKLAHOMA

Southern Oklahoma Development Association, P.O. Box 3125, Ardmore, Oklahoma 73401, Phone: (404) 389-5488. Executive Director: J. J. Gigoux; President: Bill Hoover. 1966.
Lawton Metropolitan Area Planning Commission, City Hall, Lawton, Oklahoma 73501, Phone: (405) 357-7100. Director of Planning: Pat Painter; Chairman: Gale Sadler. 1961.
Eastern Oklahoma Economic Development District, P.O. Box 1367, Muskogee, Oklahoma 74401, Phone: (918) 682-6337. Executive Director: Joseph L. McAuliff; Chairman: Col. M. A. Hagerstrand, Ret.
Association of Central Oklahoma Governments, 221 Plaza Court Building, Oklahoma City, Oklahoma 73102, Phone: (405) CE5-9651. Executive Director: L. Douglas Halley; President: William S. Morgan. 1966.
Central Oklahoma Economic Development District, 612 Federal National Bank Building, Shawnee, Oklahoma 74801, Phone: 273-6410. Executive Director: Earl V. Price; President: Bill R. Wright. 1966.
Indian Nations Council of Governments, 700, 9 East Fourth Building, Tulsa, Oklahoma 74103, Phone: (918) 587-1537. Staff Director: Donald E. Osgood; Chairman: Mayor John Hess (Sand Springs). 1967.
Tulsa Metropolitan Area Planning Commission, 700, 9 East Fourth Building, Tulsa, Oklahoma 74103, Phone: (918) 587-1537. Director: Donald E. Osgood; Chairman: Ed Dubie.
Northeast Counties of Oklahoma Development District, P.O. Box 591, Vinita, Oklahoma 74301, Phone: (918) 256-4478. Executive Director: L. B. Earp; President: Lynn C. Barnes. 1966.
Kiamichi Economic Development District, Eastern Oklahoma State College, Wilburton, Oklahoma 74578, Phone: (918) 465-2367. Executive Director: Wesley W. Watkins; President: James L. Wright.
Ark-Homa Regional Planning Commission (see Fort Smith, Arkansas).

OREGON

Linn-Benton Association of Governments, P.O. Box 490, Albany, Oregon 97321, Phone: (503) 926-4264. Director: Gary Holloway; Chairman: Frank E. Kikel. 1967.
Central Lane Planning Council, 1643 Oak Street, Eugene, Oregon 97401, Phone: (503) 342-1757. Executive Secretary: Lawrence A. Rice; Chairman: Kenneth H. Kohnen. 1945.

Medford, Oregon (In the process of forming a Council), Honorable Earl M. Miller, County Judge, Jackson County Courthouse, Medford, Oregon.

Columbia Region Association Of Governments, 424 S.W. Main Street, Portland, Oregon 97204, Phone: (503) 226–4331. Executive Secretary: Homer C. Chandler; Chairman: Mayor E. G. Kyle. 1966.

Central Umpqua Regional Planning Council, Douglas County Courthouse, Roseburg, Oregon 97470, Phone: (503) 672–3311, Ext. 29. Coordinator: J. E. Slattery; Chairman: Pete Serafin. 1966.

Columbia County Mayors Planning Commission, City Hall, Saint Helen's, Oregon 97051, Phone: (503) 397–0111. Chairman: M. E. McMichel.

Mid-Willamette Valley Council of Governments, Marion County Courthouse, Salem, Oregon 97301, Phone: (503) 364–4401. Executive Secretary: Wesley Howe; Chairman: Pat McCarthy. 1959.

PENNSYLVANIA

Turnpike Planning And Development Commission, 500 Eighth Avenue, Altoona, Pennsylvania 16602, Phone: (814) 944–4415. President: Robert A. Halloran; Executive Director: Robert J. Heckel.

Lehigh Valley Congress Of Governments, 420 Main Street, Bethlehem, Pennsylvania 18018. President: Mayor Leonard G. Witt. 1967.

North Central Economic Development Association, 315 East Market Street, Clearfield, Pennsylvania 16830, Phone: (814) 765–7000. Chairman: Fred L. Rhoades; Executive Director: Flanders M. O'Neal.

Cambria County Planning Commission, Court House, Ebensburg, Pennsylvania 15931, Phone: (814) 472–7885. Director: T. Frederick Young; Chairman: Elvin W. Overdorff. 1965.

Erie County Planning Commission, Courthouse, Eire, Pennsylvania 16501, Phone: (814) 456–6560. Planning Director: Christopher Capotis; Chairman: Allen H. Kuhn.

Tri-County Regional Planning Commission (Cumberland, Dauphin and Perry Counties), 341 South Cameron Street, Harrisburg, Pennsylvania 17011, Phone: (717) 234–2639. Director: Oliver M. Fanning; Chairman: L. V. Holcombe. 1965.

Blair County Planning Commission, Courthouse Annex No. 2, Hollidaysburg, Pa. 16648, Phone: (814) 695–5084. Director: Richard C. Sutter; Chairman: John L. Wike. 1964.

Conemaugh Valley Conference, Bren Lyn Village, Johnstown, Pennsylvania 15904, Phone: (814) 266–3757. Chairman: Wayne C. Corle.

Lancaster County Planning Commission, 900 East King Street, Lancaster, Pennsylvania 17602, Phone: (717) 393–1721. Director: Ellis W. Harned; Chairman: H. Clifford Kreisle. 1958.

Lancaster Inter-Municipal Committee, Municipal Building, Lancaster, Pennsylvania 17604, Phone: (717) 397–3501. Chairman: Councilman Edward C. Kraft (Lancaster). 1967.

Joint Planning Commission of Lehigh and Northampton Counties, Post Office Box 2087, Lehigh Valley, Pennsylvania 18001, Phone: (215) 264–4544. Director: Michael N. Kaiser; Chairman: Robert K. Young. 1961.

Susquehanna Economic Development Association, Route 1, Lewisburg, Pennsylvania 17837, Phone: (717) 523–1109. President: Robert O. Brouse; Director: Rolland D. Berger.

Northwestern Pennsylvania Regional Planning and Development Commission, Room 14, Seneca Building, Seneca Street, Oil City, Pennsylvania. Executive Director: Howard Irons; Chairman: Arden L. Moyer.

Delware Valley Regional Planning Commission, 1317 Filbert Street, Philadelphia, Pennsylvania 19107, Phone: (215) LO8–3211. Director: Walter K. Johnson; Chairman: Paul N. Ylvisaker. 1967.

Regional Conference of Elected Officials, Inc., 1317 Filbert Street, Philadelphia, Pennsylvania 19107, Phone: (215) LO3–3662. Executive Director: Chester A. Kunz; Chairman: Mayor James H. J. Tate (Philadelphia). 1967.

Allegheny Seminar, 905 Bruce Hall, University of Pittsburgh, Pittsburgh, Pennsylvania 15213, Phone: (412) 621–3500, Ext. 441. Executive Secretary: Dr. Joseph A. James; Chairman: William J. Herge. 1963.

Southwestern Pennsylvania Economic Development District, Park Building, Pittsburgh, Pennsylvania 15222, Phone: (412) 391–1240. Chairman: Hiram Milton; Executive Director: Lewis A. Vidic.

Southwestern Pennsylvania Regional Planning Commission, 564 Forbes Avenue, Pittsburgh, Pennsylvania 15219, Phone: (412) 391–4120. Director: William R. B. Froelich; Chairman: A. B. Kenney. 1962.

Berks County Planning Commission, Berks County Courthouse, Reading, Pennsylvania 19601. Director: George Fasic.

Lackawanna County Planning Commission, Courthouse Annex, Scranton, Pennsylvania 18503, Phone: 346–7421. Planing Director: John F. Radkiewicz; Chairman: John F. Murphy. 1959.

Tocks Island Regional Advisory Council, 612 Monroe Street, Stroudsburg, Pennsylvania 18360, Phone: (717) 421–9841. Executive Director: Frank W. Dressler; Chairman: Denton Quick. 1965.

Northern Tier Economic Development Association, Incorporated, 11 Main Street, Towanda, Pennsylvania 18848, Phone: (717) 265–9103. Planning Director: Martin Maier; Chairman: Walter Bell. 1967.

Northern Tier Regional Planning Commission, 111 Main Street, Towanda, Pennsylvania 18848, Phone: (717) 265–9103. Executive Director: Martin Maier; Chairman: Joseph Kesnow. 1965.

Economic Development Council of Northeastern Pennsylvania, 403 First National Bank Building, Wilkes-Barre, Pennsylvania 18701, Phone: (717) 824–7801. Director: Raymond R. Carmon; President: John S. Davidson. 1964.

Luzerne County Planning Commission, Luzerne County Courthouse, Wilkes-Barre, Pennsylvania 18702, Phone: (717) 825–2749. Director of Planning: Edward H. Heiselberg; Chairman: John A. Hourigan, Jr. 1957.

York County Planning Commission, 1320 West Market Street, York, Pennsylvania 17404, Phone: (717) 843–9954. Director: Reed J. Dunne, Jr.; Chairman: Norman R. Wolf. 1959.

RHODE ISLAND

Statewide Comprehensive Transportation and Land Use Planning Program, 36 Kennedy Plaza, Suite 300, Providence, Rhode Island 02903, Phone: (401) 521–7100, Ext. 656. Director: Richard J. Bouchard; Chairman: Angelo A. Marcello. 1964.

Warwick, Rhode Island (In the process of forming a Council), Mayor Philip Noel, City Hall, Warwick, Rhode Island 02886.

SOUTH CAROLINA

Lower Savannah Economic Development District, Farmers and Merchants Bank, Aiken, South Carolina 29801, Phone: (803) 245–2653. Chairman: Eugene P. Brabham.

Berkeley-Charleston Regional Planning Commission, The Center, Room 301, Charleston, South Carolina 29403, Phone: (803) 723–1671. Director of Planning: Hosey H. Hearn; Chairman: Joe Griffith.

Richland and Lexington Counties Joint Planning Commission, City Hall, Columbia, South Carolina 29201, Phone: (803) 765–1041. Director: Sidney F. Thomas, Jr.; Chairman: Dr. A. C. Flora.

Greenville-Pickens Regional Planning Commission, 18 Thompson Street, Greenville, South Carolina 29601, Phone: (803) 233–3944. Secretary: Bert A. Winterbotton; Chairman: A. D. Asbury.

South Carolina Appalachian Advisory Commission, 18–A Thompson Street, Greenville, South Carolina 29601, Phone: (803) 239–4801. Executive Director: James S. Konduras; Chairman: R. E. Hughes.

Greenwood Metropolitan District Planning Board, Davis & Floyd, Inc., P.O. Box 428, Greenwood, South Carolina 29646, Phone: (803) OR9–8811; Chairman: P. Floyd.

Upper Savannah Economic Development District, City Hall, Greenwood, South Carolina 29646, Phone: (803) 678–336. Chairman: Clayto L. Roberts.

Pee Dee Economic Development District, Olanta, South Carolina 29114, Phone: (803) 396–4283. President: Martin Green.

Low County Resource Conservation and Development Project, P.O. Box 493, Walterboro, South Carolina 29488. Phone: (803) 541–2424 (City Supervisor). Project Coordinator: Murry Wood; Chairman: Charles Frazier.

Piedmont Resource Conservation Development Project, c/o Soil Conservation Service, Agricultural Building, York, South Carolina, Phone: (803) 684–4821. Executive Director: J. T. Brannen; Chairman: John Carriker.

Augusta SMSA Council of Local Governments (see Augusta, Georgia).

SOUTH DAKOTA

Siouxland Interstate Metropolitan Planning Commission (see Sioux City, Iowa).

TENNESSEE

Bristol, Tennessee-Virginia, Joint Planning Commission, 317 Professional Building, Bristol, Tennessee 37620, Phone: (615) 764–3135. Planning Director: Louis J. Smith; Chairman: C. B. Kearfott, Jr. 1966.

Chattanooga Area Regional Council of Governments, Municipal Building, Chattanooga, Tenessee 37402. Phone: (615) 267–6618, Ext. 224. Acting Director; Charles A. Rose; President: Mayor Ralph H. Kelley (Chattanooga) 1967.

Andersen County Regional Planning Commission, County Courthouse, Clinton, Tennessee 37716. Phone: (615) 457–2544. Chairman: Max Lawson.

First Tennessee-Virginia Economic Development District, East Tennessee State University, P.O. Box 2779, Johnson City, Tennessee 37601, Phone: (615) 928–0224. Executive Director: Jack B. Strickland; Chairman: Wesley L. Davis. 1966.

East Tennessee Economic Development District, 1810 Lake Avenue, Knoxville, Tennessee 37916, Phone: (615) 974–2386. Executive Director: John W. Anderson; Chairman: Judge Harvey Sproul (Loudon County). 1966.

Knoxville Area Council of Local Governments, 1810 Lake Avenue, Knoxville, Tennessee 37916. Phone: (615) 974–2386. Executive Director: John W. Anderson, Jr.; Chairman: L. D. Word. 1967.

Miss-Ark Tenn Council of Governments, 424 City Hall, Memphis, Tennessee 38103, Phone: 527–6611. Ext. 307. Executive Director: Neil G. Smith; Chairman: J. W. Ramsey. 1967.

Mid-Cumberland Council of Governments, 226 Capitol Boulevard Building, Nashville, Tennessee 37219. Phone: (615) 256–1552. Executive Director. Frank W. Ziegler; President: John S. Ridley. 1967.

Upper Duck Regional Planning Commission, Courthouse, Shelbyville, Tennessee 37160, Phone: (615) 684–7820. Executive Secretary: Claybourne Ross.

TEXAS

West Central Texas Council of Governments, P.O. Box 60, Abilene, Texas 79604, Phone: (915) 673–3781. Executive Director: Wendell H. Bedichek; President: Mayor Ralph N. Hooks (Abilene).

Amarillo Council of Governments (in process of formation), City Hall, Amarillo, Texas 79101. Contact: Arthur Storey.

North Central Texas Council of Governments, P.O. Box 888 (El Patio East), Arlington, Texas 76010, Phone: (817) 275–2651. Executive Director: William J. Pitstick; Chairman: Mayor C. P. Waggoner. 1966.

Austin-Travis County Organization For Regional Planning, % Department of Planning, P.O. Box 1088, Austin, Texas 78767, Phone: (512) GR 7–6511. Planning Coordinator: L. Wayne Golden; Chairman: David B. Barrow. 1966.

Brazos Valley Economic Development District, P.O. Box 3067, Bryan, Texas 77801, Phone: (713) 823–5970. Executive Director: Glenn J. Cook; Chairman: William Albert Miller. 1966.

Coastal Bend Economic Development District, 4225 South Port Avenue, Corpus Christi, Texas 78415, Phone: (512) 852–5651, Ext. 266. Executive Director: Jack L. Jones; President: Lew Borden.

Coastal Bend Regional Planning Commission, P.O. Box 2350, Corpus Christi, Texas 78403, Phone: (512) 854–2606, Ext. 266. Executive Director: Lon R. Starke; Chairman: Homer Dean. 1966.

Deep East Texas Economic Development District, 205 North Temple Drive, Diboll, Texas 75941, Phone: (713) 829–4216. Director: C. A. Neal Pickett; President: Honorable Charles Wilson, Senator.

El Paso Council of Governments, Suite 600, Electric Building, El Paso, Texas 79901, Phone: (915) 533–1659. Executive Director: Leslie G. Smyth; President: E. L. Hurt.

Houston-Galveston Area Council, 430 Lamar Avenue, Houston, Texas 77002, Phone: (713) 228–5351. Executive Director: Gerard H. Coleman; President: Edward Schreiber.

South Texas Council of Governments and Southwest Texas Regional Economic Development District, P.O. Box 1365, Laredo, Texas 78040, Phone: (512) 722–4641. Director: Emilio J. Gutierrez; Chairman: Roberto A. Benavides.

Lubbock Metropolitan Council of Governments, City Hall, Lubbock, Texas 78205, Phone: (806) 762–6411. Chairman: W. D. Rogers, Jr.; President: Rodrick L. Shaw. 1967.

Lower Rio Grande Valley Development Council, 1st National Bank Building, Suite 411, McAllen, Texas 78501, Phone: (512) 682–3481. Director: Robert A. Chandler; President: Mayor Paul G. Veale. 1967.

Midland-Odessa Council of Governments (in the process of formation), City Hall, Midland, Texas 79701; Mayor: Hank Avery.

Orange County Council of Governments, Orange County Courthouse, Orange, Texas 77630, Phone: (713) 883–4814. Chairman: Judge Charlie G. Grooms (Orange County).

Concho Valley Council of Governments, P.O. Box 1751, San Angelo, Texas 76901, Phone: (915) 655–9149. President: James Ridge.

Alamo Area Council of Governments, 422 Three A Life Building, San Antonio, Texas 78205, Phone: (512) 223–5564. Chairman: Blair Reeves; Executive Director: Robert E. Jamison. 1966.

Sherman-Dennison Metropolitan Area Regional Planning Commission, 600 North Highland Avenue, Sherman, Texas 75090, Phone: (214) 8992–8111. Chairman: Harry M. Shytles, Jr., MD.

Ark-Tex Council of Government, P.O. Box 1967, Texarkana, Texas 75501, Phone: (214) 792–8237. Director: Bob Shawn; Chairman: Neal Courtney. 1966.

North East Texas Economic Development District, Inc., P.O. Box 1967, Texarkana, Texas 75501, Phone: (214) 792–8237, Ext. 46. Director: Sylvin R. Lange; President: Cameron McElroy. 1966.

Smith County-Tyler Area Council of Governments, P.O. Box 2039, Tyler, Texas 75702, Phone: (214) LY 4–6651. Executive Director: A. A. Arnold.

Golden Crescent Council of Governments, P.O. Box 1758, Victoria, Texas 77901. President: C. A. Dickerson, Jr.; Director: Jerald A. Keith.

Central Texas Economic Development District, Connally Technical Institute, Waco, Texas 76705, Phone: (817) 799–3991. Director: Preston M. Hays; Chairman: J. Howard English.

Heart of Texas Council of Governments, 410 Professional Building, Waco, Texas 76703, Phone: (817) PL 2–1001. Director: William D. Ringo; Chairman: P. M. Johnston.

North Texas Regional Planning Commission, P.O. Box 1431, Wichita Falls, Texas 76307, Phone: (817) 322–5611, Ext. 276. Director: Edwin B. Daniels; Chairman: J. R. Mowery. 1966.

UTAH

Davis County Community Correlation Council, 211 Davis County Courthouse, Farmington, Utah 84025, Phone: (801) 867–2211, Ext. 62. Planning Director: Rodney F. Sutton; Chairman: Stanley M. Smoot.

Salt Lake County Council of Governments, 5461 South State Street, Murray, Utah 84107, Phone: (801) 262–2421. President: William E. Dunn. 1967.

Weber Area Council of Governments, 714 Municipal Building, Ogden, Utah 84401, Phone: (801) 399–8201. Planning Director: Graham F. Shirra; Chairman: Albert C. Bott. 1967.

Utah County Planning Council, Provo, Utah 84601, Phone: (801) 373–5510. Director: Mr. Ira Snell; Chairman: Floyd Harmer.

Central Utah Resource Development Association, Courthouse, Richfield, Utah 84701, Phone: (801) 896–5473. Executive Director: Rudolph S. Pace; President: J. H. Springer. 1966.

VERMONT

Franklin Regional Planning Commission, Bakersfield, Vermont 05441. Chairman: Mr. Valentine Bonk.

Windham Regional Planning Commission, P.O. Box 446, Brattleboro, Vermont 05301. Chairman: Jeremy Freeman.

Central Vermont Regional Planning Commission, Calias, Vermont 05648. Chairman: Jonathan Brownell.

Rutland Regional Planning Commission, Castleton, Vermont 05735. Chairman: John Mulligan.

Bennington Regional Planning Commission, P.O. Box 104, Manchester Center, Vermont 05255. Chairman: Henry Van Loon.

Addison Regional Planning Commission, P.O. Box 42, R.D. No. 1, Middlebury, Vermont 05753. Chairman: Arthur Gibb.

Chittenden Regional Planning Commission, 11 Wealthy Avenue, South Burlington, Vermont 05401. Chairman: William Kellner.

South Winsor Regional Planning Commission, 23 East Lane, Springfield, Vermont 05156. Chairman: John Howland.
Mad River Regional Planning Commission, P.O. Box 398, Waterbury, Vermont 05091. Chairman: L. Samuel Miller.

VIRGINIA

Lenowisco Regional Planning and Economic Development Commission, 413 Wood Avenue, Big Stone Gap. Virginia 24219, Phone: (703) 523-0121. Acting Director: William P. Sage, Jr.; Chairman: William B. Fraziar. 1965.
Rock Bridge Area Regional Planning Commission, 2138 Sycamore Avenue, Buena Vista, Virginia 24416, Phone: (703) 261-2500. Executive Director: Edward F. Parcha; Chairman: H. B. Chittum. 1966.
Eastern Shore of Virginia Regional Planning and Economic Development Commission, P.O. Box 38, Capeville, Virginia 23313, Phone: (703) 331-2857. Chairman: George F. Parsons.
Reservoirs Regional Planning Commission, P.O. Box 456, Chatham, Virginia 24531, Phone: (703) 432-8927. Executive Director: Walter J. Haberer; Chairman: Dan Callahan. 1961.
Jackson River Valley Regional Planning Commission, 441 E. Ridgeway Street, Clifton Forge, Virginia 24422, Phone: (703) 862-4264. Chairman: Norman C. Scott.
Rappahannock Regional Planning Commission, P.O. Box 863, Fredericksburg, Virginia 22401, Phone: (703) 373-2890. Executive Director: S. L. Lewis, Jr.; Planning Director: Francis S. Kenny.
Southside Regional Planning Commission, P.O. Box 555, Halifax, Virginia 24558, Phone: (703) 476-6433. Chairman: R. L. Lacy.
Brunswick County Regional Planning and Development Commission, 101 East Church Street, Lawrenceville, Virginia 23868, Phone: (703) 848-3200. Executive Director: Robert W. Baker; Chairman: E. G. Dawson. 1965.
Tri-County Regional Planning Commission, Madison, Virginia 22727, Phone: (703) 948-4321. Chairman: Joseph R. Drake.
Peninsula Regional Planning Commission, 118 Main Street, Newport News, Virginia 23601, Phone: (703) 596-3011. Secretary: Albert J. Stodghilt; Chairman: C. E. Johnson.
Southeastern Virginia Regional Planning Commission, 110 West Plume Street, Norfolk, Virginia 23510, Phone: (703) 622-5301. Executive Director: Robert F. Foeller; Chairman: James F. Hope. 1967.
Richmond Regional Planning Commission, 1015 East Main Street, Richmond, Virginia 23219, Phone: (703) 644-8587. Director: Richard B. Robertson; Chairman: B. Earl Dunn. 1956.
Roanoke Valley Regional Planning Commission. 4841 Williamson Road, N.W., Roanoke, Virginia 24012, Phone: (703) 362-3777. Planning Director: Thomas D. Smedley; Chairman: Edward B. Lassiter. 1956.
Middle Peninsula Regional Planning Commission, Urbanna, Virginia 23175, Phone: (703) PL8-7771. Executive Director: R. Atwell Taylor; Chairman: W. T. Bareford.
Northern Neck Regional Planning And Economic Development Commission, Weems, Virginia 22576. Executive Secretary: Charles Wiley; Chairman: Walther B. Fidler.
Bristol, Tennessee-Virginia, Joint Planning Commission (see Bristol, Tennessee).
First Tennessee-Virginia Economic Development District (see Johnson City, Tennessee).
Metropolitan Washington Council of Governments (see District of Columbia).

WASHINGTON

Skagit County Planning Commission, Rt. 3, Box 457, Anacortes, Washington 98221, Phone: (206) 336-3366. Executive Secretary: Thomas Thompson, Jr.; Chairman: Arnell Johnson.
Western Skagit Governmental Conference, P.O. Box 547, Anacortes, Washington 98221, Phone: (206) 293-2151. Recording Secretary: Gordon Logan; Chairman: Jesse W. Ford.
Whatcom County Regional Planning Council, Whatcom County Courthouse, Bellingham, Washington 98225, Phone: (206) 733-9431. Planner-in-charge: Harry R. Fulton; Chairman: Mayor James Van Andal. 1966.

Benton-Franklin Governmental Conference, 906 Fadwin Avenue, Richland, Washington 99352, Phone: (509) 943–9185. Director: Donald P. Morton; Chairman: Lawrence Scott.

Puget Sound Governmental Conference, Seattle Ferry Terminal, Pier 52, Seattle, Washington 98104, Phone: (206) 623–1200. Director: Robert R. McAbee; Chairman: A. L. Krekow. 1957.

Spokane Regional Planning Conference, Room 309, City Hall, Spokane, Washington 99201, Phone: (509) 624–4341, Ext. 334. Planning Director: Vaughn P. Call; Chairman: James A. Blodgett. 1966.

Pacific Northwest River Basins Commission, 110 East 13 Street, Vancouver, Washington 98660, Phone: (206) 862–5421, Ext. 606. Director of Planning: Ray E. Holmes; Chairman: Charles W. Hodde.

Chelan-Douglas Regional Planning Council, Chelan County Courthouse, Wenatchee, Washington 18801, Phone: (509) NO3–2101. Director: George Volker; Chairman: Homer J. Trefry.

Yakima County Conference of Governments, Room 417, Courthouse, Yakima, Washington 98901, Phone: (509) 248–2521, Ext. 279. Executive Secretary: Warren Q. Sutliff; Chairman: Cliff Onsgard.

Columbia Region Association Of Governments (see Portland, Oregon).

WEST VIRGINIA

Southern West Virginia Economic Development Corporation, P.O. Box 936, Bluefield, West Virginia 24701, Phone: (304) 327–8159. Executive Director: Thomas Gannaway; Chairman: Samuel Lajfer.

Kanawha County Planning And Zoning Commission, 410 Kanawha Boulevard East, Charleston, West Virginia 25301, Phone: (304) 346–6711. Chairman: Frederick L. Thomas.

Regional Development Authority, 301 Kanawha Banking and Trust Building, Charleston, West Virginia 25301, Phone: (304) 344–8197. Executive Secretary: Clarence E. Moran; President: Ernest J. Gilbert.

Hinton-Summers County Planning And Zoning Commission, 604 Greenbrier Road, Hinton, West Virginia 25951. President: Edward E. Eackles.

Cabell County Planning Commission, Cabell County Courthouse, Huntington, West Virginia 25701, Phone: (304) 525–7754 (County Clerk). Chairman: Lake Polan.

Kyova Interstate Planning Commission, P.O. Box 1833, Huntington, West Virginia, Phone: (304) 522–7366; Chairman: Michael R. Prestera.

Petersburg-Grant County Planning Commission, Petersburg, West Virginia 26847. Chairman: Roswell H. Alt.

Grafton-Taylor County Planning Commission, P.O. Box 47, Rosemont, West Virginia 26424. President: Brannon Riffle.

St. Mary's-Pleasants County Planning Commission, St. Mary's, West Virginia 26170. Chairman: Arthur G. Olds.

Steubanville-Weirton Area Transportation Study, 3023 Pennsylvania Avenue, Weirton, West Virginia 26062. Chairman: Norman D. Ferrari.

Weston-Lewis County Planning Commission, Weston, West Virginia 26452. Chairman: John G. Davisson.

Belmont, Ohio and Marshall Metropolitan Planning Committee, % Waterhouse and Rogers Realty, 148 East Cove, Wheeling, West Virginia, Phone: (304) 242–1440. Chairman: H. William Rogers.

Wheeling-Ohio County Planning Commission, c/o City-County Building, Wheeling, West Virginia 26003, Phone: (304) 232–0320. Chairman: Jack Waterhouse.

Mingo County Planning Commission, Office of the City Clerk, P.O. Box 1517, Williamson, West Virginia 25661. Chairman: G. D. Poole.

WISCONSIN

Fox Valley Council of Governments, 12th Floor, Zuelke Building, Appleton, Wisconsin 54911, Phone: (414) 739–6156. Executive Director: Eugene E. Franchett; Chairman: Laurel K. Heaney. 1967.

Northeastern Wisconsin Regional Planning Commission, 2111 North Richmond, Appleton, Wisconsin 54911, Phone: (414) 739–4250. Director: William E. Morris; Chairman: Gordon Buboll. 1962.

Green Bay-Brown County Regional Planning Commission, 100 North Jefferson Street, Green Bay, Wisconsin 54301, Phone: (414) 437–7611. Planning Director: Ralph M. Bergman; Chairman: Nathain Malcove. 1962.

Mississippi River Regional Planning Commission, Courthouse, Room 101, La Crosse, Wisconsin 54650, Phone: (608) 784-4888. Chairman: John Thomas; Secretary: Lawrence Weber. 1964.

Intergovernmental Cooperation Council, c/o Village of Whitefish Bay, 801 East Lexington Boulevard, Milwaukee, Wisconsin 53217, Phone: (414) 962-6690. Secretary: Thomas G. Drought; Chairman: Henry F. Mixter.

Southeastern Wisconsin Regional Planning Commission, 916 North East Avenue, Waukesha, Wisconsin. Phone: (414) 542-8083. Director: Kurt W. Bauer; Chairman: George C. Berteau. 1960.

COUNCILS CONFIRMED AFTER THE DIRECTORY WAS SENT TO PRINTER

Arkansas Valley Association of Local Governments, City Hall, Salida, Colorado 81201. Chairman: Mayor Ed Touber (Salida).

Dukane Valley Council, Northern Illinois Gas Company, Aurora, Illinois 60507. Chairman: Marvin Chandler.

Centre Regional Council of Governments, c/o Frederick E. Fisher, Borough Manager, State College, Pennsylvania 16801.

Golden Crescent Council of Governments, City Hall, Victoria, Texas 77901. Executive Secretary: Gerald A. Keith.

Lewis Regional Planning Council, County Courthouse, Chehalis, Washington 98532.

APPENDIX L

List of Public Laws Reflecting Use of Systems Technology

As the U.S. Congress has been confronted with the need to pass legislation concerning the various public problem areas, these bills and resolutions have begun to include some mention of analytical planning processes, the encouragement of new technologies, automatic data processing, and terminology designed to promote some (often vague) facet of the systems approach. Specific reference to and discussion of a number of these laws is contained in section V.D. of this report.

The great majority of the public laws listed below was passed during the 88th, 89th, and 90th Congresses, but in certain instances, earlier legislation which received favorable action has been included.

The selection parameters were two: laws oriented to a specific social or community problem; and an indication that the authors were attempting to utilize innovative thinking, techniques, and devices in the improvement of the problem. The problem areas where public laws of this latter characteristic were passed include: urban planning, environmental pollution, transportation, housing, law enforcement, health services, education, recreation, manpower, and the creation of new agencies.

The public laws are arranged in chronological order. The synopses were prepared originally for inclusion in the *Digest of Public General Bills*, published on a biweekly basis (during congressional sessions) by the Legislative Reference Service of the Library of Congress.

PUBLIC LAW 81–171—HOUSING ACT OF 1949

S. 1070. Introduced by Mr. Ellender and others on February 25, 1949

Housing Act of 1949.—Establishes a national housing policy based on: (1) Encouragement of private enterprise to serve as large a part of the total need as possible, and (2) governmental assistance, where feasible; to enable private enterprise to serve still more of the total need; to eliminate substandard housing through clearance of slums; to provide decent housing for low-income urban and rural nonfarm families and decent farm dwellings for farm owners lacking means to do so for themselves. Contains specific congressional charge to the Housing and Home Finance Agency and others to encourage and assist: (1) The production of housing of sound standards of design and construction, at reduced costs, (2) the use of new and standardized methods, materials, and equipment, (3) the development of well-planned communities, and (4) the stabilization of the housing industry at a high annual volume of residential construction.

Title I.—Provides for temporary and long-term Federal loans and capital grants to local public agencies to assist communities in eliminating slums and redeveloping project areas by private enterprise.

To finance these loans the Administrator may issue obligations of $25 million for the first year, $225 million for the second year, and $250 million for each of the next 3 years, total authorization not to exceed $1 billion. The Administrator may contract to make capital grants totalling $100 million per year for 5 years, total authorization not to exceed $500 million. With respect to a project on open unplatted land, no grants shall be made, but long-term loans may be made for acquisition of land (to be repaid in 40 years) and temporary loans may be made for public buildings and facilities necessary to support the new uses of the land (to be repaid in 10 years). The aggregate of capital grants to a local public agency shall not exceed two-thirds of the aggregate costs of the projects, the remaining third to be covered by local grants-in-aid.

Title II.—Authorizes Federal contributions for low-rent housing of $85 million for the first year, additional $55 million in each of the next 3 years and $58 million for the fifth year (total authorization not to exceed $308 million per year) to localities which assume community responsibility and initiative in estimating their own needs and demonstrating that their needs cannot be fully met by private enterprise without Federal aid, 10 percent of which, for the first 3 years, shall be available only for rural nonfarm projects. Provides for preference to displaced families and to World War II veterans. Federal aid shall be given only where there is a gap of at least 20 percent between the upper rental limits for admission to the proposed low-rent housing and the lowest rents at which private enterprise is providing a substantial supply of decent housing. Provides higher room cost limitations under the U.S. Housing Act of $1,750 a room (now $1,000 and $1,250 depending on size of city) and allows an additional limitation of $750 under certain emergency conditions. Authorizes Federal aid for the rehabilitation of existing dwellings. Makes special provision for large families of low income. Authorizes construction of 135,000 dwelling units per year for 6 years beginning July 2, 1949, not to exceed 810,000 units. Requires local contributions equal to 20 percent of the Federal contributions, to be made either in cash or tax exemptions of real and personal property [amending U.S.C. 42: 1401–1430].

Title III.—Directs the Housing and Home Finance Administrator to undertake a program of research (where possible, in cooperation with industry, labor, and other Federal and local agencies) to improve building methods, etc., and to encourage localities to make studies of their own housing needs [amending Public Law 901 § 301, 80th Cong.].

Title IV.—Provides for Federal aid for farm housing through loans by the Secretary of Agriculture, through the Farmers Home Administration, to farm owners to provide decent dwellings for themselves, their tenants, share croppers or laborers. Authorizes secretary to finance such loans by issuing obligations (upon security of obligations of borrowers) not to exceed $25 million the first year, and additional $50 million, $75 million, and $100 million during the second, third, and fourth year, respectively. Authorizes annual contributions of $500,000 with additional contributions not to exceed an aggregate of $1 million, $1,500,000, and $2 million per annum for the second, third, and fourth years, respectively. Authorizes $1 million for grants for minor improvements of farm buildings, to remove hazards, which amount shall be increased by further amounts of $2,500,000, $4 million, and $5 million in the second, third, and fourth years, respectively, in addition to the loan and contribution program. Contains preference for veterans of all wars and families of deceased servicemen.

Title V.—Authorizes Administrator to appoint advisory committees and a Deputy, and directs inclusion of Secretary of Labor in the membership of the National Housing Council in the Housing and Home Finance Agency. Directs the Director of the Census to take a census of housing in each State in 1950 and decennially thereafter, as a part of the population inquiry of the decennial census. Makes necessary technical amendments to and repeals of other statutes.

PUBLIC LAW 85–767—REVISES, CODIFIES, AND ENACTS INTO LAW, TITLE 23 OF THE UNITED STATES CODE, ENTITLED "HIGHWAYS"

H.R. 12776. Introduced by Mr. Fallon on June 3, 1958

Revises, codifies, and enacts into law, title 23 of the United States Code, entitled "Highways".

Continues the three Federal-aid system (the primary, secondary, and interstate) and sets forth the procedures for the designation of roads to be included in each.

Authorizes the Secretary of Commerce to deduct up to 3¾ percent of the sums authorized for apportionment to a State for the administration of the provisions of law pertaining thereto.

Sets forth the method for apportioning funds among the States.

States desiring to participate in benefits may do so by submitting a program of proposed projects for the utilization of funds apportioned.

Permits the Secretary to acquire lands on request of a State required in connection with the Interstate System. Sets forth certain standards which must be met before State plans may be approved.

Limits the Federal share payable for any project on the primary or secondary system to 50 percent of the cost (except under certain conditions).

Requires any State receiving funds for the Interstate System to limit by law the weight and width of vehicles using such system.

Declares it to be the national policy to regulate the erection of advertising signs within 660 feet of the right-of-way of the Interstate System.

Provides for the authorization and apportionment of funds for forest highways, forest development roads, etc.

Sets forth certain general provisions concerning the administration of the Act. Repeals numerous Acts relating to Highways.

PUBLIC LAW 86–380—ADVISORY COMMISSION ON INTERGOVERNMENTAL RELATIONS

H.R. 6904. Introduced by Mr. Fountain on May 6, 1959

Establishes an Advisory Commission on Intergovernmental Relations composed of 24 members, appointed from various segments of the Federal, State, and local governments and private citizens. Sets forth the function of the Commission to: (1) Bring together representatives of the Federal, State, and local governments to consider common problems, (2) provide a forum to discuss programs requiring intergovernmental cooperation, (3) give attention to the conditions and controls involved in the administration of Federal grant programs, (4) make technical assistance available for the review of proposed legislation, (5) encourage discussion and study of public problems likely to require intergovernmental cooperation, and (6) recommend the most desirable allocation of governmental functions and responsibilities among the several levels of government.

PUBLIC LAW 87–415—PROVIDES FOR A PROGRAM OF OCCUPATIONAL TRAINING AND RETRAINING FOR THE NATION'S LABOR FORCE

S. 1991. Introduced by Mr. Clark and others on May 29, 1961

Provides for a program of occupational training and retraining of the Nation's labor force. Directs the Department of Labor to make studies and then develop training and skill development programs utilizing the resources of industry, labor, educational institutions, and State and local agencies. Grants weekly retraining allowances equal to unemployment compensation and relocation allowances to persons offered jobs in other areas. Provides for on and off the job training and vocational training for the unemployed.

PUBLIC LAW 88–365—URBAN MASS TRANSPORTATION ACT

S. 6. Introduced by Mr. Williams and others on January 14, 1963

Establishes a long-range Federal program to assist State and local bodies and mass transportation companies to provide mass transportation facilities necessary for the orderly growth and development of urban communities. Authorizes Federal grants for mass transportation facilities and equipment on the basis of a comprehensive plan for the development of the urban area. Permits grants of up to two-thirds of the cost which cannot be financed from estimated revenues, with the other one-third to be provided by local grants. Provides for Federal loans for projects where such loans cannot be obtained privately on reasonable terms.

Establishes a 3-year emergency program with less strict planning requirements but with the Federal share limited to one-half of the unfinanced cost. Makes the additional one-sixth of the amount available if full planning requirements are met within 3 years.

Provides for a $500 million contract authorization over a 3-year period.

Replaces the present $25 million demonstration grant authority with a research, development, and demonstration program and makes the $25 million authorization available for the broader program. Permits the use of an additional $30 million of the new authority for the broader program.

Requires an adequate relocation program for families displaced by projects.

Provides for the coordination of Federal assistance for highways and mass transportation facilities. Permits States to enter into interstate compacts for comprehensive planning for the growth of urban areas and for carrying out mass transportation programs.

Requires projects receiving aid to be carried out in accordance with the Davis-Bacon Act (prevailing wage rate). Requires consideration of air pollution control before facilities and equipment may be acquired.

PUBLIC LAW 88-443—HOSPITAL AND MEDICAL FACILITIES AMENDMENTS

H.R. 10041. Introduced by Mr. Harris on February 20, 1964

Hospital and Medical Facilities Amendments—Authorizes $5 million for 1965 and $10 million for each of the next 4 fiscal years to enable the Surgeon General to make grants to public or nonprofit private agencies for special projects in the area-wide planning of health and related facilities. Limits grants to 66⅔ percent of the cost of the project.

Authorizes for fiscal 1965 and each of the next 4 fiscal years $70 million for grants for construction of public or other nonprofit facilities for long-term care; $20 million for grants for construction of public or other diagnostic or treatment centers; $10 million for grants for construction of public or other nonprofit rehabilitation facilities; for grants for construction of public or other nonprofit hospitals and public health centers and for grants for modernization of such facilities $150 million for 1965, $160 million for 1966, $170 million for 1967, and $180 million for the next 2 fiscal years.

Sets forth State allotments and requirements for State plans.

Authorizes Surgeon General to make a loan of funds to assist projects for construction or modernization of hospital and other medical facilities.

Authorizes Surgeon General to insure any mortgage for the purpose of financing construction or modernization of a private nonprofit hospital or other medical facility. Establishes a Medical Facilities Mortgage Insurance Fund.

Provides that the Surgeon General shall consult with a Federal Hospital Council consisting of 12 members appointed by the Secretary of Health, Education, and Welfare.

Authorizes the Surgeon General to conduct studies and demonstrations relating to coordinated use of hospital facilities.

PUBLIC LAW 88-444—H.R. 11611. INTRODUCED BY MR. HOLLAND ON JUNE 16, 1964

Establishes the National Commission on Technology, Automation and Economic Progress composed of 14 members to: (1) Identify and assess the past effects and the current and prospective role and pace of technological change; (2) identify and describe the impact of technological and economic change on production and employment which are likely to occur during the next 10 years; (3) define those areas of unmet community and human needs toward which application of new technologies might be devoted; (4) assess the most effective means for channeling new technologies into promising directions, and (5) recommend administrative and legislative steps to be taken by the Federal Government to support and promote technological change in the interest of continued economic growth.

Establishes a Federal Interagency Committee to advise the Commission and maintain liaison with the resources of certain Departments and agencies.

Public Law 88-452—Economic Opportunity Act

S. 2642. Introduced by Mr. McNamara and Others on March 16, 1964

Economic Opportunity Act.—Provides for mobilization of human and financial resources of the Nation to combat poverty in the United States.

Title I, Youth Programs.—Establishes a Job Corps within the Office of Economic Opportunity and authorizes the Director of the Office to enter into agreements with any Federal, State, or local agency or private organization for the provision of facilities and service, to provide education and vocational training to enrollees in the Corps, provide for programs of work experience for enrollees, and to establish certain standards and regulations for the program.

Provides that the Corps shall be composed of male individuals between 16 and 21 years of age. Provides that participation in the Corps shall not relieve enrollees of draft obligations. Provides for allowances and maintenance of enrollees and for a readjustment allowance not to exceed $50 for each month of satisfactory participation.

Establishes work-training programs for unemployed youths, male and female, in cooperation with State and local agencies and authorizes the Director to enter into agreements providing for payment of part or all of the cost of a program which meets certain requirements.

Provides for work-study programs to promote the part-time employment of students in institutions of higher education who are from low-income families and are in need of earnings to pursue their education.

Title II, Urban and Rural Community Action Programs.—Provides grants for financial and technical assistance, research, training, and demonstrations to provide stimulation and incentive for urban and rural communities to mobilize public and private resources to combat poverty through community action programs. Provides for participation of State agencies in such programs.

Title III, Special Programs to Combat Poverty in Rural Areas.—Authorizes grants not to exceed $1,500 and loans not to exceed $2,500 to low-income rural families to help effectuate an increase in their income. Authorizes cooperation with public or private nonprofit corporations having as their object the improvement of the productivity and income of low-income farmers and authorizes loans to local cooperative associations furnishing services and facilities to such families.

Title IV, Employment and Investment Incentives.—Provides loans, participations, and guarantees to private borrowers when it is determined that such financial assistance will result in stable employment for persons not already employed by the borrower, and who will be recruited from among long-term unemployed persons. Authorizes loans to small business where such loans will assist in providing employment and investment incentives.

Title V, Family Unity Through Jobs.—Provides payments for experimental, pilot, and demonstration projects to help unemployed fathers and other members of needy families secure and retain employment.

Title VI, Establishes in the Executive Office of the President the Office of Economic Opportunity Headed by a Director.—Establishes an Economic Opportunity Council to consult with and advise the Director and National Advisory Council to review the operations of the Office. Authorizes $962,500,000 for the purposes of this act.

PUBLIC LAW 88-560—HOUSING ACT OF 1964

S. 3049. Introduced by Mr. Sparkman on July 29, 1964

Title I, Mortgage Insurance Programs.—Permits the Federal Housing Commissioner to provide additional relief for home mortgagors in default due to circumstances beyond their control.

Enables the Commissioner to extend aid to distressed homeowners who, after relying upon Federal Housing Administration construction standards and inspections, find structural or other major defects in their properties purchased with FHA-insured loans.

Removes the requirement that the Commissioner find that the property for which a home improvement loan outside of urban renewal areas is sought to be "economically sound" and substitutes requirement that it be an "acceptable risk"; authorizes payment of lender insurance claims in cash.

Permits the insurance of morgages on nursing home facilities sponsored by private nonprofit corporations or associations; requires the Surgeon General to certify that a State agency has certified as to the need and minimum standard of licensing and operation of nursing home facilities.

Title II, Urban Renewal and Growth: Code Enforcement.—Requires that beginning 3 years after enactment no workable program will be certified or recertified unless the locality has had in effect for at least 6 months a minimum standards housing code and the locality is carrying out an effective program of enforcement to achieve compliance. Permits the cost of enforcement activities to be included as a part of eligible project costs.

Permits the Housing and Home Finance Administrator to enter into a single loan contract with a local public agency to provide the temporary financing necessary to carry out any two or more urban renewal projects being undertaken by it at any one time.

Increases the aggregation amount of obligational authority for urban renewal grants from $4 to $4.85 billion.

Requires local public agencies to assure that there will be available adequate housing for individuals as well as families displaced from an urban renewal area.

Permits the disposal of urban renewal land at a special reduced price for use in the construction of housing for low- or moderate-income individuals as well as families.

Requires the Administrator to set a single interest rate (based on the "going Federal rate" on the date of the authorization of the contract) for each urban renewal contract for loans or advances authorized after the effective date hereof.

Expands existing urban renewal powers with respect to projects involving the use of air rights to provide housing, and related facilities and uses, for low- or moderate-income families.

Authorizes additional federally reimbursed relocation payments for low- or moderate-income families and elderly individuals, and to small independent businesses displaced from urban renewal areas.

Increases from $5 to $10 million the grant limitation for the urban renewal demonstration grant program and makes various amendments to the program. Makes all counties, regardless of population, eligible for comprehensive planning assistance and permits Indian reservations to be eligible. Increases by $30 million the authorization for planning grants.

Permits nonresidential projects of urban renewal in the District of Columbia.

Title III, Housing for Low-Income Families.—Makes single low-income displaced persons eligible for admission to low-rent housing regardless of age or disability status.

Authorizes a special subsidy of up to $120 per dwelling per year where a dwelling unit is occupied by an eligible displaced family.

Permits acceptance of certifications by local governing bodies that they have complied with the equivalent elimination (of substandard dwellings) requirements.

Increases the limit on annual contributions by $36 million. Establishes for the low-rent program the same basic requirements for relocating displaced families and individuals as those which are applicable under the urban renewal program. Provides for relocation payments to families, individuals, businesses and nonprofit organizations displaced from low-rent housing project sites.

Increases from $5 to $10 million the amount authorized for grants for low-income housing demonstration programs.

Title IV, Community Facilities.—Provides that instrumentalities of one or more States and instrumentalities of municipalities or other political subdivisions in one or more States are not precluded from receiving assistance under the public facility loans program if otherwise eligible.

Makes any community of 150,000 eligible for a public facility loan if it has been designated as a redevelopment area and permits assistance to any public agency or instrumentality serving one or more municipalities, without regard to the aggregate population of the communities which it is serving, so long as each is within the existing population limits of the program.

Requires all repayments and other receipts after June 30, 1964, in connection with advances made under advance planning programs to be made into a revolving fund established hereby. Authorizes an additionl $20 million to the fund as may be necessary to carry out the advance planning program.

Title V, Mortgage insurance Procedural Amendments.—Places claims by lenders for insurance payments (on FHA-insured title I property improvement loans) certified for payment prior to December 31, 1957, on the same basis as is now provided for those certified for payment subsequent to that date. Authorizes optional cash payment of insurance benefits on any claim.

Increases the dollar limit on the amount of mortgage which can be insured under the program for low-cost housing in outlying areas from $9,000 to $11,000.

Provides for certain technical changes in FHA insurance benefits and simplifies payment procedures. Eliminates mandatory acquisition or foreclosure within 1 year of multifamily project in default. Establishes new room count limits for FHA multifamily projects.

Permits individual elderly persons to occupy housing for low- and moderate-income families. Permits servicemen to obtain housing that is being provided under FHA mortgage insurance program for homes for low- or moderate-income families.

Authorizes FHA to insure mortgage loans made by private lenders to the purchasers of the properties without regard to any of the limitations or requirements that would otherwise apply.

Broadens experimental housing program by providing insurance if the housing involves the utilization or testing of new design, materials, or similar items.

Amends and broadens mortgage insurance program for condominiums.

Title VI, Participations in FNMA Pool of Mortgages.—Vests fiduciary powers in the Federal National Mortgage Association in order to facilitate the financing of its own and other mortgages through the sale to private investors of beneficial interests or "participations" in such mortgages. Extends authority to the pooling of Veterans' Administration mortgages.

Title VII, Rural Housing.—Extends the rural housing programs until September 30, 1965. Authorizes an additional $150 million for loans under title V rural housing direct loan program.

Redefines the term "domestic farm labor" to make immigrant farm laborers eligible to occupy insured housing.

Authorizes grants to States or political subdivisions, or public or private nonprofit organizations, to assist in providing housing and related facilities for domestic farm labor.

Title VIII, Miscellaneous.—Permits FNMA, under its secondary market operations, to purchase, sell, and deal otherwise in participations in insured and guaranteed mortgages.

Increases from $50 to $75 million the amount for grants under the open-space program. Increases amounts for direct loan elderly housing from $275 to $350 million.

Permits college housing loans to be made to eligible nonprofit corporations without requiring the educational institution to cosign the note for college housing loan whenever State law prevents the educational institution from acting as cosigner.

Permits the Secretary of Defense to acquire FHA insured rental housing projects.

Authorizes $500,000 annually for a 3-year period beginning July 1, 1964 for fellowships for city planning and urban studies.

Increases the lending area to 100 miles from a savings and loan associations home office.

Raises the limitation on investments by a Federal savings and loan association in first liens on improved real estate without regard to limitation of the Home Owners' Loan Act and investments in participating interest in first liens on real property of the types on which it may make loans under its general lending authority and without regard to area restrictions. Permits a Federal savings and loan association to invest not more than 5 percent of its assets either in direct investments on real property located within urban renewal areas or in obligations secured by a first lien on real property so located. Permits investment not exceeding 2 percent of their assets in the stock of corporations organized under the laws of the State in which the association's home office is located if the entire capital stock of the corporation is available for purchase only by savings and loan associations of that State and by federally chartered associations having their home offices in that State.

Provides that no home mortgage shall be accepted as collateral security for an advance by a Federal home loan bank if, when the advance is made, the home mortgage loan secured by it has more than 30 years to run to maturity.

Permits national banks to invest in real estate loans up to 80 percent of the appraised value of the real estate and for a term no longer than 30 years.

PUBLIC LAW 88–578—LAND AND WATER CONSERVATION FUND ACT

H.R. 3846. Introduced by Mr. Aspinall on February 18, 1963

Establishes a program to improve outdoor recreation facilities by providing Federal assistance to States, and by Federal acquisition of certain land and water areas. Sets up a special fund with revenues from (1) entrance and user fees of recreation areas, (2) surplus property sales, and (3) motorboat fuels tax. Such moneys are to be allocated 60 percent for State purposes, 40 percent for Federal purposes.

Starting the third year in which the fund is in operation, $60 million is authorized each fiscal year for 8 years.

Provides for apportionment of available sums among the States. Requires States to pay 50 percent of the costs of planning projects, and 70 percent of acquisition or development costs. Requires a comprehensive statewide outdoor recreation plan. The Secretary of the Interior must approve State projects prior to payment.

Authorizes the President to allocate funds for Federal purposes under this act, when such acquisition is authorized by law.

PUBLIC LAW 89–4—APPALACHIAN REGIONAL DEVELOPMENT ACT OF 1965

S. 3. Introduced by Mr. Randolph and others on January 6, 1965

Title I, Appalachian Regional Commission.—Establishes an Appalachian Regional Commission composed of one Federal member appointed by the President, with the consent of the Senate, and one member from each State in the Appalachian region. Provides for the Commission to develop comprehensive and coordinated plans and programs to promote economic development; to conduct and sponsor investigations, research, and studies; to formulate interstate compacts and other forms of interstate cooperation; to encourage the formation of local development districts; to encourage private investment; to serve as a coordinating unit; and to provide a forum for consideration of regional problems.

Title II, Special Appalachian Programs.—Authorizes the Secretary of Commerce to assist in the construction of development highways and access roads to link with interstate highways in the Appalachian area.

Authorizes the Secretary of Health, Education, and Welfare to make grants for the construction and operation of multicounty demonstration health facilities.

Authorizes the Secretary of Agriculture to make grants to assist in the improvement and development of pastureland for livestock in the region, and to provide technical assistance in the organization and operation of timber development organizations. Authorizes financial assistance to seal and fill abandoned areas, to extinguish underground fires, and to expand wildlife restoration projects.

Authorizes preparation of a comprehensive plan for the development and utilization of water resources in the area. Authorizes the Secretary of Health, Education, and Welfare to make grants for construction of school facilities under the provisions of the Vocational Education Act without regard to ceilings or allotments among the States.

Authorizes additional grants for sewage treatment works. Permits supplements to grants-in-aid programs.

Title III, Administration.—Provides for the designation of "local development districts" and authorizes grants for the administrative expenses (up to 75 percent) of such districts.

Title IV, Appropriations and Miscellaneous Provisions.—Authorizes $237,200,000 through fiscal 1966 to carry out the act. Places employees on projects receiving aid under the prevailing wage and hour requirements (Davis-Bacon Act).

PUBLIC LAW 89-10—ELEMENTARY AND SECONDARY EDUCATION ACT

Introduced by Mr. Perkins on January 12, 1965

Title I, Financial Assistance to Local Educational Agencies for the Education of Children of Low-Income Families.—Declares it to be the purpose of Congress to provide Federal financial assistance to educational agencies in areas with concentrations of low-income families to improve the educational facilities of the area.

Authorizes the Commissioner of Education to make payments to State educational agencies for basic and special incentive grants to local agencies from July 1, 1966, to June 30, 1968.

Provides a formula for determining the amount of Federal assistance to a State. This formula to be based on a computation using the Federal percentage of the average per pupil expenditure multiplied by the number of children in low-income families in the State or area in which the grant is to be used but the Federal contribution to an agency is not to exceed 30 percent of that agency's budget for the fiscal year ending June 30, 1966.

Establishes a criteria for determining the eligibility of a local educational agency. Requires that there be in the school district served by the agency at least 100 children in low-income families or that more than 3 percent of the total number of children served by the agency come from such families.

Defines "Federal percentage" as 50 percent and "low-income" as $2,000 or less. In determining the number of children in low-income families, only those from 5 to 17 are to be counted.

Provides for special incentive grants to agencies with amounts to be determined by previous agency expenditures.

Requires the local agency to apply to the State for such grants and provides standards for determining eligibility.

Requires State educational agencies to provide the Commissioner with satisfactory assurance that proper disbursement will be made;

that this act will be complied with; and that annual reports will be made to the Commissioner.

Provides standards and methods of payments by the Commissioner to the States.

Requires that laborers and mechanics on construction projects receiving assistance under this title be paid in accordance with the standards set forth in the Davis-Bacon Act (prevailing wage and hour rate).

Provides for the withholding of funds by the Commissioner in the event of noncompliance with the requirements above. Permits judicial review of the Commissioner's holdings.

Requires a report from the Commissioner to the Secretary of Health, Education, and Welfare on the operation of this title.

Title II, School Library Resources and Instructional Materials.— Authorizes $100 million for the fiscal year 1966 for use in the acquisition of school library resources and instructional material for use in public and private nonprofit elementary and secondary schools in the States. Sets forth standards for determining allotments to be made to the States under this title.

Requires those States desirous of allotments to submit to the Commissioner plans which: Designate an agency to administer the plan; set up a program for expenditures of the funds; set forth criteria for allocating and selecting the materials; set forth procedures to insure that the Federal assistance will not supplant but supplement State funds; provide for proper disbursement and accounting procedures to assure proper disbursement; and provide for reports deemed necessary by the Commissioner. Requires the Commissioner to approve any State plan which complies with the above provisions.

Where there is no State agency which provides library and instructional materials the Commissioner is authorized to provide for an equitable distribution of the State's funds and to charge the cost of distribution out of that State's allotment.

Requires the Commissioner to give a State opportunity for a hearing before finally disapproving a State plan and provides for judicial review of the Commissioner's final action.

Title III, Supplementary Educational Centers and Services.— Authorizes the appropriation of $100 million for the fiscal year 1966 and such sum for 3 succeeding years for supplementary educational centers and services.

Establishes a formula for apportionment among the States. Provides for reapportionment of unused or allotments found to be unnecessary by the Commissioner to other States.

Sets forth the uses to which the Federal funds may be put to include construction or acquisition of equipment to expand elementary and secondary education programs, guidance counseling and programs to encourage children to reenter or remain in school, comprehensive academic services, developing and conducting exemplary educational programs, specialized instruction and equipment for students in advanced courses, educational radio and television, and special educational and related services for rural areas.

Sets up requirements for receiving grants and standards for the Commissioner to follow in making grants.

Provides that payments made under this title may be made in installments, in advance, or by way of reimbursement.

Establishes an eight member Advisory Committee on Supplementary Educational Centers and Services to advise the Commissioner on matters relating to this title.

Provides for the recovery of payments made under this title if within 20 years the owner of the facilities assisted by Federal funds ceases to be a State or local agency or the facilities cease to be used for the purposes intended without the Commissioner's assent.

Requires that labor standards at construction sites be in accordance with the Davis-Bacon Act.

Title IV, Cooperative Research Act.—Educational Research and Training. Authorizes the Commissioner to make grants for research, surveys, and demonstrations in education and to make findings available to interested parties. Sets forth the standards required for issuance of a grant. Provides for transfers of funds to other Federal agencies engaged in such work.

Requires the Commissioner to submit annual progress reports on such research to Congress.

Authorizes $100 million for use over a 5-year period for the construction of regional facilities for research and related purposes. Sets up standards to be used by the Commissioner in using such funds. Requires that the wage rate at such construction sites as are approved by the Commissioner be in accordance with the Davis-Bacon Act as amended.

Title V, Grants to Strengthen State Departments of Education.—Authorizes $10 million for such purpose as the name of the title implies for fiscal year 1966 and such funds for the succeeding 4 fiscal years as may be necessary.

Sets forth a formula for apportionment of 85 percent of such appropriated funds among the States and reserves 15 percent for grants for special projects.

Grants are to be made upon approval by the Commissioner of a State application which must set forth the use to be made of such funds. The application is required to show that the grant will be used for such purposes as: Educational planning on a statewide basis; collection, analysis, and dissemination of educational data; programs fostering or aiding educational research programs and projects; publication and distribution of curricular materials; improvement of teacher preparation; education financing; pupil achievement tests; training of educational personnel; and providing consultative and technical assistance and related service to local educational agencies and schools.

Restricts the amount of Federal participation in such programs to not more than 66 percent and not less than 50 percent of the total to be expended on such programs.

Sets forth the standards which the Commissioner must use in approving a State application for grants authorized under this title.

Provides for the interchange of personnel between the States and the Office of Education. Sets forth standards for the treatment of exchange employees as to their status within the Office and the treatment to be given to such employees in regard to such matters as pay, leave, retirement, and insurance.

Requires the Commissioner to give reasonable time for a hearing before finally disapproving a State plan and provides for judicial review of the Commissioner's decision regarding approval or disapproval of a State plan.

Authorizes the establishment of a 12 member Advisory Council on State Departments of Education to review the administration of grants made under this title and to make recommendations for improvement of such administration.

Authorizes the Secretary to engage technical and nontechnical assistance for the Council and requires the Council to report on its findings and recommendations to the Secretary who will in turn transmit such report to the President and Congress.

Title VI, General Provisions.—Authorizes the Commissioner to appoint, with the approval of the Secretary, advisory committees to advise and consult with him. Authorizes the Commissioner to delegate authority to employees of the Office and to utilize the facilities of other Federal agencies. Prohibits any manner of Federal control education.

PUBLIC LAW 89–15—MANPOWER ACT OF 1965

S. 974. Introduced by Mr. Clark and others on February 3, 1965

Empowers the Secretary of Labor to conduct studies in evaluation of and research in the unemployment problem through the use of grants or contracts, to establish programs financed by grants to alleviate unemployment and facilitate reemployment and to provide for studies of on-the-job training.

Requires the Secretary to conduct and engage in pilot projects to increase the mobility of unemployed workers and authorizes the use of grants and loans for such purposes.

Authorizes $500,000 over a 2-year period to assist in the placement of persons seeking employment.

Provides for the payment of training allowances for a period of up to 104 weeks. Such payments not to exceed $10 more than the average unemployment compensation rate. Provides for increases in payments in case of dependents for a supplemental training allowance to certain persons.

Provides for payment of subsistence expenses where the trainee does not live within community distance of his training facilities.

Limits the availability of training allowances to those persons with 2 years of gainful employment and not in a family whose head is employed.

Limits to 25 percent the number of persons under 22 who may receive such allowances.

Limits Federal payments for training allowances to 90 percent of total after 1966.

Provides for supplementary programs in redevelopment areas.

PUBLIC LAW 89–42.—AUTHORIZING FUNDS FOR COMPREHENSIVE RIVER BASIN WORKS FOR FLOOD CONTROL AND NAVIGATION PURPOSES

H.R. 7655. Introduced by Mr. Jones of Alabama on March 25, 1965

Authorizes additional funds not to exceed $263 million for prosecution of projects in certain comprehensive river basin plans for flood control and navigation.

PUBLIC LAW 89-57—FEDERAL AID HIGHWAY ACT OF 1966

S. 3155. Introduced by Mr. Randolph on March 29, 1966

Authorizes appropriations for the construction of Federal-aid-highways and the Interstate Highway System.

PUBLIC LAW 89-80—WATER RESOURCES PLANNING ACT

S. 21. Introduced by Mr. Anderson and others on January 5, 1965

Title. I.—Establishes a Water Resources Council composed of the Secretaries of Interior; Agriculture; Army; and Health, Education, and Welfare to maintain a continuing study of the adequacy of water supplies and to establish comprehensive regional or river projects. Sets forth standards to be used in evaluating such plans and requires that they be submitted, with recommendations, to the President for his review and transmittal to Congress.

Title II.—Authorizes the President, upon the request of the Governor of one or more of the affected States, to create a river basin water resources commission to serve as the principal agency for the development of water and related land resources in its region. Sets forth provisions for the appointment of the commission, its duties, and powers.

Title III.—Provides for financial assistance to the State for the comprehensive planning required by this act. Authorizes $5 million for the fiscal year beginning after enactment hereof and $5 million for each of the 9 succeeding fiscal years. Sets forth a formula for allocating these funds to the States. Sets forth standards which State programs must meet in order to be entitled to such funds, and the formula for determining the Federal share in the State programs.

PUBLIC LAW 89-117—HOUSING AND URBAN DEVELOPMENT ACT OF 1965

H.R. 7984. Introduced by Mr. Patman on May 6, 1965

Title I.—Establishes a program of rent subsidy payments to moderate income families who are elderly or handicapped or who were displaced from their homes by public action or occupy substandard housing. Extends the program to low income families who are eligible for public housing. Authorizes rehabilitation grants to low income home owners in urban renewal areas to finance necessary repairs and home improvements.

Title II.—Authorizes the Federal Housing Authority to provide mortgage insurance on loans to private land developers and extends existing FHA insurance programs for 4 years. Establishes a program of special FHA home mortgage insurance for veterans who have not received a GI loan.

Title III.—Extends Federal urban renewal programs for 4 years authorizing an additional $2.9 billion for urban renewal grants through fiscal 1969.

Title IV.—Prescribes uniform land acquisition procedures to be followed in connection with the acquisition of land by eminent domain under Federally-assisted housing and urban development programs.

Title V.—Increases the authorization for college housing loans under the Housing Act of 1950 over a 4-year period.

Title VI.—Authorizes the Housing and Home Finance Administrator to make grants to local and public agencies for basic water and sewer facilities, neighborhood health and recreation centers and similar community facilities.

Title VII.—Increases by $1.625 million the amount of special assistance which the President can authorize the Federal National Mortgage Association to provide for housing and community development.

Title VIII.—Increases present grant levels for preservation of open-space land and removes the current $75 million contract authority substituting authority for appropriation of needed amounts. Establishes a program of grants for provision of open-space land in built up urban areas and urban beautification and improvement. Authorizes $5 million for grants of up to 100 percent of the cost of beautification projects useful as demonstrations of new techniques or facilities.

Title IX.—Extends existing rural housing programs for 4 years. Authorizes the Secretary of Agriculture to make loans to farmers and rural residents to buy previously occupied dwellings and farm service buildings. Authorizes the Secretary to insure rural housing loans and sell insured housing loans at discounts to private investors, if necessary.

Title X.—Removes present dollar limits on authorizations of funds for urban planning grants, Federal-State urban affairs training programs and public works planning programs. Authorizes the Housing Administrator to make loans to private nonprofit corporations to finance water purification systems and small community sewage systems.

PUBLIC LAW 89–136—PUBLIC WORKS AND ECONOMIC DEVELOPMENT ACT OF 1965

S. 1648. Introduced by Mr. Douglas and Others on April 1, 1965

Broadens the authority of the Secretary of Commerce to make grants and loans for needed public works and public service or development facilities in redevelopment areas. Establishes a new program of supplemental grants to assist with local shares of Federal grant-in-aid programs. Authorizes $250 million annually for development facility grants and $170 million for business and development facility loans.

Authorizes the Secretary to make industrial and commercial loans, to guarantee loans for working capital made to private borrowers by private lending institutions in connection with projects in redevelopment areas, and to contract to pay to or on behalf of businesses entities locating in redevelopment areas a portion of interest costs incurred in financing their expansion from private sources.

Establishes an economic development fund for loans, collections, and repayments under the act.

Authorizes Federal grants-in-aid for technical assistance, economic planning, and administrative expenses to appropriate nonprofit State, area, district, or local organizations.

Establishes standards for the designation of redevelopment areas and provides for an annual review of the eligibility of areas by the Secretary. Authorizes the Secretary to designate new multicounty economic development districts and to increase development facility grants for projects in redevelopment areas within designated districts by an amount equal to 10 percent of the cost of the project assisted. Limits financial assistance to economic development centers and for 10 percent bonuses to redevelopment areas within designated districts to $50 million annually.

Provides for the establishment of multistate regional planning commissions and authorizes the Secretary to provide technical assistance to the Commission through other agencies, private contractors, or, through grants-in-aid.

Provides for the appointment by the President, with the advice and consent of the Senate of an Administrator of Economic Redevelopment in the Department of Commerce.

Directs the Secretary to appoint a National Public Advisory Committee on Regional Economic Development consisting of 25 representatives of labor, management, and the general public.

Provides criminal penalties for falsification of information by any applicant for financial assistance.

PUBLIC LAW 89-170—TO PROVIDE FOR STRENGTHENING AND IMPROVING THE NATIONAL TRANSPORTATION SYSTEM

H.R. 5041. Introduced by Mr. Harris on February 24, 1965

Permits the Interstate Commerce Commission to make cooperative agreements with the various States to enforce economic and safety requirements.

Permits a State to require motor carriers operating in such State to register its certificate of public convenience and necessity.

Permits persons injured by motor carriers operating in violation of Commission regulations to initiate court action to require compliance. Increases the penalty for failing or refusing to comply with regulations of the Commission.

Sets forth new requirements and procedures for recovery of reparations and requires actions for such recovery to be filed within 2 years.

PUBLIC LAW 89-174—TO ESTABLISH A DEPARTMENT OF HOUSING AND URBAN DEVELOPMENT

H.R. 6654. Introduced by Mr. Reuss on March 23, 1965

Establishes an executive department known as the Department of Housing and Urban Development headed by a Secretary appointed by the President with the consent of the Senate. Provides for the Department to advise the President with respect to Federal programs relating to housing and urban development, develop policies for fostering urban development, coordinate Federal activities, provide technical assistance and information, and encourage comprehensive planning by State and local governments.

Provides for three Assistant Secretaries, a General Counsel, and an Under Secretary.

Transfers all the functions and duties of the Housing and Home Finance Agency of the Federal Housing Administration and the Public Housing Administration in that agency to the new agency. Transfers the Federal National Mortgage Association to the Department.

388

Public Law 89-182—State Technical Services Act of 1965

S. 949. Introduced by Messrs. Magnuson, Byrd of West Virginia, and Ribicoff on February 2, 1965.

Provides the Federal support of State and regional centers to facilitate communication and the flow of information between these centers and American enterprise.

Sets forth the standards and requirements that a State must meet to receive grants.

Provides for Federal support of interstate programs and establishes an advisory committee to review and evaluate such programs.

Provides for a review of the plans, programs, and administration of same every 5 years by a public committee.

Sets forth the situations in which the Secretary of Commerce may terminate aid.

Public Law 89-196—Authorizing $1.5 Million for Expenses of Both the Commission on Law Enforcement and the D.C. Commission on Crime and Law Enforcement

S.J. Res. 102. Introduced by Mr. Ervin on August 5, 1965.

Authorizes funds for the Commission on Law Enforcement and Administration of Justice and the District of Columbia Commission no Crime and Law Enforcement.

Public Law 89-197—To Improve the Quality of State and Local Law Enforcement and Correctional Personnel

H.R. 6508. Introduced by Mr. Celler on March 18, 1965

Law Enforcement Assistance Act.—Authorizes grants by the Attorney General to any public or private nonprofit agency for the purpose of improving the quality of State and local law enforcement and correctional personnel, and employees in programs for the prevention or control of crime.

Provides for studies by the Attorney General with respect to matters relating to law enforcement organization, techniques, and practices; and prevention or control of crime.

Authorizes the Attorney General to appoint advisory and technical committees to assist him.

Public Law 89-220—Authorizing Funds for Research and Development in High-Speed Ground Transportation To Improve the National Transportation System

S. 1588. Introduced by Messrs. Magnuson, Pell, and Pastore on March 22, 1965

Authorizes the Secretary of Commerce to undertake research and development in high-speed ground transportation to improve the national transportation system.

Public Law 89-234—Water Quality Act of 1965

S. 4. Introduced by Mr. Muskie and others on January 6, 1965

Establishes a national policy for the prevention, control, and abatement of water pollution. Creates within the Department of

Health, Education, and Welfare a Federal Water Pollution Control Administration to administer all functions of the Department relating to water pollution.

Authorizes $20 million for Federal grants to States and municipalites for one-half of the cost of projects determined to be new and improved methods of controlling pollution.

Authorizes the Secretary, after public hearing and consultation, to establish water quality standards to aid in preventing, controlling, and abating water pollution.

Increases the limitation on grants for the construction of sewage treatment works. Provides for an additional 10 percent increase in grants for projects which conform to comprehensive metropolitan or regional planning.

PUBLIC LAW 89–272—PROVIDING FOR RESEARCH PROGRAMS RELATIVE TO CONTROLLING AIR POLLUTION RESULTING FROM FUMES FROM GASOLINE AND DIESEL-POWERED VEHICLES

S. 306. Introduced by Mr. Muskie and Others on January 7, 1965

Provides for research programs relative to controlling air pollution resulting from gasoline and diesel-powered vehicles and establishes a Federal Air Pollution Control Laboratory to carry out such programs. Provides for the appointment of a technical committee to encourage the development of low-cost techniques to reduce offensive byproducts of sulfur combustion.

Authorizes grants to municipalities for the elimination of air pollution resulting from solid waste disposal if such municipalities have prohibitions on the open burning of solid wastes with effective enforcement procedures.

Requires all new gasoline-powered vehicles to be equipped with blowby systems to keep hydrocarbon emissions down to certain percentages. Prohibits the importation of any gasoline vehicles not meeting the standards herein. Authorizes the Secretary of Health, Education, and Welfare to make grants to certain State agencies designated to carry out the policies herein. Declares it a policy of the United States to insure the inspection and maintenance of the means required to control the emission of air pollutants from gasoline-powered vehicles.

PUBLIC LAW 89–329—HIGHER EDUCATION ACT OF 1965

H.R. 9567. Introduced by Mrs. Green of Oregon on June 30, 1965

Title I, Community Service Programs.—Strengthens the educational resources of colleges and universities and provides financial assistance for students in postsecondary and higher education. Authorizes $5 million for fiscal year 1966 and such sums as are needed for 4 additional years for the purpose of assisting in the solution of community educational problems.

Requires that States, to receive the allotted funds, must submit a plan showing how these funds are to be used and either designate or create an agency for the application of the plans. The State plan must vest in the agency sole authority to administer the funds, provide a comprehensive program, show procedure and programs to be used, show that the Federal funds will not be used to supplant State or local funds, and assure proper disbursement of the Federal funds.

Establishes a National Advisory Committee on Extension and Continuing Education to advise the Commissioner on policy matters. Provides for review of administration of the programs by enabling the Commissioner to establish a 12-man Review Council on Extension and Continuing Education.

Authorizes the Secretary of Health, Education, and Welfare to engage technical assistance as required by the Council and requires the Council to make reports of its findings. Provides for compensation of Council members at the rate of $100 per day.

Title II, College Library Assistance and Library Training and Research, Part A, College Library Resources.—Authorizes $50 million for fiscal 1966 and such funds as may be necessary for 4 additional years for grants to institutions of higher learning to assist in acquiring library materials.

Provides for the expenditure of 75 percent of the above sum in basic grants not to exceed $5,000 to each institution for the above purposes. To obtain such a grant, institutions must apply to the Commissioner and provide in the application that the funds will be expended in the year given, that proper disbursement and accounting procedures will be used, and that reports on such use will be submitted to the Commissioner.

Provides for supplemental grants by the Commissioner not to exceed $10 for each full-time student.

Provides for the distribution of the remaining 25 percent of the funds allotted by special purpose grants which must be used for library materials.

Requires that the Commissioner establish an Advisory Council on College Library Resources consisting of eight members, to establish criteria for the granting of special-purpose grants.

Requires those institutions receiving such funds to be nationally accredited. Provides that the funds are not to be used in connection with any school or department of divinity.

Part B, Library Training and Research.—Authorizes $15 million for fiscal year 1966 and such sums as may be needed for 4 additional years for grants to institutions of higher learning to assist them in training persons in librarianship and for research and demonstration projects relating to librarianship. Authorizes the Commissioner to appoint panels to evaluate this research.

Part C, Strengthening College and Research Library Resources.—Authorizes a 5-year program for acquiring all library materials currently published throughout the world and providing catalog information for such material. Authorizes $5 million for fiscal 1966 and funds as needed to reimburse the Library of Congress for subsequent years.

Title III, Strengthening Developing Institutions.—Authorizes $30 million in fiscal 1966 and such sums as may be necessary for 4 additional years for the purpose of assisting in the raising of the academic quality of colleges (developing instititions). Defines "developing institutions."

Provides for the establishment of an Advisory Council on Developing Institutions to advise the Commissioner with respect to policy matters concerning the administration of this title.

Authorizes the Commissioner to make grants to developing institutions to pay for the expense of cooperative agreements designed to strengthen such institutions. Requires that such grants be applied for and that such application be approved by the Commissioner by the implementation of standards set forth in the act.

Authorizes the Commissioner to award fellowships of up to 2 years to encourage individuals to teach at these developing institutions. Such fellowships must be applied for by the institutions and approved by the Commissioner.

Title IV, Student Assistance, Part A, Educational Opportunity Grants.—Authorizes appropriations for 6 additional years for the purpose of providing loans to students and to institutions of higher education to make educational opportunity grants.

Sets forth the manner and mode in which the Commissioner may allocate funds to the States for achieving the purposes of this title.

Authorizes the Commissioner to enter into contracts not exceeding $100,000 to encourage full utilization of educational talent and to identify youths of exceptional financial need and to encourage them to complete secondary and postsecondary educational training. Authorizes such funds as may be necessary to accomplish this purpose.

Part B, Federal, State, and Private Programs of Low-Interest insured Loans to Students in Institutions of Higher Education.—Authorizes $1 million and further sums if necessary to establish a student loan insurance fund to insure eligible lenders against losses on student loans.

Authorizes an additional $17,500,000 for advances for reserves for State student loan insurance programs. Limits outstanding loans to $700 million for fiscal 1966 and sets forth a formula for computing amounts to which States are entitled. Provides for increases in subsequent years of the total amount of insured loans.

Limits insurance to loans not exceeding $2,000 per academic year per student and to $7,500 total per student in case of graduate or professional students and to $5,000 total for other students. Provides that the insurance shall be on 100 percent of the unpaid balance.

Sets forth the eligibility requirements of students obtaining loans and the terms upon which the student loans must be made in order to be covered by the insurance.

Provides for Federal payments to reduce student costs.

Sets forth the requirements for the issuance of certificates of insurance by the Commissioner and provides that the effective date of the insurance shall be the date of the certificate.

Provides for a procedure to be followed by eligible lenders upon the default, death, or disability of a student borrower.

Establishes a student loan insurance fund to be available to the Commissioner and sets forth procedure for the Commissioner to follow when the fund is insufficient to pay defaulted loans.

Vests in the Commissioner the powers to carry out the duties assigned to him. Provides that he may sue and be sued in any State court of general jurisdiction and in any Federal district court.

Prohibits any Federal control of education.

Part C, College Work-Study Program Extension and Amendments.— Provides for the transfer of educational activities under the Economic

Opportunity Act to the Commissioner of Education and authorizes funds for such program through fiscal 1970.

Title V, Amendments to Higher Education Facilities Act.—Authorizes additional funds for aid in the construction of public junior colleges and technical institutes and graduate facilities.

Title VI, Advisory Council to House Committee on Education and Labor.—Authorizes the chairman of the Committee on Education and Labor of the House of Representatives to establish an advisory council to make studies and recommendations with respect to programs established hereunder.

Public Law 89–404—To Promote a More Adequate Program of Water Research

S. 22. Introduced by Mr. Anderson and Others on January 6, 1965

Authorizes the appropriation of an additional $85 million over a 10-year period beginning in fiscal 1967, for grants to public and private institutions for water research projects under the provisions of the Water Resources Act (42 U.S.C. 1961b). Requires all contracts be submitted to Congress 60 days in advance of execution for review.

Public Law 89–454—To Provide for Expanded Research in the Oceans and the Great Lakes, and To Establish a National Oceanographic Council

S. 944. Introduced by Mr. Magnuson and Others on February 2, 1965.

Marine Resources and Engineering Development Act.—Declares it the policy of the United States to develop, encourage, and maintain a comprehensive long-range program in marine sciences to assist in protecting health and property, enhance commerce, transportation and national security.

Establishes a National Council on Marine Resources and Engineering Development composed of the Vice President, members of the President's Cabinet, and other executive department officials, to advise and assist the President on ocean surveys, developing comprehensive oceanographic programs, providing for cooperation between the various segments of Government dealing with oceanographic and marine sciences activities.

Establishes a 15-member Commission on Marine Sciences composed of experts in various disciplines to assist the President and the Council.

Public Law 89–553—To Broaden the Scope of the Small Reclamation Projects Act

S. 602. Introduced by Mr. Moss and Others on January 19, 1965

Broadens the scope of the Small Reclamation Projects Act to increase the limitation on the costs of projects (from $5 million to $6.5 million), and funds authorized therefor (from $100 to $200 million). Changes the formula used for determining the interest rate for the interest-bearing portions of loans under this act from a formula based on yield to a formula based on coupon rates. Permits advancement of up to half the cost of the project.

PUBLIC LAW 89–560—AUTHORIZING SECRETARY OF AGRICULTURE TO COOPERATE WITH STATE AND LOCAL AGENCIES IN PLANNING FOR CHANGES IN THE USE OF AGRICULTURAL LAND IN RAPIDLY EXPANDING URBAN AREAS

S. 902. Introduced by Mr. Ellender on February 1, 1965.

Authorizes the Secretary of Agriculture to cooperate with States and other public agencies by providing soil surveys and other technical assistance in planning for use changes in rapidly expanding areas including farm and nonfarm areas.

PUBLIC LAW 89–561—TO AUTHORIZE FEASIBILITY INVESTIGATIONS OF CERTAIN WATER RESOURCE DEVELOPMENT PROPOSALS

S. 3034. Introduced by Mr. Jackson on March 7, 1966.

Authorizes the Secretary of the Interior to engage in feasibility investigations of certain water resource development proposals.

PUBLIC LAW 89–562—TO AUTHORIZE FUNDS FOR CONTINUED ASSISTANCE IN DEVELOPMENT OF URBAN MASS TRANSPORTATION FACILITIES

S. 3700. Introduced by Mr. Williams of New Jersey on August 8, 1966

Amends the Urban Mass Transportation Act to: (1) Authorize additional appropriations in 1968 and 1969 for grants made under the act; (2) authorize the Secretary of Housing and Urban Development to, (a) make grants to States and local public bodies for the planning, engineering, and designing of urban mass transportation projects and for studies concerning the nature, utilization, and economic feasibility of facilities, equipment, and systems, (b) provide for 100 graduate training fellowships annually, for managerial, technical and professional personnel in the mass transportation field, and (c) make grants to institutions of higher learning to assist them in carrying out research in urban transportation problems; and (3) direct the Secretary of Housing and Urban Development to prepare long-range programs for modernizing urban mass transportation facilities and providing national leadership to State, local, and private organizations for planning, financing, and operation of future urban mass transportation systems.

PUBLIC LAW 89–563—TRAFFIC SAFETY ACT OF 1966

S. 3005. Introduced by Mr. Magnuson on March 2, 1966 (approved September 9, 1966 (Public Law 89–563))

Directs the Secretary of Commerce to establish Federal minimum safety standards for all "motor vehicles" (defined as any mechanically powered vehicle manufactured primarily for use on public streets, except those operated exclusively on rails).

Directs the Secretary to issue initial standards on or before January 31, 1967, and new and revised standards on or before January 31, 1968. Provides that all standards prescribed shall be effective within 180 days or 1 year after publication, unless the Secretary directs otherwise.

Allows the Interstate Commerce Commission to continue to require certain higher standards of performance based on its authority under existing law.

Provides for judicial review by the Court of Claims upon petition within 60 days of publication of the standard by a person adversely affected by such standard.

Provides for establishing a National Motor Vehicle Safety Advisory Council, a majority of which shall be representatives of the general public, including State and local governments, and the remainder shall include representatives of the motor vehicle industry. Directs the Secretary to consult with such Council on safety standards under this act.

Provides that the Secretary, both directly and through grants, shall conduct research, testing, development, and training for motor vehicle safety.

Provides that, initially, safety standards shall not apply to used autos. Directs that a study of safety standards and inspection requirements of the States, relating to used autos be made, a report on the findings filed, and not later than 1 year from the date such report is filed, uniform safety standards with respect to such used vehicles shall be set by the Secretary.

Requires all manufacturers of motor vehicles and equipment to notify any purchaser by certified mail of any safety-related defect, including an evaluation of the risk involved, and to notify the Secretary of all reports, documents, etc., and communications regarding any safety defect. Provides for onsite inspections by the Secretary. Requires that all vehicles and equipment be certified by the manufacturer as complying with the Federal safety standards.

Requires manufacturers or distributors to repurchase defective vehicles or equipment at the price paid plus not less than 1 percent of the original price for each month of delay in replacement.

Provides penalties for violations of safety standards, inspection, certification, and notification requirements of up to $1,000 for each violation to a maximum of $400,000 for a related series of violations.

Directs the Secretary to administer the provisions of this act through a National Traffic Safety Agency to be established within the Department of Commerce.

Directs the Secretary to establish Federal minimum safety standards for pneumatic tires, provide for conspicuous labeling on tires of safety-related information, and require that motor vehicles be equipped with tires meeting such standards.

Authorizes the Secretary to make a complete investigation of the need for facilities to conduct research, development, and testing in traffic and auto safety.

Authorizes the Secretary to maintain a register listing the names of all individuals who have had their driver's license revoked, except for revocations for less than 6 months based on nonmoving violations. Provides that such information shall be available only at the request of a Federal, State, or local government, and only with respect to an application for a driver's license or permit.

PUBLIC LAW 89–564—PROVIDING FOR A COORDINATED NATIONAL HIGHWAY SAFETY PROGRAM

S. 3052. Introduced by Messrs. Randolph and Cooper on March 8, 1966

Highway Safety Act.—Provides for a coordinated national highway safety program through financial assistance to the States to accelerate highway traffic safety programs.

Requires States to have highway safety programs approved by the Secretary designed to reduce highway accidents and deaths, injuries, and property damage resulting therefrom. Provides for a reduction of 10 percent of Federal-aid highway funds and directs the Secretary not to apportion funds to a State not having such a program by January 1, 1969. Provides that these programs are to be in accordance with uniform standards established by the Secretary, and such standards are to be such as to improve driver performance, provide for an effective record system of accidents, accident investigations, vehicle registration, operation and inspection, highway design and maintenance, traffic control, vehicle codes, and laws, surveillance of traffic for detection and correction of high or potentially high accident locations, and emergency services.

Requires the Secretary to make a study of the relationship between consumption of alcohol and its effect on highway safety which is to cover review and evaluation of State and local laws and enforcement methods and procedures relating to driving under the influence of alcohol, State and local programs for the treatment of alcoholism, and such other aspects of the problem as may be useful.

Establishes a National Highway Safety Advisory Committee, also composed of the Secretary or his designee as Chairman, the Federal Highway Administrator, and 29 members, not more than four of whom shall be Federal officers, or employees. Provides that such members shall be selected from among representatives of State and local governments, of public and private institutions contributing to, affected by, or concerned with highway safety, and other public and private agencies, organizations, or groups demonstrating an active interest in highway safety, as well as research scientists and other individuals expert in this field.

Requires that in approving programs for projects on the Federal-aid system, the Secretary shall give priority to those projects which incorporate improved standards and features with safety benefits.

Requires that the Secretary make a detailed estimate of the cost of carrying out the provisions of this Act and submit a report thereon to the Congress not later than January 10, 1968.

PUBLIC LAW 89–574—FEDERAL-AID HIGHWAY ACT

S. 3155. Introduced by Mr. Randolph on March 29, 1966

Authorizes an increase of $4.9 billion in the appropriations for the Interstate Highway System. Includes authorizations for an additional year until June 30, 1972, and authorizes use of the apportionment factor contained in the 1965 cost estimate of completion in making appropriations of interstate funds for fiscal years 1968 and 1969.

Authorizes appropriations for the Federal-aid primary and secondary highway systems and extensions thereof within urban areas for fiscal 1968 and 1969.

Authorizes appropriations for other highways in Federal domain areas for fiscal 1968 and 1969.

Increases the authorization for reconstruction of disaster-damaged highways for fiscal 1967 and thereafter from $30 to $50 million annually, Provides for a 2-year carryover of unexpended authorizations for that purpose, and provides for reimbursement from the general fund of the Treasury to cover the Federal share of the cost of reconstruction of roads not on the Federal-aid systems.

Modifies the prohibition against use of the highway trust fund for making appropriations for highway beautification to limit this prohibition to appropriations made during fiscal 1966, and provides contract authority thereafter for such purposes as are now applicable to the regular Federal-aid highway program.

Public Law 89–599—Granting Congressional Consent to Compact Between Missouri and Kansas Creating Kansas City Area Transportation District and Authority

S. 3051. Introduced by Mr. Long of Missouri and others on March 8, 1966

Grants the consent of Congress to the compact between Missouri and Kansas creating the Kansas City Area Transportation District and the Kansas City Area Transportation Authority.

Public Law 89–670—Department of Transportation Act

H.R. 15963. Introduced by Mr. Holifield on June 27, 1966

Establishes a Department of Transportation at the Cabinet level and transfers to that Department jurisdiction over the following agencies, offices and responsibilities: (1) Bureau of Public Roads, (2) Federal Aviation Agency, (3) Civil Aeronautics Board, (4) Maritime Administration, (5) Coast Guard, (6) Office of the Under Secretary of Commerce for Transportation, (7) the Federal Safety functions pertaining to rail, motor, and pipelines presently under the Interstate Commerce Commission.

Creates within the Department the Federal Highway, Railroad, Maritime, and Aviation Administrations.

Authorizes the Secretary of Transportation to develop national transportation policies and to promote research in the transportation field, including noise abatement.

Provides for establishment of a 5-member National Transportation Safety Board to investigate accidents.

Authorizes the Secretary to develop economic criteria and data for use in the investment of Federal funds for transportation.

Public Law 89–672—Authorizing the Secretary of the Interior To Enter into Contracts for Scientific and Technological Research into Problems Confronting His Department

S. 3460. Introduced by Mr. Jackson on June 6, 1966

Authorizes the Secretary of the Interior to enter into contracts with educational institutions, public or private agencies, or individuals for scientific and technological research into any aspect of the problems related to programs of his Department.

Public Law 89–675—Clean Air Act Amendments

S. 3112. Introduced by Mr. Muskie and others on March 21, 1966

Authorizes grants under the Clean Air Act to air pollution control agencies for maintenance of air pollution control programs in addition to present authority for grants to develop, establish, or improve such programs. Makes the use of appropriations under the act more flexible by consolidating the appropriation authorizations under the act and deleting the provision limiting the total of grants for support of air pollution control programs to 20 percent of the total appropriation for any year. Extends the duration of the programs authorized by the act.

PUBLIC LAW 89–749—COMPREHENSIVE HEALTH PLANNING AND PUBLIC HEALTH SERVICES AMENDMENTS

S. 3008. Introduced by Mr. Hill on March 2, 1966

Promotes and assists in the extension and improvement of comprehensive health planning and public health services by providing grants to States for such planning and services, grants to nonprofit institutions for training, studies, etc., in connection with such programs and the interchange of HEW personnel with States to provide for more effective discharge of Federal responsibilities in the field of health.

PUBLIC LAW 89–753—TO IMPROVE AND MAKE MORE EFFECTIVE PROGRAMS UNDER THE FEDERAL WATER POLLUTION CONTROL ACT

S. 2947. Introduced by Mr. Muskie and others on February 18, 1966

Clean Rivers Restoration Act.—Provides for supplementing the existing water pollution control program for planning and construction of treatment works on a river basin basis. Further amends the Federal Water Pollution Control Act by: (1) Providing for a 6-year, $6 billion program of grants to municipalities for sewage treatment construction, with the Federal Government paying up to 50 percent of the cost, provided: (a) the Governor agrees to abolish water quality standards for all rivers and streams in the State, and (b) the State provides 30 percent of the total cost; (2) removing the present limits on individual project grants; (3) providing incentives for State participation in the program by offering a 10-percent bonus on grants in cases where the project conforms with a comprehensive plan for a metropolitan area; (4) providing for a long-term, low-interest loan program to assist communities in States where State funds are inadequate; (5) providing for an increase from $5 million to $10 million in grants for State programs to assist States in improving their own programs; and (6) authorizing a $25-million-a-year program of grants for the demonstration of advanced waste treatment and water purification methods.

PUBLIC LAW 89–754—DEMONSTRATION CITIES AND METROPOLITAN ACT OF 1966

S. 3708. Introduced by Mr. Muskie on August 9, 1966

Title I, Comprehensive City Demonstration Programs.—Provides for Federal assistance to enable cities to: (1) Rebuild urban slum areas; (2) provide public facilities to improve the quality of life in such areas; and (3) participate in and coordinate other activities aided under the existing Federal programs in connection with such "demonstration" programs.

Provides for grants to be made by the Secretary of Housing and Urban Development, after approval of a locally prepared comprehensive city demonstration program, amounting to: (1) 80 percent of the cost of planning and developing such a program; (2) 80 percent of the cost of administering such a program; (3) 100 percent of the cost of relocations necessitated by such a program; and (4) 80 percent of non-Federal contributions required to be made to other programs or activities assisted by Federal grant-in-aid programs which are undertaken in connection with such a program.

Title II, Planned Metropolitan Development.—Authorizes the Secretary of Housing and Urban Development to make grants for development projects in metropolitan areas, such projects to include sewer, water, and sewage treatment facilities; highway, mass transit, airport, and other transportation facilities; and recreation and other open-space facilities. Requires metropolitan areas to have an adequate system for coordinating local public development policies and activities on the basis of metropolitan-wide comprehensive planning. Requires that in order to be eligible for grants, projects must be consistent with the total metropolitan plan of their area. Establishes a time limit within which grants may be made. Limits grants to 20 percent of the total cost of the project.

Title III, Urban Information and Technical Assistance Services.—Authorizes Federal financial assistance to States and metropolitan area agencies to enable them to: (1) Make available information and data on urban needs and assistance programs through centers established for such purpose, and (2) provide technical assistance to small communities.

Public Law 89-774—To Create the Washington Metropolitan Area Transit Authority for Improvement of Transportation Facilities

S. 3488. Introduced by Mr. Robertson and others on June 9, 1966

Grants the consent of Congress for the States of Virginia and Maryland and the District of Columbia to amend the Washington Metropolitan Area Transit Regulation Compact to establish an organization empowered to provide transit facilities in the National Capital Region.

Public Law 90-148—Air Quality Act of 1967

S. 780. Introduced by Mr. Muskie and others on January 31, 1967

Broadens and extends the Clean Air Act. Expands the authority to conduct and assist in research relating to fuels and vehicles to eliminate pollution of air by preventing and controlling the discharge into the air by various types of pollutants.

Permits the Secretary to regulate and require registration of fuel additives. Provides a civil penalty of $1,000 for each and every day of the continuance of violations of such regulations.

Authorizes the Secretary of Transportation to impose as a condition to approval of the highway safety program of any State a requirement that such program include procedures to assure the adequacy of performance of systems or devices in motor vehicles for control of emissions from them. Directs the Secretary from time to time to establish emission standards for heavy industries which because of their nature are sources of substantial amounts of pollutants and can with reasonable efforts and expenditures be prevented or substantially reduced. Permits the Secretary to issue cease and desist orders and permits judicial review from adverse decisions.

Authorizes the establishment of regional air quality commissions which shall encompass such areas which in his judgment have such characteristics as to warrant treatment as a unit for air pollution control purposes. Directs such commissions to establish air quality standards for its air quality region, taking into consideration the various relevant aspects thereof, such as the concentration of industry, other commercial establishments, and population and the technological and economic feasibility of achieving such quality, as well as standards for pollutant emissions in order to achieve or preserve such air quality. Such commissions shall prepare a system of alerts to avert situations in which there may be imminent and serious danger to the public health or welfare from air pollutants. Sets forth penalties for violations and provides for subpena powers.

PUBLIC LAW 90–174—PARTNERSHIP FOR HEALTH AMENDMENTS

H.R. 6418. Introduced by Mr. Staggers on March 1, 1967

Increases the authorized appropriations for grants for comprehensive health planning and public health services under the Public Health Service Act.

Provides for assisting each health care facility in the participating States to develop a program for capital expenditures or replacement, modernization, and expansion consistent with the overall State plan.

Authorizes the Secretary of Health, Education, and Welfare to make grants and contracts for projects for the conduct of research, experiments, or demonstrations aimed at developing new methods or improvement of existing methods of organization, delivery, or financing health facilities and services. Provides guidelines for this program.

Authorizes the Secretary to enter into agreements providing for cooperative planning between the Public Health Service and communities to cope with health problems resulting from disasters or other health emergencies of such nature as warrant Federal assistance.

Authorizes the Secretary to accept volunteer services and to enter into agreements and arrangements with schools of medicine or health hospitals and other health care facilities for the mutual use and interchange of facilities, resources, and services.

Authorizes the Secretary to provide medical, surgical, dental treatment, and hospitalization for Federal employees at remote medical facilities of the Public Health Service where other medical care and treatment are not available. Requires payment for this care and treatment when its users are not entitled to it under any other provision of law.

Clinical Laboratories Improvement Act—Requires clinical laboratories which deal with the health of man to obtain a license to operate if they intend to operate in interstate commerce. Provides that the license be issued under standards found necessary by the Secretary to carry out the purpose of this act.

Provides standards for revoking, limiting, or suspending licenses, and gives an aggrieved party a right to a hearing and judicial review. Makes it a misdemeanor to violate this act and provides for a fine of up to $1,000 and/or imprisonment for not more than 1 year for each violation. (Adds 42 U.S.C. 264.)

PUBLIC LAW 90–219—ESTABLISHES A FEDERAL JUDICIAL CENTER IN THE ADMINISTRATIVE OFFICE OF THE UNITED STATES COURTS

H.R. 5385. Introduced by Mr. Celler on February 15, 1967

Establishes a Federal judicial center in the administrative office of the U.S. Courts. Directs the Center to aid in the stimulation, development, and conducting of programs of continuing education and training for personnel in the judicial branch of the Government. Provides that the activities of the Center be suprervised by a Board composed of the Chief Justice of the United States, two judges of the U.S. courts of appeals, three judges of the U.S. district courts, and the Director of the Administrative Office of the U.S. courts. Authorizes the employment of necessary personnel and necessary appropriations.

PUBLIC LAW 90–220—AUTHORIZING THE PROSECUTION OF A TRANSIT DEVELOPMENT PROGRAM FOR THE NATIONAL CAPITAL REGION

H.R. 11395. Introduced by Mr. Whitener on July 12, 1967

Authorizes under the National Capital Transportation Act the prosecution of a transit development program for the National Capital region so as to further the objectives of the act of July 14, 1960. (Amends 40 U.S.C. 681.)

PUBLIC LAW 90–274—TO ESTABLISH JUDICIAL MACHINERY FOR THE RANDOM SELECTION OF FEDERAL JURIES

S. 989. Introduced by Mr. Tydings on February 16, 1967

Jury Selection and Service Act.—Establishes judicial basic procedures for the random selection of Federal grand and petit jurors so as to eliminate discrimination in such selection. Requires such juries to be composed of a cross selection of the community. Provides that sources of names other than voter lists may be used to supplement voter lists.

Provides that no citizen shall be excluded from service as a grand or petit juror on account of race, color, religion, sex, national origin, or economic status. Provides that prejudice in selecting jurors is established when a party shows that a jury selection system fails to comply substantially with this act.

Qualifies a person for jury service unless he: (1) Is not a 21-year-old citizen of the United States who has resided for a period of 1 year within the judicial district; (2) is unable to read, write, and understand the English language with a degree of proficiency sufficient to fill out satisfactorily the juror qualification form; (3) is unable to speak the English language; (4) is incapable, by reason of mental or physical infirmity, to render satisfactory jury service; or (5) has a charge pending against him for the commission of or has been convicted in a State or Federal court of record of a crime punishable by imprisonment for more than 1 year and his civil rights have not been restored by pardon or amnesty.

Provides a procedure for challenging the noncompliance with the provisions for jury selection established by this act in both civil and criminal cases. Allows any person or the United States to pursue other criminal or civil remedies to prevent or vindicate illegal discrimination in the jury selection process. Provides for the maintenance and inspection of records.

PUBLIC LAW 90–222—ECONOMIC OPPORTUNITY AMENDMENTS OF 1968

S. 2388. Introduced by Mr. Clark on September 12, 1967

Authorizes the appropriation of $2.258 billion, during fiscal year 1968 to carry out the provisions of the Economic Opportunity Act.

Title I, Job Corps program.—(1) Increases and expands the methods of screening and selecting underprivileged individuals for participation in the Job Corps program; (2) authorizes the Director to provide for the support of enrollees and to pay them a personal allowance of from $35 to $65 per month; (3) authorizes the establishment of stringently enforceable standards of conduct for enrollees at Job Corps centers; (4) prohibits payments to any individual or organization solely as compensation for the referral of names of candidates for the Job Corps; (5) permits the enrollment of applicants on probation or parole where the applicant's release from the supervision of probation or parole officials is mutually satisfactory to those officials and to the Director and will not violate applicable laws or regulations; (6) limits Job Corps enrollment to 2 years except as the Director may authorize in special cases; (7) provides that Job Corps enrollment shall not relieve any individual of his military obligation; (8) requires that Job Corps center programs provide each enrollee education, vocational training, work experience, planned avocational and recreational activities, physical rehabilitation, and development and counseling; (9) permits enrollees to obtain a high school equivalency certificate with the concurrency of the Secretary of Health, Education, and Welfare; (10) requires the Director to assure that Job Corps activities do not displace presently employed workers or impair existing contracts; (11) provides each former enrollee with a readjustment allowance not to exceed $50 for each month of Job Corps participation. An enrollee, however, is not entitled to that portion of the readjustment allowance which is not paid to his dependents unless he has remained in the program for 6 months, except where an enrollee has remained in the program for 3 months and is expected to complete his training in less than 6 months. Provides for the payment of an enrollee's readjustment allowance at the public employment service office nearest to his home or to the community in which he intends to reside after termination of training. Permits up to $25 per month of the readjustment allowance to be paid during an enrollee's service directly to his spouse, child, or other dependent relative; (12) requires that an enrollee's pertinent records including counseling and testing data be made available to the Department of Labor and the Office of Economic Opportunity; (13) requires careful and systematic evaluation of the Job Corps program directly or through independent contracts, including consultation with other agencies to compare the relative effectiveness of Job Corps and other programs; (14) requires the Director to enter into agreements with State educational agencies establishing and operating model community vocational education schools and skill centers to be centrally located in urban areas having high dropout rates, large numbers of unemployed youths and a need for such school and center (the project must include a job survey of the area, a training program reflecting the needs of the job market as projected by the survey and the establishment of a community advisory committee, arrangements must also be worked out with schools in the area and the administrator of the skill center for maximum utilization of the center both during school and after school hours); (15) authorizes agreements with States to assist in the operation of State-operated programs to carry out the purpose of the Job Corps; (16) requires the submission of a plan setting forth the proposed establishment of a Job Corps center to the Governor of the State in which the center is to be located and gives the Governor a veto over such plan; (17) provides, with certain specified exceptions, that Job Corps enrollees are not to be considered Federal employees or subject to laws relating to Federal employment including those regarding hours of work, rates of compensation, leave, unemployment compensation, and Federal employee benefits (the three enumerated exceptions relate to services performed by enrollees for purposes of the Internal Revenue Code and Social Security Act, compensation for work injuries, and tort claims against the United States); (18) permits adjustment and settlement of claims for damage to persons or property resulting from Job Corps operations in amount not exceeding $500 when such a claim is a proper charge against the United States and not cognizable under section 2672 of title 28, United States Code; (19) limits Job Corps enrollment during fiscal year 1968 to 45,000 enrollees; (20) requires that on or before June 30, 1968, at least 25 percent of the Job Corps enrollees and residents receiving training be women; (21) limits the direct operating cost of Job Corps centers in operation for more than 9 months to a maximum of $7,300 per enrollee; and (22) prohibits inquiries concerning the political affiliation or beliefs of any enrollee and requires that all disclosures concerning such matters be ignored, except as to membership in political parties or organizations which constitute by law a disqualification for Government employment. Prohibits discrimination by any Government employee against or in favor of any enrollee or applicant because of his political affiliations or beliefs. Prohibits officers, employees, and enrollees of the Job Corps from taking any active part in political management or campaigns except as provided by or pursuant to statute. Prohibits such persons from using their official position or influence for the purposes of interfering with an election or effecting the result thereof, although all such persons retain the right to vote as they may choose and to express in their private capacities opinions on all political subjects and candidates.

Work and training for youth and adults.—(1) States that the purpose of this program is to provide useful work and training opportunities with related services and assitance to assist low-income youths to continue or resume their education and to help unemployed or low-income youths and adults to obtain and hold regular competitive employment with maximum opportunities for local initiative in developing programs which respond to local needs and problems;

(2) requires the Director to designate community program areas for the planning and conduct of comprehensive community work and training programs; (3) provides for the designation of a prime sponsor, which may be a public or private nonprofit agency, to receive comprehensive work and training program funds and to be responsible for planning, administering, coordinating, and evaluating a comprehensive work and training program in each respective program area; (4) authorizes financial assistance for numerous work and training programs such as: (a) The Neighborhood Youth Corps which provides part-time employment, on-the-job training, and useful work experience for low-income students in the ninth through 12th grades and who need earnings to permit them to resume or maintain their attendance in school, and (b) programs to provide work activities for the chronically unemployed poor with poor employment prospects who are unable because of age, lack of employment opportunity, or otherwise to secure appropriate employment or training assistance under other programs. (Such programs must enable participants to participate in projects involving community betterment or beautification including activities contributing to the management, conservation, or development of natural resources, recreational areas, parks, highways, and other lands.); (5) requires that by July 1, 1968, all work and training programs be consolidated into the comprehensive work and training program and that financial assistance for such components be provided to the prime sponsor unless the director grants an extension of time or provides financial assistance to organizations other than a prime sponsor to carry out component programs when, after considering the views of the prime sponsor, if any, the director determines that such assistance would enhance program effectiveness or acceptance on the part of persons served and would serve purposes of title I; (6) prohibits financial assistance for programs unless the following specified special conditions are met: (a) No participants will be employed on projects involving political parties or activities related to sectarian instruction or religious worship; (b) the program will not result in the displacement of employed workers or impair existing contracts for services; (c) the rates of pay and conditions of employment will be appropriate and reasonable in light of the type of work, geographical region, and proficiency of the participant; and (d) the program will, to the maximum extent feasible contribute to the occupational development or upward mobility of individual participants; (7) authorizes financial assistance to State agencies for the provision of technical assistance and training with emphasis upon such services to rural areas, to assist in coordinating State activities, to operate work and training programs in communities where a prime sponsor is not yet established, and to provide work and training opportunities on State projects and in State agencies; (8) limits Federal financial assistance to programs to 90 percent of the cost of such programs, giving the Director discretion to exceed such percentage where he determines it necessary.

Revises and extends the special impact program which is designed to establish special programs directed to the solution of critical problems existing in particular communities or neighborhoods within urban areas having especially large concentrations of low-income persons and rural areas having substantial outmigration to eligible urban areas and which are of sufficient size and scope to have an appreciable impact in such communities and neighborhoods in arresting tendencies toward dependency, chronic unemployment, and rising community tensions. Programs may include economic and business development programs including those which provide financial and other incentives to business to locate in or near the areas served so as to provide employment opportunities for residents of such areas, community development activities which create new training and employment activities and contribute to an improved living environment, and manpower training programs which support and complement economic business and community development programs.

Extends and revises those provisions of law dealing with urban and rural community action programs which are designed to assist communities in opening opportunities which enable low-income persons to achieve self-sufficiency. Includes the following provisions with respect to such programs: (1) Requires that communities be encouraged and aided to plan and conduct community action programs which shall be designed to: (a) Provide services and assistance including innovative approaches to enable low-income persons to achieve economic independence, improve their living conditions, and increase their participation in community activities; (b) stimulate agencies and institutions which provide services to low-income persons to expand, modify, and improve their programs; and (c) mobilize, utilize, and coordinate relevant public and private resources; (2) encourages the establishment of housing development and services organizations as delegate agencies to focus on housing needs or low-income families and individuals. Such organizations may be nonprofit housing development corporations and may become sponsors of housing under existing programs of specialized housing agencies but may not insure mortgages or duplicate the long-term capital financing functions of programs now administered by specialized housing agencies; (3) requires that after July 1, 1968, each community action agency adopt a systematic approach to the achievement of purposes and to utilization of funds; (4) requires that 50 percent of the funds authorized and appropriated be used to finance "component programs" and "national emphasis programs" which are locally selected to respond to particular community needs; (5) lists the following "national emphasis programs": (a) Headstart, (b) follow-through, (c) a legal services program, (d) a comprehensive health services program (e) upward bound, (f) project fund, and (g) a family planning program; (6) authorizes training research and technical assistance in connection with national emphasis programs; (7) limits Federal financial assistance in community action agencies to 90 percent of the approved cost of assisted programs with limited authority in the Director to exceed such percentage.

Changes the rural areas programs to a rural loan program designed to meet some of the special needs of low-income rural families by establishing a program of loans to assist in raising and maintaining their income and living standards.

Revises the migrant program and makes the following provisions: (1) States that the purpose of the program is to assist migrant and seasonal farmworkers and their families to improve their living conditions and develop skills necessary for a productive and self-sufficient life; (2) authorizes financial assistance to State and local agencies and nonprofit institutions and cooperatives to carry out migrant assistance programs; (3) authorizes such programs to include projects or activities in the fields of day care for children, education, health, improved housing, and sanitation, including the provision and maintenance of emergency and temporary housing, legal advice and representation, consumer training and counseling, the promotion of increased community acceptance of migrant and seasonal farmworkers and their families, and equipping of unskilled migrant workers through education and training to meet the changing demands in agricultural employment brought about by technological advancement. Extends through 1968 the present program for indemnity payments to dairy farmers.

Revises the program providing for employment and investment incentives in order to assure that in connection with loan assistance, emphasis is placed upon the preservation or establishment of small business concerns located in urban areas of high concentration of poverty, and that management training programs are of sufficient scope and duration to provide reasonable opportunity for individuals served to develop entrepreneurial and managerial self-sufficiency.

Establishes a program of financial assistance to day-care projects which provide day care for children from low-income families or from urban and rural areas having high concentrations or proportions of low-income persons in order to enable the parents or relatives of such children to undertake or continue vocational training, basic education, or gainful employment.

Authorizes grants to public and private agencies to pay not in excess of 90 percent of the cost of day-care projects.

Makes various amendments which concern the administration and coordination of the Economic Opportunity Act programs.

Provides a new formula for the treatment of income for certain public assistance purposes.

Continues the VISTA program in essentially its present form, but provides, among other things, authority for the establishment of full time programs.

Amends the Manpower Development and Training Act to provide that the training allowance for all youths shall be at the same rate, eliminating the difference in existing law between the training allowance for youth coming out of other training programs and youth not attending such program.

Title II, Emergency Employment Act.—Provides for financial assistance to designated urban and rural areas which contain a high proportion of low-income families and individuals which have severe problems of unemployment and underemployment.

Permits such assistance to include services and supporting facilities in such fields as health, public safety, education, recreation, streets, parks, and municipal maintenance, housing and neighborhood improvement, conservation and rural development, beautification, and other fields of human betterment and public improvement.

Requires reports to the Congress on the progress of these programs. Authorizes the appropriations of $1 billion for fiscal year ending June 30, 1968, and $1,500 million for the fiscal year ending June 30, 1969.

PUBLIC LAW 90–351—SAFE STREETS AND CRIME CONTROL ACT

H.R. 5037. Introduced by Mr. Celler on February 8, 1967

Provides for a 5-year program of planning and program grants to States and local governments to carry out the provisions of this act.

Bases this act on the principle that the States and local governments must plan and execute programs to improve their police, courts, and correctional systems.

Establishes a sliding scale of grants to implement the act which will provide up to 90 percent of the cost to communities which qualify for assistance; up to 60 percent of the cost of executing action programs will be borne by the Federal Government; up to 50 percent of the cost of certain new construction will be assumed by the Federal Government; finally, up to 100 percent of the cost of certain research will be assumed by the Federal Government.

Provides that the Attorney General be assisted in its administration by a director appointed by the President with the advice and consent of the Senate.

Authorizes the Attorney General to conduct research and evaluation studies, to collect, evaluate, publish, and disseminate statistics and other information on the condition and progress of law enforcement and criminal justice in the several States.

PUBLIC LAW 90–391—VOCATIONAL REHABILITATION AMENDMENTS

H.R. 16770. Introduced by Mr. Daniels and others on April 24, 1968

Vocational Rehabilitation Amendments.—Extends and expands generally the provisions of the Vocational Rehabilitation Act. Increases generally the authorized appropriations provided under the act. Allows additional State allotments for the innovation of vocational rehabilitation services when the predetermined fixed allotment is insufficient.

Permits the Secretary upon the request of an agency to authorize share funding and administrative responsibility in order to permit a joint program to provide services to handicapped individuals. Permits counseling, guidance, adjustments, training maintenance, physical restoration, placement, and followup services to the handicapped.

Redefines the term "rehabilitation facility" under the act so as to mean a rehabilitation center, workshop, or other facility operated for the primary purpose of assisting handicapped individuals and providing evaluation and work adjustment services for disadvantaged individuals and provides singly or in combination one or more of the following services for handicapped individuals: (1) Management, medical, psychological, social, and vocational services; (2) testing, fitting, or training in the use of prosthetic devices; (3) prevocational conditioning or recreational therapy; (4) physical and occupational therapy; (5) speech and hearing pathology; (6) psychological and social services; (7) evaluation; (8) personal and work adjustment; (9) vocational training; (10) evaluation or control of special disabilities; and (11) extended employment for the severely handicapped who cannot be readily absorbed in competitive labor market.

PUBLIC LAW 90–396—AUTHORIZING SECRETARY OF COMMERCE TO ARRANGE FOR THE COLLECTION OF STANDARD REFERENCE DATA FOR THE BENEFIT OF SCIENTISTS AND THE GENERAL PUBLIC

H.R. 6297. Introduced by Mr. Millen of California on February 28, 1967

Standard Reference Data Act.—Authorizes and directs the Secretary of Commerce to provide or arrange for the collection, compilation, critical evaluation, publication and dissemination of reliable standardized scientific and technical reference data for the benefit of the Nation's scientists and engineers and the general public.

Authorizes the Secretary in carrying out this program, to utilize the reference data services and facilities of other agencies and instrumentalities of the Federal Government and of State and local governments, persons, firms, institutions, and associations, with their consent and in such a manner as to avoid duplication of these services and facilities. Requests that all agencies and instrumentalities of the Federal Government exercise their duties and functions in such manner as will assist in carrying out the purpose of this act.

Provides, in order to effect integration and coordination of standard reference data activities, that the Secretary, in consultation with other interested Federal agencies, shall prescribe and publish in the Federal Register such standards, criteria, and procedures for the preparation and publication of standard reference data as may be necessary to carry out the provisions of this act.

Provides that copyright may be obtained on behalf of the United States in standard reference data compiled and evaluated by the National Bureau of Standards, and that standard reference data may be sold by the Secretary or by a person or agency designated by him, and at prices which reflect the cost of collection, compilation, evaluation, publication, and dissemination of the data, including administrative expenses.

PUBLIC LAW 90–422—EXTENDING AUTHORIZATION FOR ACTIVITIES UNDER THE STATE TECHNICAL SERVICES ACT

S. 3245. Introduced by Mr. Magnuson on March 27, 1968

Extends for an additional year the authorization of appropriations under the State Technical Services Act of 1965.

PUBLIC LAW 90–423—EXTENDING PROGRAM OF RESEARCH AND DEVELOPMENT IN THE FIELD OF HIGH-SPEED GROUND TRANSPORTATION

H.R. 16024. Introduced by Mr. Staggers on March 18, 1968

Extends for 2 years the act of September 30, 1965, directing the research and development of high-speed ground transportation and replaces the Secretary of Commerce with the Secretary of Transportation as directing the program. Authorizes $16,200,000 for fiscal year 1969 and $21,200,000 for fiscal year 1970. Authorizes the Secretary to contract for the construction of two suburban rail stations in furtherance of a demonstration program.

Public Law 90-440—To Create a District of Columbia Air Pollution Control Board

S. 1941. Introduced by Mr. Tydings and others on June 13, 1967

District of Columbia Air Pollution Control Act.—Directs the District of Columbia to prescribe regulations to control the emissions of any kind of substance into the atmosphere and to protect and improve air quality in the District and sets forth standards for these regulations.

Directs the Commissioner of the District of Columbia to prepare and carry out a comprehensive program for the control and prevention of air pollution in the District, and sets forth minimum requirements for this program.

Public Law 90-448—Housing and Urban Development Act of 1968

S. 3497. Introduced by Mr. Sparkman on May 15, 1968

Title I, Low Income Housing.—Establishes a homeownership assistance program for low-income families in the form of periodic payments to reduce interest costs on a market-rate home mortgage or a cooperator's share of a cooperative association's mortgage; and makes eligible to receive such payments, nonprofit organizations sponsoring projects as well as the individual lower income purchasers, broadly revising the National Housing Act.

Limits the amount of mortgage to $15,000 ($17,500 in high-cost areas) except that for families of five or more persons the limits are to be $17,500 and $20,000; and these mortgages are to be insured under the new special risk insurance fund.

Allows assistance payments for poth urban and rural areas, and assigns administration of the payments in rural areas to the Secretary of Agriculture.

Authorizes the Secretary to provide budget debt management, and related counseling to homeowners with mortgages under the act.

Authorizes mortgage insurance for families of low and moderate income who cannot qualify for mortgage insurance under regular FHA programs because of credit histories but who the Secretary finds are "reasonably satisfactory credit risks" and capable of homeownership.

Establishes a special risk insurance fund out of which claims would be paid on mortgage insured under the new act and authorizes $5 million to be advanced from the general insurance fund to establish the new fund.

Allows a low- or moderate-income purchaser to buy an individual family unit and undivided interest in the common areas and facilities of the project, which could be financed with a mortgage with an interest rate no less than the below-market interest rate under the National Housing Act.

Authorizes the Secretary of HUD to undertake a program of assistance to nonprofit sponsors of low- and moderate-income housing, including provisions of information and technical assistance with respect to construction, rehabilitation, and operation of low- and moderate-income housing.

Creates a National Homeownership Foundation to provide technical and limited financial assistance to public and private organizations who provide increased homeownership and housing opportunities for lower income families. Provides that the Foundation would be administered by an 18-member board. Authorizes an appropriation of $10 million to carry out the Foundation's functions.

Directs the Secretary of HUD to institute a program under which public and private organizations shall submit plans for lower income housing using new and advanced technologies. Directs the Secretary to approve not more than five of such plans as most promising considering such factors as: (1) adaptability to large-scale construction at moderate cost; (2) maintenance of environmental quality; (3) the possibility of mass production of the technology; and (4) the financial soundness of the organization submitting the plan.

Authorizes the Secretary of HUD, in cooperation with the private insurance industry, to develop a plan for establishing an insurance program to enable homeowners to meet their monthly mortgage payments in time of personal economic adversity. Directs the Secretary to make a report on his actions along with his recommendation for establishing such a program within 6 months following enactment of this act.

Establishes a National Advisory Commission on Low-Income Housing to undertake a comprehensive study and investigation of the resources and capabilities in the public and private sectors of the economy which may be used to fulfill more completely the objectives of the national goal of "a decent home and suitable living environment for every American family" particularly as such goal relates to low-income families. Directs the Commission to make an interim report to the President and Congress by July 1, 1969, and a final report by July 1, 1970.

Title II, Rental Housing for Lower Income Families.—Establishes an assistance program for rental and cooperative housing for lower income families in the form of periodic payments to the mortgagee on behalf of the mortgagor, serving to reduce interest costs on a market rate project market; and allows the Secretary to reimburse the mortgagee for its expenses in handling the mortgage.

Increases the authorization for rent supplement program payments by $40 million for fiscal year 1970, and by an additional $100 million for fiscal year 1971.

Increases the authorization for annual contributions contracts for the low-rent public housing programs by $100 million on the date of enactment and by $150 million in 1969 and 1970.

Authorizes the Secretary of HUD to make grants to public housing agencies to assist in financing tenant services for families living in low-rent housing projects, with preference given to programs providing for maximum feasible participation of the tenants in the development and operations of such services.

Permits local housing authorities to sell any low-rent housing unit to a tenant if suitable for individual ownership without regard to the present limitations.

Permits public aid to Indian families without regard to the present limitation that such a family may not live on property located on or adjacent to their farm.

Provides an additional subsidy under section 2(2) of the United States Housing Act of 1937 for large families and families of unusually low income.

Title III, FHA Insurance Operations.—Permits payment of FHA mortgage insurance premiums by the Secretary of Transportation in cases where a serviceman assumed home mortgage previously insured under any other provision of the National Housing Act.

Directs the Secretary of Defense or the Secretary of Transportation to continue making the premium payment on behlaf of the widow of a serviceman who dies in service for 2 years after his death or until she sells the house.

Requires the Secretary of HUD to approve a request for the extension of time for curing a default on any FHA-insured mortgage on multifamily housing or for a modification of the terms of such a mortgage only pursuant to regulations prescribed by him.

Authorizes the Secretary of HUD to establish the interest rates on all FHA mortgage insurance programs except the land development program and the below-market interest rate program.

Extends the sales housing program for two, three, and four family residences to all low and moderate income families, and removes dividend restrictions for nondwelling facilities.

Authorizes the Secretary of HUD to insure supplemental loans made by financial institutions approved by him to multifamily projects and group practice facilities covered by FHA-insured mortgages. Defines "supplemental loan" as a loan made to finance improvements or additions to a project or facility. Prescribes qualifications of the supplemental loans making them eligible for insurance.

Increases the maximum loan limitation for home improvement loans of $3,500 to $5,000 and the maximum maturity from 5 years and 32 days to 7 years and 32 days, and also increases the maximum finance fees.

Extends the term of FHA land development mortgages from 7 to 10 years. Permits the extension of maturity of land development mortgages beyond 10 years if such an extension is determined warranted by unusual or unforeseen circumstances.

Permits the Secretary of HUD to establish interest rates for new mortgage insurance programs at such rates as he believes necessary to meet the market.

Permits the rehabilitation and sale of individual units in a multifamily structure, and permits the blanket mortgage to cover four or more units (now five or more units).

Increases the maximum mortgage obligation under the insurance program for mortgages on single-family housing in outlying areas from $12,500 to $13,500. [Amends 12 U.S.C. 1709(i).]

Authorizes FHA to insure mortgages on seasonal homes not exceeding $15,000 and 75 percent of the appraised value on an acceptable risk basis.

Title IV, Guarantees for Financing New Community Land Development.—New Communities Act—Authorizes the Secretary of HUD to guarantee the bonds, debentures, notes, and other obligations to help finance new community development projects.

Sets forth the requirements to make a new community development eligible for assistance, and any obligation eligible to ge guaranteed under the Act, limiting to $50 million any obligation for any single new community development project.

Authorizes the Secretary to establish guarantee fees and to make such other charges as he considers reasonable and to establish a guarantee revolving fund and make any guarantee by the Secretary conclusive evidence of the eligibility of the obligations.

Requires the Secretary to adopt measures to encourage small builders, and requires payment of prevailing rates as determined by the Secretary of Labor, in accordance with the Davis-Bacon Act, to laborers and mechanics employed in land development.

Authorizes the Secretary to make supplementary grants to States to local public bodies carrying out new community assistance projects.

Title V, Urban Renewal.—Authorizes the Secretary of HUD to provide financial assistance to local public agencies to assist them in carrying out neighborhood development programs.

Increases the amount of funds available for urban renewal by $1,400 million and for urban renewal in model cities area by $350 million and increases the rehabilitation grant to low-income homeowners from $1,500 to $3,000.

Removes the present limitation on the acquisition and rehabilitation of residential properties by a local urban renewal agency. Authorizes the Secretary of HUD to make grants for property determined to be uninsurable after inspection in order to make it meet reasonable underwriting standards imposed by a statewide property insurance plan approved by the Secretary under title XII of the National Housing Act.

Permits capital grants for low and moderate housing open-land projects, limiting the amount, but permits a local public agency to borrow funds to finance project undertakings on the private market at an interest rate in excess of the Federal lending rate set out in its loan contract with the Government.

Permits payment of capital grants for urban renewal projects and deems such projects completed upon a determination by the Secretary of HUD that not more than 5 percent of the acquired land remains to be disposed of, that the local public agency does not expect to be able to dispose of such land in the near future, that the agency has agreed to retain such land for use in accordance with the urban renewal plan, and that all other project activities are completed.

Extends the rehabilitation loan program from October 1, 1969, to June 30, 1973, and authorizes such loans: (1) in areas which are definitely planned for rehabilitation and where the property is an owner-occupied residential structure in violation of local housing codes, and (2) for property determined to be uninsurable after inspection pursuant to a statewide insurance plan approved under title XII of the National Housing Act. Limits rehabilitation residential loans to persons whose income does not exceed locally applicable limits for occupants of projects financed with below-market interest rate mortgages insured under section 221(d)(3) of the National Housing Act [12 U.S.C. 1715(d)(3)]; but exempts from the above limitation urban renewal and code enforcement projects now receiving financial assistance under title I of the Housing Act of 1949, which contemplate the use of these loans.

Authorizes the Secretary of HUD to make grants for demolition of nonresidential structures that are harborages or potential harborages of rats.

Permits the carrying out of air rights urban renewal projects and the construction of necessary foundations and platforms to provide educational facilities.

Requires that a majority of housing units in each community's urban renewal projects to be developed for predominately residential uses be for low- and moderate-income families, and that at least 20 percent of the total units in such projects in each community be for low-income families, except that the Secretary of HUD may waive the 20 percent requirement to the extent that units for low-income families are determined to be unnecessary.

Defers the effective date for meeting the code enforcement requirement for certification of workable programs for community improvement in the case of Indian tribes until January 1, 1970. [Amends 42 U.S.C. 1451(c).]

Authorizes the Secretary of HUD to contract to make grants, in an aggregate amount not to exceed $15 million in any fiscal year, to cities and other municipalities or counties to assist in taking interim steps to alleviate harmful conditions in any slum and blighted area of the community which is planned for rehabilitation or federally assisted code enforcement in the near future.

Provides for the utilization of local private nonprofit agencies for rehabilitation grants in code enforcement areas.

Provides for an additional relocation payment for displaced families, handicapped individuals, and individuals 62 years of age or over, of not to exceed $500 in the first 12 months and $500 in the second 12 months (present payment of not to exceed $500 may be made only during the first 5 months) after displacement. Authorizes a relocation payment of not to exceed $5,000 to a displaced individual who is the owner of real property occupied by him acquired for a project, if he purchases and occupies a dwelling within 1 year subsequent to displacement.

Title VI, Urban Planning and Facilities.—Authorizes the Secretary of HUD to make planning grants to State planning agencies for assistance to district planning agencies for rural and other non-metropolitan areas, and makes planning grants directly to tribal planning councils or other bodies for planning on Indian reservations.

Broadens the definitions of comprehensive planning to include planning for the provision of governmental services and for the development and utilization of human and natural resources but not to include planning aimed at encouraging businesses to reloca te from another area.

Changes the conditions for acquisition of land.

Provides that any funds authorized but not appropriated for the basic water and sewer facilities, neighborhood facilities, and the advance acquisition of land programs will remain available for appropriation through fiscal year 1970. Authorizes an appropriation of $115 million for fiscal year 1970 for grants for water and sewer projects.

Provides funding provision for open space land program.

Title VII, urban mass transportation.—Increases the amount appropriated under the Urban Mass Transportation Act by $190 million.

Broadens the definition of "mass transportation" to allow greater flexibility and scope and eliminates the present requirement that such transportation must serve the "general public" and operate "over prescribed routes."

Extends the emergency program from November 1, 1968, to July 1, 1970.

Permits private transit companies to furnish up to 50 percent of the local share of the net project cost of a mass transit project or, in cases of an applicant's (State or local public body) financial inability to put up any portion of the local share, private companies would be permitted to put up 100 percent of such share.

Title VIII, Secondary Mortgage Market.—Amends the Federal National Mortgage Association Charter Act to divide the heretofore existing FNMA into two separate corporations: (1) a Federal National Mortgage Association to continue its present secondary market operations as a Government-sponsored private corporation; and (2) a Government National Mortgage Association, to remain in the Government, which will exercise the special assistance and management and liquidating functions formerly exercised by the FNMA.

Permits the Government National Mortgage Association as a trustee under trusts created for sales of participation certificates to issue such certificates for refinancing purposes without regard to the requirement of appropriation act authority.

Allows the FNMA to issue and sell securities backed by a set-aside portion of its portfolio of mortgages.

Authorizes an additional $500 million for the purchase of mortgages by the Government National Mortgage Association in its special assistance function.

Makes certain transitional provisions regarding the capital stock of FNMA and provides for the board of directors during a transitional period.

Title IX, National Housing Partnerships.—Authorizes the creation of private corporations for profit, not to be Government agencies, with one to be called the Corporation, which shall have a 15-member board of directors, 12 elected by stockholders, and three appointed for 3-year terms by the President with the advice and consent of the Senate.

Authorizes the President to create additional corporations and outlines the powers of the corporation as well as process of organizations and financing of it.

Authorizes the corporation to form, as a separate organization, a limited partnership called the "Partnership" under the D.C. Uniform Limited Partnership Act. Permits the stockholders of the corporation and other persons to become limited partners in the partnership. Provides that the partnership may engage in any of the activities authorized for the corporation.

Requires the corporation to submit an annual report to the President for transmittal to Congress and to make audits annually.

Title X, Rural Housing.—Authorizes direct and insured loans in rural areas (areas with 5,500 population or less) to low- and moderate-income families, and provides for aid for rental or cooperative housing where no other aid is available.

Authorizes financial and technical aid to provide, in rural areas, housing and related facilities for rural trainees and their families enrolled in federally-aided training courses; but requires the Secretary of Agriculture to first consult with the Secretaries of Labor, HEW, and HUD, and Director of OEO to ascertain that no other aid is available.

Requires a labor area survey and full coordination among all Federal, State and local government agencies administering related programs before training and housing sites can be selected.

Requires that advances for land purchasers be payable within periods not longer than 33 years and to bear interest but other advances would be nonrepayable or repayable without interest.

Broadens the eligible purposes of domestic farm labor housing loans to include the purchase of necessary land for building sites.

Authorizes the Secretary of Agriculture to make grants to or contracts with public and private nonprofit organizations for the development, administration, or coordination of programs to aid low-income individuals and families in carrying out mutual or self-help housing efforts. Authorizes the Secretary to establish the Self-Help Housing Land Development Fund for such purposes, and to make loans to such individuals for the acquisition of land and building materials in order to construct decent dwellings under mutual or self-help programs. Authorizes not to exceed $5 million for each of the fiscal years 1969 through 1973. Prohibits the making of grants or contracts after June 30, 1973, except pursuant to a commitment or other obligation entered into before that date.

Title XI, Urban Property Protection and Reinsurance.—Urban Property Protection and Reinsurance Act—Establishes an Advisory Board, consisting of 19 members, to be appointed by the Secretary of HUD, representing the general public, the insurance industry, State and local governments, and the Federal Government, to advise the Secretary with respect to the following programs, which will terminate in 1973 (except for the reinsurance program, which will be liquidated in 1976):

Part A creates: (1) statewide plans to assure fair access to insurance requirements called "FAIR" plans, (2) an all-industry placement facility, (3) industry cooperation and plan evaluation programs.

Requires every insurer reinsured under this program to cooperate with the State insurance authority in each State in which it acquired reinsurance in carrying out statewide plans to assure fair access to insurance requirements, and requires these "FAIR" plans to be designed to make essential property insurance more readily available in urban areas.

Requires all plans to include an all-industry placement facility authorized to do business with every insurer in the State participating in the plan, helping agents and brokers place insurance up to full insurable value of a property, and to distribute risks equitably among insurers.

Requires every insurer participating in a plan to file a statement pledging its full participation and cooperation in carrying out the plan with the State insurance authority and with the corporation.

Provides for the State insurance authority to transmit copies of a plan and any amendments, to the Secretary and to advise him concerning the operation of the plan, and the need to adopt other programs to make essential property insurance more readily available.

Authorizes the Secretary to modify the criteria for plans if he finds that such action is necessary to carry out the purpose of the program.

Part B provides: (1) Reinsurance coverage by specifying reinsurance of losses from riots or civil disorders, (2) conditions for reinsurance agreements, and (3) recovery of premiums.

Authorizes the Secretary to offer to any insurer, subject to conditions of the Act, reinsurance against property losses from riots or civil disorders.

Authorizes the Secretary to enter into reinsurance contracts for riot losses, and to establish the amount of losses from riots or civil disorders each insurer must retain before the Secretary will reimburse the insurer for its losses.

Sets forth various conditions under which the Secretary will terminate existing reinsurance and provides that the Secretary will not offer reinsurance in any State if the State itself does not assume a portion of the responsibility for assisting the Secretary to reinsure against losses resulting from riots or civil disorders within 1 year after the enactment of the Act.

Authorizes the Secretary to recover any unpaid premiums for reinsurance, and imposes a 5-year statute of limitations on such recoveries and on the recovery by an insurer of excess premiums paid to the Secretary.

Part C sets forth general requirements for: (1) Claims and judicial review, (2) fiscal intermediaries and servicing agents, (3) a National Insurance Development Fund, (4) records, annual statements, and audits, (5) study of reinsurance and other programs, (6) services and facilities of other agencies—utilization of personnel, services, facilities, and information, (7) advance payments, (8) taxation, (9) appropriations, (10) financing, and (11) creation of the position of Federal Insurance Administrator in the Department of HUD.

Authorizes and directs the Secretary to conduct a study of reinsurance to help assure an adequate supply of burglary and theft and other property insured in urban areas and to help to assure adequate availability of surety bonds for construction contractors in urban areas and to undertake other studies.

Title XII, District of Columbia Insurance Placement Act.—Directs the establishment by all insurers licensed in the District of Columbia of an Industry Placement Facility to formulate and administer a program, subject to disapproval by the Commissioner of the District of Columbia, to apportion equitably insurance of property which is insurable and whose owners request the aid of the Facility to procure insurance.

Requires the Facility to assure all property owners fair access to basic property insurance through the normal insurance markets by submission of appropriate regulations to the Commissioner, who may adopt such of the regulations as he approves.

Authorizes the Commissioner to establish a joint underwriting association, to consist of all licensed insurers in the District of Columbia writing basic property insurance, if he finds, after notice and hearing, that such association is necessary to carry out the purposes of this title. Prescribes basic requirements for the operation of the association, and provides for the examination and inspection of the association's records by the Commissioner. Requires the association to submit an annual report to the Commissioner. Provides for appeal to the Commissioner after any final decision of any inspection bureau, the Facility, or the association, and provides for judicial review of final orders of the Commissioner. Authorizes the assessment of each insurer licensed to do business in the District of Columbia in order to make available to insurers reinsurance against losses to property resulting from riots or civil disorders.

Title XIII, National Flood Insurance.—National Flood Insurance Act—Establishes a program of Federal assistance for flood insurance, to be related to a unified national program for flood plain management. Authorizes companies to form pools, in accordance with conditions set out by the Secretary of HUD and under his supervision. Provides for Federal assistance in the form of premium subsidies and reinsurance coverage to compensate for heavy losses, with contributions of risk capital by participating insurance companies. Entitles the companies to a certain percentage of profits, and imposes on them the payment of a reinsurance premium to the Government. Allows other insurance companies to participate on other than a risk-sharing basis. Provides that the property owner will bear part of the cost of flood insurance in the form of a premium.

Authorizes an alternative program where the private insurance companies would act as fiscal agents for the Government, with the Government taking all financial risks involved, in the event that the program of private industry risk participation cannot be carried out.

Limits coverage to $17,500 for single-family dwellings, with a total of $30,000 coverage for any single structure occupied by two to four families. Insures contents for $5,000 per dwelling unit. Provides subsidized coverage of $30,000 for business and other properties which would also be eligible for a certain amount of unsubsidized coverage.

Provides that the Federal subsidies to the insurance pools represent the difference between the actuarial rate and the chargeable premium rate, both to be determined by the Secretary.

Gives preference in providing coverage to those areas which give assurance that they will adopt effective land use and flood control regulations by June 30, 1970, and those which show a positive interest in the program. Prohibits the writing of policies both after June 30, 1970, unless permanent land use regulations with effective enforcement provisions have been adopted, and at any time for property declared to be in violation of State or local land development ordinances.

Prohibits the granting of Federal disaster assistance with respect to property which is covered or could under certain circumstances have been covered by flood insurance under this act.

Authorizes the Secretary to identify within 5 years all flood plain areas and within 15 years to establish flood risk zones and make estimates of flood-caused losses in those zones.

Title XIV, Interstate Land Sales.—Interstate Land Sales Full Disclosure Act—provides for the regulation of sales of interests in subdivisions in commerce or through the mails. Requires a statement to include: The name and address of each person having an interest; a description of the property; the condition of the title to the land; the terms and conditions of the proposed disposal of the land; the condition of access to utilities; any blanket encumbrances; a copy of articles of incorporation; a copy of the deed; a copy of council or title insurance policy; copies of forms of conveyance to be used in selling parcels; and other information as may be deemed necessary.

Exempts certain property transactions from coverage, unless the method of disposition is adopted for the purpose of evasion of this act.

Title XV. Mortgage Insurance for Nonprofit Hospitals.—Authorizes the Secretary of HUD to insure a mortgage covering a new or rehabilitated hospital, including operational equipment, if the mortgage involves a principal obligation of not more than $25 million and does not exceed 90 percent of the estimated replacement cost of the property or project. Permits the Secretary also to insure a mortgage which provides permanent financing or refinancing of existing mortgage indebtedness in the case of a hospital presently lacking permanent financing which was completed after 1965 and the date of enactment of this act. Provides that the aggregate principal balance of all insured mortgages covering hospitals in need of permanent financing or refinancing shall not exceed $20 million at any one time.

Title XVI, Housing Goals and Annual Housing Report.—Directs the President to make a report to the Congress, not later than January 15, 1969, containing a 10-year plan for the national housing needs, along with legislative recommendations for fulfilling these needs. Requires annual reports to be made by the President on January 15, 1970, and on each succeeding year, showing the progress made under the plan and the reasons why, if any, the goals set forth in the plan have not been reached along with estimates of the need for the following year. Requires a final report to be submitted by January 15, 1979.

Title XVII, Miscellaneous.—Increases funds appropriated for planning assistance for model cities by $12 million for the fiscal year 1969.

Increases funds appropriated for model cities program financial, technical, and relocation assistance by $1 billion for the fiscal year 1970.

Permits grants for developing and testing urban renewal demonstration grant program and authorizes funds necessary for urban information and technical assistance program, and studies of advances in technology in housing and urban development.

Establishes a new program of annual grants to the existing college housing direct loan program.

Makes limited profit sponsors eligible for loans to provide housing for the elderly (only private nonprofit corporations now eligible), but limits such loans to 90 percent of development costs (other borrowers receiving up to 98 percent of development cost).

Expands the Federal-State training programs in the field of housing and community development.

Permits the Secretary of HUD to include the study of self-help in the construction, rehabilitation, and maintenance of housing for low-income persons and families in the low-income housing demonstration program.

Clarifies the authority of HUD concerning international housing.

Extends eligibility for rent supplement payment and allows consolidation of low-rent public D.C. housing projects, and extends time for the earthquake studies.

Makes other technical amendments to the National Housing Act.

Expands and clarifies the provisions of law relating to the establishment and administration savings accounts by savings and loan associations. [Amends 12 U.S.C. 1464(b)]

Authorizes Federal savings and loan associations to invest in time deposits or certificates of deposit in banks insured by the FDIC. [Amends 12 U.S.C. 1464(c)]

Broadens the authority of a Federal savings and loan association to invest up to 1 percent of its assets in loans guaranteed by the Agency for International Development to help finance housing projects or home financing institutions in developing nations outside of Latin America.

Permits a Federal savings and loan association to make loans for the construction of new structures related to residential use of the property under the existing exception applicable to property improvement loans.

Authorizes a Federal savings and loan association to invest in loans to federally supervised financial institutions secured by investments in which the association has statutory authority to invest directly.

Authorizes Federal home loan banks, subject to regulations by the Federal Home Loan Bank Board, to purchase AID-guaranteed housing loans and to sell participations therein to any bank member.

Amends the Federal Reserve Act to change the time limit on construction loans to 36 months (now 24 months), to permit national banks to continue to purchase participations in existing mortgages, and to provide that loans by national banks shall not be considered real estate loans where a bank looks primarily for repayment out of income of borrowers or security other than real estate (even though a bank may take a mortgage on real estate as additional security for a loan).

Eliminates the present corporate status of HUD's public housing entity and provides for the retirement of its $1 million capital stock.

Extends the study of the savings and loan industry authorized by the Second Supplemental Appropriation Act, 1966 until 1969.

Extends the provisions of the Small Business Act to the Trust Territory of the Pacific Islands.

PUBLIC LAW 90–454—AUTHORIZING COOPERATION WITH THE STATES IN PROTECTING AND DEVELOPING ESTUARINE AREAS OF THE COUNTIES WHICH HAVE SPORTING, SCENIC, OR RECREATIONAL VALUE

H.R. 25. Introduced by Mr. Dingell on January 10, 1967

Directs the Secretary of the Interior—in consultation with the States, the Secretary of the Army, and other Federal agencies—to conduct a 2-year study and inventory of the Nation's estuaries and the waters of the Great Lakes. Requires the Secretary to submit, not later than January 30, 1970, to the Congress through the President, a report of the study, together with legislative recommendations, including recommendations on the desirability of establishing a nationwide system of estuarine areas, the terms, conditions, and authorities to govern such system, and the designation and acquisition of any specific area which he believes should be acquired by the United States. Provides that no lands could be acquired unless authorized by a subsequent act of Congress.

Authorizes the Secretary to enter into an agreement with the State of New York or any political subdivision or agency thereof, for the permanent management, development, and administration of certain publicly owned lands on Long Island, N.Y., which were studied by the Secretary in conjuction with the State of New York in 1961 and 1965, the costs of which would be shared in an equitable manner. Directs the Secretary to study publicly owned areas in other States with a view toward recommending the desirability of authorizing the Secretary to enter into similar agreements for the administration management, and development of those areas.

PUBLIC LAW 90–515—TO ESTABLISH A NATIONAL WATER COMMISSION TO REVIEW WATER RESOURCES PROBLEMS AND PROGRAMS

S. 20. Introduced by Mr. Jackson and others on January 11, 1967

Establishes a National Water Commission to provide for a comprehensive review of national water resources problems and programs.

Provides that the seven members who shall make up the commission are to be appointed from outside the Federal Government by the President and serve at his pleasure.

PUBLIC LAW 90–547—TO INCREASE AUTHORIZATIONS FOR WATER RESOURCES PLANNING ACTIVITIES

S. 3058. Introduced by Mr. Jackson on February 29, 1968

Increases the authorization of appropriations for administering the Water Resources Planning Act.

PUBLIC LAW 90–575—HIGHER EDUCATION AMENDMENTS OF 1968

S. 3769. Introduced by Mr. Morse on July 11, 1968

Extension of Education Programs; Higher Education Act of 1965.— The Community Service and Continuing Education Programs. Extends for 4 years and authorizes appropriations of $50 million each for the fiscal years 1969 and 1970 and $60 million each for the fiscal years 1971 and 1972.

Extends the authorization for appropriations for college library resources for 4 years by authorizing appropriations of $50 million for the fiscal year 1969, $75 million for the fiscal year 1970, and $90 million each for the fiscal years 1971 and 1972.

Extends for 4 years the appropriation authorization for library training and research at a level of $15 million for the fiscal year 1969, $28 million for the fiscal year 1970, and $38 million each for the fiscal years 1971 and 1972.

Extends for 4 years the authorization of appropriations to the Commissioner of Education to enable him to transfer funds to the Library of Congress for purposes of strengthening college and research library resources through the Library of Congress, at the level of $10 million for fiscal year 1969 and $11,100,000 for each of the next 3 fiscal years.

Extends the appropriation authorization for the developing college program for 4 years at the level of $55 million for the fiscal year 1969, $70 million for the fiscal year 1970, $91 million for the fiscal year 1971, and $96,500,000 for the fiscal year 1972.

Extends for 4 years the appropriation authorization for educational opportunity grants at the level of $70 million for fiscal year 1969, $100 million for fiscal year 1970, and $140 million each for the fiscal years 1971 and 1972.

Extends for 4 years the authorization for Federal insurance of student loans.

Extends for 4 years the terminal dates for making student loans on which the Commissioner may make payments to reduce student interest costs under the insured loan program.

Extends for 4 years the appropriations authorizations for the college work-study program, at the level of $200 million for the fiscal year 1969, $255 million for the fiscal year 1970, and $285 million each for the fiscal years 1971 and 1972.

Extends for 2 fiscal years, i.e., through fiscal year 1972, the specific dollar appropriations authorizations of education professions development.

Extends for 4 years (i.e., through fiscal year 1972) the program of equipment for higher education at the current authorization level of $60 million per year for equipment other than television equipment, and $10 million per year for instructional television equipment.

Extends for 4 years, at the current authorized level of $5 million per year, the program which authorizes grants or contracts for higher education workshops or institute (1) for training in the use of educational media equipment in institutions of higher education or (2) for specialists in educational media or librarians or other specialists using such media.

Higher Education Facilities Act—Grants for Undergraduate Facilities.—Extends for 3 years the authorization for appropriations for grants for construction of academic facilities under title I of the act at the present authorization level of $936 million per year.

Extends for 3 fiscal years, the current authorization for appropriations of $7 million per fiscal year for administration of State plans.

Extends for 3 fiscal years, graduate facilities at the present level of $120 million per fiscal year.

Extends for 3 fiscal years, loans for facilities through the fiscal year 1972, at the current annual level of $400 million, the authorization for appropriations for payments into the revolving fund for loans for construction of academic facilities.

Permits disaster assistance to public institutions of higher education if the institution is located in an area which suffered a major disaster before July 1, 1972.

National Defense Education Act.—Extends for 4 years the authorization for specific appropriations for capital contributions to student loan funds, at the level of $250 million for fiscal year 1969, $275 million for the fiscal year 1970, and $300 million each for fiscal years 1971, and 1972, for student loans.

Extends for 4 years the authorizations for appropriations and allotments for assistance for acquisition of equipment and materials and for grants for administrative expenses and supervisory services of State educational agencies.

Extends for 4 years the national defense fellowship program and would continue through that period the present annual limit of 7,500 fellowship awards.

Extends for 4 years State programs of guidance, counseling, testing, etc., by authorizing appropriations of $30 million for fiscal year 1969,

$40 million for fiscal year 1970, and $54 million each for fiscal years 1971 and 1972.

Extends for 4 years the language development programs and authorizes appropriations of $19 million for fiscal year 1969, $30 million for fiscal year 1970, and $38,500,000 each for fiscal years 1971 and 1972.

Continues for 4 fiscal years the appropriation authorization for educational media at the level of $5 million per year.

Extends for 4 years the authorization for appropriations for improvement of statistical services of State educational agencies.

Continues for 4 years the authorization for appropriations for institutes for teachers in specified subjects or fields and for educational media specialists.

International Education Act.—Extends for 3 years the fiscal year 1969 appropriation ceiling of $90 million contained for international education.

National Vocational Student Loan Insurance.—Extends for 4 fiscal years the various time-limited authorizations under the National Vocational Student Loan Insurance Act of 1965.

National Foundation on the Arts and the Humanities.—Extends for 4 fiscal years the authorization for annual appropriations of $500,000 to the Commissioner of Education for grants to State educational agencies, or loans to nonprofit private schools, for equipment suitable for use in providing education in the humanities and the arts.

Student Assistance Programs: Assistance for State Educational Opportunity Grant Programs.—Provides that upon request of a State Governor not to exceed 15 percent of the State's apportionment (not including any reapportionments) under this part for any fiscal year shall, instead of being distributed to institutions by the Commissioner, be paid to the State for educational opportunity grants to such students if (1) the State provides an equal amount for this purpose, (2) the amount provided by the State represents an increase in expenditures for this purpose by the State over the amount expended by it for this purpose for the previous fiscal year, and (3) such educational opportunity grants are made on the same basis of need, and are substantially the same in other respects, as educational opportunity grants by institutions of higher education under this part of the act.

Talent Search Provisions.—Authorizes the Commissioner to make grants, as well as to enter into contracts, for carrying out this section, and by making it a purpose of such grants or contracts to identify qualified youths of "financial or cultural need" with "an exceptional potential for postsecondary educational training," rather than "exceptional financial need" alone, and to encourage them to complete secondary school and to take such training.

Special Services for Disadvantaged Students.—Authorizes the Commissioner to make grants to or contracts with institutions of higher education and other public or private nonprofit organizations for the planning, development, or carrying out of programs or projects of remedial and other special services for students with an academic potential who are enrolled or accepted for enrollment at the institution concerned and who by reason of deprived education, cultural or economic background, or physical handicap are in need of such services to assist them to initiate, continue, or resume their postsecondary education.

Authorizes to be appropriated $15 million for the fiscal year 1970, and $80 million each for the fiscal years 1970 and 1971.

Student Loan Insurance Programs.—Authorizes deferment of repayment of State or privately insured loan during attendance of student borrower at eligible institution or during military, Peace Corps, or VISTA service; and authorizing Federal payment of all interest accruing during and such attendance or service.

Increases the interest payments made by the Commissioner from 3 percent per annum to the total amount of interest which accrues on the loan during any period during which payment of principal is deferred because the borrower is attending an eligible institution or is a member of the Armed Forces, or a volunteer in the Peace Corps or VISTA.

Authorizes deferral of repayment of any such loan from a State loan program, or insured under a State or private nonprofit program, during such attendance or service; and provides for the Commissioner to pay interest on any such outstanding loan (whether made under a State loan program, or insured under a State, Federal, or private nonprofit program) during any such attendance or service.

Provides for coordination between non-Federal and Federal programs with respect to maximum amounts of individual loans insured, length of installment obligations, and minimum amounts of repayment installments on such loans.

Provides that the maximum amount of loans to any one student outstanding at any one time that may be insured by State and private nonprofit loan insurance agencies is $7,500.

Provides that the maximum amount of loans to undergraduate (as well as graduate or professional) students that may be insured by the Commissioner in an academic year or its equivalent would be $1,500 rather than $1,000, and that the aggregate amount of loans that may be outstanding at any one time for such students would be $7,500 rather than $5,000.

Provides that the State or private nonprofit guarantee agency must require the borrower, after commencement of the repayment period, to make repayments, with respect to the total of his loans insured under the programs covered by this part, at a rate of not less than $360 per year.

Provides for Federal guarantee of student loans insured by States or nonprofit private institutions or organizations, and Federal advances to reserve funds of student loans insurance programs of such States, institutions, or organizations.

Authorizes the Commissioner to enter into a guarantee agreement with any State or nonprofit private insurer under the program to guarantee on behalf of the United States, 80 percent of each student loan insured by such insurer. No guarantee premium would be charged.

Provides for increase of Federal advances to reserve funds of student loan insurance programs of States or nonprofit private insittutions or organizations.

Provides for increases of maximum interest rate under student loan insurance programs and administrative costs.

Increases from 3 to 4 percent, during the loan repayment period, the Federal interest subsidy that is payable with respect to loans to students (having an adjusted family income of less than $15,000) which are made under the Federal or a non-Federal student loan insurance program or under a State student loan program covered by the act.

Enables the Commissioner to pay to lenders, from interest subsidy appropriations, a temporary "administrative cost allowance" not to exceed the rate of 1 percent of unpaid balances of the principal of insured student loans made by such lenders, in States which have a statutory interest limitation (including any such limitation in a statute establishing a State student loan insurance program) that does not permit an interest rate of 7 percent, if the Commissioner determines that such statutory limitation threatens to impede the carrying out of the purpose of the act. Provides that this allowance, when authorized by the Commissioner, may be paid on loans insured, and installments paid, on or after the date of enactment of the bill, but may not continue beyond the 4th month after adjournment of the State's first regular legislative session adjourning after January 1, 1969. Permits this allowance to be paid without regard to the student's adjusted family income.

Provides that the amendments made by this section shall not apply to (1) any loan made or contracted for prior to the date of enactment of the bill, or (2) any loan, or part thereof, made to consolidate a loan made or contracted prior to that date.

Provides that each eligible lender that is entitled to have the Federal Government pay it interest on behalf of student borrowers could elect to have the Federal Government retain and accumulate the Federal interest payments on student loans, and to pay over such accumulation (with interest at a rate determined by the Treasury in the light of the average yield of U.S. obligations of 5-year maturity) when the final payment of principal falls due.

Increases from one-fourth of 1 percent to one-half of 1 percent the maximum premium that may be charged by the Commissioner for Federal loan insurance.

Permits Federal savings and loan associations to invest in loans for vocational education. Makes pension funds eligible for lending if approved by the commissioner for this purpose.

Loan Forgiveness and Payments on Insured Loans.—Provides for increases in loan forgiveness for teachers in schools serving children from low-income families, and authorization of such forgiveness for service in Armed Forces.

Provides for loan cancellation for service in the Armed Forces of the United States at the rate of 25 percent of the total amount of the loan (plus interest), for each complete year of consecutive service, and to provide for increasing the rate of cancellation from 15 to 20 percent for each complete academic year of service as a full-time teacher in a public or other nonprofit elementary or secondary school which is in the school district of a local educational agency which is eligible in such year for assistance under the Elementary and Secondary Education Act of 1965, and which has been determined by the Commissioner, pursuant to regulations and after consultation with the State educational agency of the State in which the school is located, to be a school in which there is a high concentration of students from low-income families.

Provides for cancellation of student loans made for service in the Armed Forces at a rate of 25 percent per annum for each complete year of consecutive service for health professions under the Public Health Service Act.

Provides for payment by the Commissioner, on behalf of the borrower, of the total amount owed by a borrower under Federal and non-Federal loan insurance programs in the case of his death, or in the case of his permanent and total disability (as determined in accordance with regulations of the commissioner).

Enables the Commissioner to make grants to proprietary schools which otherwise qualify as an "institution of higher education" to conduct work-study programs whereby students could be employed by public or other nonprofit institutions to perform work in the public interest.

Provides that the Federal share of the compensation of students employed in work-study programs will not exceed 90 percent under the Economic Opportunity Act.

Cooperative Education.—Establishes cooperative education program. Authorizes to be appropriated $8 million for fiscal year 1970 and $10 million each for fiscal years 1971, and 1972, for grants under section 472 by the Commissioner of Education to institutions of higher education for planning, establishing, expanding, or carrying out programs of cooperative education that alternate periods of academic study with full-time public or private employment which will give the student, so far as practicable, work experience related to his academic or occupational objective.

Authorizes appropriations of $500,000 for the fiscal year 1969, and $750,000 for each of the next 3 fiscal years, for making such grants or contracts for the training of persons in the administration of cooperative education and for research into methods of improving the use of cooperative education.

Prohibits the use of funds appropriated under this part for paying the compensation of employed students.

Requires that each application: (1) set forth programs or activities for which a grant is authorized; (2) provide for maintenance or program expenditures to at least the level maintained during the fiscal year preceding the year of application; (3) provide for the making of reports, in such form and containing such information as the Commissioner may reasonably require, and the keeping of such records and affording such access thereto as the Commissioner may find necessary; (4) provide for such fiscal control and fund accounting procedures as may be necessary; and (5) includes such other information that may be necessary.

Provides that from funds appropriated, the Commissioner is authorized to (1) make grants to or contracts with institutions of higher education, or combinations of such institutions, and (2) make grants to other public or private nonprofit agencies or organizations, or contracts with public or private agencies or organizations when such grants, or contracts will make an especially significant contribution to attaining the objective of this section.

General Provisions Concerning Student Assistance.—Provides a statement of congressional intent on appropriate governmental assistance for universal educational opportunity at the post-secondary level; and provisions for a study and plan with respect thereto by stating that it is the intent of Congress that universal educational opportunity at the postsecondary level be made available through appropriate governmental assistance.

Provides for the appointment of a Presidential commission to be known as the Universal Educational Opportunity Commission. Requires the commission to submit a plan or alternative plans for providing universal educational opportunity at the postsecondary level, to be developed on the basis of a study or studies which take into account among other things various existing financial aid programs, income tax devices, and such factors as the cost effectiveness and the immediate and longrun economic impact of alternative plans.

Provides that no grant, award, or loan, to any student under any act amended by the bill shall be considered in duplication of benefits for the purpose of section 1781 of title 38, United States Code.

Provides that nothing in this bill, or any act amended by the bill, shall be construed to prohibit any institution of higher education from refusing to award, continue, or extend any financial assistance under any such act to any individual because of any misconduct which in its judgment bears adversely on the individual's fitness for such assistance.

Provides that for purposes of programs assisted under titles I, IV, X, XIV, XVI, or XIX of the Social Security Act, no grant or loan to an undergraduate student which is made or insured under any program administered by the Commissioner shall be considered income or resources.

Authorizes to be appropriated $150,000 to be used by way of grant to or contract with public or private organizations to develop, within a year after enactment, improved testing and other procedures for (1) determining more accurately the capabilities of students with varying social and economic backgrounds for postsecondary education and (2) for making such determinations initially while the student is in his first or second year of secondary education.

Provides for payment to institutions of higher education operating fellowship programs, of an amount consistent with prevailing practices under similar federally supported programs.

Amendments to Other Titles of the Higher Education Act of 1965.— Community service and continuing education programs. Provides that the Federal share of costs of State plans under this program, which dropped to 50 percent after fiscal year 1967, will again be 75 percent for fiscal years 1969 and 1970, and will drop to 65 percent for the fiscal year 1971 and 60 percent for the fiscal year 1972.

College Library Assistance, Research, and Training.—Library assistance for new colleges. Makes eligible, in accordance with criteria prescribed by regulation, new institutions of higher education in the fiscal year preceding the first year in which students are to be enrolled.

Authorizes the Commissioner to make grants to institutions of higher education and other public or private agencies, institutions, and organizations, for the planning or development of programs for the opening of library or information science schools, or of programs intended to lead to the accreditation of such existing schools.

Enables the Commissioner of Education to transfer funds to the Librarian of Congress for specified purposes by: (1) making it clear that the purpose of clause (1) of the section will be met through acquisition, by the Librarian of Congress so far as possible, of "copies of" all library materials currently published throughout the world which are of value to scholarship; (2) authorizing the Librarian of Congress to compile and distribute catalog and general bibliographic information about library materials forming a part of the collection of the Library whether or not such materials were recently acquired under the authority of section 231; and (3) adding authority to enable the Librarian of Congress to pay administrative costs of cooperative arrangements for acquiring library materials published outside of the United States, its territories and possessions, and not readily obtainable outside of the country of origin, for institutions of higher education, combinations thereof, or other public or private nonprofit research libraries.

Strengthening Developing Institutions.—Increases the share for junior colleges in title III of Higher Education Act of 1965. Authorizes the Commissioner of Education to award grants to professors retired from active duty at institutions of higher education that do not qualify for assistance under title III, to encourage such professors to teach and conduct research at developing institutions.

Education Professions Development.—Extends these programs to poverty areas. Extends the provisions of medical insurance coverage to Teachers Corps not otherwise covered.

Allocation of Fellowships Under Title V–C of Higher Education Act of 1965.—Provides that in determining what is an equitable distribution of fellowships the Commissioner may take into account such factors as the respective numbers of children aged 3 to 17, the undergraduate student enrollments in institutions of higher education, and the numbers of lower income families, in the States. Provides that the basic distribution rule shall not apply to the extent that the Council determines that an urgent need for a certain category of education personnel is not likely to be met without a preference in favor of that category, in which event the Commissioner may give preference to programs designed to meet that need, but that in no case may such preferred programs constitute more than 50 percent of the total number of fellowships awarded in a year.

Requires the Commissioner, in making grants and contracts for programs or projects to seek to achieve an equitable geographical distribution of training opportunities throughout the Nation, taking into account such factors as the relative numbers of children aged 3 to 17 and numbers of low-income families in the States.

Instructional Equipment.—Provides that in order to promote coordination of Federal programs for assistance in the purchase of special equipment for education in the natural or physical sciences, the Commissioner shall consult with the National Science Foundation and other agencies in developing general policy in respect of such equipment.

General Provisions Amendments.—Conforms definitions of institution of higher education in Higher Education Act of 1965 and National Defense Education Act of 1958. Establishes an Advisory Council on Graduate Education chaired by the Commissioner of Education, to include—in addition to one representative each from the Office of Science and Technology in the Executive Office of the President, the National Science Foundation, the National Foundation on the Arts and Humanities—members appointed by the Commissioner from among leading authorities in the field of education, except that at least one of the appointed members would have to be a graduate student.

Provides that the Council shall advise the Commissioner on matters of general policy arising in his administration of programs relating to graduate education.

Authorizes the Commissioner to furnish to the Council such technical assistance and to make available to it such secretarial, clerical, and other assistance and such pertinent data as the Council may require.

Amendments to Higher Education Facilities Act.—Provides that construction of a facility may be considered to result in the expansion of the institution's student enrollment capacity (which is a basic condition for eligibility under title I of the Act) if the Commissioner finds that the student enrollment capacity of an institution would decrease if an urgently needed academic facility is not constructed.

Authorizes the Commissioner to enter into contracts with institutions of higher education or with higher education building agencies to make annual grants to reduce costs of securing loans from sources other than title III for the construction of academic facilities.

Requires such contracts to be entered into where the Commissioner finds that: (1) The applicant was unable to secure the necessary funds for construction of the facilities upon reasonable terms and conditions, and (2) the construction would be undertaken in an economical manner and would not be of elaborate or extravagant design or materials.

Amendments to the National Defense Education Act.—Directs the Commissioner to seek to achieve an equitable geographical distribution of graduate programs supported under title IV based on such factors as students enrollment in institutions of higher education and the number of faculty members who have not attained the degree of doctor of philosophy or its equivalent as compared with the number of members, who have such a degree.

New Educational Programs.—This section of the bill would amend the Higher Education Act of 1965 by redesignating title VIII as title XIII, by making appropriate changes in section numbering, and by inserting after title VII the following new titles:

Networks for Knowledge.—Authorizes a program of project grants to stimulate colleges and universities to share their technical and other educational and administrative facilities and resources through cooperative arrangements while maintaining their institutional identities, and to test and demonstrate the effectiveness of a variety of such arrangements, preferably on a multi-institutional basis. Such grants might be made directly to the colleges or universities involved or, when deemed more effective, to other established public or nonprofit private agencies or organizations.

Education for the Public Service.—Provides grants and contracts to strengthen and improve education for the public service. Provides that grants or contracts could be made only upon application to the Commissioner that sets forth an authorized purpose for which such grant or contract may be made and relates such purpose to the program described in any application of such applicant.

Authorizes the use of payments for part of the compensation of students employed in non-Federal public service as part of a program approved for assistance under this section, and also encourage agencies of the Federal Government to enter into arrangements with institutions of higher education for the employment of students enrolled in approved programs.

Public Service Fellowship.—Authorizes the Commissioner to award graduate or professional fellowships, not to exceed 3 academic years, to persons who plan to pursue careers in public service.

Directs the Commissioner to allocate fellowships under title IX among institutions of higher education with programs approved under the provisions of part B, for the use of persons accepted into such programs, so as most nearly: (1) To provide an equitable distribution of such fellowships throughout the United States; and (2) to attract recent college graduates to pursue public service careers.

Directs the Commissioner to approve a program under part B upon application by the institution and his findings that the program is significantly aimed at education persons for the public service, or in a field for whose practitioners there is a significant and continuing need in the public service; that the program is or may be expected to be of high quality; that the application describes the relationship of the program to the program described in any application of such institution submitted; that the application contains satisfactory assurance that the institution will recommend fellowship awards only for persons of superior promise who have demonstrated a serious intent to enter the public service, and will make reasonable continuing efforts to encourage fellowship recipients to enter the public service upon completing an approved program.

General Provisions.—Requires the Commissioner in administering title IX to give primary emphasis to the assistance of programs and activities not otherwise assisted by the Federal Government.

Authorizes payments to be made in installments, and in advance or by way of reimbursements, with necessary adjustments on account of overpayments or underpayments.

Prohibits grants, contracts, or fellowships awarded to, or for study at, a school or department of divinity, which term is defined as in the Higher Education Facilities Act of 1963 and the Higher Education Act of 1965.

Authorizes the Commissioner in administering title X, to utilize the services of any agency of the Federal Government and of any other public of nonprofit agency or institution, on a reimbursable basis or otherwise in accordance with agreements between the Commissioner and the head thereof.

Improvement of Graduate Program.—Strengthens and improves the quality of graduate programs leading to a doctoral or professional (other than medical) degree, and increases the number of such quality programs.

Provides that there be authorized to be appropriated $5 million for the fiscal year ending June 30, 1970, and $10 million for each of the 2 succeeding fiscal years, for grants by the Commissioner to institutions of higher education having programs leading to the Ph. D. or comparable professional or other degree to pay part of the cost of carrying out projects or activities.

Provides that payments under this title may not exceed 90 percent of the total cost of the project or activity; nor may they exceed 50 percent of the cost of the purchase or rental of books or equipment or other materials (after taking into account sums received as Federal financial assistance under other programs); nor be used for sectarian instruction or religious worship or in connection with any part of the program of a school of divinity, which term is defined as in the other titles of the Higher Education Act of 1965.

Requires the Commissioner, insofar as practicable and consistent with the other purposes of the title, to give weight to the objective of having an adequate number of graduate and professional schools of good quality within each appropriate region.

Requires the Commissioner to consult with the National Science Foundation, the National Foundation on the Arts and the Humanities, and the Federal Judicial Center for the purpose of promoting the coordination of Federal programs bearing on the purposes of this title.

Law School Clinical Experience Programs.—Authorizes contracts with law schools accredited by a nationally recognized accrediting agency or association approved by the Commissioner to pay not to exceed 90 percent of the cost (but not more than $75,000 in any fiscal year to any one law school) of programs to provide their students with clinical experience in the practice of law, and preferably with experience in the preparation and trial of cases. Federal assistance to establish or expand such programs might cover necessary expenditures for planning, equipment, training of faculty members, salary for additional faculty members, travel and per diem for faculty and students, reasonable stipends to students for public service rendered outside the academic year as part of such a training program, and any other items allowed pursuant to regulations issued by the Commissioner.

United States Foreign Service Corps.—Establishes a United States Foreign Service Corps consisting of students and Government employees selected for admission under the provisions of this title, and enrolled in a program of education, training, or research, or a course of study approved by the Board of Trustees established hereunder.

Public Law 90–576—Vocational Education Amendments of 1968

H.R. 18366. Introduced by Mr. Perkins and other on July 8, 1968

Requirement with respect to enforcement of compulsory attendance laws.—Requires the appropriate official in each school district to certify, as a condition of obtaining any funds from any program under this act, that the compulsory attendance laws are being enforced in his district. Subjects a person making a false certification to a criminal penalty of not more than $10,000 or imprisonment of not more than 5 years or both.

Directs the Secretary to prepare recommendations to Congress which shall indicate the minimum level of educational attainment that he feels essential to meet the needs of today.

Elementary and Secondary Education Act amendments.—Provides for preschool programs for children of low-income families and for the allotment of funds to the States in equal proportions based on (1) the relative number of public assistance recipients in each State as compared to all States; (2) the average number of unemployed persons as compared with the Nation; and (3) the relative number of children living with families with incomes of less than $1,000 in each State as compared to all States.

Provides the necessary requirements for State plans, authorizes the Commissioner to approve or disapprove the State plans, and provides for judicial review of his actions.

National School Lunch and Child Nutrition Acts Amendments.— Provides for temporary emergency assistance to provide nutritions to needy children in school and in other group activities outside of school.

Vocational Education Amendments, General Provisions.—Authorizes appropriations of $575 million for fiscal year 1970, $750 million for fiscal year 1971, and $750 million each for fiscal years 1971 and 1972. Further provides that of the sums appropriated for any fiscal year, 70 percent shall be for grants for State vocational education programs, 10 percent for grants for research and training in vocational education, 10 percent for grants for exemplary programs and projects, and 10 percent for grants for State special emphasis programs. Authorizes for each fiscal year through 1973, the appropriation of such sums as may be necessary to pay the cost of administering State plans and of the activities of advisory councils under this title and of evaluation and dissemination activities required by the title.

Provides for allotting the sums appropriated for any fiscal year on the basis of per capita income and the number of persons in the various age groups, so that, as multiplied by the State's allotment ratio, 55 percent is based on the population aged 15 through 19, 30 percent on the population aged 20 through 24, and 15 percent on the population aged 25 through 65, so that each State's allotment ratio is determined by subtracting from 1.00 the product of 0.50 multiplied by the quotient obtained by dividing the State per capita income by the national per capita income.

Allows no State to have an allotment ratio higher than 0.60 or lower then 0.40, and the allotment ratio for Puerto Rico, Guam, American Samoa, the Virgin Islands, and the Trust Territory of the Pacific Islands shall be 0.60.

Creates a National Advisory Council on Vocational Education consisting of 21 members appointed by the President. Includes persons representative of labor and management, persons representative of new and emerging occupational fields, persons familiar with manpower problems and administration of manpower programs, persons knowledgeable about the administration of State and local vocational education programs, persons experienced in the education and training of handicapped persons, persons familiar with the special problems and needs of individuals disadvantaged by their socio-economic backgrounds, persons having special knowledge of postsecondary and adult vocational education programs, and (constituting at least one-third of the total membership) persons representative of the general public, including parents and vocational students.

Directs the National Council: (1) To advise the Commissioner concerning the administration of, general regulations for, and operation of, vocational education programs; (2) to review the administration and operation of this title, including the effectiveness of such programs in meeting their purposes; (3) to review and advise the Commissioner on the approval of each State plan and application for assistance under the title; (4) to review, evaluate, and transmit at least annually to Congress and the President all reports, evaluations, and materials which the State advisory councils are required to submit to the National Council; (5) to conduct independent evaluations of programs and disseminate the results thereof; (6) to make recommendations for the improvement of administration and operations, including recommendations for legislative changes; (7) and to prepare an annual report on its evaluations, recommendations, and other activities and on vocational education and its effectiveness in meeting the need for vocational education throughout the Nation and submit such report to the Congress and the President before January 31 of each year.

Provides that any State desiring to receive a grant under this title for any fiscal year shall establish a State advisory council.

Directs each State advisory council to include as members persons who are familiar with the vocational education needs and the problems of management and labor in the State, who represent State industrial and economic development and planning agencies; who are representative of community and junior colleges, area vocational schools, technical institutes, and other post-secondary or adult education agencies or institutions which may provide programs of vocational or technical education and training, and other institutions of higher education, who are familiar with the administration of State and local vocational education programs.

Provides that the State advisory council shall evaluate vocational education programs, services, and activities and shall prepare and submit an annual evaluation report through the State board to the Commissioner and to the National Advisory Council on Vocational Education.

Authorizes the Commissioner to delegate any of his functions (except the making of regulations) under the Vocational Education Act of 1963 to any subordinate in the Office of Education. Authorizes the Commissioner to utilize the services and facilities of any Federal or public nonprofit agency or institution pursuant to an agreement, and to pay for such services in advance or by way of reimbursement as agreed upon. Directs the Commissioner to consult with the Department of Labor and other Federal departments and agencies administering programs which may be coordinated with those carried out under the Vocational Education Act of 1963, and to the extent practicable to coordinate such programs on the Federal level with those of other Federal departments and agencies and to require State and local authorities to coordinate their programs and projects of other public or private programs with the same or similar purposes.

Provides that if appropriations for any fiscal year do not exceed $300 million, 90 percent of the sums appropriated for that fiscal year shall be for grants to States under the State plan program and 10 percent for grants and contracts under the research and training authority, and unless the sums appropriated for a fiscal year are less than the total amount of appropriations for grants under the Vocational Education Act and the George-Barden Act for fiscal year 1969, the amount allotted to any State for the State plan programs shall not be less than the amount alloted to such State under such acts for fiscal year 1969.

States that nothing in the Vocational Education Act of 1963, shall be construed to authorize any payment for religious worship or instruction, or for the construction, operation, or maintenance of any facility to be used for such purposes. Requires that preference in federally compensated work under such programs or projects be given to students from low-income families. Provides that such program or project will not displace employed workers or impair existing contracts for services, and provides that no Federal compensation may be paid to a student for work for profitmaking employers, except that such employers may be reimbursed for the reasonable cost of work-experience programs when the work performed by students is manifestly unequal to the compensation paid to such students. Forbids such facilities to be used in such a manner as to result in racial segregation.

State Vocational Education Programs.—Authorizes the Commissioner to make grants to the States to assist them in conducting vocational education programs for persons of all ages in all communities of the States which are designed to insure that educational training programs for career vocations are available to all individuals who desire and need such education and training.

Provides that grants to States may be used for the following purposes: (1) exploratory occupational education programs; (2) vocational education programs for high school students which is designed to prepare them for advanced or highly skilled postsecondary vocational and technical education; (3) vocational education for other students attending high school; (4) vocational education for persons who have already entered the labor market and who desire or need training or retraining to achieve stability; (5) or to provide such persons with opportunities for continuing vocational education; (6) vocational education for persons who have completed or left high schools and who are available for study in a program of postsecondary vocational or technical education; special programs designed to meet the special vocational education needs for persons who have academic or socioeconomic disadvantages which hinder them in other vocational education programs which may be available, or who are handicapped; and (7) ancillary services and activities to assure quality in all vocational education programs, such as teacher training and supervision, guidance and counseling, program evaluation, special demonstration, and experimental programs, and development of instructional materials.

Provides that funds may be used to provide training through contractual arrangements with private vocational training institutions if such institutions can make a significant contribution to attaining the objectives of the State plan, and if such institutions can provide substantially equivalent training at a lesser cost or can provide equipment or services not available in public institutions.

Includes in the State plan requirements, provisions requiring a long-range program plan and requiring an annual program plan. Requires that the annual program plan, to be submitted each year with the State plan, would describe the content of, and the allocation of funds to, programs, services, and activities to be carried out under the State plan for the fiscal year ahead (whether or not with Federal funds) and indicate how and to what extent these programs, services, and activities (and allocations of funds thereto) will carry out the objectives set forth in the long-range program plan, and indicate how and to what extent allocations of Federal funds allotted to the State take into consideration the criteria for allocation of Federal funds.

Adds to the Vocational Education Act of 1963, a particular reference to new and emerging manpower needs and job opportunities on the local, State, and national levels.

Provides that Federal funds will not be allocated to local educational agencies in a manner, such as the matching of local expenditures at a uniform matching ratio throughout the State, except to the extent that the considerations required in the above-mentioned criteria are uniform throughout the State. Requires that applications from local educational agencies for funds to be developed in consultation with persons representative of the educational and training resources available to the area to be served by the applicant, show promise of providing the persons to be served with education programs designed to make substantial progress toward preparing such person for a career, include a comprehensive plan for meeting the vocational education needs in the area or community served by such agency, and indicate how and to what extent the vocational education programs, services, and activities proposed in the application will meet the needs set forth in such comprehensive plan. Provides that no local educational agency which is making a reasonable tax effort, as defined by regulations, will be denied funds for the establishment of new vocational education programs solely because the local educational agency is unable to pay the non-Federal share of the cost of such new programs.

Requires that the State plan provide minimum qualifications for teachers, teacher-trainees, supervisors, directors, and other personnel having responsibilities for vocational education in the State and the policies and procedures developed to improve the qualifications of such personnel and to insure that such qualifications continue to reflect a direct relationship with the need for personnel in vocational education programs carried out under the State plan.

Requires the State plan to provide for cooperative arrangements with other agencies, organizations, and institutions concerned with manpower needs and job opportunities, such as business and labor organizations, institutions of higher education, and community action agencies.

Restates the requirement, of the Vocational Education Act of 1963, that the State plan shall include policies and procedures for allocations of Federal funds which insure, among other things, that Federal funds will be so used as to supplement, and, to the extent practical, increase, but not supplant, State or local funds that would otherwise have been made available for the purposes for which funds may be used under the State plan. Adds a new provision providing that final action with respect to any application for funds under this part shall not be taken without first affording the local educational agency reasonable notice and opportunity for a hearing. Requires that, of the sums available to each State under the State plan, at least 15 percent shall be used for vocational education for persons who have completed or left high school and who are available for study in a program of postsecondary vocational or technical education, at least 15 percent shall be used for special programs designed to meet the special educational needs for persons who have academic or socioeconomic disadvantages which hinder them in other vocational education programs which may be available, and at least 10 percent shall be used for special programs designed to meet the special vocational education needs for persons who are physically or mentally handicapped.

Provides that the Commissioner shall not approve a State plan until he has made specific findings as to the compliance of such plan with the requirements of this part and he is satisfied that adequate procedures are set forth to insure that the assurances and provisions of such plan will be carried out. Requires that the Commissioner submit his findings to the Senate Committee on Labor and Public Welfare and the House Committee on Education and Labor at least annually. Provides that, if the Commissioner finds that the requirement for reserving at least 15 percent of the funds available under the State plan for postsecondary vocational and technical education will result in a State's inability to use the 15 percent reservation for that purpose in any fiscal year, he may waive that requirement, upon application by that State. Requires the Commissioner to include in the findings which he submits to the congressional committees supporting evidence respecting the justification of such waiver.

Provides that whenever the Commissioner, after opportunity for hearing has been provided, finds that there has been a failure to comply substantially with any requirement in the plan of that State, or with any requirement in the application of a local educational agency, further payments shall not be made to the State under this title, or, if so determined by the Commissioner, the State may not make further payments to the local educational agencies affected by the failure, until he is satisfied that there is no longer any such failure to comply.

Contains provisions for judicial review of the Commissioner's action with respect to State plans.

Provides for judicial review of the final action of a State board if the local educational agency is dissatisfied with the final action of the State board with respect to approval of an application by the local agency.

Provides that the Commissioner shall pay to each State an amount equal to 50 percent of the State and local expenditures in carrying out its State plan with the following exceptions:

(1) Allows the 50 percent limitation with respect to that part of the amount available to any State under this part for any fiscal year which exceeds 110 percent of the amount paid to the State during the preceding fiscal year if the State expenditures for carrying out the State plan for that fiscal year are at least 110 percent of such expenditures for the preceding fiscal year to be waived by the Commissioner. The Commissioner must find that the State is making a reasonable effort to increase its expenditures for carrying out the State plan. The Commissioner may waive the 50 percent matching requirement under the above-mentioned conditions only to the extent that the amount for which such matching would not be required will be used for new or expanded programs or projects which would not otherwise be carried out because of the lack of State and local matching funds and if he finds that such programs are designed to meet urgent vocational education needs.

(2) Allows the matching requirement with respect to so much of the amount available to any State which exceeds the amount paid to the State during fiscal year 1969, as will be used for new vocational education programs to be carried out by one or more local educational agencies within that State having an urgent need for establishing new vocational education programs be waived by the Commissioner. Requires that the Commissioner must find that such local educational agency or agencies are making a reasonable effort and cannot otherwise provide the matching funds to establish such new vocational education programs.

Provides that no payments shall be made to any local educational agency or to any State unless the Commissioner finds that the combined fiscal effort of that local agency and the State with respect to the provision of vocational education by that agency for the preceding fiscal year was not less than such combined fiscal effort for the second preceding fiscal year; and no payments shall be made to any State unless the Commissioner finds that the fiscal effort of that State for vocational education for the preceding fiscal year was not less than it was for the second preceding fiscal year.

Research and Training in Vocational Education.—Provides that, from sums available to each State for grants under this part, the Commissioner is authorized to make grants to and contracts with institutions of higher education, public and private agencies and institutions, State boards, and, with the approval of the appropriate State board, to local educational agencies in that State to encourage research and training in vocational education and the development of vocational education programs designed to meet special vocational education needs of youths and to provide education for new and emerging careers and occupations.

Provides that the Commissioner may, for any fiscal year, reserve up to 50 percent of the allocations for projects which are of national or regional importance or interstate projects. Requires that if such reservation is made for any fiscal year, special consideration must be given in approving applications for the remaining funds to applications (1) from State boards, and (2) from institutions of higher education to support research coordination units and projects approved by research coordinating units.

Authorizes grants and contracts to be used for research in vocational education; training programs designed to familiarize persons involved in vocational education with research finding and successful pilot and demonstration projects in vocational education; experimental, developmental, and pilot programs and projects designed to test the effectiveness of research findings; demonstration projects; the development of new vocational education curriculums; and projects in the development of new careers and occupations.

Includes in projects in the development of new careers and occupations, research and experimental projects designed to identify new careers in such fields as mental and physical health, crime prevention and corrections, welfare, education, municipal services, child care, and recreation, requiring less training than professional positions and to delineate within such careers roles with the potential for advancement from one level to another; training and development projects designed to demonstrate improved methods of securing the involvement, cooperation, and commitment of both the public and private sectors toward the end of achieving greater coordination and more effective implementation of programs for the employment of persons in the fields described above, including programs to prepare professionals (including administrators) to work effectively with aides, and projects to evaluate the operation of programs for the training, developments, and utilization of public service aides, particularly their effectiveness in providing satisfactory work experiences, and in meeting public needs.

Provides that such an application for a project must be reviewed by a panel of experts who are not employees of the Federal Government before the Commissioner may approve it.

Provides that the Federal share of the amount expended by an applicant to carry out an approved project, may not exceed 90 percent, except that the Federal share of projects funded out of the reservation for national or regional research projects shall be determined in accordance with regulations.

Exemplary Programs and Projects.—Authorizes the Commissioner to make grants to or contracts with State boards of vocational education or local educational agencies for exemplary programs and projects and with other public or private agencies or institutions when such a grant or contract would make an especially significatnt contribution to carrying out the purposes of this act. Authorizes the Commissioner to make grants or contracts pursuant to this new part to pay all or part of the cost of (1) planning and developing, or (2) establishing, operating, and evaluating programs and projects to broaden the occupational aspirations and opportunities of youths, especially those with academic, socioeconomic, or other handicaps.

Requires that a grant to a State or local educational agency be made only upon a determination by the Commissioner (1) that there is satisfactory assurance of participation of manpower agencies and, as appropriate, persons broadly representative of employers, labor organizations, community action organizations, and other community institutions in the planning, establishment, and carrying out of the program or project, and (2) that there will be participation of students enrolled in nonprofit private schools in the area to be served by the program or project to the extent consistent with the number of such students with educational needs of the type to be met by the program or project.

Provides that no grant or contract (other than a grant to or contract with a State board) may be made unless the program or project involved has been submitted to the State board in the State where it is to be conducted and has not been disapproved by the board within 60 days or within such longer period as the Commissioner may allow pursuant to regulations.

Provides that the Federal share may not exceed 90 percent of the amount expended by the applicant in carrying out the approved project.

State Special Emphasis Programs.—Provides that, from the sums available for grants, the Commissioner shall carry out a program of making grants to assist States in the establishment of special emphasis programs to meet special educational needs in the States.

Provides that grants may be used in accordance with approved applications for planning and developing programs designed to provide special vocational educational activities, the establishment, maintenance, and operation of new vocational education programs (including the lease or construction of necessary facilities and the acquisition of necessary equipment), designed to meet the needs of individuals in the State and to offer a diverse range of educational experience to persons of varying talents and needs by providing, especially through new and improved approaches, special vocational education programs. Includes in such programs:

(1) New and expanded vocational education programs and services specifically designed for persons and students who have academic, social, economic, cultural, or other handicaps, and who reside in areas with concentrations of persons so disadvantaged;

(2) Cooperative work-study programs, designed to combine a meaningful work experience with formal education;

(3) Residential vocational education schools;

(4) Development of curriculum materials;

(5) Acquisition of vocational library resources;

(6) Strengthening vocational and technical education at the post-secondary level in junior and community colleges;

(7) Comprehensive guidance and counseling;

(8) Comprehensive vocational guidance and counseling for adults;

(9) Specialized instruction and equipment for students interested in studying vocational education subjects which are not taught in the local schools or which can be provided more effectively on a centralized basis, or for persons who are handicapped;

(10) Making available modern vocational education equipment and specially qualified personnel;

(11) Developing, producing, and transmitting radio and television programs for classroom and other vocational education use;

(12) Providing special educational and related services for persons who are in or from rural areas;

(13) Providing support or services for the comprehensive and compatible recording, collecting, processing, analyzing, interpreting, storing, retrieving, and reporting of State and local vocational education data, including the use of automated data systems;

(14) Research and training programs and experimental, developmental, or pilot programs designed to meet the special vocational education needs of youth;

(15) Cooperative arrangements for the training or retraining of experienced vocational education personnel;

(16) Encouraging community involvement in vocational education programs; and

(17) Programs designed to prevent school dropouts and assist all students in bridging the gap between school and the world of work.

Provides that the Commissioner may approve an application for a grant under this part only upon his determination that each of the proposed projects, programs, and activities for which the application is approved, will make a significant contribution to strengthening the ability of the State and local educational agencies in that State to participate effectively in meeting the vocational education needs of the State, and the application has been developed in consultation with the State advisory council and sets forth the procedures by which the proposed activities will be evaluated by such advisory council.

Homemaking Education.—Authorizes appropriations for purposes of this section of $25 million for fiscal year 1970, $35 million for fiscal year 1971, and $50 million for fiscal year 1972.

Requires any State participating in this section to have in effect, an approved State plan and to submit a supplementary plan which: (1) Designates the State board as the sole agency for administration, or supervision of the administration, of the State plan; (2) sets forth a program under which Federal funds will be expended solely for (A) homemaking education for persons who have entered, or are preparing to enter, the work of the home, and (B) ancillary services and activities; (3) sets forth fiscal control and fund accounting procedures; and (4) provides for submitting reports, keeping records, and affording the Commissioner access to such records.

Provides that the Commissioner shall pay 50 percent of the amount expended for homemaking education and ancillary services and activities, but in no case more than a State's allotment.

Cooperative Vocational Education Programs.—Provides financial assistance for personnel to coordinate cooperative work-study programs and to provide instruction related to the work experience.

Authorizes appropriations of $25 million for fiscal year 1970, $50 million for fiscal year 1971, and $75 million for fiscal year 1972.

Provides that, in order to participate in the program, a State must, through its State board, submit a State plan to the Commissioner setting forth policies and procedures to be used in establishing cooperative work-study programs through local educational agencies and public and private employers.

Provides for a Federal share of not more than 90 percent of the State's expenditures under its State plan for the cooperative work-study program.

Miscellaneous Provisions.—Requires the Secretary of Health, Education, and Welfare to transmit to the committees having legislative jurisdiction and the appropriations committees of each House of the Congress not later than March 31 of each year a report evaluating programs and projects assisted under such programs, which shall include his recommendations relating thereto, including legislative recommendations.

Provides that if Congress has not passed or formally rejected legislation extending the authorization for any legislation during the regular session in which a comprehensive report is required, the authorization is automatically extended for one fiscal year beyond its expiration date and at the same level of the last year's authorization.

Provides that, notwithstanding any other provision of law, unless expressly in limitation of these provisions, funds appropriated in any fiscal year to carry out any of the programs administered by the Office of Education shall remain available for obligation until the end of such fiscal year.

Early Education of Handicapped Children.—Authorizes the Commissioner to arrange by contract or grant with appropriate public agencies and private nonprofit organizations for developing and carrying out experimental preschool and early education programs for handicapped children.

Limits the Federal share to 90 percent of the cost of developing, carrying out, or evaluating the program.

Provides that the Commissioner shall conduct either directly or by contract with independent organizations a thorough and continuing evaluation of the effectiveness of each program.

Authorizes appropriations of $5 million for fiscal year 1969, $10 million for fiscal year 1970, and such sums as may be authorized for fiscal year 1971.

Provides that the Commissioner of Education (1) shall prepare and disseminate to all appropriate State and local agencies and institutions and others concerned with education, complete information on programs of Federal assistance; (2) shall inform the public on federally supported programs for education by providing information to communications media; (3) shall develop, on both formal and informal bases, a close liaison for interchange of ideas and information with representatives of American business and with service, labor, or other organizations, both public and private, to advance American education; (4) shall collect data and information on programs qualifying for assistance under programs administered by him for the purpose of obtaining objective measurements of the effectiveness achieved in carrying out the purposes of such programs; (5) may upon request provide advice, counsel, technical assistance, and demonstrations to State educational agencies, local educational agencies, or institutions of higher education undertaking to inititate or expand programs in order to increase the quality of depth or broaden the scope of such programs, and shall inform such agencies and institutions of the availability of assistance pursuant to this clause; (6) shall prepare and disseminate to State educational agencies, local educational agencies, and other appropriate agencies and institutions an annual report setting forth developments in the utilization and adaptation for programs carried out pursuant to this title; and (7) may enter into contracts with public or private agencies, organizations, groups, or individuals to carry out these provisions.

Provides that, upon request from a State educational agency, the Commissioner shall provide counseling and technical assistance to elementary and secondary schools in rural areas, as defined by the Commissioner, of such State in determining benefits available to such agencies and schools under Federal laws, and in preparing applications and meeting other requirements for such benefits. Technical assistance may also be provided to institutions of higher education in rural areas.

Requires the Commissioner of Education to: (1) prepare a catalog of all Federal assistance programs in the area of education, which must include the name of the program, the authorizing statute, the specific Federal administering officials and a brief description of such program; (2) set forth the availability of benefits and eligibility restrictions in each such program; (3) set forth the budget requests for each such program, past appropriations, obligations incurred, the average assistance provided under each such program, and pertinent financial information indicating the size of each such program for selected fiscal years, and any funds remaining available; (4) set forth the prerequisites, including the cost to the recipient of receiving assistance under each such program, and any duties required of the recipient after receiving benefits; (5) identify appropriate officials, in Washington as well as in each State and locality to whom application or reference for information for each such program may be made; (6)

set forth the application procedures; and (7) contain a detailed index designed to assist the potential beneficiary to identify all education assistance programs related to a particular need or category of potential beneficiaries; and contain such other programs information and data as the Commissioner deems necessary.

Authorizes appropriations of such sums as may be necessary for each of fiscal year 1970 through 1972, for the collection and dissemination of information section.

Provides that the Commissioner of Education shall not in any manner effect or agree to the consolidation of any programs which will result in the commingling at the Federal, State, or local level of funds derived from different appropriations; nor shall he transfer funds derived from appropriations to the use of any program not covered by that appropriation; nor shall he enter into any agreement with any State educational agency which would have the effect of requiring or providing for the approval of programs involving funds from different appropriations on any basis other than that provided for in the law which authorizes the appropriation of funds for each such program; nor shall the making of any grant or contract derived from any appropriation to the Office of Education be conditional upon the receipt of any grant or contract from any other such appropriation.

Provides that the Commissioner of Education shall make a study of the feasibility of consolidation of education programs in order to provide for more efficient use of Federal funds at the local level and to simplify application procedures for such funds.

PUBLIC LAW 90–577—INTERGOVERNMENTAL COOPERATION ACT OF 1968

S. 698. Introduced by Mr. Muskie and Others on January 26, 1967

Authorizes full information for the Governors on grants made to their States and provides for more uniform administration of Federal grant funds to the States. Revises the scheduling of fund transfers to the States and permits the States to budget Federal grant funds in a manner similar to that in which other revenues are budgeted.

Provides for congressional review of future grant programs to insure that such programs are systematically reexamined and reconsidered in light of changing conditions.

Authorizes the Federal departments and agencies to render technical assistance and training services to State and local governments on a reimbursable basis.

Establishes a coordinated intergovernmental urban assistance policy, which requires local government review of certain applications for Federal aid in urban programs.

Prescribes a uniform policy of procedure for urban land transactions and use undertaken by the General Services Administration, by requiring consistency of that Agency's policies with local zoning regulations and development objectives.

Authorizes the President to submit to Congress plans for the consolidation of individual categorical grants and to effect the interagency transfer of administrative responsibility for grant programs, subject to the type of congressional veto proviso governing reorganization plans.

Provides a policy of uniform treatment for individuals who are annually affected by such Government projects as urban renewal and highway construction, by establishing uniform relocation payments, and advisory assistance programs for those displaced by Federal and federally assisted programs. Requires compliance with the relocation requirements as a condition of Federal grants to State and local governments. Imposes on all federally assisted programs the present urban renewal requirement that no property acquisition project may proceed until there is assurance of available standard housing for those displaced.

Provides full Federal reimbursement of relocation payments up to a maximum of $25,000; and above that, Federal sharing according to the project's cost-sharing formula.

Establishes a uniform policy for the acquisition of real property by Federal Government agencies and by State agencies using Federal funds for public improvement programs.

PUBLIC LAW 90–636—S. 2938. INTRODUCED BY MR. CLARK AND OTHERS ON FEBRUARY 7, 1968

Extends various expiring provisions under the Manpower Development and Training Act as labor mobility demonstration projects, trainee placement assistance demonstration projects, and correctional institutions.

SELECTED REFERENCES

The literature which treats the subject of systems technology applied to social and community problems is diverse and often hard to locate. In addition to the books, magazine articles, and professional journal monographs which touch on some aspect of these problems, useful material also appears in congressional hearings and reports, industrial and university proposals, State and local government planning documents, program and project progress reports, and specially commissioned studies by foundations.

This selective listing is divided for the purpose of expediting its use into 20 categories:

A. Urban Planning
B. Environmental Pollution
C. Transportation
D. Housing Redevelopment
E. Law Enforcement and the Administration of Justice
F. Health Services
G. Manpower and Unemployment
H. Planning-Programming-Budgeting Systems
I. Federal Aid Programs
J. Regional Information Systems
K. Government—Local and State
L. Government—State
M. Government—Local
N. Aerospace Technology in the Urban Area
O. Books
P. Congressional Documents
Q. Serial Publications
R. Bibliographies
S. Symposia
T. Miscellaneous

A. URBAN PLANNING

A–1 Alcott, James. Technology and urban needs. A statement from the Engineering Foundation Research Conference on the social consequences of technology. Kansas City, Missouri, Midwest Research Institute, 1966. 24 p.

A–2 Banfield, Edward C. Why government cannot solve the urban problem. *In* Daedalus, Fall 1968. p. 1231–1241.

A–3 Campbell, Robert D., *and* Hugh L. LeBlanc. An information system for urban planning. Prepared for the Maryland National Capital Park and Planning Commission. Washington, D.C., U.S. Department of Housing and Urban Development, [1967] 96 p.

A–4 Carroll, James D. Science and the city: the question of authority. *In* Science, v. 163, February 28, 1969, p. 902–911, references.

A–5 Carter, Luther J. Systems approach: political interest rises. In Remarks in the Senate by the Honorable Gaylord Nelson on Systems analysis. *In* Congressional record, v. 112, part 21, October 21, 1966. p. 28434–28435.

A-6 Chartrand, Robert L. Modern management analysis and traditional management analysis: a survey of the impact. In Remarks in the Senate by the Honorable Gaylord Nelson on Systems analysis. *In* Congressional record [daily edition], November 21, 1967, p. S16879–S16883.

A-7 Chartrand, Robert L. The systems approach in social legislation. In Extensions of remarks in the Senate by the Honorable Hugh Scott. *In* Congressional record [daily edition], June 11, 1968, p. E5158–E5160.

A-8 Concepts of an urban management information system, a report to the City of New Haven, Connecticut. Yorktown Heights, New York, Advanced Systems Development Division, International Business Machines Corporation, [no date] 45 p.

A-9 Crane David A. Fort Lincoln New Town. Technologies study: the application of technological innovation in the development of a new community. Prepared for Edward J. Logue, . . . District of Columbia Redevelopment Land Agency, National Capital Planning Commission, and District of Columbia Government. Washington, D.C., December 1968. 246 p. + appendixes.

A-10 Devaney, F. John. The use of computers in city and regional planning. *In* Traffic quarterly, v. 20, October 1966. p. 511–524.

A-11 Dewey, Arthur W. Electronic data processing of growth elements in suburban municipalities. University of Connecticut, Storrs, Agricultural Experiment Station, November 1966. 16 p. (Research Report 17)

A-12 The diamond interchange—a mathematical model. *In* SDC magazine, v. 11 no. 5, May 1968. p. 18–19.

A-13 DiCesare, F., *and* J. C. Strauss. Simulation of an urban transportation transfer point. *In* Socio-economic planning sciences, v. 1, no. 3, July 1968. p. 405–414.

A-14 Doctors, Samuel I. Project management in urban redevelopment. *Reprinted from* MSU business topics, Summer 1968. p. 19–23.

A-15 Doxiadis, C. A. Ekistics; a scientific approach to the problems of human settlements. Presented to the Committee on Science and Astronautics, U.S. House of Representatives, 10th meeting, February 4, 1969. 91st Congress, 1st session. Washington, D.C., U.S. Govt. Print. Off. 1969. 4 p. (Committee Print)

A-16 Energy and the new civilization: a systems approach to the cities. *In* Da review, May 1, 1967. p. 2–4. [Interview with C. A. Doxiadis.]

A-17 Grimm, Sergei N. The scientific urban planning system. *In* Law and computer technology, v. 2, no. 1, January 1969. p. 19–25.

A-18 Herrman, Cyril C. Systems approach to city planning: San Francisco applies management techniques to urban redevelopment. *In* Harvard business review, v. 45, September/October 1966. p. 71–80.

A–19 House, Peter. The simulated city: the use of second generation gaming in studying the urban system. Presented at the Annual symposium of the ACM on "The application of computers to the problem of urban society," October 18, 1968. p. 1–2.

A–20 Lanham, Richard. People, problems and planning. *In* SDC magazine, v. 10, no. 10, October 1967. p. 1–15.

A–21 Management analysis newsletter. Edited by the Division of Methods, Data Processing, and Office Services, State House Providence, Rhode Island, Issue no. 2, October 1968. 12 p.

A–22 Martin, A. E. Environment, housing, and health. *In* Ekistics, v. 24, no. 144, November 1967. p. 395–396. (Reprint of the conclusions of an article appearing in Urban studies v. 4, no. 1, February 1967, p. 1–21.)

A–23 Mitchel, William H. Urban planners and information. Fourth annual conference of planners, sociologists, and computer specialists. *In* Datamation, v. 12, no. 10, October 1966. p. 82–84, 86.

A–24 Morse, F. Bradford. Private responsibility for public management. Special Report. *Reprinted from* Harvard business review, v. 45, no. 2, March–April 1967. 8 p.

A–25 Muskie, Edmund S. The role of congress in promoting and controlling technological advance. *In* George Washington law review, v. 36, no. 5, July 1968. p. 1138–1149.

A–26 Nelson, Gaylord. A space age trajectory to the great society. Remarks in the Senate. *In* Congressional record, v. 111, part 20, October 18, 1965. p. 27242–27248.

A–27 Rosenberg, Gerhard. City planning theory and the quality of life. *In* Ekistics, v. 24, no. 144, November 1967. p. 411–414.

A–28 Science, engineering and the city. A symposium sponsored jointly by the National Academy of Sciences and the National Academy of Engineering. Washington, D.C., National Academy of Sciences, 1967. 142 p. (Publication 1498)

A–29 Scott, Hugh. Congressional interest in systems approach spurs legislation for National Commission. *In* Aerospace management, Fall-Winter 1966. p. 11–13.

A–30 Smith, Robert G. The systems approach and the urban dilemma. Washington, D.C., George Washington University, July 1968. 45 p.

A–31 Socio-economic planning sciences, v. 1, no. 3, July 1968. [This issue contains a collection of papers presented at the Annual symposium of the Association for Computing Machinery, New York City, November 10, 1967.]

A–32 Sweeney, Stephen B., *and* James C. Charlesworth. Governing urban society: new scientific approaches. Philadelphia, Pennsylvania, the American Academy of Political and Social Science, 1967. 254 p.

A–33 Taylor, Harold H., *and* Milton C. Hallberg. Use of computers in the storage and retrieval of severance effect information. Institute for Research on Land and Water Resources, Pennsylvania State University, 1966. 35 p.

A–34 Touche, Ross, Bailey and Smart. Automated data processing planning project. Report to the Governor of the State of Washington, January 1967. 180 p.

A–35 Uhl, Edward G. The urban crisis: industry's resources, capabilities, and responses. [no date] 15 p.

A–36 U.S. *Advisory Commission on Intergovernmental Relations*, Washington, D.C. Urban and rural America: policies for future growth. Washington, D.C. U.S. Govt. Print. Off., April 1968. 186 p.

A–37 U.S. *Advisory Commission on Intergovernmental Relations*, Washington, D.C. State legislative and constitutional action on urban problems in 1967. Washington, D.C,. April 1968. 29 p. refs.

A–38 U.S. *Department of Housing and Urban Development*, Washington, D.C. Solving urban problems through urban information and technical assistance. A program guide for urban information and technical assistance, Title IX, demonstration cities and metropolitan development act of 1966. Washington, D.C., [1968] 46 p.

A–39 U.S. *Department of Housing and Urban Development*, Washington, D.C. Office of Intergovernmental Relations and Planning Assistance. Community development training and urban information and technical assistance: 100 outstanding programs. Supported under Title VIII of the 1964 Housing Act and Title IX of the 1966 Demonstration Cities and Metropolitan Development Act. Washington, D.C., June 1968. 48 p.

A–40 Williams, W. K. Computers in urban planning. *In* Socio-economic planning sciences, v. 1, no. 3, July 1968. p. 297–308.

B. Environmental Pollution

B–1 AVCO water resources program. AVCO Economic Systems Corporation, [1960] 10p.

B–2 Carlson, Jack W. "What price—a quality environment." Given before the 23rd annual meeting of the Soil Conservation Society of America, University of Georgia, Athens, Georgia, August 20, 1968. 26 p.

B–3 Cleary, Edward J. The Orsanco story: water quality management in the Ohio valley under an interstate compact. Baltimore, Maryland, Johns Hopkins Press (published for Resources for the Future, Inc.), 1967. 335 p.

B–4 Design, development, and demonstration of a pollution control planning simulation. Abt Associates. 1967. 25 p.

B–5 Effective water management for the nation's future. Task Force on the Functions of Federal, State, and Local Governments. Washington, D.C , Republican National Committee. June 1966. 7 p.

B–6 Effective water management for the nation's future. Task Force on the Functions of Federal, State, and Local Governments. Washington, D.C., Republican National Committee. June 1966. 7 p.

B–7 Hodgson, James B. Analysis of the National Research and Development Management System for Water Resources. Proposed study to perform a study under the provisons of Title II of the Water Resources Research Act of 1964 as amended. February 15, 1967. various paging.

B–8 Jeglic, J. M., *and* G. D. Pence. Mathematical simulation of the estuarine behavior and its applications. *In* Socio-economic planning science, v. 1, no. 3, July 1968. p. 363–389.

B–9 Legislative analysis; the water pollution control bill. 89th Congress, 2nd session. Washington, D.C., American Enterprise Institute for Public Policy Research. July 29, 1966. 37 p.

B–10 Ludwig, John H. Air pollution control technology; research and development on new and improved systems. *In* Law and contemporary problems, air pollution control, v. 33, Spring 1968. p. 217–226.

B–11 National Academy of Sciences. National Research Council. Waste management and control. A report to the Federal Council for Science and Technology, Committee on Pollution. Washington, D.C., 1966. 257 p. (Publication 1400)

B–12 Stanford L. Optner and Associates. Phase one report on the sanitation management information system to the City Administrative Officer and the Board of Public Works. Los Angeles, California, March 1966. 1 v.

B–13 Proceedings of the IBM scientific computing symposium on water and air resources management. White Plains, New York, IBM Data Processing Division, 1968. 329 p.

B–14 Savas, Emanuel S. Computers in urban air pollution control systems. *In* Socio-economic planning science, v. 1, 1967. p. 157–183. refs.

B–15 Systems analysis techniques and concepts for water resource planning and management. Ann Arbor, Michigan, Technology Planning Center, Inc., [no date] 46 p. refs.

B–16 Ulbrich, E. A. Adapredictive air pollution control for the Los Angeles basin. *In* Socio-economic planning sciences, v. 1, no. 3, July 1968. p. 423–440. refs.

B–17 U.S. *Congress. House. Committee on Science and Astronautics.* The adequacy of technology for pollution abatement. Hearings before the Subcommittee. 89th Congress, 2nd session. v. 1. Washington, D.C., U.S. Govt. Print. Off., 1966. 604 p.

B–18 U.S. *Congress. House. Committee on Science and Astronautics. Subcommittee on Science, Research, and Development.* Environmental pollution—a challenge to science and technology. 89th Congress, 2nd session. Washington, D.C., U.S. Govt. Print. Off., 1966. 60 p. (Committee Print)

B–19 U.S. *Congress. Senate. Committee on Interior and Insular Affairs. And the House. Committee on Science and Astronautics.* Congressional white paper on a national policy for the environment. 90th Congress, 2nd session. Washington, D.C., U.S. Govt. Print. Off., October 1968. 19 p. (Committee Print)

B-20 U.S. *Congress. Senate. Committee on Interior and Insular Affairs. And the House. Committee on Science and Astronautics.* Joint House-Senate colloquium to discuss a national policy for the environment. Hearing before the Committees. 90th Congress, 2nd session. Washington, D.C., U.S. Govt. Print. Off., July 17, 1968. 233 p.

B-21 U.S. *Department of Health, Education, and Welfare. Public Health Service, Consumer Protection and Environmental Health Service, National Air Pollution Control Administration.* Report for consultation on the San Francisco bay area air quality control region. December 1968. 46 p. refs.

B-22 U.S. *National Commission on Urban Problems,* Washington, D.C. Building the American city. Report of . . . to the Congress and to the President of the United States. Part VI, Improvement of the environment. 91st Congress, 1st session. Washington, D.C., U.S. Govt. Print. Off., [1969] p. 487-500. (House Document No. 91-34)

B-23 Water management decisions. Technology Planning Center, Inc. 45 p.

B-24 Water quality management data systems guide. Joint Committee on Water Quality Management Data. The Pennsylvania Department of Health. July 1, 1966. 69 p.

C. TRANSPORTATION

C-1 Breuning, S.N., *and* T.N. Harvey. Implications of integrated transportation engineering design systems for computer software and hardware. *In* Proceedings of the 22nd national conference, Association for Computing Machinery. Washington, D.C., Thompson Book Company, 1967. p. 255-262.

C-2 Freeway controls. *In* Public automation, v. 4, no. 5, May 1968. p. 1.

C-3 Hanson, Marke, *editor.* Project METRAN; an integrated evolutionary transportation system for urban areas. Cambridge, Massachusetts, M.I.T. Press, 1966. 262 p.

C-4 Kevany, Michael J. An information system for urban transportation planning: the BATSC approach. Technical memorandum (TM-38920/000/01). Santa Monica, California, System Development Corporation. 56 p.+appendix.

C-5 Meeting the electronic data processing needs of the Minnesota Highway Department: an evaluation/planning study of present and expanded computer systems. Minnesota Department of Highways, Management Control Division. October 1966. 95 p.

C-6 Murray, Roger J. Marriage of digital computer and hybrid master controller. J.F.K.I.A. signal and surveillance system. [New York] Port of New York Authority, [no date]. 9 p.

C-7 Mushkin, Selma J., *and* Robert Harris. Transportation outlays of states and localities: projections to 1970. The Council of Governments. May 1965. 47 p.

C-8 Orchanian, Paul L. Stopping scofflaws. *In* Output; monthly feature of Public automation, v. 4, no. 5, May 1968. 3 p.

C-9 Parsons Brinckerhoff-Tudor-Bechtel. San Francisco Bay area rapid transit district demonstration project. Technical reports. Number 4, v. II, Rapid transit propulsion systems, 157 p., appendix. Number 8, Acoustics studies, 157 p. Number 9, Adhesion characteristics, 1 v. San Francisco, 1968. 3 nos.

C-10 Perilla, Oscar. Towards development of scientific approaches to transportation and urban design. [New York] Port of New York Authority, December 1967. 35 p. (Report 67-6)

C-11 Schofer, Ralph E., and Franklin F. Goodyear. Electronic computer applications in urban transportation. In Proceedings of the 22nd national conference, Association for Computing Machinery. Washington, D.C., Thompson Book Company, 1967. p. 247-253.

C-12 Seifert, William W. Transportation development—a national challenge. Presented to the Committee on Science and Astronautics, U.S. House of Representatives, Panel on Science and Technology, 10th meeting, February 5, 1969. 91st Congress, 1st session Washington, D.C., U.S. Govt. Print. Off., 1969. 18 p. (Committee Print)

C-13 Taylor, Harold H., and Milton C. Hallberg. Use of computers in the storage and retrieval of severance effect information. University Park, Pennsylvania, Institute for Science and Engineering, and the Institute for Research on Land and Water Resources, The Pennsylvania State University, 1966. 35 p.

C-14 Transportation in modern America. Task Force on the Functions of Federal, State and Local Governments. Washington, D.C., Republican National Committee. June 1966. 22 p.

C-15 Troubles with mobility. An analysis, a model and a data center. In SDC magazine, v. 11, no. 5, May 1968. p. 1-19.

C-16 U.S. Congress. Senate. Metropolitan area pilot transportation study. Study implementing Senate Resolution 250. 90th Congress, 2nd session. Washington, D.C., U.S. Govt. Print. Off., July 19, 1968. 139 p. (Senate Document No. 117)

C-17 U.S. Library of Congress. Legislative Reference Service, Washington, D.C. Urban mass transportation: current trends and prospects, by Elizabeth M. Heidbreder. May 20, 1966. 26 p. (E-151)

C-18 Utilization study of methods to obtain improved results from existing transportation technologies. Gaithersburg, Maryland, Federal Systems Division, International Business Machines Corporation, January 3, 1967. 49 p.

C-19 Webber, Melvin M., and Shlomo Angel. The social context for transport policy. Presented to the Committee on Science and Astronautics, U.S. House of Representatives, Panel on Science and Technology, 10th meeting, February 5, 1969. 91st Congress, 1st session. Washington, D.C., U.S. Govt. Print. Off., 1969. 18 p. (Committee Print)

C-20 Young, Dennis. Scheduling a vehicle between an origin and a destination to maximize traveler satisfaction. *In* Proceedings of the 22nd national conference, Association for Computing Machinery. Washington, D.C., Thompson Book Company, 1967. p. 233–245.

D. Housing Redevelopment

D-1 Cogswell, Arthur R., Werner Hausler, *and* C. David Sides, Jr. A computer based building cost analysis and design system. North Carolina Fund Low Income Housing Demonstration Office, Chapel Hill, North Carolina, March 1967. 29 p. + data forms.

D-2 Hodge, Patricia Leavey, *and* Philip M. Hauser. The challenge of America's metropolitan population outlook—1960 to 1985. Washington, D.C., U.S. Govt. Print. Off., 1968. 90 p. (Research Report No. 3) Prepared for the consideration of the National Commission on Urban Problems.

D-3 Legislative analysis; housing and urban development bills. 89th Cong., 2nd sess. Washington, D.C., American Enterprise Institute for Public Policy Research. July 7, 1966. 25 p. (Analysis No. 9).

D-4 Manvel, Allen D. Local land and building regulation. How many agencies? What practices? How much personnel? Washington, D.C., U.S. Govt. Print. Off., 1968. 48 p. (Research Report No. 6) Prepared for the consideration of the National Commission on Urban Problems.

D-5 The meandering path to six million homes. *In* Progressive architecture, June 1968. p. 95–153.

D-6 Republican Coordinating Committee. Housing and urban development. Task Force on the Functions of Federal, State and Local Governments. Washington, D.C., Republican National Committee, June 1966. [9] p.

D-7 Shirk, George H. The Pei plan—total downtown renewal. *In* Proceedings of the Mid-America conference on urban design. Kansas City, Missouri, Midwest Research Institute, [no date] p. 71–79.

D-8 U.S. *Library of Congress. Legislative Reference Service*, Washington, D.C. The rent supplement program: pro and con arguments, by Tom F. Lord. August 25, 1965, revised July 14, 1966. 7 p. (E–110 rev.)

D-9 U.S. *National Commission on Urban Problems*, Washington, D.C. Building the American city. Report of . . . to the Congress and to the President of the United States. Introduction and summary, p. 1–31. Part I, the urban setting: population, poverty, race, p. 40–55. Part II, Housing programs, p. 56–197. Part III, Codes and standards, p. 199–321. Part V. Reducing housing costs, p. 417–485. 91st Congress, 1st session. Washington, D.C., U.S. Govt. Print. Off., [1969] (House Document No. 91–34)

D-10 U.S. *President's Commission on Urban Housing*, Washington, D.C. A decent home; the report of the President's Committee on Urban Housing. Washington, D.C., U.S. Govt. Print. Off., December 1968. 252 p.

E. Law Enforcement and the Administration of Justice

E-1 Administration of justice: Los Angeles Police Department. *In* SDC magazine, v. 8, no. 10, October 1965. p. 17–21.

E-2 Administration of justice: New York City Police Department. *In* SDC magazine, v. 8, no. 10, October 1965. p. 25–27.

E-3 Administration of justice: New York State Identification and Intelligence System. *In* SDC magazine, v. 8, no. 10, October 1965. p. 22–25.

E-4 Ball, Vaughn C. The impact of data processing technology on the legal profession. *In* Computers and automation, v. 17, no. 4, April 1968. p. 43–47.

E-5 Bermant, O. I. Digital computers in traffic control—the experience and the significance. *In* Socio-economic planning science, v. 1, no. 3, July 1968. p. 415–421.

E-6 Chartrand, Robert L. Systems technology and judicial administration. *In* Judicature, v. 52, no. 5, December 1968. p. 194–198.

E-7 Cincinnati/Hamilton County regional computer center. Annual report 1967. 123 p.

E-8 City of New Orleans. Traffic court—management information systems. [no date]. 13 p.

E-9 Columbus, E. G. Automatic data processing: a practical police tool. White Plains, N.Y., Data Processing Division, International Business Machines Corporation, 1967. 77 p.

E-10 Gallati, Robert R. J. Criminal justice systems and the right to privacy. *In* Governmental ADP: the practitioners speak. Selected readings from Public automation, compiled by Public Automated Systems Service, 1313 East Sixtieth Street, Chicago, Illinois. p. 104–105.

E-11 Halloran, Norbert A. The advance of court automation. *In* Governmental ADP: the practitioners speak. Selected readings from Public automation, compiled by Public Automated Systems Service, 1313 East Sixtieth Street, Chicago, Illinois. p. 96–99.

E-12 Halloran, Norbert A. Removing obstacles to court automation. *In* Governmental ADP: the practitioners speak. Selected readings from Public automation, compiled by Public Automated Systems Service, 1313 East Sixtieth Street, Chicago, Illinois. p. 100–103.

E-13 Hirsh, Lisa. The anatomy of crime. *In* City, v. 2, no. 6, November–December 1968. p. 18–23.

E-14 Mahoney, W. Michael. Measuring the effectiveness of criminal rehabilitation programs. Bethesda, Maryland, RMC, February 1968. 10 p. (RMC Research Document RD–015) (Originally published as Occasional Paper No. 5 by the U.S. Department of Health, Education, and Welfare.)

E-15 Miller, Robert J. Can EDP win the war on crime? *In* Datamation, v. 14, June 1968. p. 81, 83–84.

E-16 Ragan, Lawrence. Chicgao's police EDP system. *In* Datamation, v. 13, July 1967. p. 52–53.

E-17 Role of police in an urban society. Highlights of the 1967 National Conference of the American Society for Public Administration. *In* SDC magazine, v. 10, no. 5, May 1967. p. 9.

E-18 Travelers Research Center, Inc., Hartford, Conn. Planning-programming-budgeting for police departments, by Frank J. Leahy. Hartford, Connecticut, April 1968. 21 p. refs. (7344-311)

E-19 U.S. President's Commission on Law Enforcement and Administration of Justice. The challenge of crime in a free society. Washington, D.C., U.S. Govt. Print. Off., February 1967. 340 p.

F. Health Services

F-1 Adams, Harold, *and* Walter Balk. An analysis of the division of vocational rehabilitation. Albany, State University of New York at Albany, 1968. 71 p.

F-2 Emergency ambulance service. City of New York, Office of the Mayor, March 8, 1968. 71 p.

F-3 Feulner, Edwin J., Jr., *and* Thomas W. Roehl. A cost-benefit analysis of the Wood County project. [Wood County, Wisconsin, Vocational Rehabilitation Administration] August 1967. 32 p. incl. summary.

F-4 Siegel, Stephen J. Harnessing ADP for hospitals. *In* Output; monthly feature of Public automation, March 1967. 3 p.

G. Manpower and Unemployment

G-1 Adelson, Marvin. The system approach—a perspective. *In* SDC magazine, v. 9, no. 10, October 1966. p. 1-9.

G-2 [Auerbach Corporation] A survey of computer-assisted placement systems. Submitted to Bureau of Employment Security, U.S. Department of Labor. Washington, D.C. n.d. 1 v. refs.

G-3 [Auerbach Corporation] Employment security and ADP systems study. Phase I. Conceptual report. Submitted to Bureau of Employment Security, U.S. Department of Labor, Washington, D.C. April 17, 1967. 1 v. (Technical report 1393-100-TR-2)

G-4 Goodwin, Robert C. Revised steering committee, task force and task group charters, and a tentative schedule of meetings for the first half of phase II of the ES data systems study. Memorandum. Washington, D.C., Bureau of Employment Security, U.S. Department of Labor. December 14, 1967. 2 p. + attachments.

G-5 Hodgson, James D., *and* Marshall H. Brenner. Successful experience: training hard-core unemployed. *In* Harvard business review, v. 46, no. 5, September-October 1968. p. 148-156.

G-6 U.S. *Department of Commerce. National Bureau of Standards, Technical Analysis Division,* Washington, D.C. The uses of systems analysis in manpower adjustment. A summary of a six-month study for the President's Commission on Automation, Technology and Economic Progress. January 25, 1966. 42 p. + appendixes.

G–7 [U.S. *Department of Labor*] Bureau of Employment Security, Washington, D.C. Draft plan for employment security automated reporting system (ESARS). Washington, D.C. April 10, 1968. 1 v.

H. PLANNING-PROGRAMMING-BUDGETING SYSTEMS

H–1 Cotton, John F. Planning-programming-budgeting systems for local government. *In* Municipal finance, v. 41, August 1968. p. 26–33.

H–2 Flood, Merril M. Operations research in the civilian sector of government. Washington, D.C., U.S. Department of Commerce, National Bureau of Standards, 1966. p. 3–6. (Miscellaneous Publication 294)

H–3 [George Washington University. State-Local Finances Project, Washington, D.C.] Introduction to report on PPBS demonstration—state, county, city. [October 1968] 40 p.

H–4 Luther, Robert A. PPBS in Fairfax County: a practical experience. *In* Municipal Finance, v. 41, August 1968. p. 34–42.

H–5 Marvin, Keith E. Statement of Keith E. Marvin . . . before the Institute on Program Planning and Budgeting Systems for Librarians at Wayne State University, Detroit, Michigan. October 2, 1968. 17 p.

H–6 Mushkin, Selma J., Harry P. Hatry, *and* Marjorie C. Willcox. Encouraging improved planning in state and local governments—the federal role. Washington, D.C., State-Local Finance Project, George Washington University, September 1968. 54 p. + appendix.

H–7 Research for Better Schools, Inc., Philadelphia, Pennsylvania. An annotated bibliography of benefits and costs in the public sector. Philadelphia, Pa., November 1968. 242 p.

H–8 Stacy, M. Joseph. Master plan for an automated financial information system for the Commonwealth of Massachusetts. M. Joseph Stacey, Comptroller. 1966. 149 p.

H–9 U.S. *Civil Service Commission*, Washington, D.C. A follow-up study of the three week residential seminar in PPBS. Washingto, D.C., 1968. 98 p.

H–10 U.S. *Library of Congress. Legislative Reference Service*, Washington, D.C. The origins and evolution of the federal planning-programming-budgeting system (PPBS), by Robert L. Chartrand. October 14, 1968. 15 p. (SP 155)

H–11 U.S. *Library of Congress. Legislative Reference Service*, Washington, D.C. The planning-programming-budgeting system: an annotated bibliography, by Robert L. Chartrand, *and* Dennis W. Brezina. April 11, 1967. 23 p. (SP 127)

H–12 Zwick, C. J. Evolution of major r & d program methodology in government. *In* Proceedings of the symposium on National r & d in the 1970's. October 18–19, 1967. Washington, D.C., National Security Industrial Association, 1968. p. 213–216.

I. Federal Aid Programs

I–1 Advisory Commission on Intergovernmental Relations, Washington, D.C. Statutory and administrative controls associated with federal grants for public assistance. Washington, D.C., U.S. Govt. Print. Off., May 1964. 108 p. (A Commission Report; A–21)

I–2 Midwest Research Institute. Federal aid program information: a survey of local government needs. Final Report. Prepared for Urban Planning Research and Demonstration Program, U.S. Department of Housing and Urban Development. Kansas City, Missouri, September 1967. 97 p. + appendixes.

I–3 U.S. *Congress. House.* Listing of operating federal assistance programs compiled during the Roth study. Prepared by the staff of Representative William V. Roth, Jr. 90th Congress, 2nd session. *Reprinted from* the Congressional record, June 25, 1968. Washington, D.C., U.S. Govt. Print. Off., 1968. 399 p. (House Document No. 399)

I–4 U.S. *Congress. House. Committee on Government Operations. Subcommittee on Government Operations.* The federal research and development programs: the decisionmaking process. Hearings before the Subcommittee. 89th Congress, 2nd session, January 7, 10 and 11, 1966. Washington, D.C., U.S. Govt. Print. Off., 1966. 212 p.

I–5 U.S. *Congress. Joint Economic Committee. Subcommittee on Economic Progress.* Federal programs for the development of human resources. 90th Congress, 2nd session. v. I. Washington, D.C., U.S. Govt. Print. Off., 1968. 326 p.

I–6 U.S. *Congress. Joint Economic Committee. Subcommittee on Economic Progress.* Federal programs for the development of human resources. 90th Congress, 2nd session. v. II. Washington, D.C., U.S. Govt. Print. Off., 1968. 358 p.

I–7 U.S. *Congress. Joint Economic Committee. Subcommittee on Urban Affairs.* A directory of urban research study centers. 90th Congress, 1st session. Washington, D.C., U.S. Govt. Print. Off., August 1967. 77 p.

I–8 U.S. *Congress. Senate. Committee on Government Operations. Subcommittee on Intergovernmental Relations.* Catalog of federal aids to State and local governments. 88th Congress, 2nd session. Washington, D.C., U.S. Govt. Print. Off., 1964. 154 p. (Committee Print) Supplement, January 4, 1965. 89th Congress, 1st session. Washington, D.C., U.S. Govt. Print. Off., 1965. 65 p. (Committee Print) Second supplement, January 10, 1966. 89th Congress, 2nd session. Washington, D.C., U.S. Govt. Print. Off., 1966. 257 p. (Committee Print)

I–9 U.S. *Library of Congress. Legislative Reference Service*, Washington, D.C. Catalog of federal programs to aid urban economic and physical development, by Joel B. Hincks. June 16, 1967. 1 v. (E–237)

I-10 U.S. *National Science Foundation,* Washington, D.C. Federal funds for research, development, and other scientific activities; fiscal years 1967, 1968, and 1969. Washington, D.C., U.S. Govt. Print. Off., August 1968. 259 p. (NSF 68–27)

J. REGIONAL INFORMATION SYSTEMS

J-1 Haak, Harold H., *and* W. Richard Bigger. A coordinated data system for metropolitan San Diego. San Diego, Public Affairs, Research Institute, San Diego State College, September 1, 1966. 47 p.

J-2 Intergovernmental cooperation act of 1968. Public Law 90–577; 82 Stat. 1098 [S. 698], October 16, 1968. 9 p.

J-3 Intergovernmental Task Force on Information Systems. The dynamics of information flow; recommendations to improve the flow of information within and among federal, state and local governments. April 1968. 31 p.

J-4 Interurban automation. *In* Public automation, v. 3, no. 2, February 1967. p. 1–5.

J-5 Isaacs, Herbert H., *and* Michael McCune. The feasibility of a central computer system for San Gabriel valley cities. Los Angeles, Herbert H. Isaacs Research and Consulting, Inc., January 31, 1967. 147 p. refs.

J-6 A national clearinghouse network for urban information. Requirements survey, network design and organizational planning. Progress report no. 1 for Department of Housing and Urban Development, September 24, 1968. Kansas City, Midwest Research Institute. 30 p.

J-7 A southwestern regional information dissemination network. A special merit proposal to the U.S. Department of Commerce Technical Services Program by the Coordinating Board, Texas College and University System, [no date.] 98 p. + appendixes.

J-8 University of Connecticut *and* The Travelers Research Center, Inc. A regional municipal information handling service for the capitol region. Report of research supported by the Connecticut Research Commission, October 1967. v. I, 44 p. + appendix (refs.); v. II, 37 p. + appendix.

K. GOVERNMENT—LOCAL AND STATE

K-1 Fite, Harry H. The computer challenge to urban planners and state administrators. Washington, D.C., Spartan Books, 1965. 142 p. (The American University Technology of Management Series, v. 2. Paul W. Howerton, General Editor.)

K-2 Governmental ADP: the practitioners speak. Selected readings from Public automation, compiled by Public Automated Systems Service, 1313 East Sixtieth Street, Chicago, Illinois, [1968]. 105 p.

K–3 Hearle, Edward F. R., *and* Raymond J. Mason. A data processing system for state and local governments. Englewood Cliffs, New Jersey, Prentice-Hall, 1963. 150 p.

K–4 Intergovernmental Task Force on Information Systems. The dynamics of information flow; recommendations to improve the flow of information within and among federal, state and local governments. April 1968. 31 p.

K–5 Isaacs, Herbert H. User-oriented information systems for state and local government. Santa Monica, California, System Development Corporation, March 5, 1965. 21 p. refs. (SP–1988)

K–6 Price, Dennis G. Automation in state and local governments. *In* Datamation, v. 13, March 1967. p. 22–25.

K–7 Rowan, Thomas C. System development for regional, state and local government. *In* SDC magazine, v. 8, no. 10, October 1965. p. 1–15.

K–8 U.S. *National Commission on Urban Problems*, Washington, D.C. Report of the National Commission on Urban Problems. Part IV. Government structure, finance, and taxation. Washington, D.C., [1969]. 1 v.

L. Government—State

L–1 Automatic data processing in state government of Texas. Report of Governor's Committee, John Connally, Governor. Austin, Texas, Executive Department, August 7, 1964. 16 p. + appendixes.

L–2 Automated data processing in state government: status problems and prospects. Public Automated Systems Service, 1965. 40 p.

L–3 Background study report for the conference on science, technology and state government. Louisvlle, Kentucky, Southern Interstate Nuclear Board, September 19–20, 1968. 53 p.

L–4 Data processing in Wisconsin state government—a five year plan: 1967–72. Department of Administration, Bureau of Systems and Data Processing, [no date]. 46 p.

L–5 Development of integrated accounting and data processing systems for the state of South Dakota. Baltimore, Maryland (Olympia, Washington), John A. Donaho and Associates, Inc., May 1967. 22 p. + appendixes.

L–6 Five year plan for the implementation and administration of the Technical Services Act for the state of Nebraska. A systems approach to economic and technological development. [University of Nebraska, University Extension Division], September 1966. 49 p.

L–7 Hoos, Ida R. Automation, systems engineering, and public administration: observations and reflections on the California experience; preliminary findings on an ongoing research study of the state of California's experience in the application of systems analysis to five designated problem areas—crime, transportation, waste management, information handling, and welfare. *In* Public administration review, v. 26, December 1966. p. 311–319.

L–8 Information management and electronic data processing within the Office of Governor, State of Illinois. [Springfield], Illinois, April 1, 1965. 24 p.

L–9 Management Information Division progress report, first six months, February 15–August 15, 1966. Springfield, Illinois Department of Finance, September 2, 1966. 44 p. refs.

L–10 Mooney, Orus M. The digital computer: its development and its use in Texas state government. *In* Public affairs comment, v. 13, September 1967. p. 1–4.

L–11 Pennsylvania's computer age; a look at EDP in the Commonwealth. [Harrisburg], Pennsylvania, Bureau of Management Information Systems, 1967. 42 p.

L–12 Scriven, Donald D. Missouri communications study. September 1968. 10 p.

L–13 State of Alaska implementation plan—Alaska information system: July 1967–July 1972. [no date]. 100 p. + attachments.

L–14 State of New Hampshire information system study report. Prepared by Internal Automation Operation, General Electric, May 1966. 84 p.

L–15 Talley, Brooks H. ADP in Kentucky state government. Frankfort, Kentucky, Legislative Research Commission, 1967. 46 p. (Research Report No. 42)

L–16 Touche, Ross, Bailey and Smart. Automated data processing planning project; report to the Governor, State of Washington. [Seattle], 1967. 180 p.

L–17 West Virginia state information system program implementation plan report. Lockheed Missiles and Space Company, December 1967. 1 v. (T–37–67–4)

M. GOVERNMENT—LOCAL

M–1 Alcott, James. Technology and urban needs. A statement from the Engineering Foundation Research Conference on the Social consequences of technology. Kansas City, Missouri, 1966. 24 p.

M–2 Amsterdam, Robert. GIST—a geographic information system for New York City, preliminary design. New York Office of the Mayor, July 1968. 16 p.

M–3 Amstutz, Arnold E. City management—a problem in systems analysis. *In* Technology review, v. 71, no. 1, October/November 1968. p. 47–52.

M–4 APOF [Automated Planning and Operations File] conceptualization study. Final report . . . by the SDC Study Team. Santa Monica, California, System Development Corporation, March 31, 1966. 121 p. (Technical Memorandum TM–(L)–2905)

M–5 Automated data processing in municipal government: status, problems, and prospects. Chicago, Illinois, Public Automated Systems Service, 1966. 34 p.

M–6 Black, Harold, *and* Edward Shaw. Detroit's data banks. *In* Datamation, v. 13, March 1967. p. 25–27.

M–7 Cincinnati/Hamilton County regional computer center. Annual report 1967.

M–8 City computers. *In* Public automation, v. 3, no. 12, December 1967. p. 3–5.

M–9 City limits? *In* SDC magazine, v. 9, nos. 7, 8, Summer 1966. p. 1–4.

M–10 SDC and the city. *In* SDC magazine, v. 9, nos. 7, 8, Summer 1966. p. 5–24.

M–11 The city responding to reason. *In* SDC magazine, v. 9, nos. 7, 8, Summer 1966. p. 25–28.

M–12 Community development training and urban information and technical assistance: 100 outstanding programs. [Washington, D.C.], Office of Intergovernmental Relations and Planning Assistance, Department of Housing and Urban Development, June 1968. 48 p.

M–13 Concepts of an urban management information system. A report to the City of New Haven, Connecticut. Yorktown Heights, New York, Advanced Systems Development Division, International Business Machines Corporation, [no date]. 45 p.

M–14 Crecine, John P. Computer simulation in urban research. *In* Public administration review, v. 28, January–February 1968. p. 66–77.

M–15 Haak, Harold H., *and* W. Richard Bigger. A coordinated data system for metropolitan San Diego. San Diego State College, September 1, 1966. 47 1.

M–16 Interim report on data processing. Prepared for the Citizens Committee on Cook County Government. December 1967. 19 p.

M–17 Isaacs, Herbert H., *and* Michael McCune. The feasibility of a central computer system for San Gabriel Valley cities. Los Angeles, California, Herbert H. Isaacs, Research and Consulting, Inc., January 31, 1967. 147 p.

M–18 Lindsay, Franklin A. Managerial innovation and the cities. *In* Daedalus, Fall 1968. p. 1218–1230.

M–19 Long range plan for systems: highlights of a report to the City Administrative Officer, Los Angeles. Los Angeles, California, Stanford L. Optner and Associates, June 1968. 28 p.

M–20 McConnell, Thomas J., *and* Picot B. Floyd. Albany, Dougherty County: electronic data processing; a feasibility study. [Athens], University of Georgia, Computer Center and Institute of Community and Area Development, 1966. 65 p.

M–21 Matteson, Robert J. M., Roger J. Herz, *and* Anna H. Clark. Electronic data processing; an evaluation of the use of computers. New York, New York (City) Temporary Commission on City Finances, 1966. 58 p. (Staff Paper 3).

M–22 Mitchel, William H. SOGAMMIS—a systems approach to city administration. *In* Public automation, April 1966. 4 p.

M–23 Mitchel, William H. Tooling computers to the medium-sized city. *In* Public Management, v 49, March 1967. p. 63–73.

M–24 Modernizing local government to secure a balanced federalism. A statement on national policy by the Research and Policy Committee. New York, Committee for Economic Development, July 1966. 77 p.

M–25 Science, engineering, and the city. A symposium sponsored jointly by the National Academy of Sciences and the National Academy of Engineering. Washington, D.C., National Academy of Sciences, 1967. 142 p. (Publication 1498)

M–26 Scriven, Donald D. Report on St. Charles County project. May 28, 1968. 1 v. incl. appendixes.

M–27 Stitelman, Leonard. Automation in government: a computer survey of the Detroit metropolitan region. Detroit, Michigan, Metropolitan Fund, Inc., November 1967. 21 p.

M–28 Tamaru, Takuji. Prospects in municipal information systems: the example of Los Angeles. *In* Computers and automation, v. 17, January 1968. p. 15–18.

M–29 University of Connecticut *and* The Travelers Research Center, Inc. A regional municipal information handling service for the capitol region. Report of research supported by the Connecticut Research Commission, October 1967. v. I, 44 p. + appendix (refs.); v. II, 37 p. + appendix.

M–30 U.S. *Department of Housing and Urban Development,* Washington D.C. Science and the city. Washington, D.C., U.S. Govt. Print. Off., 1967. 43 p. (HUD MP–39)

M–31 U.S. *Department of Housing and Urban Development,* Washington, D.C. Solving urban problems through urban information and technical assistance. A program guide for urban information and technical assistance. Washington, D.C., U.S. Govt. Print. Off., 1968. 46 p.

M–32 Vincent J. Richard. Municipal information systems' concepts. [Storrs] Institute of Public Service, University of Connecticut, 1967. 58 p. (Connecticut. University. Institute of Public Service. MITP Series 2)

M–33 Weiner, Myron E. Cooperation as a model for municipal use of computers. [Storrs] Institute of Public Service, University of Connecticut, 1968. 15 p. (Connecticut. University. Institute of Public Service. MITP Series 3)

M–34 Weiner, Myron E. Information, technology and municipal government. [Storrs] Institute of Public Service, University of Connecticut, 1967. 38 p. (Connecticut. University. Institute of Public Service. MITP Series 1)

M–35 Wright, J. Ward. Local government and the computer age. Remarks before the Subcommittee on Census and Statistics. June 29, 1966. 18 p.

M–36 Yavitz, Boris, *and* Thomas M. Stanback, Jr. Electronic data processing in New York City; lessons for metropolitan economics. New York, Columbia University Press, 1967. 159 p. (With an introduction by Eli Ginzberg.)

N. Aerospace Technology in the Urban Area

N–1 Black, Ronald P., *and* Charles W. Foreman. Technological innovation in civilian problem areas. Falls Church, Virginia, Analytic Services, Inc., September 1967. 95 p. + appendixes.

N–2 Chartrand, Robert L. Modern management analysis and traditional management analysis: a survey of the impact. *In* extension of remarks of the Honorable Gaylord Nelson. Congressional record (daily ed.), v. 113, November 21, 1967. p. S16879–S16883.

N–3 Gilmore, John S., John J. Ryan, *and* William S. Gould. Defense systems resources in the civil sector: an evolving approach, an uncertain market. Denver, University of Denver Research Institute, July 1967. 201 p.

N–4 Gregory, William H. Aerospace and social challenge. Part 1: industry probes socio-economic markets. *In* Aviation week and space technology, v. 88, no. 24, June 10, 1968. p. 39–57. Part 2, June 17, 1968. p. 43–55. Part 3, July 1, 1968. p. 38–51.

N–5 Greenberg, D. S. Defense research: Senate critics urge redevelopment to urban needs. *In* Science, v. 160, April 26, 1968. p. 400–402.

N–6 Midwest Research Institute. A national clearinghouse network for urban information. Kansas City, Missouri, September 24, 1968. 30 p. (Progress Report No. 1).

N–7 Henderson, Hazel. Should business tackle society's problems? *In* Harvard business review, 1963. p. 77–85.

N–8 Rittenhouse, C. H. The transferability and retraining of defense engineers. Prepared for U.S. Arms Control and Disarmament Agency. Washington, D.C., U.S. Govt. Print. Off., 1967. 126 p. (ACDA/E–110)

N–9 Stitelman, Leonard. Automation in government: a computer survey of the Detroit metropolitan region. Detroit, Michigan, Metropolitan Fund, Inc., November 1967. 25 p.

O. Books

O–1 Bellush, Jewel, *and* Murray Hausknecht. Urban renewal: people, politics, and planning. Garden City, New York, Anchor Books, Doubleday, 1967. 542 p.

O–2 Berry, Brian J. L., *and* Jack Meltzer, *editors.* Goals for urban America. Englewood Cliffs, New Jersey, Prentice-Hall, Inc., 1967. 152 p.

O–3 Boguslaw, Robert. The new utopia; a study of system design and social change. Englewood Cliffs, New Jersey, Prentice-Hall, Inc., 1965. 213 p.

O–4 Boulding, Kenneth E., *and others.* Environmental qualtiy in a growing economy; essays from the sixth RFF Forum. Baltimore, Maryland, Johns Hopkins Press for Resources for the Future, Inc., 1966. 173 p.

O–5 Campbell, Robert D., *and* Hugh L. LeBlanc. An information system for urban planning. Washington, D.C., George Washington University, Urban Planning Data System Project, U.S. Govt. Print. Off., 1965. 96 p.

O–6 Carter, Launor F., *and others.* National document-handling systems for science and technology. New York, John Wiley, 1967. 344 p. (Information Sciences Series)

O–7 Chartrand, Robert L., Kenneth Janda, *and* Michael Hugo. Information support, program budgeting, and the Congress. New York, Spartan Books, 1968. 231 p.

O–8 Cleary, Edward J. The Orsanco story: water quality management in the Ohio valley under an interstate compact. Baltimore, Maryland, Johns Hopkins Press (published for Resources for the Future, Inc.), 1967. 335 p.

O–9 Cornog, Geoffrey Y., *and others, editors.* EDP systems in public management. Chicago, Illinois, Rand McNally and Company, 1968. 212 p. (Proceedings of the Conference on applications of electronic data processing: state and local government, cosponsored by the University of Georgia and System Development Corporation.)

O–10 Crawford, Roger James, Jr. Utility of an automated geocoding system for urban land and analysis. [Seattle], Urbana Data Center, University of Washington, 1967. 134 1. (Research Report No. 3)

O–11 Eldredge, Hanford Wentworth, *editor.* Taming megalopolis. v. 1: What is and what could be. v. 2: How to manage an urbanized world. Garden City, New York, Doubleday, 1967. 2 v. (Anchor Books)

O–12 Ewald, William R., Jr., *editor.* Environment and policy; the next fifty years. Bloomington, Indiana, Indiana University Press, 1968. 459 p. (Based on papers commissioned for the American Institute of Planners' two-year consultation, part II: the Washington conference, October 1–6, 1967.)

O–13 Ewald, William R., Jr., *editor.* Environment for man; the next fifty years. Bloomington, Indiana, Indiana University Press, 1967. 308 p. (Based on papers commissioned for the American Institute of Planners' two-year consultation, part I: optimum environment with man as a measure, Portland, Oregon, August 14–18, 1966.)

O–14 Faltermayer, Edmund K. Redoing America; a national report on how to make our cities and suburbs livable. New York, Harper and Row, 1968. 242 p.

O–15 Fite, Harry H. The computer challenge to urban planners and state administrators. Washington, D.C., Spartan Books, 1965. 142 p. (The American University Technology of Management Series, v. 2. Paul W. Howerton, *general editor.*)

O–16 Greenwood, Frank. Managing the systems analysis function. [New York], American Management Association, Inc., 1968. 137 p.

O–17 Hearle, Edward F. R., *and* Raymond J. Mason. A data processing system for state and local governments. Englewood Cliffs, New Jersey, Prentice-Hall, Inc., 1963. 150 p. refs.

O–18 Heilbroner, Robert L. Most notorious victory; man in an age of automation. New York, Free Press, 1966. 441 p.

O–19 Jantsh, Erich. Technological forecasting in perspective. (A framework for technological forecasting, its techniques and organization; a description of activities and annotated bibliography.) Paris, Organisation for Economic Cooperation and Development, 1966. 493 p. (Working Document DAS/SPR/66.12)

O–20 Johnson, Thomas F., James R. Morris, *and* Joseph G. Butts. Renewing America's cities. Washington, D.C., Institute for Social Science Research, 1962. 130 p.

O–21 Kahn, Herman, *and* Anthony J. Wiener. The year 2000; a framework for speculation on the next thirty-three years. New York, Macmillan Co., 1967. 431 p.

O–22 Landers, Richard R. Man's place in the dybosphere. Englewood Cliffs, New Jersey, Prentice-Hall, Inc., 1966. 266 p.

O–23 MacBride, Robert. The automated state; computer systems as a new force in society. Philadelphia, Pennsylvania, Chilton Book Co., 1967. 407 p.

O–24 Meyerson, Martin. Face of the metropolis. New York, Random House, 1963. 249 p.

O–25 Meyerson, Martin, Barbara Terrett, *and* Paul N. Ylvisaker, *editors*. Metropolis in ferment. *In* Annals of the American Academy of Political and Social Science, v. 314, November 1957 (entire issue). p. 1–164; index, p. 227–231.

O–26 Murphy, Brian, The computer in society. London, England, Anthony Blond Ltd., 1966. 128 p. refs.

O–27 Nelson, Richard R., Merton J. Peck, *and* Edward D. Kalachek. Technology, economic growth, and public policy. Washington, D.C., Brookings Institution, 1967. 238 p.

O–28 Sackman, Harold. Computers, system science, and evolving society; the challenge of man-machine digital systems. New York, John Wiley, 1967. 638 p. refs.

O–29 Schnore, Leo F., *and* Henry Fagin, *editors*. Urban research and policy planning. v. 1: Urban affairs annual reviews. Beverly Hills, California, Sage Publications, 1967. 638 p.

O–30 Toward the year 2000: work in progress. *In* Daedalus, v. 96, no. 3, Summer 1967 (entire issue). p. 639–994.

O–31 Weaver, Robert C. Dilemmas of urban America. Cambridge, Massachusetts, Harvard University Press, 1965. 138 p.

P. Congressional Documents

P–1 U.S. *Congress. House. Committee on Government Operations. Subcommittee on Government Operations.* The federal research and development programs: the decisionmaking process. Hearings before the Subcommittee. 89th Congress, 2nd session, January 7, 10 and 11, 1966. Washington, D.C., U.S. Govt. Print. Off., 1966. 212 p.

P–2 U.S. *Congress. House. Committee on Science and Astronautics. Subcommittee on Science, Research, and Development.* Policy issues in science and technology. Review and forecast. 3rd progress report of the Subcommittee. 90th Congress, 2nd session. Washington, D.C., U.S. Govt. Print. Off., 1968. 54 p.

P–3 U.S. *Congress. House. Committee on Science and Astronautics.* Technology assessment seminar. Proceedings before the Subcommittee on Science, Research, and Development of the . . . 90th Congress, 1st session, September 21 and 22, 1967. Washington, D.C., U.S. Govt. Print. Off., 1967 (revised August 1968). 184 p.

P–4 U.S. *Congress. House. Committee on Science and Astronautics. Subcommittee on Science, Research, and Development.* Environmental pollution: a challenge to science and technology. 89th Congress, 2nd session. Washington, D.C., U.S. Govt. Print. Off., 1966. 60 p.

P–5 U.S. *Congress. Joint Economic Committee.* Employment and manpower problems in the cities: implications of the report of the National Advisory Commission on Civil Disorders. 90th Congress, 2nd session. Washington, D.C., U.S. Govt. Print. Off., September 16, 1968. 23 p. (Senate Report No.-1568)

P–6 U.S. *Congress. Joint Economic Committee.* Impact of the property tax: its economic implications for urban problems. Supplied by the National Commission on Urban Problems to the Committee. Washington, D.C., U.S. Govt. Print. Off., May 1968. 48 p. (Joint Committee Print)

P–7 U.S. *Congress. Joint Economic Committee. Subcommittee on Economic Progress.* Automation and technology in education. 89th Congress, 2nd session. Washington, D.C., U.S. Govt. Print. Off., August 1966. 11 p. (Joint Committee Print)

P–8 U.S. *Congress. Joint Economic Committee. Subcommittee on Economic Progress.* Federal programs for the development of human resources. 90th Congress, 2nd session. v. I. Washington, D.C., U.S. Govt. Print. Off., 1968. 326 p.

P–9 U.S. *Congress. Joint Economic Committee. Subcommittee on Economic Progress.* Federal programs for the development of human resources. 90th Congress, 2nd session. v. II. Washington, D.C., U.S. Govt. Print. Off., 1968. 358 p.

P–10 U.S. *Congress. Joint Economic Committee. Subcommittee on Economic Progress.* Technology in education. Hearings before the Subcommittee. 89th Congress, 2nd session, June 6, 10, and 13, 1966. Washington, D.C., U.S. Govt. Print Off., 1966. 273 p.

P–11 U.S. *Congress. Joint Economic Committee. Subcommittee on Urban Affairs.* A directory of urban research study centers. 90th Congress, 1st session. Washington, D. C., U. S. Govt. Print. Off., August 1967. 77 p.

P–12 U.S. *Congress. Joint Economic Committee. Subcommittee on Urban Affairs.* Urban America: goals and problems. Hearings before the Subcommittee. 90th Congress, 1st session, September 27, 28, October 2, 3, and 4, 1967. Washington, D.C., U.S. Govt. Print. Off., 1967. 239 p.

P–13 U.S. *Congress. Joint Economic Committee. Subcommittee on Urban Affairs.* Urban America: goals and problems. 90th Congress, 1st session. Washington, U.S. Govt. Print. Off., August 1967. 303 p.

P–14 U.S. *Congress. Senate. Committee on Government Operations. Subcommittee on Executive Reorganization.* Federal role in urban affairs. 89th Congress, 2nd session—90th Congress, 2nd session, [August 15, 1966—1968] Washington, D.C., U.S. Govt. Print. Off., [1966–1968] 21 parts+appendix to part 1.

P–15 U.S. *Congress. Senate. Committee on Government Operations. Subcommittee on Intergovernmental Relations.* Creative federalism. Part I, the federal level. Hearings before the Subcommittee. 89th Congress, 2nd session, November 16. 17, 18, and 21, 1966. Washington, D.C., U.S. Govt. Print. Off., 1966. 467 p.

P–16 U.S. *Congress. Senate. Committee on Government Operations. Subcommittee on Intergovernmental Relations.* Intergovernmental personnel act of 1967. Intergovernmental manpower act of 1967. Hearings before the Subcommittee. 90th Congress, 1st session, April 26, 27, and 28, 1967. Washington, D.C., U.S. Govt. Print. Off., 1967. 314 p.

P–17 U.S. *Congress. Senate. Committee on Government Operations. Subcommittee on Intergovernmental Relations.* Catalog of federal aids to state and local governments. 88th Congress, 2nd session. Washington, D.C., U.S. Govt. Print. Off., 1964. 154 p. (Committee Print) Supplement, January 4, 1965. 89th Congress, 1st session. Washington, D.C., U.S. Govt. Print. Off., 1965. 65 p. (Committee Print) Second supplement, January 10, 1966. 89th Congress, 2nd session. Washington, D.C., U.S. Govt. Print. Off., 1966. 257 p. (Committee Print)

P–18 U.S. *Congress. Senate. Committee on Government Operations. Subcommittee on Intergovernmental Relations.* Intergovernmental cooperation act of 1967 and related legislation. Hearings before the Subcommittee. 90th Congress, 2nd session, May 9, 10, 14, 15, 16, 21, 22, 28, and 29, 1968. Washington, D.C., U.S. Govt. Print. Off., 1968. 536 p.

P–19 U.S. *Congress. Senate. Committee on Labor and Public Welfare. Special Subcommittee on the Utilization of Scientific Manpower.* Scientific manpower utilization, 1965–66. Hearings before the Special Subcommittee. 89th Congress, 1st and 2nd sessions, November 19, 1965; May 17 and 18, 1966. Washington, D.C., U.S. Govt. Print. Off., 1966. 213 p.

P–20 U.S. *Congress. Senate. Committee on Labor and Public Welfare. Special Subcommittee on the Utilization of Scientific Manpower.* Scientific manpower utilization, 1967. Hearings before the Special Subcommittee. 90th Congress, 1st session, January 24, 25, 26, 27; March 29 and 30, 1967. Washington, D.C., U.S. Govt. Print. Off., 1967. 377 p.

P–21 U.S. *Congress. Senate. Committee on Labor and Public Welfare. Subcommittee on Employment, Manpower, and Poverty.* The impact of federal research and development policies upon scientific and technical manpower. Washington, D.C., U.S. Govt. Print. Off., December 1966. 69 p.

P–22 U.S. *Congress. Senate. Committee on Employment and Manpower.* Impact of federal research and development policies on scientific and technical manpower. Hearings before the Subcommittee. 89th Congress. 1st session, June 2, 3, 4, 7, 8, 9, 10, and July 22, 1965. Washington, D.C., U.S. Govt. Print. Off., 1965. 954 p.

P–23 U.S. *Congress. Senate. Select Committee on Small Business.* Automatic data processing and the small businessman. Prepared by the Science Policy Research Division, Legislative Reference Service, Library of Congress, Washington, D.C., U.S. Govt. Print. Off., June 3, 1968. 148 p. (Senate Document No. 82).

P–24 U.S. *Congress. Senate. Select Committee on Small Business. Subcommittee on Science and Technology.* Policy planning for technology transfer. Hearings before the Subcommittee. Prepared by the Science Policy Research Division, Legislative Reference Service, Library of Congress. Washington, D.C., U.S. Govt. Print. Off., April 6, 1967. 184 p.

Q. Serial Publications

Q–1 City. Bi-monthly review of urban America. Published by Urban America, Inc., 1717 Massachusetts Avenue, NW., Washington, D.C. 20036.

Q–2 Ekistics. Reviews on the problems and science of human settlements. Published by the Athens Center of Ekistics of the Technological Institute, P.O. Box 471, Athens, Greece.

Q–3 Journal of housing. Published 11 times per year by the National Association of Housing and Redevelopment Officials, 1413 K Street, NW., Washington, D.C. 20005.

Q–4 Journal of the American Institute of Architects. Published monthly by The Octagon, 1735 New York Avenue, NW., Washington, D.C. 20006.

Q–5 Journal of the Institute of Planners. Published bimonthly by the American Institute of Planners, 917 15th Street, NW., Washington, D.C. 20005.

Q–6 Land economics. Published quarterly by the University of Wisconsin, Social Science Building, Madison, Wisconsin 53706.

Q–7 SDC magazine. Published monthly by Corporate Communications, System Development Corporation, 2500 Colorado Avenue, Santa Monica, California 90406.

Q–8 Urban affairs quarterly. Published by Sage Publications, Inc., 275 South Beverly Drive, Beverly Hills, California 90212.

R. Bibliographies

R–1 Harrison, Annette. Bibliography on automation and technological change and studies of the future. [Santa Monica, California, Rand Corporation], March 1968. 34 p. (P–3365–3).

R–2 Public Administration Service. Public Automated Systems Service. Automation in the public service; an annotated bibliography. Chicago, Public Administration Service, [1966]. 70 p.

R–3 U.S. *Library of Congress. Legislative Reference Service*, Washington, D.C. Air pollution; a selected, annotated bibliography, by Maria H. Grimes. [January 1968]. 16 p. (SP 162)

R–4 U.S. *Library of Congress. Legislative Reference Service*, Washington, D.C. The planning-programming-budgeting system: an annotated bibliography, by Robert L. Chartrand, *and* Dennis W. Brezina. April 11, 1967. 23 p. (SP 127)

R–5 U.S. *Library of Congress. Legislative Reference Service*, Washington, D.C. Slums and slum clearance: a list of references, by Tom F. Lord. December 20, 1965. 4 p. (E–124)

R–6 U.S. *Library of Congress. Legislative Reference Service*, Washington, D.C. The social and psychological impact of urban redevelopment, selected references, 1959–1965, by Tom F. Lord. November 2, 1966. 3 p. (E–191)

R–7 U.S. *Library of Congress. Legislative Reference Service*, Washington, D.C. The social responsibility of business, a selected bibliography, by George K. Brite. March 11, 1969. 23 p. (E 319 R)

R–8 U.S. *Library of Congress. Legislative Reference Service*, Washington, D.C. Urban affairs: selected references, 1955–1966, by Tom F. Lord. December 20, 1966. 5 p. (E–200)

R–9 U.S. *Library of Congress. Legislative Reference Service*, Washington, D.C. Urban planning: a bibliography, 1960–1968, by Marion Schlefer. August 13, 1968. 42 p. (E–297)

R–10 U.S. *Library of Congress. Legislative Reference Service*, Washington, D.C. Water pollution, a selected bibliography, by Migdon Segal. June 27, 1967. 5 p. (SP 132)

S. Symposia

S–1 Annual symposium of the Association for Computing Machinery, held on November 10, 1967 in New York City on Application of Computers to the Problems of urban society, edited by S. N. Levine. *In* Socio-economic planning sciences, v. 1, no. 3, July 1968. p. 201–442.

S–2 Council of State Governments. Interstate conference on automated data processing; its impact on public policy and service, May 25–27, 1965, Lansing, Michigan. 168 p.

S–3 EDP systems for state and local governments. Proceedings of the conference, September 30–October 2, 1964, New York. New York University in association with System Development Corporation. Santa Monica, Corporate Communications, [1964] 166 p. (BRT–42/000/00)

S–4 Golovin, Nicholas E. Reflections on the AIAA/ORSA Forum on Systems analysis and social change. April 2, 1968. 28 p.

S–5 Hoos, Ida R. Systems analysis as a technique for solving problems—a realistic overview. New York, Annual symposium, Association for Computing Machinery, "Applications of computers to the problems of urban society," October 18, 1968. 11 p.

S–6 Laughery, Kenneth R., Theodore E. Anderson, *and* Edwin A. Kidd. A computer simulation model of driver-vehicle performance at intersections. *In* Proceedings of the 22nd national conference, Association for Computing Machinery. Washington, D.C., Thompson Book Co., 1967. p. 221–231.

S–7 Macy, B., J. M. Bednar, *and* R. E. Roberts. The impact of science and technology on regional development—final report prepared for the Office of Regional Development

Planning, U.S. Department of Commerce. Kansas City, Missouri, Midwest Research Institute, 1967. 168 p.

S–8 The Development Administrators Training Program, Institute of Public Service, University of Connecticut, *and* Ad Hoc Committee on Management Analysis in State and Local Government. Management analysis and the administration of critical government programs. Papers from the third annual conference on management analysis in state and local government, held October 24–25, 1966, Buck Hills Falls, Pennsylvania. 1 vol.

S–9 Proceedings of the IBM scientific computing symposium on water and air resource management. White Plains, New York, IBM Data Processing Division, 1968. 392 p.

S–10 Proceedings of the Mid-America conference on urban design, March 30 and 31, 1966, Kansas City Art Institute. 114 p.

S–11 Proceedings of the 22nd national conference, Association for Computing Machinery. Washington, D.C., Thompson Book Company, 1967. 599 p. (A.C.M. Publication No. P-67)

S–12 Public systems management—concepts and applications. Proceedings, October 19–20, 1967. Los Angeles, Management Education Conferences, University of California Extension, [1967]. 194 p.

S–13 Science, engineering, and the city. A symposium sponsored jointly by the National Academy of Sciences and the National Academy of Engineering. Washington, D.C., National Academy of Sciences, 1967. 142 p. (Publication 1498)

S–14 Science, technology, and state government. Proceedings of the NSF–SINB conference. Presented by the National Science Foundation and the Southern Interstate Nuclear Board, September 19–20, 1968, Louisville, Kentucky. Louisville, Kentucky, Kentucky Science and Technology Commission, February 1, 1969. 238 p.

S–15 Southern Interstate Nuclear Board, [Atlanta, Georgia] Background study report for the conference on science, technology and state government, September 19–20, 1968, Louisville, Kentucky. 53 p.

S–16 The urban challenge: the management and institutional response. A conference sponsored by the George Washington University, North American Rockwell Corporation, and the System Development Corporation, Airlie House, Warrenton, Virginia, June 19–21, 1968. [Remarks of General Bernard A. Schriever.]

S–17 U.S. *Department of Housing and Urban Development.* Conference on new approaches to urban transportation, November 27, 1967, Washington, D.C. 104 p.

S–18 Vaughn, Charles L., *editor.* Systems approach to social problems. Proceedings of the sixth Management conference on marketing in the defense industries. Sponsored jointly by the Defense Marketing Group of the American Marketing Association, Boston Chapter *and* the Bureau of Business Research, Boston College. Chestnut Hill, Massachusetts, Boston College Press 1968, 138 p.

T–1 Adelson, Marvin. The systems approach—a perspective. *In* SDC magazine, October 1966. p. 1–9.

T–2 Beshers, James M., *editor.* Computer methods in the analysis of large-scale social systems. Cambridge, Massachusetts, Joint Center for Urban Studies, Massachusetts Institute of Technology *and* Harvard University, 1965. 207 p. (Proceedings of a conference held on October 19–21, 1964 in Cambridge.)

T–3 Chartrand, Robert L. Systems technology and social imagination. In Extensions of remarks of the Honorable F. Bradford Morse. *In* Congressional record [daily edition], v. 113, November 15, 1967. p. A5629–A5631.

T–4 Hare, Van Court, Jr. Systems analysis: a diagnostic approach. New York, Harcourt, Brace and World, Inc., 1967. 544 p.

T–5 Meyer, Jerome B. Systems approach to national affairs. *In* Data, September 1966. p. 59–64.

T–6 Michael, Donald N. Some long-range implications of computer technology for human behavior in organizations. *In* The American behavioral scientist, v. 9, no. 8, April 1966. p. 29–35.

T–7 Mitre Corporation. Definition of a planning and design study for a national center for education technology. Washington, D.C., May 1968. 60 p.

T–8 Quade, E. S., *and* W. I. Boucher, *editors.* Systems analysis and policy planning: applications in defense. New York, American Elsevier Publishing Company, Inc., 1968. 453p.

T–9 Seaborg, Glenn T. Time, leisure and the computer: the crisis of modern technology. Remarks at a conference on "The university in a changing society," celebrating the centennial of Howard University. Washington, D.C., March 1, 1966. 17 p.

T–10 Seiler, John A. Systems analysis in organizational behavior. Homewood, Illinois, Richard D. Irwin, Inc., *and* Dorsey Press, 1967. 219 p.

T–11 A strategy for planning. A report to the National Governors Conference by the Committee on State Planning. October 18, 1967. 14 p.

T–12 Uhl, Edward G. The urban crisis: industry's resources, capabilities, and response. 15 p.

T–13 U.S. *Department of Health, Education, and Welfare,* Washington, D.C. Toward a social report. Washington, D.C., U.S. Govt. Print. Off., January 11, 1969. 101 p.

T–14 U.S. *National Aeronautics and Space Administration,* Washinton, D.C. Office of Technology Utilization. Applications of systems analysis models; a survey. Washington, D.C., U.S. Govt. Print. Off., 1968. 69 p. refs. (NASA SP–5048)

T–15 U.S. *National Commission on Technology, Automation, and Economic Progress.* Technology and the American economy. v. 1. Washington, D.C., U.S. Govt. Print. Off., February 1966. 115 p.

AUTHOR INDEX

FOREWORD

In 1965 the Committee on Labor and Public Welfare established a Special Subcommittee on the Utilization of Scientific Manpower under the chairmanship of Senator Gaylord Nelson. The special subcommittee was asked to look into the related problems of the use of highly trained scientific and technical personnel in meeting the Nation's domestic needs and the application of the systems analysis techniques to these problems.

The result was an extensive set of hearings, and now this report on the application of systems technology to social and community problems, prepared by the Science Policy Staff of the Library of Congress' Legislative Research Service.

The concerns of the Special Subcommittee have been assumed by the standing Subcommittee on Employment, Manpower, and Poverty, which is printing this report.

At the time the hearings began there was in some quarters a belief that the problem solving techniques (systems analysis) and the technicians that had performed such wonders in space and advanced weapons technology could move with equal effect to "solve" our domestic dilemmas.

In the meantime 3 years experience with Program Budgeting (PPBS) in Washington and extensive efforts to apply systems technology to domestic social programs has convinced most observers that the systems analysts were accurate when they assigned a more modest role to themselves. We now recognize that the key elements in domestic dilemmas are political, having to do with choice of direction and commitment of resources. And we know that the hard information, crucial to the effective functioning of systems techniques, is simply not available in relation to many of our domestic problems.

However, in those areas where vast amounts of detailed facts must be considered the full use of systems techniques should be made.

This report details the interest taken in systems approaches by the Congress and the administrative agencies. It documents the growing use of systems approaches at the state and local level through the analysis of questionnaires sent to all 50 states, 22 cities and a number of regional planning commissions. It also contains the discriptions of a number of state and local applications of systems techniques to transportation and water resource planning, to health services, education, environmental pollution, and urban housing problems.

This vast amount of data, together with the glossary of terms used in this field, should make the report a valuable document to students, Congressmen, and local officials all across the Nation.

RALPH YARBOROUGH,
Chairman, Committee on Labor and Public Welfare.

(475)

LETTER OF TRANSMITTAL

U.S. SENATE,
COMMITTEE ON LABOR AND PUBLIC WELFARE,
Washington, D.C., June 27, 1969.

Hon. RALPH YARBOROUGH,
Chairman, Committee on Labor and Public Welfare,
Senate Office Building, Washington, D.C.:

DEAR MR. CHAIRMAN: This letter transmits to you a report on "Systems Technology Applied to Social and Community Problems."

In 1965 the Special Subcommittee on the Utilization of Scientific Manpower was set up to examine the present and potential use of the justly famed techniques of problem analysis and solution commonly called "systems analysis" developed in our space and advanced weapons programs for the solution of domestic problems. The results of that investigation are contained in the volumes of hearings conducted in Washington and Los Angeles by the subcommittee in 1966 and 1967, and in the document I am transmitting to you today.

The report was drawn up by the staff of the Legislative Reference Service's Science Policy Research Division under the direction of Charles S. Sheldon II. Mr. Robert Chartrand and Mrs. Louise G. Becker of that staff deserve special thanks for their imaginative and tireless efforts in compiling this report over the past year.

It is our expectation that the report will be of great interest not only to Congressmen seeking information on this important topic, but also to State and local officials seeking some grasp of the promise of the systems approach.

The systems approach is no substitute for the political commitment to devote the resources necessary to meet our domestic crisis. But the systems approach is already of decisive help in isolating solvable problems and working out paths to their solution.

The Subcommittee on Employment, Manpower, and Poverty will give careful consideration to the recommendations of this report to increase the value of systems technology for the solution of domestic problems.

Sincerely,

GAYLORD NELSON, *U.S. Senator.*

LETTER OF TRANSMITTAL

THE LIBRARY OF CONGRESS,
LEGISLATIVE REFERENCE SERVICE,
Washington, D.C., April 11, 1969.

Hon. GAYLORD NELSON,
Committee on Labor and Public Welfare,
U.S. Senate, Washington, D.C.

DEAR SENATOR NELSON: In your position as chairman of the Special Subcommittee on the Utilization of Scientific Manpower, during the 89th and 90th Congresses, you requested that a report be prepared which would analyze the hearings conducted by the special subcommittee on S. 430, the Scientific Manpower Utilization Act, and S. 467, to create a National Commission on Public Management. In addition, it was indicated that a more comprehensive study of the potential of systems tools and techniques in coping with nondefense, nonspace public problems would be useful. In response to this request, I am pleased to transmit the report "Systems Technology Applied to Social and Community Problems," prepared by the Science Policy Research Division of the Legislative Reference Service.

This report presents an analysis of the issues involved in Federal, State, and local use of innovative technology in dealing with such problems as environmental pollution, transportation planning, housing redevelopment, law enforcement, and health services. Past public legislation reflecting encouragement of research, development, and utilization of new equipments and processes is reviewed. The activities of congressional committees and subcommittees related to the examination of the benefits and limitations of systems technology are discussed, as are the programs and projects undertaken by executive branch departments and agencies. An analysis of the series of hearings convented by the Senate Special Subcommittee on the Utilization of Scientific Manpower is featured, together with a study of the questionnaires sent by you in 1966 and 1968 to the Governors of all 50 States, the mayors of 22 large cities, and responsible officials of certain regional development commissions. Examples of State and local activities reflecting the use of systems tools and techniques are included. Finally, questions are set forth which may warrant further discussion and action by the Congress.

Mr. Robert L. Chartrand, specialist in information sciences, is the principal author and director of the study. He was assisted by Mrs. Louise Giovane Becker, who prepared section VII and certain appendixes; Dr. Warren H. Donnelly, who contributed to section III; and Mrs. Mauree W. Ayton, who helped prepare the extensive selected references. All three are of the Science Policy Research Division. The manuscript was reviewed by Dr. Charles S. Sheldon II, Chief of the Science Policy Research Division, and the Assistant Chief,

Richard C. Carpenter. Valuable guidance and support were provided by William B. Cherkasky, William J. Spring, and Miss Linda Billings of your staff. I would also like to acknowledge the cooperation of numerous executive branch agencies, universities, foundations, and industrial elements in providing selected data utilized during the preparation of the report.

Sincerely,

LESTER S. JAYSON, *Director.*